INTRODUCTION TO SPACE SYSTEMS

SPACE TECHNOLOGY LIBRARY
Published jointly by Microcosm Press and Springer

For further volumes:
http://www.springer.com/series/6575

Miguel A. Aguirre

Introduction to Space Systems

Design and Synthesis

Springer

Miguel A. Aguirre
Velazquez 87
Piso 7 Izq (7th Floor Left)
Madrid 28006
Spain

ISBN 978-1-4614-3757-4 ISBN 978-1-4614-3758-1 (eBook)
DOI 10.1007/978-1-4614-3758-1
Springer New York Heidelberg Dordrecht London

Library of Congress Control Number: 2012936646

Printed on acid-free paper

Springer is part of Springer Science+Business Media (www.springer.com)

Foreword

The definition of all space systems starts with the establishment of their fundamental parameters: requirements to be fulfilled, overall system and satellite design, analysis and design of the critical elements, development approach, and cost and schedule. There are only a few texts that cover the early design of space systems, and none of them have been specifically dedicated to it. Furthermore, all existing space engineering books focus on analysis. None of them deal with space system synthesis, that is, with the interrelations between all the elements of the space system. This book aims to understand the interaction between all the forces, both technical and nontechnical, which influence the definition of a space system. It refers to the entire system: space and ground segments, mission objectives, as well as to cost, risk, and mission success probabilities.

The early design of a space system is a small but important phase of space system implementation. This is the phase where the most fundamental decisions are taken. It deserves its own text. The early design also requires understanding how the main space system constituents fit together. To have this understanding, it is necessary to have not only an analytical perspective but also a synthetic perspective. The analytical approach is well covered by many texts, but the synthetic perspective is not. Books with a synthetic perspective are necessary.

This book is aimed at space system engineers. It is their job to design and then implement the space system. They are responsible for ensuring fulfillment of the identified users' needs, end-to-end performance, and coherence between all the system elements, and they need to have an overall perspective of the space system. Indeed, they need to have a synthetic view. It will also be of interest to managers who require an interdisciplinary understanding of space systems and how each element interacts with the others. Experts in specific engineering disciplines reading this book would gain from understanding the overall flow of design and how the different engineering disciplines interact with each other. It will also be of interest to advanced students of space system engineering as an overview of spacecraft systems engineering interactions. The book will be of use to anybody who has some professional experience and is interested in acquiring a generalist perspective over a space system as an end-to-end system with all its richness of interactions, both technical and programmatic. On the other hand, this book will not be adequate for students of space engineering who have not gone through a previous course on space systems analysis and technologies.

Quite exceptionally for a book on space system design, this book has a single author. I have devoted most of my professional life to the early design of space systems, and I have written this book from a generalist perspective, focusing on the interactions of the conflicting factors that drive the design. This emphasis on the "synthetic general perspective" is what justifies the single authorship of the book. I am not an expert in any of the specialized engineering areas described in the book; nevertheless, through my experience in space system overall design, I have become a specialist in generalities and a specialist in the interactions of all the engineering areas. This generalist perspective is going to be the perspective of the book.

This book, written as it is by a single author, is inevitably biased by my own experience. That is, it is richer in examples drawn from Earth Observation missions than from other types of space missions. I have, however, included numerous examples from Space Sciences and some from Space Communications missions to illustrate various points of interest. For this same reason, manned missions have not been treated. Earth Observation missions, on the other hand, arguably provide the best framework for gaining a thorough end-to-end perspective of space systems. While often addressing complex scientific issues, they always involve the delivery of mission products that elucidate or resolve issues of importance to society. End users of these products frequently utilize the acquired data to make decisions of importance to wide sectors of society. Meteorological or environmental monitoring missions are clear examples of this. Hence, these kinds of missions must be considered from a myriad of angles beyond those strictly concerned with satellite design.

Acknowledgments

This book was produced with the help of a large number of European Space Agency (ESA) colleagues, who reviewed the work and provided many valuable suggestions. Special thanks must go to all my colleagues at the Future Earth Observation Division – headed by Pierluigi Silvestrin – all of whom provided much needed encouragement and helpful input. Special thanks are also due to David Simpson from Astrium UK (retired), who reviewed the entire text in detail. This allowed me to benefit from his vast experience of space system design.

Contents

List of Figures

List of Tables

Chapter 1

Introduction

Keywords Space Systems Analysis, Space Systems Synthesis, Space Systems Design, Space Systems Stakeholders, System Engineering, System Architecting, Space Project Phases

This chapter is an overall introduction to the whole book. The chapter starts with a description of the aims of the book and provides a justification for its existence. It continues with the actors and roles involved in the process of starting a new space system. The next section deals with the generalist "architectural" perspective that is necessary to define a space system. The following two sections deal with the evolution of the design process. The first one discusses the design and implementation of a space system and the second recalls the names and aims of the phases that are normally used to define a space system. The next section presents what constitutes a space system. The chapter finishes with a brief glossary of terms that are used repeatedly in the book and with a list of recommended supplementary texts.

In addition to the present introduction, this book is structured as follows:

- The first part of the book describes and analyzes space system design and implementation. It is divided into the following chapters:
  - One chapter is on the science of space system design and implementation. It is a recompilation of the technical/scientific disciplines needed for the proper design and implementation of space systems.
  - It continues with two chapters on analysis of system requirements and constraints.
  - The actual design process is then considered in two chapters. The first chapter deals with the different factors that influence the design. The second chapter explains in detail the iterative steps necessary to arrive at the final architecture.
- A second part of the book analyzes concrete aspects of the space system design process. It concentrates on the interactions between design decisions and uses continuously past design examples to illustrate these interactions. This second part is divided into the following chapters:
  - One introductory chapter divides the design process into domains.
  - The satellite observables and instruments domain chapter.
  - The orbit, space environment, and attitude domain chapter.
  - The satellite configuration domain chapter.

M.A. Aguirre, *Introduction to Space Systems: Design and Synthesis*, Space Technology Library 27,
DOI 10.1007/978-1-4614-3758-1_1, © Springer Science+Business Media New York 2013

  - The satellite operations data flow domain chapter.
  - The instrument output data flow domain chapter.

- The final chapter analyzes alternative and low-cost design and implementation approaches.

1.1. Aim of the book

This book has two aims:

- To describe and analyze the process needed for the proper design of a space system, in order that the reader can understand what is a good system design and how to arrive to it.
- To describe and analyze space systems not from the point of view of their elements but from the point of view of the interrelations between them. That means not from the point of view of system analysis but from the point of view of system synthesis.

The design process spans from the highest to the most detailed level but this book deals with the early design phases. The design process that is covered by this book begins with a vague statement of needs by users or sponsors and ends with the delivery of a comprehensive documentation set. The aim of the documentation shall be to enable an informed decision on the actual large-scale industrial project approval or rejection. Meyer Rechtin[1] defined system architecture as "the set of information that defines a system value cost and risk sufficiently for the purpose of the sponsor." Using this definition of "system architecture," this book aims to explore the process of creating the space system architecture. The output of this architecture definition process shall include:

- A clear definition of the mission aims including the identification of users and the requirements that will be addressed by the designed system
- A set of formal documents defining the functional and performance system requirements, both qualitative and quantitative
- A description of the proposed system with supporting documentation to demonstrate that the described system will fulfill the identified requirements
- A description of the design and development approach in sufficient detail to provide credibility to the cost and schedule predictions
- Cost and schedule predictions based on the design selected and on the development and implementation approach chosen
- Mathematical models allowing the derivation of the end-to-end system performance in terms understandable for the users

The process described above occurs before the approval of a fully fledged industrial program and a small team does it. However, by the time the process is finished approximately 90 % of the system costs have been effectively decided and locked[2]

---

[1] The Art of Systems Architecting, Mark Maier and Eberhardt Rechtin, CRC Press 2000.

[2] NASA Systems Engineering Handbook, SP-610S 1995.

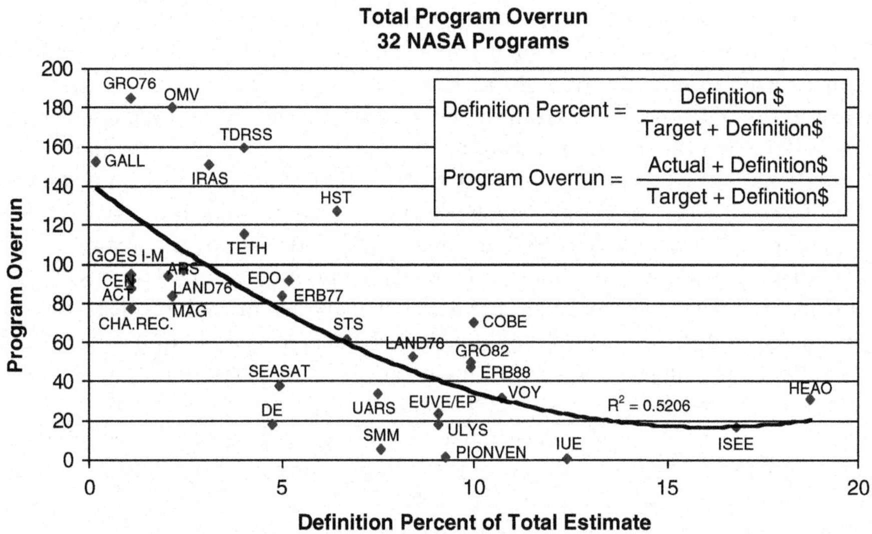

Figure 1.1. Design Cost Versus Overall Cost (Presentation by Werner Gruhl Office of the Comptroller NASA HQ 1985).

into the total. It is a small proportion of the system definition and implementation effort but it is an inordinately important one.

In Chapter 5 of the television series produced by Tom Hanks *From the Earth to the Moon*,[3] the heroes are the engineers. Specifically the script is built around Thomas Kelly and his team from Grumman, designing and implementing the Lunar Module. Nevertheless, at the very beginning and at the very end of the episode there are two brief scenes where T. Dolan and J. Houbolt appear. These are the people, who, in 1959, proposed the concept of a Lunar Module and Lunar Rendezvous. Originally the whole Apollo Module was designed for landing on the Moon and returning to Earth. This original concept was scrapped due to mass and financial constraints. Another cheaper and more feasible solution was sought, which led to the more risky but affordable Lunar Module and Lunar Rendezvous concepts. This ability to think around problems, work within constraints, and propose workable solutions made the Apollo program possible. Houbolt and Dolan deserve their, small but important, appearance in the television episode and this book deserves to be written.

To dedicate due attention to the early phases of a project is the safest way to avoid large problems later. Figure 1.1 provides a summary of cost overrun as the percentage of effort dedicated to the early definition. Projects where enough effort has been dedicated to early definition have clearly smaller cost overruns.

This book is complementary to all the other space system engineering books. It concentrates in the interrelations and synthesis that the other books normally do

[3] *From the Earth to the Moon*, Fifth Chapter, Tom Hanks, Executive Producer, based on *A Man on the Moon* by Andrew Chaikin.

not cover and does not include information that the other books cover properly. For example, this book is not a good source from which to learn about the different methods used to provide the right thermal environment for a satellite; nevertheless, it is a good source to learn how the thermal design, the attitude control system, and the pointing requirements of the instruments and the platform could drive in conflicting directions the attitude control, configuration, and structural stability of the satellite. The tool that this book uses for this learning has been the comparative analysis of previous designs. In line with this desire for a complementary, generalist perspective, our analysis of the different engineering domains will be brief and will forego the formulas and detailed analytical procedures necessary for the actual design. To compensate for this lack, extensive analysis of the factors and interrelations driving the design are included, illustrated with multiple examples from both current and past designs. For example, Section 9.7 offers an exhaustive analysis of the geometry of the celestial sphere around the satellite, including the time-varying orientations and dimensions of the Sun, the Earth, observation targets, and the surrounding deep space. What are not included are the formulae – fundamentally orbital mechanics and spherical trigonometry – needed to calculate these time-varying geometries. This lack of formulae is in line with the complementary nature of this book and with its generalist perspective; it does not contain the mathematical formulas that are already included in previous books but concentrates in understanding the overall context where the mathematical formulae will be applied.

This book considers the space mission design process as an interactive one. The textbook situation where a clear set of user needs generates a clear specification with a design, which answers the specification, unfortunately does not exist. The users of the data to be provided by the space mission have always a wide, and sometimes confusing and contradictory, set of desires that can be partially or totally fulfilled by different designs answering different sets of final performances. These designs will have attached widely different price, risk, and schedule tags and would require different programmatic approaches. Furthermore, the scientists, who use, and the political sponsor, who pays, will be different entities. The sponsor could have explicit or implicit objectives, and being the source of the funds, these will become increasingly important to meet. An example of these objectives could be the development and use of a specific technology or the execution of the work in a specific country or within a specific programmatic framework.

The beginning of this book defined its aim as to arrive to a "proper" space system design – but what does this mean? A "proper" design is one which balances all the conflicting needs coming from the user community and provides a system which generates reasonable scientific mission output within the affordability and programmatic constraints established by the program sponsor. The scientific output will be defined as reasonable by providing answers to an important set of user needs. The system design will be made optimal by the correct allocation of functions and performances to the different elements making up the overall system and by the use of technologies, designs, and development approaches that fits in the risk and cost profile defined by the sponsor. This implies that the scientific requirements, the development risks, and the programmatic constrains shall be considered as a whole and given the same priority by the designer of the system architecture. This approach is totally in accordance with the approach outlined in the introduction of Wertz[4] *Space Mission Analysis and Design* that deserves to be quoted literally:

> Space mission analysis and design begins with one or more broad objectives and constraints and then proceeds to define a space system that will meet them at the lowest possible cost. Broad objectives and constrains are the key to this process. Procurement plans for space systems too often substitute detailed numerical requirements for broad mission objectives, To get the most performance for the money spent, we must require of the system only what it can reasonably achieve.

The previous paragraphs establish four principles that are of fundamental importance:

- To consider requirements establishment as part of the iterative design process and not as a separate step. Requirements establishment is not completed prior to the design process but is a part of it
- To design a system that is affordable by prioritizing and negotiating unduly penalizing requirements.
- To consider not only mathematical analysis and detailed numerical outputs but also to understand nonnumerical factors
- To reach a balance of requirements, constraints, and implementation difficulties

The need for "balance" is acknowledged also by NASA[5]

> A cost-effective system must provide a particular kind of balance between effectiveness and cost.

Indeed this whole book strives toward the understanding of the balance that exits in a well-designed system.

In the previous paragraphs the need to consider requirements as "flexible" and part of the overall design process has been repeatedly emphasized. Nevertheless, N. Augustine rightly pointed out[6] that during the fully fledged industrial implementation phases the requirements must remain as stable as possible. This is not a contradiction. The mutual adjustment between design and requirements during the space system architectural definition phases of a space mission must ensure that the meshing of identified requirements and resulting design is as robust as possible. This will minimize, ideally reduce to zero, the need to change requirements during the subsequent fully funded industrial implementation phases.

1.2. Roles in the architecture definition process

The last section made reference to two participating entities or actors: the user and the financial sponsor. The distinction of the roles played by the different stakeholders in the creation of a space system is key to achieve a proper understanding of the

[4] Space Missions Analysis and Design, James Wertz and Wiley Larson, Space Technology Library, Springer 1999.

[5] Nasa Systems Engineering Handbook, SP-610S 1995.

[6] Augustine's Laws Norman Augustine, AIAA 1983.

creation process. The European Space Standard ECSS-P-001[7] uses the following definitions:

- Consumer is the actor that will use the data output of the space project. It is the scientific or operational community using the data.
- Customer is the actor that behaves as the agent of the consumer, transforms his needs into specifications, leads the resulting project, and accepts the delivery of the space system. For example, this could be a space agency.
- Supplier is the actor that actually builds the space system. It is the space industry.

Wertz in his *Space Missions Analysis and Design* uses a different division with three actors:

- Sponsor is the institution that pays for the space project.
- Developer is the institution that builds it.
- User/Operator is the institution that uses the output data of the space project.

The differentiation presented by Wertz between sponsor and user is quite correct. In most space programs the stakeholder that pays for the system is not the stakeholder that uses and this is an important distinction. This duality is not exclusive to the space domain; it also exists in the implementation of many government-funded goods and services. This duality makes the process of establishing a space program also unavoidably "political." This does not mean that the designer has to be political; indeed it is better if he does not try. A good designer is normally a bad politician. However, the designer has to be aware of the political environment to be able to design a system that will eventually be launched. It must not be forgotten that as stated by Brand Forman in 1990 "if the politics do not fly the hardware never will."

In Europe, once a project is approved and once the industrial organization has been established, the sponsor does not play a very active role. Nevertheless, in the USA the yearly budgeting process by Congress makes the involvement of the sponsor a recurrent event during the whole duration of the mission and possible project cancelation by the sponsor is a permanent danger.

On the other hand, the already mentioned ECSS standards divide the figure of the developer into the pair customer and supplier. This division is necessary because almost all space projects are done by the interaction between a space agency that represent the users and a space industry that actually builds the space system. Therefore, this text is going to consider the existence of four actors in the process of designing the architecture of a space mission. The specific terms to be used are going to be the ECSS-recommended terms, but with the addition of Wertz's proposed sponsor. The actors to be considered in this text will be the following:

- The sponsor, the entity that pays for the system and establishes the constraints, cost ceiling, desired schedule, programmatic framework, and any set of appropriate guidelines, to be applied to the implementation of the space system.

[7] Glossary of Terms European Cooperation for Space Standardization, ECSS-P-001, ESA 1997.

- The consumer, the community who uses the output of the system and who establishes requirements on the quantity, quality, and mode of delivery of the data to be produced by the space system.
- The customer, the entity who represents the interest of the consumer with respect to the supplier, who specifies the system, oversees its implementation, accepts and certifies it, delivers the system to the consumer, and is responsible to the sponsor for finishing on time and within the cost ceiling.
- The supplier, the entity who interacts with the customer, helps in the preliminary design, performs subsequent detailed definition and development, builds the system, delivers it to the customer, and is paid by him.

This role of the customer is actually the role of the "classical architect" that interprets and defends the needs of the consumer – and the sponsor – with respect to the supplier. It must be noted that in some situations, the number of actors on the process can be smaller. In almost all the privately funded space projects, the sponsor and the consumer can be one and the same entity. A private company will buy a space system to fulfill a clear set of company needs. The company, acting as both sponsor and consumer, will define the required performances and the financial envelope for the program. In any project where the needs are clear and they can be fulfilled with an already available and verified system, the customer, that means the space agency, may not be needed. The customer is the interpreter and defender of the interest of the consumer with respect to the supplier and he becomes less necessary when the level of novelty of the system is low. **A standard system does not need to be the subject of the architectural definition discussed in this book; it just needs to be purchased**. For the purposes of this text, the emphasis shall be placed on space projects that have all the actors. Space projects are complex, involve conciliation between conflicting needs, and involve advancement of the state of the art. This is the reason they need an entity that interacts with both the sponsor and the supplier. This necessary entity is the customer. In the early 1960s, Europe created the European Launcher Development Organisation (ELDO) to develop a launcher. This organization did not have managerial or technical capabilities to fulfill the duties of a customer. ELDO tried 11 times to launch a rocket and it failed every time.[8] This proves that it is necessary to have an entity with managerial and technical know-how, which can act as a link between all the actors who participate in a space mission. This is the role of the customer.

There is one situation where there are two customers: one of them federating and interacting with the consumers and another with the supplier. This is the case of operational meteorological missions, where there is a research agency: NASA or ESA, and an operational agency: NOAA or Eumetsat. The operational agency establishes the requirements after strong interaction with the consumers and transfers these requirements to the research agency that imposes them on the customer. This approach has demonstrated its value for operational meteorological missions, which are complex, expensive, and built to last many years. The operational agency is also normally in charge of the day-to-day operation of the certified in-orbit space system.

[8] ESA Achievements, BR-250, ESA 2005.

The heavy investment and the complex and long-lasting interfaces with the consumers required by these missions justify the division of the customer's role into these two parts.

Similar situations and approaches have existed for centuries: the building of the harbor in Malaga in 1550 was a major endeavor for its time. The construction of the harbor had:

- The consumer: the traders and fishermen of Malaga
- The sponsor: the Spanish Crown which paid for the project and was represented by a "Corregidor"
- The customer/architect: provided the overall design, certified the work, and authorized payments termed "Veedor"
- The supplier: made up of the contractors and subcontractors, who actually built the harbor

As in any self-respecting present-day space project the supplier for the Malaga harbor was organized with a prime contractor, called "sobrestante mayor" and with subcontractors, the "maestros de obras." As with many modern projects this project experienced delays and cost overruns.[9] It is not the case that there is nothing new under the Sun,[10] but there are not as many new things under it as they are often advertised.

1.3. The perspective during the system architecture definition phases

This book is written from the central perspective of the customer. The customer interprets the desires and needs of the consumer and transforms them into a set of specifications with a progressive level of detail and formality. The customer, with the help of the supplier, establishes the system architecture, allocating functions to system elements and breaks down the overall performance requirements, allocating them to designated system elements, which contributes to the overall performance. In this process, the customer refines the needs of the consumer and interacts between the supplier and consumer until user needs, system design, performances breakdown, and functions allocation are balanced and coherent. In parallel the customer will take care of the fulfillment of the programmatic constraints of the sponsor; for example, overall cost caps, acceptable risk, schedule, and technologies to be used. Once all the conflicting interests of the actors have been balanced the overall system architecture can be frozen and the first comprehensive formal system-level specification can be issued. The overall architecture and the system specification will derive the detailed designs and set of specifications for the main system elements. The level of detail necessary for the definition of the system architecture will depend upon the aim of arriving at "the set of information that defines a system's value, cost and risk

---

[9] Ingenieria en el tiempo de Felipe II Patrimonio Nacional 2003.

[10] Eclesiastes, Cohelet, date of writing unknown.

sufficiently for the purpose of the sponsor." This information, which will be delivered as documentation to the sponsor and the customer, will allow a well-informed decision regarding the implementation of a fully fledged and funded industrial development for a space project. This step shall and must occur before the project is approved. The customer, acting as a link between the consumer and the supplier, behaves as a classical architect during the early design phases. The establishment of the architecture of a space system is concerned with proving the feasibility as well as the desirability of the mission for the consumer and sponsor before it becomes a reality. The architecture definition effort strives to provide a balance and compromise between the consumer's needs, the supplier's capabilities, and the sponsor's ability to pay for the system. Compromise can only be assured by an interplay of activities, which are detailed enough to ensure a well-informed decision to allow the project's final approval. The architectural design is paramount to achieving a decision to approve a project and it is critical to the overall success of the space system. All these interrelations are summarized in Figure 1.2.

As far as detailed design and allocation of functions and performances to the different system elements is concerned, the generalist perspective of the architect is also fundamental. He is independent of the specialists in charge of the design of the system elements. The system architect can act as an arbitrator between the specialists contributing to the design process. Therefore, during the early stages of a space project, the customer has to act with the generalist perspective of the architecture of the system. The extent to which this system architect's perspective differs from that of the system engineer is explored in the prologue of the quoted Maier and Rechtin book, which compares and differentiates the roles of the system architecture definition and the system engineering. Quoting again literally from that book:

> The early stages of a space project are indeed a period where situations are often unknown and goals are not totally defined. In these stages, there is more synthesis than analysis, the work is performed by small teams and the interactions of the Customer with the User and Sponsor are strong.

This means, that the early stages of a project tend more toward "System Architecture Definition" than toward "System Engineering." According to Table 1.1, which is also taken literally from the Maier–Rechtin book:

The "art" element indicated in the table is absolutely pertinent. The optimal allocation of functions and the optimal identification of key performance criteria cannot be established by science and mathematics alone. At the beginning of the design process,

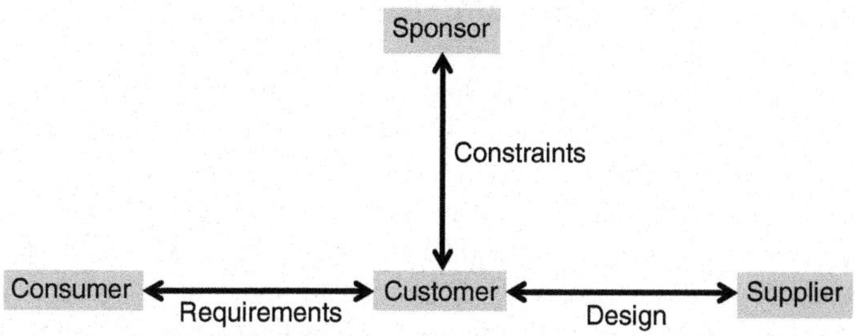

Figure 1.2. Interactions Between the Key Actors During the Space System Design and Implementation Process.

Table 1.1. System Architect Versus System Engineer Perspectives.

The system architecting/system engineering continuum			
Characteristics	Architecting	Both	Engineering
Situation	Ill structured	Constrained	Understood
Goals	Satisfaction	Compliance	Optimization
Method	Heuristics	™	Equations
	Synthesis	™	Analysis
	ART and science	Art AND science	SCIENCE and art
Interfaces	Focus in misfits	Critical	Completeness
System integrity maintained through	Single mind	Clear objectives	Disciplined methodology
Management issues	Working for sponsor/ consumer	Working with sponsor/ consumer	Working for supplier
	Conceptualization and certification	Whole waterfall	Meeting requirements
	Confidentiality	Conflict of interest	Profit versus cost

the design environment is so large that heuristics, a sense of balance and artistic beauty and creative flare emanating from this balance, would carry more truth than complex numerical optimization algorithms. The existence of the "political" needs of the sponsor – he holds the purse strings – makes the exclusive use of mathematical equations to arrive to the optimal architecture even more inappropriate.

It can be said that a good architecture:

- Fulfills needs.
- Is well balanced.
- Is affordable.
- Is elegant.

One of the aims of this book is to develop the sensitivity to recognize a good architecture.

1.4. Design and implementation as a evolving process

In the previous sections, the design process has been presented as an exercise of balancing conflicting needs. This section will provide an introduction to the problem of how the need to balance conflicting needs evolve over time.

The space system engineer needs to strive for balance during all the design and implementation phases as do the specialist engineers within their areas of interest. Therefore, the same need recurs at progressively more detailed levels of design and an iterative process ensues. This is the reason that figures such as the cascade or spiral (Figure 1.3) have been produced in the past to illustrate the design process of complex systems.

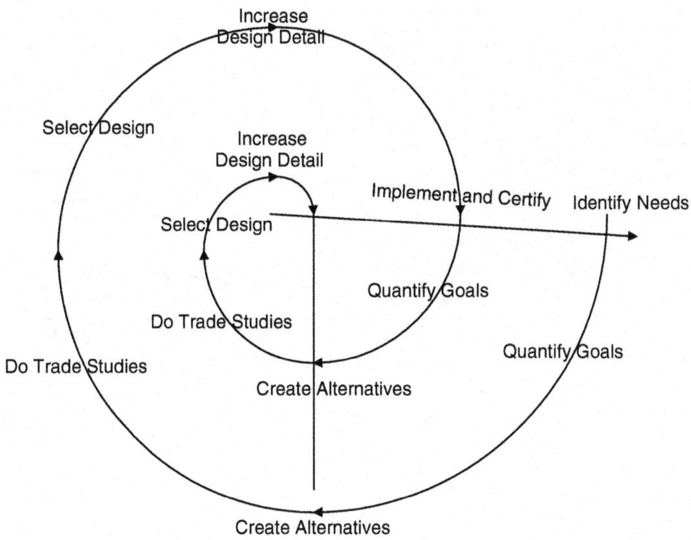

Figure 1.3. Design as a Recursive Process.

The highest level of design starts from general statement of needs and constraints, which will lead to the overall mission concept, the payload or observation concept, the broadest design features, and the highest-level set of specifications and plans for the project. Using this output as the starting point for the next design cycle, progressively more detailed designs and specifications will then be provided. The earlier design deals with the key requirements and the higher-level description of the mission but because this design underwrites the decision for project go-ahead, that means, for the system to be actually built, the higher-level description must also include very well described and analyzed critical subsets, because reliable estimation of project cost, performance, and risk will require fairly detailed analysis and design of these critical subsets. Each design at level n provides information needed to derive the specification and the starting point for the design at level $n+1$. That means at each different level of design the same problem of design by balancing conflicting needs reappears. At the beginning of the process, the design starts from a functional model of the highest level of abstraction. This level of abstraction is progressively reduced and more concrete concepts and models are provided. This progressive design process includes:

- Functional decomposition
- Physical building blocks with increased level of design granularity
- Overall design where all the critical elements have been progressively detailed
- Detail design at all levels

Once that the customer and supplier have finalized the design, they control the manufacturing, assembly building, integration, verification, and final acceptance of

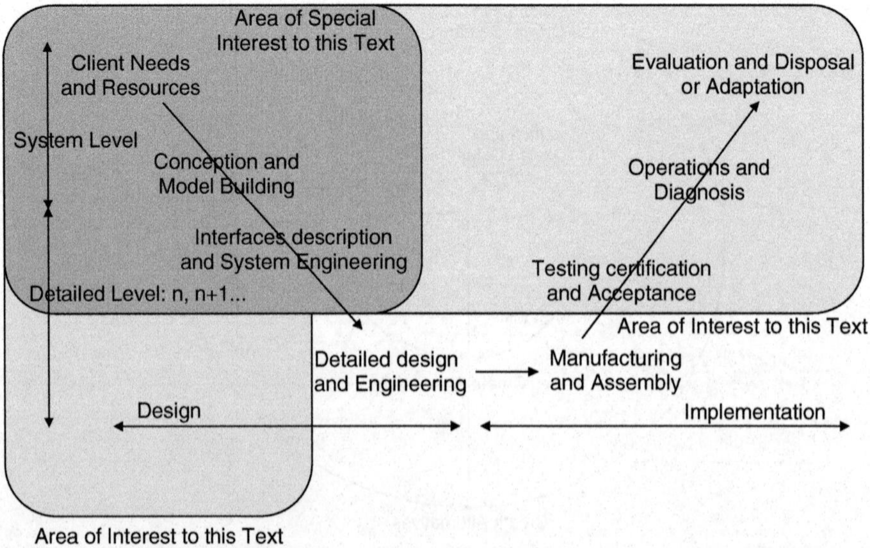

Figure 1.4. System Implementation as a V-Process and Areas of Interest to this Text.

progressively larger assemblies until the end-to-end system is finally certified and accepted as ready for use by the consumer. That means the design process is top-down, from the whole to the detailed, but the implementation is bottom-up, from the detailed to the whole, and the overall evolution in time of the system design and implementation can be depicted as a V-shaped process[11] (Figure 1.4). The descending branch is a decomposition and definition sequence and the ascending branch is an integration and verification sequence.

A sad reality associated with the V-process is that the first thing to be designed is the last thing to be verified and certified as being designed correctly. Indeed the earliest and the latest stages of a space system implementation have the end-to-end perspective in common. Once the project is approved, the responsibilities of implementing the space segment and the ground segment are normally in different projects and under the control of different hierarchies. The satellite system and the ground segment system engineers will concentrate on the very demanding task confronting each one of them: to produce their part of the system in line with specifications, on schedule, and within the cost budget. Nevertheless, the system architect acting during the earlier phases of the project has to produce a design where ground and space elements fit nicely together, functionalities are correctly allocated, and where the division of responsibilities is clear. Also, once the satellite is launched, the operations engineer handling the in-orbit satellite has to work with the ground segment and the space segment as a single entity. That means, it is at the beginning and at the end of the project when the consideration of the space and ground elements as part of a larger whole is

[11] System Engineering for Faster Cheaper Better, Forsberg and K. Mooz, Incose 1991.

the only correct one. This is the reason why this is a book about space system design, not about satellite system design.

During the design and implementation process described above, the interest in the balance of conflicting needs and in interactions between the system elements, which are the main subjects of this book, will be pertinent to:

- Activities related to highest-level system engineering during the whole design and implementation process, i.e., to activities located in the upper part of Figure 1.4
- Activities related to design at all levels, i.e., activities located in the left part of Figure 1.4
- And especially, to activities related to system-level design, i.e., activities located in the upper left quarter of Figure 1.4

1.5. Project phases and project reviews

Space projects are divided into distinct phases, which are designed to advance the system from one natural development step to the next. Each phase is dominated by a main activity and finished with a review of the work done.

ECSS defines the following phases of a space project:

- Phase 0: Mission and needs identification, which establishes mission aims and defines functions to be performed
- Phase A: Feasibility, which establishes system functional requirements, and concept
- Phase B: Preliminary definition
- Phase C: Detailed definition
- Phase D: Manufacturing assembly and testing
- Phase E: Launch and operations of the system
- Phase F: Disposal

The main object of this book is the study of the "architecture definition" phases that occur before fully funded project approval. A more detailed analysis of the contents of these phases will be provided in Chapter 6.

The transition of one phase to the following one is controlled by reviews that certify that the previous phase was successfully completed and that the next one is ready to start. The process described above is strictly sequential: requirements, design manufacturing, launch, and disposal. The lower part of the figure reproduces at a small scale the same V-diagram that Figure 1.5 depicted and that describes design from the general to the detailed and assembly and verification from the detailed to the general.

Depending on the mission and on the degree of novelty of the different components of the system, the implementation of the mission will include higher or lower levels of overlapping between design, manufacturing assembly, and verification. The system will include many larger or smaller components already existing and with proven records that would be used as ready off-the-shelf items. On the other hand, the satellite will also include totally new components that will require comprehensive detailed design and testing. To reduce mission risk, the design, manufacturing, and

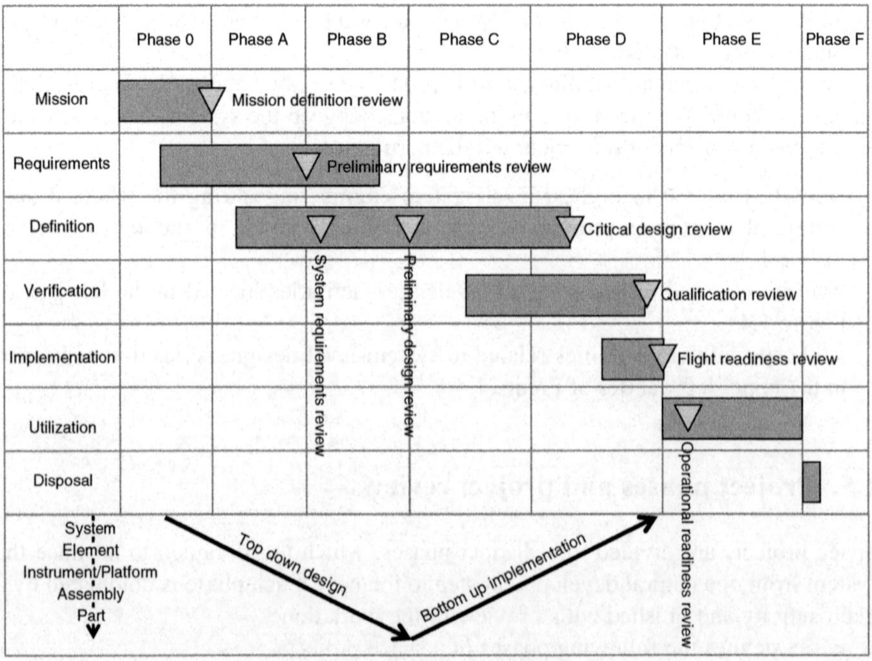

Figure 1.5. Project Phases and Milestones of a Typical Space Project According to the European Cooperation for Space Standardization (ECSS) (Space Project Management, "Project Phasing and Planning" ECSS-M-30-A Issued. Published by the European Cooperation for Space Standardization. ESA 1996).

test of these new and most critical components shall have started well before the official beginning of the detailed design phase C or verification phase D.

1.6. What is a space system?

A system is a set of interrelated or interacting elements that together fulfill a function larger than the direct addition of the functions of its elements. The system is separated from the environment and other external systems by an imaginary surface. The system interacts through that surface, which defines the system and its context. With that definition a system could be literally anything; nevertheless the elements of a meaningful and properly defined system work together to meet a clear set of objectives. In this case a meaningful space system shall be one whose elements work together to meet a space-related objective. Recalling the terminology of the previous sections, these objectives shall be to satisfy the requirements of the consumers within the constraints established by the sponsors. Through the boundaries, the space system will interact with different external elements including: the sponsor, the consumer, the external environment, and other relevant external systems; for instance, the data produced by other satellites built by other agencies but addressing similar needs.

1.7. Terminology

Some terms are going to be used in this book repeatedly and with a very specific meaning. Some of them have been already introduced. These important terms are the following:

Endeavor. Generic term to indicate the effort needed to achieve something. It is always used with a general meaning and it does not imply a clear beginning or end, or a clear line of authority.

Program. It is a long-duration, large-size endeavor, which includes a related series of rather independent undertakings and is focused to pursue a goal. It is under the control of a specific authority and it is normally long duration.

Project. A unique endeavor consisting of a set of coordinated and controlled activities, undertaken to achieve an objective conforming to specific requirements and constraints. It assumes a single authority and a clear beginning and end.

System. It is a set of interrelated or interacting elements. It is normally the object of a project. A system has a single authority.

Consumer. The user of the data produced by the space system.

Sponsor. The source of the funds for the space system.

Customer. The representative of the consumer and sponsor with respect to the supplier.

Supplier. The builder of the space system that delivers it to the customer.

Architecture. The set of information that defines a system design, value cost, and risk sufficiently for the purpose of taking a decision on the fully funded project go-ahead.

Architecture Definition Phases. The interlinked set of activities associated to the creation of an architecture. These phases are normally performed before the approval of a fully funded industrial project. Those phases are the main area of study of this book.

Implementation Phases. The interlinked set of activities associated with putting into orbit and successfully operating a space system, which has been previously defined in the definition phases. These phases are normally performed after the approval of a fully funded industrial project.

Architect. The person or set of persons evaluating, negotiating, and sometimes adjudicating conflicting legitimate technical, programmatic, and scientific needs by keeping an end-to-end perspective.

Requirements. Formal and unambiguous performances criteria against which the system will be verified.

Constraints. Formal or informal guidelines that limit the freedom of the system design or implementation.

Element. Higher-level part of the system, for example, the platform, the instrument the ground station.

Component. Lower-level parts of the system, for example, the attitude control subsystem of the satellite, the communication antenna.

1.8. Recommended supplementary reading

This text concentrates in space system synthesis and assumes previous knowledge on space system analysis. Furthermore the approach to this text is strongly based on system architecting paradigms and it requires some knowledge of system engineering and space system engineering theory. Further readings that will be necessary for a better understanding on the issues described in this book and that have strongly influenced its creation are listed in this section.

The synthesis and architecture perspective owns much to *The Art of Systems Architecting*, Mark Maier and Eberhardt Rechtin, CRC Press, 2000."

NASA Systems Engineering Handbook SP-610S 1995 provides an overall view of the special problems associated to space system engineering

Space Missions Analysis and Design, James Wertz and Wiley Larson, Space Technology Library, Springer, 1999 is the reference text on space system analysis and a tool to be used as a step-by-step guide to design a space system

Spacecraft System Engineering by Fortescue and Swinherd published by Wiley is an excellent guide for everything related to satellite technology.

A very comprehensive book on space system engineering is the *Spacecraft Techniques and Technology* produced by CNES and published by Cepadues Editions. This book is especially rich in information related to payload that perhaps the previous texts do not cover with the necessary level of detail.

The series of booklets produced by the European Cooperation for Space Standardization (ECSS) are extensively used and quoted along the chapters of this book.

The space missions analysis text mentioned above provide enough information to understand orbital mechanics. There are also specialized software packages used to calculate orbits and to define the visibility patterns of the satellite with respect to: the Sun, Earth, areas of observational interest, other satellites, and ground stations. Satellite Tool Kit (STK) from AGI is extensively used. The orbital analysis tool SaVoir from Taitus Engineering has been used to perform most of the orbital analysis necessary to create the figures of this book.

Chapter 2

Space Disciplines

Keywords System Engineering, System Architecting, Space Engineering Disciplines, Project Management, European Cooperation for Space Standardization (ECSS)

The fundamental aim of this chapter is to introduce the space engineering disciplines in the context of the architectural definition of a new space mission. This introduction describes briefly the roles and responsibilities of the different engineering disciplines during all the phases of the mission implementation but it concentrates in the role they play during the architectural definition phases. It is recalled that space system architectural definition is a rigorous process as outlined in the previous chapter. Small teams of the supplier and customer organizations, which have to keep a generalist perspective, direct the architectural definition. These small teams perform their activity interacting with specialists in the different space engineering disciplines. To understand these interactions it is necessary to introduce these engineering disciplines. This chapter has two objectives:

- To describe in detail two engineering disciplines: space system engineering and space system architecting. They are described in detail because they both require the generalist perspective adopted by this book.
- To briefly introduce all the other engineering disciplines relevant to the process of establishing the architecture of a space system. This introduction is done emphasizing the roles of the different disciplines during the space system architectural definition phases.

Engineering disciplines are divided into four categories:

- Satellite engineering disciplines
- Payload or instruments engineering disciplines
- Engineering support disciplines, which include common transversal activities such as: cost analysis, verification engineering, or quality assurance
- Ground segment engineering

To study in detail the classical engineering fields the reader must refer to the existing books that address these topics. Fortescue book[1] is especially thorough in the treating of the classical satellite engineering domains and CNES book[2] in the treatment

[1] Spacecraft Systems Engineering, Fortescue, Stark et al., John Willey and Son 1991.

[2] Space Techniques and Technology, CNES, Cepadues 2005.

M.A. Aguirre, *Introduction to Space Systems: Design and Synthesis*, Space Technology Library 27, DOI 10.1007/978-1-4614-3758-1_2, © Springer Science+Business Media New York 2013

not only of satellite engineering but also of instrument engineering and of the space engineering support disciplines and ground segment.

This chapter uses the European Cooperation for Space Standardization (ECSS) set of standards as the fundamental guideline for the description of the functions performed by the different engineering disciplines. Nevertheless, this chapter also recalls the existence of some disciplines, not considered by ECSS when discussing space engineering: one is system architecture engineering and the second one is the science and engineering involved in the generation of the final mission data products.

2.1. Space system engineering

According to NASA *System Engineering Handbook*[3], the objective of system engineering is to see that the system is designed, built, tested, and operated so that it accomplishes its purpose in the most cost-effective way possible. This shall be done considering performance, cost, schedule, and risk (Figure 2.1).

The system engineering process includes the following different functions:

- Integration and control
- Interfaces management
- Requirements engineering
- System analysis
- Design and configuration definition
- Verification

Of these six system engineering activities, interfaces management and integration and control occur during all the phases of implementation of a space mission. The

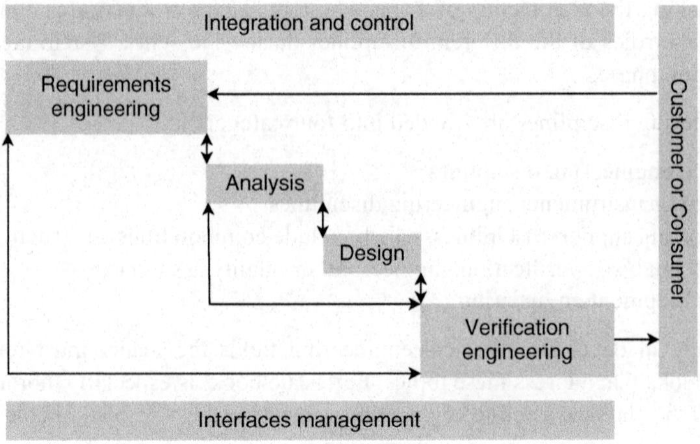

Figure 2.1. The System Engineering Process (Space Engineering Policy and Principles, ECSS-E-00, ESA 1996).

[3]NASA System Engineering Handbook, SP-610S, NASA 1995.

other four: requirements engineering, analysis, design, and verification occur more or less sequentially.

2.1.1. Integration and control

System engineering integration and control ensures the integration of the various disciplines and participants throughout all the project phases in order to optimize the total definition and realization of the system. It provides corresponding and essential inputs to project management.

The integration and control functions are strongly related to project management and include: the support to the project manager required to allow the overall management and planning, including the production and maintenance of databases, the control of information transfers, the management of interfaces, the creation and maintenance of technical budgets, and the technical support necessary to allow the project manager to perform correctly all the activities related with risk assessment and risk reduction.

Integration and control system engineering comprises the following activities:

- Planning and managing of the system architecture.
- Managing engineering databases.
- Planning document data exchange.
- Internal and external interface management.
- Environment engineering.
- Human factors.
- System budgets and margin philosophy.
- Technology risk assessment and risk alleviation.
- Ensure cost-effective procurement.
- Provide technical inputs to change management and assessment of the system engineering capability of the providers at all levels.

During the early phases before the approval of the project, the volume of information produced is moderate, the number of participants is small, and it is possible to keep the work logically structured without heavy formal procedures. Because of the modest number of documents, the space system architect can dispense with the provision of a formal management plan and system engineering plan. Nevertheless, formal risk management and risk reduction actions are necessary fundamentally to identify technology risks that shall be reduced before the approval of the project. It is also necessary to establish all the plans necessary to provide a correct estimation of cost and risk, including the overall development approach for the final implementation of the project, addressing model philosophy and technology readiness level for all elements of the system down to whatever level the technology risk arises.

2.1.2. Interfaces management

The system engineer also acts as an interface with project management team and the engineering support disciplines such as configuration control, product assurance,

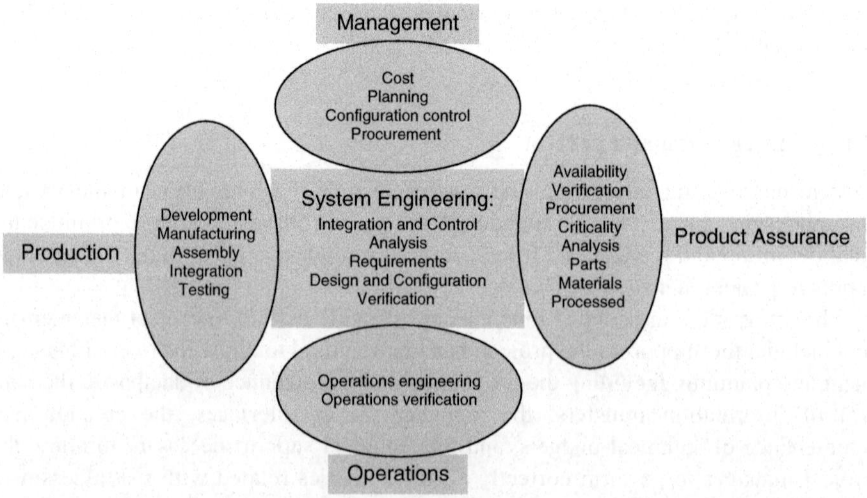

Figure 2.2. Space System Engineering Interfaces (System Engineering, ECSS-E-10, ESA 1996).

or production engineering. The system engineer reports to the project manager and provides all the technical information necessary for accurate estimation of schedule, cost, and risk as necessary for correct decision making. The system engineer interfaces with product assurance in the overall process of system verification and assessment of the design solution. The system engineer also interfaces with production and logistics including all the aspects related to manufacturing, assembling, and testing: test design, execution and reporting, and test assessment. Of special interest is the interface between the satellite system engineer and the in-orbit operations of the satellites. This will require the establishment of satellite operational procedures prior to the launch, and the activities related to satellite in-orbit commissioning, final in-orbit verification, and routine in-orbit operation (Figure 2.2).

2.1.3. Requirements engineering

Requirements engineering is connected with the duty of the system engineer to ensure that the system accomplishes its purpose. This will require the establishment, discussion, negotiation, and maintenance of documents listing the requirements on the system, eventually at each level of system integration, together with a specifications tree establishing the hierarchical relations between all the specifications. The system engineer will play a central role, being responsible with respect to the mission's consumers for providing a system able to fulfill their needs. The consumers will establish their needs that the system engineer will discuss and negotiate. After this, the system engineer will translate these highest-level consumer needs into a set of formal documentation that will apply to the design to be provided by the supplier. The supplier derives more and more detailed requirements applicable to the lower-level components of the satellite that will be imposed upon the different organizations within the industrial team who provide equipment and services to the prime industrial contractor. The supplier will accept or reject the delivered equipment and services on the basis of the corresponding requirements and in turn will deliver the system to the customer. The system engineer on behalf of the supplier and in turn the customer will accept or reject system on the basis of the applicable formal requirements as defined in the customer's System Requirements Document. Chapter 3 will study in detail the different classes of specifications, who owns them, and to whom they apply. It is evident that the cost and risk associated to any space system are functions of their performances as formalized in their requirements. The preparatory stages of any space project shall ensure that the requirements on the system are firm enough so that when the formal project is started, the changes to the original set of requirements are minimized. Requirements engineering will be further examined in Chapter 3.

2.1.4. System analysis

System analyses can be defined as: making the appropriate enquiries to assist decision makers in choosing a preferred future course of action by:

- Systematically examining and re-examining the relevant objectives and assessing alternative policies and strategies for achieving them
- Comparing, quantitatively where possible, the cost-effectiveness and risk of the alternatives

Analysis of the system performances during the early stages of a project enables attainment of the system requirements and the optimal allocation of the overall performance to the different contributing elements; for example, a pointing stability figure will be a function of the performance of the structure, the thermal control, and the attitude control subsystems, and the analysis of the overall performance will allow an investigation of alternative allocation scenarios with different apportioning to the different error sources until the right balance between them has been struck. Another aspect of the system analysis is functional analysis. This can be defined as "who has to do what?" Functional analysis is also a tool used to define the baseline architecture and the design at different levels. According to the already mentioned reference *System Engineering*, ECSS-E-10, before project approval the following system engineering analysis shall be performed:

- Analysis of mission requirements and derivation of high-level mission and satellite requirements
- Mission analysis and orbital calculations
- Operational constraints
- Functional analysis of space and ground segments
- Establishment and evaluation of possible mission concepts and derivation of their critical aspects
- Technology assessment versus design choices and establishing a preliminary approach for the technologies to be developed by the mission
- System margins approach and trade-offs between risk and performance
- Cost, risk, and schedule estimation

2.1.5. Design and configuration definition

Design and configuration definition is the process of defining the physical and logical hierarchies of the hardware and software from the highest to the lowest level of complexity of the system. Configuration is defined here, as the physical concept of the satellite, and configuration definition is the effort necessary to control the actual physical configuration of the satellite. Design and configuration control includes:

- The maintenance of the product tree, which defines all the products to be delivered and their relationships to each other
- Maintenance of budgets, such as: mass, pointing, power, computer loads, etc.
- Assessment of manufacturing, assembly, verification, and operations impact on the system's design
- The control of the "Design Justification Files," which provide the rationale supporting the design choices
- The analysis and test data, which shows that the design meets all requirements

During the earliest stage of a space mission implementation, which is the focus of this book, the volume of documentation will be moderate and normally it will not be necessary to establish a complex process of control of documentation configuration. Nevertheless, this is not true for the maintenance of budgets. Budgets maintenance is fundamental from the earliest phases because:

- Errors made in budget estimates at an early phase may propagate and remain undetected for some time, adversely affecting design decisions at several levels.
- Each budget allows the breakdown of the corresponding higher-level requirement into its contributing parts making possible the derivation of the lower-level requirements for each contributor.

There is an enhanced probability of making errors in critical system budgets early in the life of a project, as the maturity of the available information is relatively low. This is usually catered for by the adoption of a suitable safety margin philosophy. However, this must not allow complacency to arise, as such errors may result in the adoption of a nonoptimum system architecture, which could prove highly detrimental for all concerned, and no margin philosophy will recover the situation once commitment to hardware has been made. Hence, there must be particularly emphasis on achieving a system architecture that ensures good design budget margins as early as possible.

2.1.6. Verification

Verification demonstrates, through a dedicated process, that the system meets the applicable requirements and is capable of fulfilling its role during the mission's lifetime. Verification addresses all constituents of the system and is incrementally performed. Verification is done bottom-up from the small and isolated to the large and integrated (Figure 1.4). The final stage of the verification is when the integrated system delivered by the supplier has been fully verified. The system engineer is the final responsible person for certifying that the satellite and the end-to-end system requirements are fulfilled. The specialists in verification and test assist him in this task; these engineering disciplines will be introduced in Section 2.6.1. Verification is much more a task that arises at the end of the project than at the beginning. Therefore, the architectural definition phases will not include many verification activities. Nevertheless, there are two aspects of verification that are important from the earliest phases of any space mission design:

- The verification of critical technologies that are necessary to reduce the risk prior to project go-ahead, i.e., necessary predevelopment.
- The understanding of the difficulties – and costs – associated with the verification of the mission. The establishment of any requirements requires thinking in detail about how it is going to be verified.

Predevelopment is analyzed in Section 6.5.1 and the importance of assessing verifiability as early as possible is studied in Section 3.5.

2.2. Space system architecting

Architecture is defined by ANSI/IEEE 1471–2000 as "The fundamental organization of a system, embodied in its components, their relationships to each other and the environment, and the principles governing its design and evolution." System architecture definition is the activity needed to establish the architecture of a system and the discipline that addresses the definition of architecture could be called system architecting. System architecting is a novel approach whose origins lie in two sources:

- The classical figure of the architect as the interface between the builder and the client
- The formalistic approach of large software creators

The classical approach places emphasis on understanding the real needs of the consumer and sponsor in the context of the possible project environment. The formalistic approach focuses on the development of methods for the systematic analysis of complex system so as to understand them better and design and implement them properly. This section presents both approaches and analyzes to what extent they are applicable to space engineering.

System engineering is a well-established engineering discipline, while system architecting is not yet so well known. Nevertheless, system architecting is especially useful for space system design in two situations:

- The classical architect approach to system architecting is quite relevant to the work prior to project approval. The classical architect perspective has been already discussed in Section 1.2 and especially in Table 1.1 and it will be revisited and expanded in Section 2.2.1.
- The software formalism approach is useful for the analysis of complex and large systems of systems. This approach will be discussed in Section 2.2.2.

Both approaches make a strong emphasis on wider perspectives, relations with the external environment, nonnumerical approaches, and interrelations between system elements.

2.2.1. The architect in the classical role

Table 1.1 demonstrated that the perspective of the system engineer in the early stages previous to the formal approval of a space project is fundamentally a classical architect perspective. According to the classical paradigm, the architect is working with, and for, the client and he/she is the communication and interpretation channel between the client and the builder. Nevertheless, the space architect shall also reconcile the conflict generated by the fact that in the overwhelming majority of space mission, there are two clients: the consumer that uses and the sponsor that pays. A large portion of the effort during the architecture definition phases of a possible new space mission is concerned with the reconciliation of the contradiction between sponsors and consumers.

The scientific tools that the architect has for the execution of his job are just those that system engineering provides to the system designer, no more, and no less. What the correct architectural design perspective adds to system engineering is not new tools but an especial perspective. This perspective requires:

- The need to consider perspectives broader than the possible space mission itself
- The need to synthesize all the pertinent sources of information
- To include nonmathematical considerations: heuristics, self-discovery, and trial and error approaches

To have a new space mission approved requires taking a perspective that is broader than the mission itself. The mission will only be approved if the – let us assume properly designed – system fits into the wider external environment. This awareness of the external environment would need the use of tools outside traditional system engineering and more in the domain of management sciences. Section 4.2 will analyze the tools for the analysis of the external environment. This external environment is outside – and also above – the control of the system designer. Nevertheless, the awareness of this environment will be necessary to arrive at an affordable and useful system design that will be finally implemented fulfilling real consumer needs.

The architect's perspective required for the proper definition of a new space system values not only knowledge and equations but also wisdom and heuristics. This somehow surprising statement has been already briefly justified in Section 1.3. Chapter 5, which analyzes the design process, also makes continuous reference to the need to develop the sensitivity to recognize proper and balanced design that is independent of purely mathematical methods. Chapter 5 also includes numerous heuristics that are adequate for space system design. Specifically, Section 5.6.2 is dedicated to the limitations of purely numerical methods for analyzing the overall performance of a space system. Indeed the whole second – and longest – part of this book, starting at Chapter 7 and finishing at Chapter 12, is based around the idea that presenting and commenting on past designs will allow the reader to develop this nonnumerical sensitivity for what a balanced and good design is.

Quoting Scott Hyers on his notes to *An Introduction to General System Thinking* by Weinberg[4]:

> Pure scientific approach says: "Clear logical thinking requires that we vary only one factor at a time" and "All other things being equal …." A system is a collection of parts no one of which can

---

[4] Quoted by Scott Hyer in a presentation to Incose.

be changed without changing the core character of the system. Therefore the study of system is not conducive to the "one factor at a time" strategy.

This would mean that proper system thinking could not be wholly "scientific." It needs a supplementary perspective. This perspective is not easy to define and a scientific system method of architecture definition or tool kit is not easy because the required perspective is: unstructured, eclectic, heuristic, and even artistic. It is a synthesis of many skills not all of which are analytical, and yet sciences are analytical.

2.2.2. Architecture definition formalisms

The fundamental aim of formalistic system architecting is the creation of unambiguous descriptions of a complex system where the functions, relations, features, and behavior of the system are clearly explained. This is performed by the development of a clear and rigid set of concepts that applied correctly will unambiguously describe the system. This description supports reasoning about the properties of the system. It defines the system building blocks and provides a plan from which products can be procured, and systems developed. It thus enables the management and development of complex systems in an optimal way. It is evident from the previous paragraph that architecture definition formalisms provide direction on how to describe architectures; the framework does not provide guidance in how to design or implement a specific architecture.

There are several approaches to arrive at these descriptions. An introduction can be provided by the approach followed by IEEE P1471,[5] which is a set of software engineering standards, but is at the same time very general and independent of any specific framework. According to this approach every system is described by several architectural descriptions. Each architectural description is called a view.

A view is a representation of a system from the perspective of related concerns or issues. The view is provided with internal logic by the viewpoint of the representation. The viewpoint is a template pattern or specification that answers the concerns of a specific type of stakeholder in the system. We can have as many different but mutually consistent description of the system as classes of stakeholders we can imagine. The "view" answers the concerns of the stakeholders and it provides the link between the classical and the modern approaches to system engineering.

The formal architectural description provides a mechanism for understanding and managing complexity. Defense systems are extremely complicated and they comprise many fairly autonomous functional and physical blocks, and the defense world has developed extensively the use of architecture definition formalisms. The American Department of Defense has produced a set of guidelines for the creation of views.

According to Figure 2.3, an architectural framework shall have three fundamental views:

- Operational view
- System view
- The technical view

[5] IEEE Recommended Practice for Architectural Description, IEEE P1471 1999.

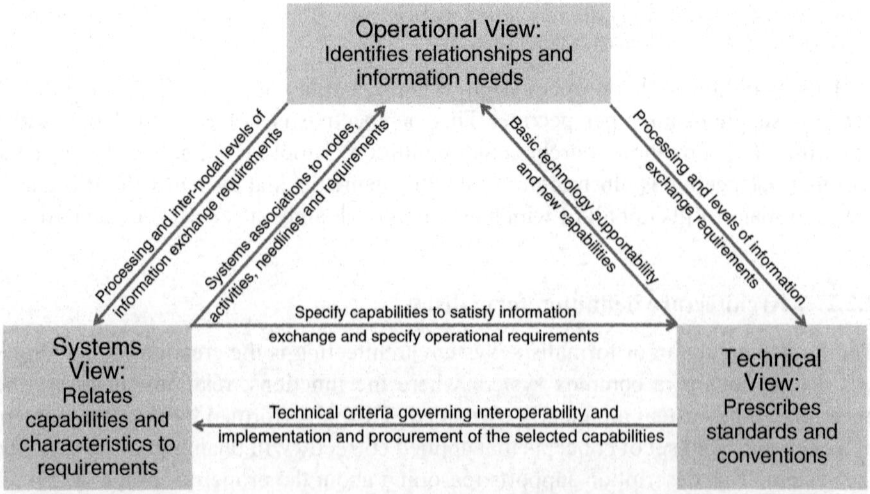

Figure 2.3. Fundamental Views as per the USA Department of Defense (C4ISR Architecture Framework, Architectures Working Group, Department of Defense, 1997).

The operational view is a description of the tasks and activities, operational elements, and information flows required to accomplish or support a military operation. It contains descriptions – often graphical – of the operational elements, assigned tasks and activities, and information flows required to support the system under consideration. It defines the types of information exchanged, the frequency of exchange, which tasks and activities are supported by the information exchanges, and the nature of information exchanges in detail sufficient to ascertain specific interoperability requirements.

The system view is a set of descriptions of systems and interconnections providing for or supporting system functions. The system architecture view shows how multiple systems link and interoperate and may describe the internal construction and operations of particular systems within the architecture. For the individual system, the system architecture view includes the physical connection, location, and identification of key nodes (including materiel item nodes), circuits, networks, platforms, etc., and specifies system and component performance parameters. The system architecture view associates physical resources and their performance attributes to the operational view and its requirements per standards defined in the technical architecture.

The technical view is the minimal set of rules governing the arrangement, interaction, and interdependence of the parts or elements whose purpose is to ensure that a conformant system satisfies a specified set of requirements. The technical architecture view provides the technical systems-implementation guidelines upon which engineering specifications are based, common building blocks are established, and product lines are developed. The technical architecture view includes a collection of the technical standards, conventions, rules, and criteria organized into profile(s) that govern system services, interfaces, and relationships for particular system architecture views and that relate to particular operational views.

2.3. Project management

For the supplier and customer, the overall authority on the space project is in the hands of a single person who ensures the success of the project with respect to it being on schedule, within the cost ceiling, and fulfilling the requirements. This is the function of the project manager. The responsibilities of project management include:

- Phasing of the project
- Cost and schedule management
- Configuration management
- Information and documentation management
- Project organization and work breakdown
- Logistics and support
- Human resources including: staffing, training, and appraisal of individual performances

A large space project is a large endeavor of the extreme complexity including scientific, technical, economic, and managerial aspects. The implementation of a space project requires the cooperation of many organizations cooperating to provide a product, which satisfies the consumer requirements while fulfilling the sponsor constraints on cost and schedule. To reach this goal, technical activities and human and financial resources shall be organized and coordinated in a structured manner in order to obtain the end product. This structure, together with related processes, constitutes a project. It implies a target (the system), a time frame, and actions to be performed under resources constraints. Project management consists of the definition, implementation, and execution of such actions including the verification of achieved results against project projections and expectations.

Project management requires careful deliberation of what shall be accomplished, laying out all the steps needed to build that future, and obtaining the resources required to carry out those steps. But most important, it requires dealing with reality, problems, delays, changes, obstacles, and, sometimes, opportunities that arise as a project takes place. That means, it implies dealing with risk in the real world. In the architectural definition phases, the leaders of the customer and supplier small teams act often as project managers as well as architects.

2.4. Satellite engineering disciplines

To design a space system involves far more than designing just the satellite, but it is the presence of the satellite that makes a system a space system, and it is evident that the satellite design will take the largest part of the effort of designing the space system. To design and build a satellite it is necessary to integrate a set of well-defined and complementary engineering disciplines. The division in satellite engineering disciplines used by ECSS includes:

- Mechanical, which includes structures, thermal control, mechanisms, and propulsion
- Electric power, electronics, and data handling
- Software

- Communications
- Satellite attitude control and orbit control

All these engineering disciplines are going to be introduced in the following sections.

2.4.1. Structure

Structure is every part of the satellite carrying loads and supporting or protecting other satellite parts. In order to carry and protect, the structure must provide strength and rigidity for withstanding environmental loads without endangering the mission's objectives. The structure must also be stable enough to avoid deformations that could degrade mission performances, for example, by mispointing the instruments. The structure also provides other functions like protection from the Sun and electrical conductivity or isolation. The structure has to provide its carrying functions during all phases of the mission: on-ground and in-orbit.

During the architecture definition phases there are two fundamental issues:

- To provide a shape – that means a configuration – to the stowed and deployed physical envelope of the satellite
- To provide an inner skeleton – a primary structure proper – to the satellite

This inner skeleton acts as the main load path for the whole satellite. The outer shape and the inner skeleton shall fulfill their structural function during the widely changing environmental situations during the entire satellite lifetime comprising the following steps:

- On ground, the structure acts as bench to mount the rest of the satellite around it. The structure's adequacy for on-ground assembly, disassembly, and transportation is also an important design factor that must be taken into account.
- During launch, the structure withstands the heavy loads generated by the firing of the rocket engines.
- In orbit the structure holds in the deployed configuration the rest of the satellite in the right location and in the right orientation. The structure shall also withstand the loads generated by satellite maneuvers.

The shape of the stowed configuration inside the launcher and the shape of the deployed satellite in orbit can be very different. This requires the use of deployment mechanisms to reconfigure the satellite once in orbit and separated from the launcher.

Structural engineering disciplines are based on stress analysis and strength of material sciences. Finite Element Analysis is the standard technique to determine stress and strength in the structure of a satellite. Structural engineering is a mature discipline with well-established procedures, well-proven structural assembly techniques, and highly reliable analytical methods. On the other hand, the correct shaping of a satellite is as much an art as it is a science. The shaping of the satellite is a true multidisciplinary activity that can only be properly performed by considering the conflicting demands of all the components of the satellite during all the satellite life stages. That means satellite configuration is based primarily on the interrelations between satellite components, which is the main focus of attention of this book, and in line with this attention, it is extensively discussed in Chapter 10. Specifically, structural engineering tools are studied in Section 10.1.1.

2.4.2. Thermal control

The thermal control components of a satellite keep the hardware within specified temperature levels, preventing them from becoming too cold or too hot. The thermal components need to fulfill this function throughout the mission's entire lifespan and during all foreseeable situations. To do so, the thermal components must be able to provide heat when the hardware becomes too cold and have the ability to dissipate heat when the hardware becomes too hot.

Heat transfer inside the satellite takes place via radiation or conduction but the interchange of heat between the satellite and the external environment always occurs through radiation. The thermal components can also be used to provide uniform thermal conditions inside the satellite reducing thermoelastic distortions and improving dimensional stability. Thermal control engineering is the technical domain that deals with the techniques necessary to ensure that all the components of the satellite are kept within adequate temperature ranges. Thermal control engineering can be divided into specialized subdomains depending of the temperature range. Thermal engineering can be divided in three main categories:

- Cryogenic, below 120 K
- High temperatures above 420 K
- Conventional between 120 and 420 K

During the early stages of the project, the thermal engineering is mainly concerned with the correct location and sizing of the radiators to dissipate the heat produced by the satellite with the orbital geometry provided by the satellite configuration, the orbit, and the attitude. That means the location of the radiators influences and is influenced by the satellite configuration. The location of the radiators is a special thermal requirement, like the need to keep an instrument within very high or low or very restricted temperature ranges. The thermal control is also a well-established engineering field and the thermal control within conventional ranges around the normal ambient temperature around 300 K can be considered mature. On the other hand, the development of dedicated thermal components for devices that require operating at very high or very low temperatures will require special developments that could require manufacturing and testing of the dedicated components prior to the project approval.

Thermal engineers use analytical techniques like finite difference analysis to determine the interchange of heat between the different parts of the satellite and between them and the space around the satellite. These analyses allow for the determination of the temperature of all the satellite elements during all the mission phases as a function of time. The design of thermal control, for satellites that do not need to dissipate large amounts of localized heat and that work at conventional temperatures, is mature, and there are proven methods of design and reliable methods of analysis. However, the design for thermal controls working at very low temperatures or dissipating large amounts of heat is complex and requires the consideration of the impact of thermal design decisions on other satellite subsystems; for example, a satellite can simplify its thermal control at the expenses of complicating its configuration. These impacts will be studied in Chapter 10. Specifically, thermal design tools and issues are revisited and expanded in Section 10.1.2.

2.4.3. Mechanisms

A mechanism is defined here as any device used to change the geometric configuration of a satellite or part of a satellite in order to ensure the satellite's correct functioning.

Mechanism reconfiguration can be internal to a specific satellite component or can introduce major and visible changes to the satellite's overall configuration. Internal mechanisms perform all the movement functions required by satellite components, for example, those allowing for the rotation of momentum wheels or the pointing of an instrument along a line of sight. External mechanisms are required for producing any temporary or permanent changes in the satellite's overall configuration. These changes are driven by the contrast between the constrained and stressed environment of the stowed satellite during launch and the unconstrained and lightly loaded environment of the satellite once in orbit. To perform these reconfiguration functions mechanisms must:

- Restrain and lock the component to be reconfigured until the moment the reconfiguration is required.
- Release the components to allow the reconfiguration to start.
- Generate forces and torques producing the movement required by the reconfiguration.
- Minimize friction during the movement.
- Measure and control the movement.
- Lock the reconfigured parts once the required movement is performed.

The change from the launch-stowed to the in-orbit-deployed configuration is of special importance.

Once the reconfiguration is finished, mechanisms become structural in functional terms because they carry loads and provide support, and because they have to provide strength and stiffness. Like structures, mechanisms shall also provide accurate positioning and dimensional stability. Mechanisms engineering uses all the structural engineering tools together with specialized knowledge in supplementary areas including:

- Electromechanical actuators to provide the energy and the movement required to change shape or to enable the movements
- Tribology, that means lubrication science and engineering, to ensure smooth movements in all conditions
- Control theory and control electronics in the very likely case that movements would be enabled and controlled by electrical actuators and sensors

Mechanisms need to perform their intended function in orbit safely and reliably. The changing thermal situation and the vacuum of the space environment make space especially unfriendly with respect to ensuring smooth movements between the different parts of the mechanism. High and low temperature will produce thermal expansion or contraction, loosening, or tightening mechanisms. The complex thermal situation and the vacuum render useless many lubricants adequate for use on Earth; the engineering discipline of developing substances that lubricate efficiently in vacuum is the well-established domain of space tribology. Mechanisms engineering also drives and can be driven by the design of the configuration of the satellite. Satellites with a large difference between stowed and deployed configurations will require many and complex mechanisms. The accurate control of pointing and alignments and the in-orbit operations of specialized mechanisms as required by the pointing of satellite instruments is an area that is normally strongly mission dependent, requiring considerable effort in the early stages of a project. To reduce mission risk and before the approval for the project, it is often necessary to design and test the most innovative and critical mechanisms required by the mission. Mechanisms will be discussed in more detail in Section 10.1.3.

2.4.4. Attitude control

To survive and properly perform their functions, orbiting satellites must point in specific directions; which is to say that satellites in orbit need to determine and control their attitude. The attitude control system is the part of the satellite, which is designed to provide control to the satellite.

Satellite control engineering is the technical discipline that ensures that the satellite's status is the desired one in the context of its variable external environment. In particular, satellite attitude control engineering deals with the control of the attitude. That means, the control of the orientation of the satellite body with respect to a reference frame. The attitude control of the satellite includes sensors and actuators acting upon the satellite. The control will also include some kind of feedback loop.

Attitude determination devices sense a satellite's attitude, which aids controlling the satellite's current attitude. The logic of the control loop can be digital or analog. In the case of a digital control loop, the control logic can be implemented by the satellite's central processor or be given its own dedicated processor. Actuators are used to provide the torques necessary for producing the required attitude changes. Satellites should be capable of assuming different attitudes with different sets of requirements for different situations. Each of these situations will define a mode for the satellite's attitude control and may require different sensors, actuators, and control logic. Attitude modes will be examined in Section 9.5.

During the architecture definition phases prior to project approval, it will be necessary to establish the control concept, the sensor and actuator tools, and the control loops. The resulting designs must be supported by analysis to demonstrate that they are in line with requirements. The selection of the satellite attitude derives directly from the needs of the events under observation, and it impacts heavily on the satellite configuration because the satellite attitude directly drives the satellite's shape. The satellite could have different attitudes depending on the different satellite operational status. In some situations, the closing of the loop will involve action from the on-ground satellite control team. This introduces a supplementary interface between attitude control and satellite operations. Attitude control is also discussed in Section 9.1.3.

2.4.5. Propulsion

Satellite orbits frequently require small and large corrections. Large corrections are necessary when the launcher releases the satellite into an intermediate orbit. Small corrections are necessary to avoid the satellite orbit drifting with respect to the nominal one due to perturbations from, for example, atmospheric drag or the gravitational effect of the Moon or Sun. Propulsion is defined as the set of components providing the satellite with thrust for autonomous orbit correction.

The propulsion components of the satellite include the propellant, propellant tanks, pipes, filters, valves, and thrusters. The propellant or propellants provide chemical energy, which is transformed into kinetic energy of the gases ejected by the thrusters. The on-board propulsion components allow the satellite to apply forces thereby controlling its orbit. On-board propulsion can be used also to apply torques to the satellite. In this case the propulsion will behave as an attitude control actuator.

One of the first and most fundamental decisions to be taken during the architectural definition phases is the orbit selection and the definition of the orbit control approaches. Being such a high-level decision, it is normally taken considering the mission as a whole with the implications on the definition of the propulsion subsystem being only one factor within many. In this case the selection of the orbit and the orbit control approach will determine if it is necessary to carry complex, simple, or no propulsion at all. The energy necessary to change the orbit will be provided by the chemical or pressure energy stored on the propellants. In the case of electric propulsion the energy will be provided by converted electric energy obtained from the Sun. The propulsion concept will be selected once that the launcher, the trajectory, and the overall orbit correction and control are established.

Propulsion and other tools of orbit determination and control are discussed in Section 9.1.2.

The design of rocket engines, both for large launcher engines and for small satellite thrusters, is highly specialized and rather disconnected from the other space engineering disciplines. The propulsion components of a satellite are almost never designed specifically for the mission but selected from within a range of already available components that appear suitable to fulfill mission needs. From this point of view propulsion engineer plays a relatively small role on the overall process of architectural definition.

The simultaneous decision regarding the launcher, orbit, and the architecture of orbital control is a fundamental one, to be taken early in the project and having major impact throughout the mission design; this means it is a decision in which the mission architect plays the leading role. The architect shall also be active in identifying the implications for the mission of design decisions made by the propulsion specialist, such as the reduction of instrument availability during orbit corrections or the choice between satellite autonomy and ground intervention for orbit determination and control, to be discussed in Chapter 11 as part of satellite operations.

2.4.6. Electrical power

Electrical power engineering is the technical domain dealing with the provision of electrical power to the satellite. The electrical power has to be: produced, stored, regulated, and distributed. In almost all cases the power is produced by solar cells, is stored in batteries, and is regulated and distributed by specialized electronic assemblies. Solar cells are located in solar arrays. These arrays are large, flat, slender structures of sufficient area to accommodate the solar cells over them. Solar arrays are often deployed to allow the collection of the required amount of solar energy. Once deployed, the solar arrays are oriented to track the variable direction of the Sun along the orbit. The power subsystem also directly controls power provision to heaters, pyro initiators, and other key elements that must be able to operate even when major components of the satellite are disabled.

Two very different aspects characterize power engineering:

- The engineering of electrical power production, control, storage, and distribution. This is an electrical engineering discipline.
- The engineering of providing deployable structures where the solar cells can be located and where they can properly track the Sun to provide maximal energy. This is a mechanical engineering discipline.

Most missions have very similar power storage, regulation, and distribution requirements and it is possible to use well-proven designs from one project to the next with small modifications. This means the design of the electrical part of power engineering is fairly standard and it does not require heavy interactions between the power engineering specialists and the rest of the architectural design team. In most cases, this will also be the situation for the "mechanical" aspects of the power production design; qualified solar arrays concepts with its hold-downs during launch, deployment, and Sun-tracking mechanisms can be used from one project to the next with small modifications. Nevertheless, the solar arrays strongly influence the satellite's configuration, and the configuration interrelates with almost all the satellite design aspects. For example, it is sometimes possible to accommodate the solar arrays as panels fixed to the spacecraft's structure, particularly if the power demand is modest and the orbit is suitable. This has the advantage of avoiding the need for solar panel hold-down, release, and deployment and if required, solar arrays drive mechanisms, all of which add cost and mass. Conversely such fixed solar panels operate with a higher temperature, as they lack the ability to radiate the absorbed solar energy from their rear faces. This is a clear example of the interactive nature of spacecraft design, where the requirements of electrical power generation must be considered as a trade-off exercise in combination with the consequences for spacecraft configuration, equipment accommodation, orbit selection, and thermal and structural design. Therefore, solar array accommodation will play an important role during the architectural design phases of space mission definition.

Power engineering from the electrical point of view is further discussed in Section 11.1.1. Solar arrays as structures are discussed in Section 10.1.4.

2.4.7. Data handling

The on-board data handling is the spacecraft's central nervous system. It monitors, stores, and controls the status, health, and performance of each satellite component. Moreover, the data handling interprets and executes the commands sent by mission control and received and decoded by the communications subsystem. Depending on the nature of these commands they will be for either immediate or delayed execution. The data handling is also capable of executing commands autonomously, as needed for the operation of the mission and depending on the design's level of autonomy. This system also records and provides time data for the satellite's components.

Satellite on-board data handling systems have two salient features:

- The hardware has to work in the harsh space environment. This forces the use of well-characterized components and also requires the use of robust and redundant designs thereby ensuring the survival of the satellite and the recovery of the data handling functionalities, even in the case of multiple failures.
- The time needed to design, implement, and fly a space mission is longer than the life cycle of data handling technologies.

These two factors combine to make a satellite's data handling technologies very conservative – even old fashioned – with respect to their on-ground equivalents. This situation is sometimes so extreme that during the early design phases of a space system it is necessary to take into account the likely probability that components used by the design under consideration could become obsolete and unavailable when, many years later, it is necessary to implement them. Besides the conservative attitude of satellite data handling engineering, the advances in information technology push continuously toward more autonomous and powerful on-board designs.

The number and style of observations and the data to be produced will be different for each new mission. Therefore, a dedicated architecture of the data handling shall be designed during the architectural definition phases of the mission. The design of the architecture requires defining the number and role of the on-board computers and dedicated data processing units that will process the information, the number, and speed of the data buses that will transfer information and the number, speed, and storage capability of the data storage devices.

In all satellites, there is always a central computer that controls the satellite as a whole, but supplementary computers could control specific areas; for example, the handling, processing, and storage of the data generated by the instruments or the control of the satellite attitude. A fully centralized data handling is normally the responsibility of a single supplier but the responsibility in decentralized data handling is normally distributed between the suppliers of the different assemblies. In many missions the architecture of the on-board data handling differentiates between components specialized in the handling of the stream of data produced by the instruments and the monitoring and control of the status and health of the whole satellite. Satellite data handling is revisited and expanded in Section 11.1.2, and the data handling of the instruments data output is further analyzed in Section 12.1.1.

2.4.8. Software

Software is defined as the set of computer programs, procedures, documentation, and their associated data. Software is clearly an independent engineering discipline with its own methods and skills. On the other hand, software resides physically in the computer and the computers are a part of data handling engineering, which has been introduced in the previous section.

The on-board software will be new for each new mission but the actual coding of the software is considered detail engineering that is implemented after the architecture definition phases of the space mission have finished. During the architecture definition phases the involvement of software specialists is minimal. At most software design is considered only at the level of functional analysis and functional blocks to be implemented. Missions advancing the state of a science will require the early development of algorithms to process the data produced by the instruments, but this is more related with the science behind the space mission than with software engineering as such.

The software related to any space mission can be divided into two:

- On-ground
- In-orbit

Taking into account the danger involved during satellite operations, in-orbit software is especially important. Critical software is defined as the software supporting a safety or dependability critical function that if incorrect or inadvertently executed can result in catastrophic or critical consequences; that means, loss of life or loss of the space mission. Critical software has specialized methods of product assurance and verification. During the architectural definition phases of the space mission it is not necessary to design the software, not even the critical software, but in order to properly define the cost of the mission, it is necessary to define the software and critical software development approaches.

2.4.9. Communications

Communications is the engineering discipline that allows the interchange of information between the satellite and ground segment.

The telemetry and telecommand data communications components allow the interchange of information between the satellite and ground segment providing the ground operations personnel with the necessary information to determine the status, health, and performance of each satellite component, and receive the commands sent from ground to control the satellite's functions, either in nominal mode or when dealing with contingencies, by receiving high-priority commands to directly address the most fundamental components of the satellite. In cases where the satellite has a single communication downlink, as occurs in many space-science satellites, the link will be used to send to Earth all the data produced by the satellite and its instruments. The information sent to the satellite generally consists of modifications to the on-board software, data to be stored on the satellite computers, or commands to change the state of the on-board components. In almost all Earth observation missions there is a second independent link that codes, encrypts, modulates, amplifies, and downlinks to Earth the data produced by the instruments. This second link is justified because the data throughput of the instruments is much larger than the data throughput of the rest of the satellite.

Almost all communication links use modulated microwave signals to transfer the data from the satellite to and from the ground, but some successful satellite-to-satellite optical communications links have been already flown. The stream of bits between the satellite and ground is coded using different digital schemes. To avoid interference between satellites and from satellites to ground communications, the frequencies and frequency bandwidths for different types of satellites is strictly regulated. The communication link performances will depend strongly on the gain of the satellite and ground antennae. Large on-ground high-gain antenna will improve the link but they are expensive as are large and pointable satellite antennae and additionally they are less reliable than low-gain omnidirectional ones.

The downlink communication needs of each new mission will depend not only on the number and style of observations and on the amount of data to be produced, but also on the number and performances of the ground stations used. The selection of the ground stations is dependent upon the orbit of the satellite. The possible use of data relay satellites may be necessary when the data flow is very high, or when it is essential to have continuity of data flow. Therefore, the design of the communications architecture requires a panoramic view of the overall mission. Normally the data flow between the satellite and ground is channeled through a limited number of ground stations, and communications are only possible during the periods when the ground stations have visibility of the satellite. For some orbits these periods can be long or even permanent as in the case of geostationary orbits (Section 9.4.4).

Telemetry and telecommand communications is further discussed in Section 11.1.3. The downlink of the data produced by the instruments is discussed in Section 12.1.2.

2.5. Instruments engineering

The justification of the existence of any space mission is their payload or instruments. The satellite is built to fly their instruments, and all the other components of the satellite are designed to be at their service. Furthermore, the structural concept, thermal approaches, power subsystems, data handling architectures, and even whole satellite platforms could be reused from one space mission to the next, but instruments typically are bespoke for each new mission. All these reasons make instruments engineering the most important part of the effort during the architecture definition phases of any new space mission, and the work done during these phases will always be fairly detailed in the definition of the instruments to be flown by the satellite. This will require a design detailed enough:

- To ensure that the performances of the instrument will be in line with the requirements. The performance of the instruments is directly related to the fulfillment of the desires of the consumer needs.
- To ensure that the risks related to the instruments have been properly identified, that the development approaches to minimize these risks have been established, and that the associated costs and schedules have been determined.

Instruments can be very varied; there are many instrument classes depending on the physical features they measure. These physical features could be an in-situ measurement of the space being traveled by the satellite. More often instruments will measure the electromagnetic radiation emitted by a remote event.

The design of space instruments requires the contribution of all the engineering disciplines listed in Section 2.4. A telescope will have its structure, thermal control, pointing mechanisms, data handling electronics, and power handling electronics. Additionally, a telescope will be an optical instrument that requires the support of optical engineers who provide its optical design and detectors engineers who provide know-how on the devices that will detect the photons collected by the telescopes. The same type of specialized knowledge will be necessary to design any other type of receiving instruments from magnetometer to radar.

Instruments can be divided into the following types:

- Passive optical instruments. These rely on electromagnetic radiation emitted or reflected by the events being observed.
- Active optical instruments. These emit and receive laser pulses.
- Passive microwave instruments. These detect microwave radiation emitted by the events being observed.
- Active microwave instruments. They emit and detect microwave signals backscattered by the events under observation.
- In-situ instruments. They do not perform remote sensing but measure the local environment around the satellite.

Depending on the nature of the instruments the nature of the dedicated knowledge to design them will be different. Some instruments are highly specialized, small, with clear interfaces, and they produce data of interest only for a narrow community. In this case they can be directly defined and implemented by the small consumer community that is going to use the data. In this case the delivery of the instruments could be contractually separated from the delivery of the rest of the satellite and the relationship between the instruments and the rest of the satellite will be controlled by specific interface rules. This approach has been very common in the past but nowadays many instruments are large, with complex interfaces with the rest of the satellite, and their data is of interest to a dispersed community that cannot be made collectively responsible for building the instruments. Consequently instruments of this second type are a distinct part – indeed the most important part – of the satellite, and the organizational arrangements to produce them are identical to the arrangements for any other part of the satellite. Chapter 8 will further study satellite instruments.

2.6. Engineering support disciplines

This text considers a third group – on top of satellite subsystem engineering and instruments engineering – that this text calls "support disciplines." These disciplines apply horizontally to all the components making a space system. These disciplines represent a considerable part of the final cost – and of the associated problems – on any space mission.

2.6.1. Manufacturing assembly, integration, verification, and testing engineering

To implement the end-to-end space system, it is necessary to perform a cascade of activities: manufacture, assembly, integrate, verify, and test all its components. This set of activities will be performed with parts, assemblies, subsystems, and elements of increasing level of complexity until the process finishes with the integration of the whole system. This set is the first part of the ascending "implementation" branch of the scheme of Figure 1.4, which depicts system design and implementation as a V-process. All these activities involve specific engineering skills – manufacturing, assembling, and testing – and all of them shall take into account the process of system architectural definition. The reasons are the following:

- During the architectural definition phases it is necessary to provide credible estimations of system implementation cost and risk, and all the activities indicated above are costly and risky.
- As indicated in the V-process scheme, the earliest and the latest phases of a space mission are logically related because they deal with the design and verification of system-level – as opposed to detailed-level – features. This will require that the approach to system-level verification is defined during the earliest system design stages.
- The definition of the architecture of the mission includes the establishment of the performance that the system is going to fulfill and the demonstration that the system is feasible and able to fulfill their performances. This requires that the approach to verify the performances is included in the architectural definition process.
- In many cases, possible new missions require the development of new techniques. To reduce the level of project risk the new techniques shall be developed before the project starts. That means, the techniques shall be developed during the architectural definition phases.

The set of the skills to perform these activities correctly are called production and verification engineering.

Production engineering covers all the preparatory work necessary to ensure that the product will be manufactured, assembled, and integrated at the lowest possible cost, in the defined timescale, and in accordance with the requirements specified. Once the project is approved, production engineering will trade-off possible production technologies, identify and optimize critical fabrication and assembly procedures, identify and design special tools, produce the necessary documentation, and finally produce and assemble the components at increasing levels of integration. The space engineering field is small when compared with: aeronautics, defense, electronics, or information technologies, and most space engineering production methods are taken from any of these mentioned engineering fields. Nevertheless, there are some production engineering areas where space is unique; for example, large, slender, deployable structures that cannot operate except in the weightlessness, and vacuum of space. It is in these specialized areas where space missions could push the state of the art in production technology. If a new space mission requires to push the state of the art, and to reduce mission risk, it is necessary to provide time and money enough to develop the necessary know-how before the mission is approved. In this case specialized technology development activities will run in parallel to the architectural definition phases of the new space mission and the architectural definition phases could not finish until the new production techniques have been matured enough to provide credibility to the estimations of performances and costs. It is extremely risky to start a new space project without having mastered all the required new production technologies. The issue of technology maturity is treated in detail in Section 4.9.3.

Verification engineering is the process of confirming that the system meets the applicable requirements. Verification engineering includes:

- The iterative checking that the consumer needs are properly captured in formal documents
- Confirmation that design constraints are respected and that all requirements are met
- Acceptance of hardware and software
- Validation of production processes, equipment and facilities, operations tools, and procedures
- Confirm integrity and performance after particular steps of the mission life cycle

As with production engineering, verification engineering is also performed in a bottom-up fashion requiring the verification of progressively larger and more integrated assemblies. The testing of large integrated assemblies will require not only the testing of their parts but also the verification of their interfaces and of the operating procedures to allow the different parts to work correctly together. Finally it will be necessary to verify the end-to-end system working simultaneously with space- and ground-located elements.

Verification engineering shall define, implement and run the tests necessary to verify the system requirements and will finally accept the whole system as fit to be accepted and paid for. Verification shall always apply to the performances in the environment that the satellite is going to have to withstand. Figure 2.4 depitcs the process necessary to arrive to the required verification approach. It is necessary to start from the requirements —what we want to verify?— going through the verification methods —how to verify?— and finishing with the actual facilities that will allow the verification —where? The selection of test methods and facilities shall take into account, not only the adequacy of the test methods and facilities but also the cost and the availability of the facilities. Proper verification requires that the tests on-ground shall properly simulate the in-orbit environment. During the architectural definition phases it is necessary to demonstrate that all the requirements are verifiable and how they are going to be verified. All missions will have especially challenging requirements whose verification will be complex and expensive. The verification methods for these requirements shall be defined at the same time and with the same dedication and detail as the space hardware. Sometimes the space mission requires the verification of subtle effects that cannot be properly observed under the gravity of the Earth or at atmospheric pressure. There are specific techniques that involve subjecting the space hardware to a simulated space environment and there are other techniques that allow Earth's gravity to be —partially— compensated enabling near-realistic testing to be performed. In some other cases test and analysis will be combined to provide a reasonable level of verification in situations that a totally realistic on-ground verification of the in-orbit performances is impossible.

Analysis and mathematical simulation are an intrinsic part of the verification process. The availability of progressively more sophisticated mathematical analysis and simulation tools and the increasing cost of complex tests are giving to analysis an increasing role during the verification process.

In any case the architectural definition phases of a new space system must establish the overall verification approach, including what is going to be verified, and how and where it can done. It is necessary to justify the logic of the verification of the space system. Development and verification approaches are discussed in Section 4.9.4.

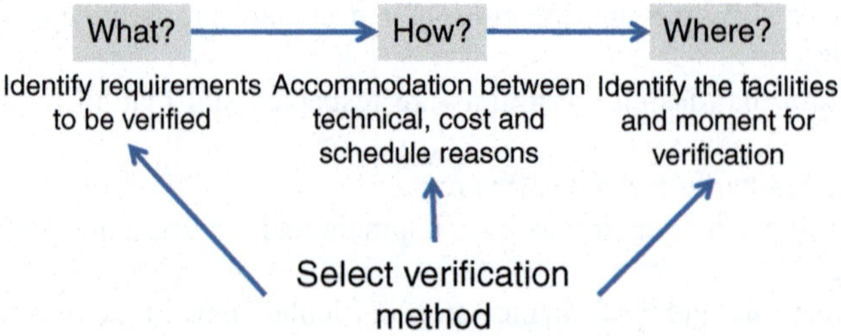

Figure 2.4. Verification Approach (System Engineering, ECSS-E-10, ESA 1996).

2.6.2. Product assurance

Product assurance is the engineering discipline devoted to the study, planning, and implementation of activities intended to assure that the design, controls, and techniques used will result in a satisfactory degree of quality. The quality will be satisfactory when the availability of the system is as desired by the consumers. Availability is defined as the ability of a system to be in a state to perform its required function.

Product assurance know-how is required during all the phases of the space mission, during the architectural definition phases before the project is approved, and during the detailed design, development, production, and operations of the space products. Product assurance engineers will be responsible for the level of quality of the system under implementation. Once the project is approved product assurance will:

- Ensure that the approach to product assurance organization, requirements, methods, tools, and resources are well defined.
- Ensure that the specified levels of quality are in line with the mission availability requirements.
- Ensure that the parts, materials, and manufacturing procedures are approved for use in a space system.
- Ensure that the verification of the end products conform to agreed procedures.

Product assurance has the dual role of first defining the rules to ensure that acceptable quality is attained and later controlling the work performed by others to certify that the defined rules of quality control are followed by all participants. The controlling aspect of product assurance is clearly a task to be performed only after the fully funded space project begins, but the definition of the rules shall be performed during the architectural definition phases. The availability requirements will directly derive of the original requirements of the consumers and it will be a driver for the mission design in aspects such as the number of satellites, numerical reliability, redundancy philosophy, or satellite life requirements. The impact of availability in space system design is discussed in detail in Section 5.5.6. All space projects include trade-off between project risk and cost, and the mission sponsor, who finally is the authority that provides the funds, will define the levels of risk avoidance and the cost targets, and will provide guidelines on cost versus risk trade-off. Different types of projects have different levels of criticality, and can tolerate different levels of risk.

Consequently they will have different levels of product assurance. Section 4.6 discusses criticality, Section 4.8 discusses project cost, and Section 4.9 discusses project risk. Risk management is the responsibility of the project manager and the role of the product assurance expert will be to clearly provide the availability implication of design decisions like elimination of redundancy or the use of lower-quality electrical components.

2.6.3. Satellite flight operations

To operate the satellite is to handle it so that it performs its intended function. All activities related to the planning, execution, and evaluation of the control of the satellite when in orbit are the domain of flight operations engineering. Flight operations imply the following set of activities:

- Monitoring of the status of the satellite and all its instruments by downlinking and inspecting information on its status.
- Programming and uplinking of "standard" or expected activities required by the satellite; for example, perform a required orbit control maneuver, a scheduled calibration that requires special satellite commanding or prearranged uplinking of updated in-orbit software.
- Establishment of observation plans and uplinking of the commands to allow the satellite to perform observations as requested by the consumers. Instruments that do not acquire data in a systematic and predictable fashion will require this.
- Analysis of anomalous situations detected in the satellite and uplinking of commands as needed to ensure the recovery of the detected nonnominal situation allowing the continuation of the space mission.
- Safe disposal of the satellite at the end of the mission.

Before the satellite is launched, operations engineering requires the preparation of the documentation to operate the satellite, the training of the personnel to perform it, the validation operations procedures and operations system, the day-to-day operations execution including the logistic support, and the disposal of the satellite. Operations engineers are also responsible for progressively activating the satellite's components – including instruments – and characterizing their in-orbit behavior. This will include the in-orbit final verification of the behavior of the satellite. Satellite operations only start when the satellite is launched. Furthermore, satellite operations also require a comprehensive definition of the satellite and a thorough understanding of its functioning. This means, a detailed definition of satellite operations can only be done rather late during the process of implementation of the space mission. Nevertheless, satellite operations engineering support is necessary from day one of the architectural definition phases of a space mission well before the mission is approved. This is necessary because when any possible new mission is considered, there are already three things that the mission consumer has already made clear:

- What quality of mission products they want?
- What quantity and frequency of data they want?
- How they want the mission products – the instrument observations – to be acquired and delivered?

The desires of the consumers on mission products acquisition and delivery will drive a cascade of requirements on the way the space mission will be operated, for example, level of satellite autonomy, and satellite modes of operation. All this requirements will be taken into account in the definition of the mission architecture and in the costing of the possible space mission. The cost of mission operations, especially if a long mission life is required, is a sizable part of the total cost of space mission implementation.

The overall operational data flow – satellite and ground – is analyzed in Chapter 11. That chapter analyzes the design decisions and the design interaction that impact the satellite in-orbit data flow, and the communications from and to the satellite. The design of the operational ground segment is discussed especially in Section 11.2.1.

2.6.4. Satellite data output processing

Most previous texts and standards on satellite engineering consider the delivery of the satellite instrument output as just another part of satellite operations engineering; nevertheless this text will always separate the processing and distribution of the data produced by the instruments. The knowledge required to operate a satellite requires a thorough understanding of the satellite and of the space environment, while the knowledge required in the processing and distribution of the data produced by the instruments is more related to the sciences and applications practiced by the consumers of the mission products. Section 7.1.4 introduces the data flows in a space mission and justifies in detail this separation between operations of the space mission and instruments output data processing and distribution. This text will analyze in Chapter 12 the instrument output data flow.

The processing and distribution of the data output of the instruments include:

- The development of processing algorithms that progressively transforms the raw data into calibrated and corrected data in instruments units; for example, the algorithms that provide calibrated and corrected radiances arriving to the instrument.
- To develop the algorithms that transform the corrected data produced by the instruments into data meaningful for the consumer of the mission; for example, into sea surface temperatures.
- To establish and maintain data networks allowing access and distribution of the data and providing interfaces to receive requests from the mission consumers, and to provide the requested data in the ways desired by the consumers.
- To have a long-term data maintenance system that would act as a repository of the mission's results, allowing continuous improvement in the science produced from the mission.

Space missions are intended to provide the information required by the consumers and most space mission ideas start as a vision in the mind of a consumer, not in the mind of a space engineer. As soon as the architectural definition phases start, one of the most important tasks is to create mathematical algorithms that relate satellite and instrument performances with the final performances required by the consumers. These algorithms will be fundamental for the specification and design of the space mission and will allow its correct design and the correct estimation of the cost and risk.

2.6.5. Cost engineering

Cost engineering is the technical discipline whose aim is to provide cost estimations. Cost estimation is defined as "the process leading to the determination of a project's expected cost."

Cost engineers provide cost estimates for specific space missions under consideration. They employ two types of methods: bottom-up and parametric. Bottom-up methods make detailed analyses of all the tasks necessary for implementing the mission. Parametric methods correlate costs with parameters like satellite mass and satellite and payload complexity. These methods are discussed in detail in Section 4.8.

Cost estimations are fundamental in the architectural definition phases of a new space project for two reasons:

- To establish the mission's financial feasibility
- To choose cost optimal solutions among the alternative architectures under consideration

Cost estimates help determine if it is feasible to fulfill the consumer's minimum requirements within the guidelines provided by the sponsor. This is a task of maximum importance for avoiding possible future cost overruns.

Cost estimates are crucial for selecting among possible alternative architectures. In order to determine the least-expensive alternative fulfilling the requirements, or the highest-performing alternative with respect to cost constraints, it is essential to be able to make use of accurate and reliable tools allowing credible cost estimates of the different alternatives under consideration. The flexibility provided by parametric methods make them especially appropriate for this purpose.

The aim of the architectural definition phases of a new space mission is to establish the desirability and feasibility of the mission itself, since no firm decision for actual implementation has yet been made. During these early phases there is still no definitive contractual agreement between the supplier and the customer, and cost quotations provided by the supplier are not normally legally binding. On the other hand early cost estimations are very important because customer and sponsor will use any cost estimations received for deciding on the mission's real desirability. These estimates will also serve for long-term financial planning. The customer's and supplier's distinct cost engineering teams frequently produce parallel cost estimations.

2.7. The consumer: the scientist behind the mission

The consumer's scientific domain will have its own body of knowledge and accepted analytical procedures necessary for properly interpreting the mission's output. The mission's ultimate aim is to satisfy the consumer, and its overall design requires some understanding of the consumer's scientific area and intentions.

The architectural definition stages of any space mission require mutual accommodation among the consumer's requirements, the sponsor's constraints, and the technical realities governing the supplier. The architect, then, will frequently interact with the consumer in these early stages. It is not the architect's obligation to become an expert in the consumer's scientific domain, but it is his duty to understand the users' needs well enough to provide them with the best design alternatives.

To provide the consumer with the best design, end-to-end mission performance simulation models must be created for the different architectures under consideration. In order to deliver mission performance in the language and with the units familiar to the consumer, these models should include modules that are based on the consumer's scientific domain. End-to-end performance simulation models will be revisited in Sections 3.4.2, 5.6, and 8.5.

Chapter 3

Requirements, Specifications, and Design

Keywords Requirements Engineering, Specifications Engineering, Nested Systems, System Layers, Value Engineering

Figure 1.2 of Chapter 1 put the customer at the center of relationships with the other three actors: with the consumer through the requirements; with the sponsor through the constraints; and with the supplier through the design, which in turn responds to the requirements and the constraints. This and the next chapter – Chapter 4 – will discuss how the customer handles the requirements and constraints. Section 1.2 defined the consumer as the end-user of the system's output, and who expresses the desired quantity, quality, and delivery mode of the data to be produced. These desires will create requirements. Requirements can be defined as needs or expectations that are formally stated and declared obligatory. Requirements are collected in specifications that can be defined as a document stating requirements. That is, needs become requirements that are collected in specifications, which become the document, used to formally verify if the consumer's needs will be correctly addressed by the system being implemented.

This chapter begins by analyzing the recursive nature of systems and how they consist of smaller subsystems and assemblies. This recursive nature demands the establishment of different levels of specifications applicable to the different levels of the system's formal breakdown – from the more general to the more detailed. The next section provides a description of the different types of requirements and the types of specifications and how the style of requirements and specifications evolve during the process of defining the mission. The next section examines requirements engineering and how to create, modify, and maintain specifications. The chapter ends by looking at the issue of value engineering; that is, how to identify the most important requirements and how to quantify the utility of the mission for the consumers in terms of those requirements.

3.1. Levels of system decomposition

Section 1.8 defined system as a set of interrelated or interacting elements, normally the object of a project and being implemented by a single authority. Each part of a system, however, can be considered as a system in itself made up of smaller parts. This process can be repeated until arriving at units that can no longer be meaningfully divided.

M.A. Aguirre, *Introduction to Space Systems: Design and Synthesis*, Space Technology Library 27,
DOI 10.1007/978-1-4614-3758-1_3, © Springer Science+Business Media New York 2013

Figure 1.3 analyzed design as a recursive process, which starts from a high level of abstraction and generality, and progressively becomes more concrete and specific.

To provide logic to this cascade of progressively more detailed decompositions, in space system engineering the following designations for distinct layers are used:

- *System*: as defined above. It is the highest-level division.
- *Segment*: a segment is a first-level division of the space system into three large blocks: space segment, ground segment, and launch segment.
- *Element*: defined in Section 1.7; represents a division of each segment into major functional blocks: instrument, satellite platform, launcher, operation center, data ground processing center, etc.
- *Subsystems*: division of the elements into sets of items grouped by engineering domains, i.e., the thermal subsystem, the structural subsystem of the satellite's platform, ...
- *Assembly*: a combination of parts that forms a functional entity; for example, an antenna pointing mechanism.
- *Parts*: the individual components of any assembly: electrical motors, bearings, mechanical or electronic parts, ...

An additional term, "component," introduced in Section 1.8, is used in this book as a generic term referring to lower-level layers of a system including subsystems, assemblies, or parts, and not to define a specific level of system's segmentation.

This logic for system segmentation goes hand-in-hand with a nested hierarchy of responsibilities, with those responsible for one layer delivering components to the layer just above while taking delivery of lower layer components from lower-layer suppliers. The number of layers employed in this hierarchy of responsibilities depends on the size of the system. Smaller systems will likely entail a smaller number of layers than larger ones, and projects may or may not use segments or subsystems layers for defining specific levels of responsibility. Figure 3.1 provides a graphical representation of the mentioned nested hierarchy of responsibilities and their corresponding layers described in the text above.

Whatever the number and names of the decomposition layers employed, each interface between layers defines a set of relationships of several supplier organizations supplying products to a costumer organization that is responsible for accepting and integrating them. Using the same terminology introduced in Section 1.3 and Figure 1.2, there is a customer to supplier relationship between each upper and lower layer of Figure 3.1.

At each interface of this figure, the customer organization, will be responsible for the integration of the products delivered by the different suppliers, until the final supplier – the industrial prime contractor responsible of the whole system – delivers it to the final customer that will accept and put the entire system at the service of the consumer. For the system to be able to adequately answer the consumer's needs, this cascade of responsibilities must produce a corresponding cascade of specifications. In line with this hierarchy of system and subsystems, the derivation of requirement and the design is always done top-down (Figure 1.5). For example, an overall requirement on instrument pointing will be apportioned between lower-level requirements on attitude control, manufacturing and assembly accuracies, materials outgassing, gravity offloading, thermal control, and thermoelastic distortion. The cascade of specifications

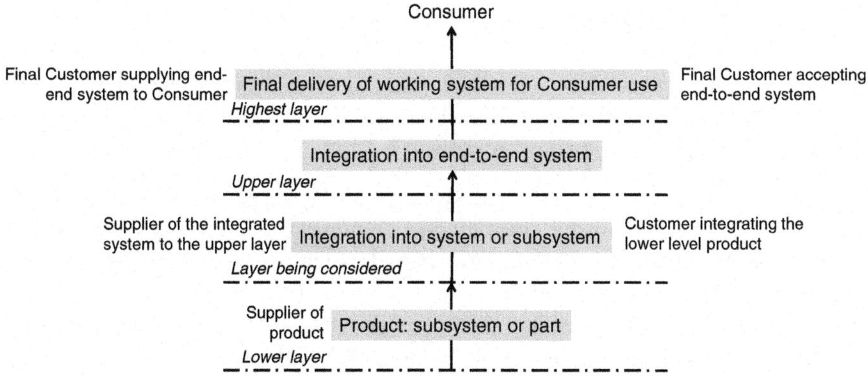

Figure 3.1. The Consumer at the Highest Level and the Hierarchical Relation Between Customers and Suppliers (System Engineering, ECSS-E-10 Part 1B, ESA 1996). The Responsibility of Each Layer Is the Customer of the Lower Layers and the Supplier of the Upper Ones.

and requirements will go hand in hand with the cascade of responsibilities and interactions depicted in Figure 3.1. At each interface between layers, those responsible for the higher level – the customer at that interface – will provide a set of requirements to those responsible for the lower level, who will be acting as his supplier

- At the top of the pyramid, the consumer is responsible for making his expectations clear. The consumer must also be aware of the constraints, (financial caps, for instance) which the sponsor has imposed on the system.
- The highest-level customer – normally the space agency – will discuss and reach agreements with the consumer about their needs and expectations and will formalize these in a mission requirement document stating the customer's commitments with respect to the consumer, and reconciling consumer's needs with the constraints imposed by the sponsor.
- The customer will create the highest-level system requirement document, which must clearly specify what requirements the system must fulfill. Once the system is implemented, this document will be used to ensure that the system is fit for purpose and can be accepted and paid for.
- The first-level supplier – the space segment prime contractor – will discuss and accept the system requirement document. Using that document as legally binding, the supplier shall design and implement the system and verify to the satisfaction of the customer that the implemented system actually fulfills the requirements.
- In turn, the first-level supplier is responsible for establishing the requirements and constraints for the second-level supplier in charge of delivering the respective components or subsystems. When the work is complete, the first-level supplier will also be responsible for validating that the finished component or subsystem meets specifications and it can be accepted and paid for.
- The second – and successive – level supplier will specify needs to lower-level suppliers, for example, the purchase of components such as space-qualified electrical motors or other electronic components for use in the satellite subsystem they are supplying. As before, the second-level supplier is also responsible for validation and payment of the delivered components.

3.2. Specification and requirements types

This section and its subsections follow the terminology and approach of the European Coordination for Space Standardization (ECSS).

The specifications serve two main purposes:

- First: to ensure that the design of the system and all its elements adequately fulfills the end-user's requirements. This process is represented by the descending **design** line of the V-scheme shown in Figure 1.5 and which advances from the general to the particular.
- Second: to confirm that each individual component and the whole system properly meet their requirements, first in isolation and later with respect to the components it interfaces with. This is the ascending **verification** line of Figure 1.5, which advances from the particular to the general.

3.2.1. Specification types

A specification is a document expressing requirements for designing, developing, and validating the system being implemented. A specification can indicate requirements associated with activities, such as tests or processes; or with product performances, such as expected numerical or nonnumerical performance parameters. This section will analyze the different kinds of specifications and how they are progressively defined as the mission's design advances.

3.2.1.1. Mission statements and mission requirements documents

A system's "mission" is the specific task, duty, or function it is designed to accomplish. In order to begin the process of definition, therefore, the system's mission must be explained. The initial document of this task, often called the "Mission Statement," outlines the needs and objectives that the system must eventually fulfill. The consumer creates the document as the collective whose initial needs gave rise to the mission and who will be the eventual end-user of the completed system. The mission statement's aim is to promote a possible new space endeavor; it is primarily a descriptive document and cannot be considered a specification, in the sense that it is not a formal collection of requirements.

As the process of definition advances, the mission statement will generate a second document usually known as the "Mission Requirements Document" where the initial vague needs become progressively more concrete – and often numeric – requirements. This document will list the end-to-end requirements necessary for implementing the system outlined by the consumers in the previous document. These requirements will establish the fundamental needs of the consumers in terms of:

1. Mission products quality, what to deliver
2. Mission products quantity, how much to deliver
3. Mission products delivery responsiveness, how to deliver

Hence, the mission requirements document formalizes and confirms the commitments obtained from the customer by the consumer.

On the other hand, the consumer does not normally pay for the system and it is the sponsor's role to provide the funds directly to the space agency, that is, the customer. Obviously, the customer will try to execute the mission requirements document to the best of its ability but the mission must be implemented according to the constraints established by the sponsor, who wields ultimate authority to cancel a project that exceeds the allocated cost. The customer's role as an intermediary between the consumer and the sponsor, also explains why the customer, and not the consumer, usually writes the mission requirements document, which, however, must be written from the consumer's perspective. If the consumer is a collective of scientists, the document shall be written from the perspective of the science to be produced. Like the mission statement, the mission requirements document plays an important part in promoting the mission. This means that the document shall contain proportionally long chapters describing and justifying the mission and explaining its purpose and advantages. These descriptive chapters will supplement other chapters that include the user's needs and the derived mission requirements.

Operational missions (Section 4.5.4) – such as operational meteorology – are promoted and financed by a single entity acting as both consumer and sponsor. This sponsor/consumer would still depend on a customer (a space agency) for the development and implementation of the system, but the consumer/sponsor would control the funds and system needs, would write its own mission requirements document, and would establish priorities. Taking into account the preliminary definition work undertaken by the customer and the supplier, the entity acting as consumer/sponsor will discuss and negotiate with the customer which elements are feasible and which are not, and this might entail changes to the mission requirements documents regarding performances, available funds, or programmatic constraints.

Neither mission statements nor mission requirements documents shall suggest or specify a design, though they almost always assume an observation technique. This is so because in order to propose a mission addressing a specific scientific or application domain, it is necessary to explain how the events of interest could be observed, that is, what techniques could be used to acquire the information required. As a result of having to define the observing technique, the mission requirements document must include not only details on the expected quality, quantity, and mode of delivery of the scientific or operational observations, but must also contain some key engineering requirements concerning the instrumentation considered compulsory by the consumer for the ultimate fulfillment of the mission's aims.

3.2.1.2. The customer requirements document as preliminary technical specification

The System Requirements Document is the highest-level specification produced by the customer. This document details specifications for the space system as a whole and is addressed to the system's prime contractor acting as the leader of the supplier organization. It describes the system's objectives, including, amongst others, technical performances, qualities, the quantity and delivery mode for the data produced, needed availability, delivery time, engineering requirements, and the duration of the mission. Some of the requirements contained in this document will be derived from the contents of the mission requirements document described above, but many will be new.

Initially, the aim of the document is not to produce a fixed and legally binding definition of the mission but to gradually reach reasonable design compromises that balance the consumer's requirements with the constraints imposed by the sponsor. Given the need to avoid a premature establishment of final requirements, the system requirements document will begin as a "Preliminary Technical Specification." This document shall contain a complete set of provisional technical requirements. The preliminary technical specification is produced early in the process of defining the mission's architecture and is used for exploring potential concepts: evaluating them and selecting preferred alternatives. Thus, it shall provide requirements with a certain degree of permissible flexibility.

The preliminary technical specification core will be the technical requirements for all the highest-level elements of the end-to-end system, with special emphasis upon the requirements of the observables and instruments. A preliminary technical specification should not offer or refer to specific design solutions, though it should begin with the observing techniques, that is, with the instrument concept proposed by the consumers in the mission requirements document. The preliminary technical specification could also include requirements related to the future project's programmatic environment, as well as general requirements derived from accepted principles expressed by the assorted engineering disciplines involved in the system's design and implementation. For example, the specification will provide: the required level of quality assurance, the need to use a specific launcher for the mission, or the safety margins required in the estimation of the satellite's mass or electrical power needs.

The preliminary technical specification is already a part of the business agreement between the customer and the supplier; so the supplier must provide well-documented concepts capable of fulfilling the requirements included in the preliminary technical specification. Nevertheless, these requirements are only used before the project has been fully approved and contain provisional, flexible requirements that are sometimes vague or qualitative; words like "optimize" or "minimize," for example, are perfectly acceptable in preliminary technical specifications. Because of this acceptable vagueness a preliminary technical specification cannot be used to validate a satellite that is being delivered.

3.2.1.3. The customer system requirements document as technical specification

A Technical Specification is a document expressing requirements for designing, developing, and eventually validating and accepting the system being implemented. The technical specification evolves directly from the preliminary technical specification and must be free of all ambiguities. This evolution is driven by the supplier's analysis of needs, including suggested possible alternative design concepts, and by the customer's continued refinement of the original concept. The customer shall balance the consumer's requirements, the sponsor's constraints, and the design concepts of the supplier. This progressive process of accommodation between ends and means and the eventual elimination of ambiguities is depicted in Figure 3.2. **The creation of the system requirements document and its gradual transformation into a document avoiding all imprecision and ambiguities is by far the most important task for the persons acting as mission architect for the customer**.

The technical specification defines the requirements for the selected solution and is a legally binding document that will form part of the final contract for the system's

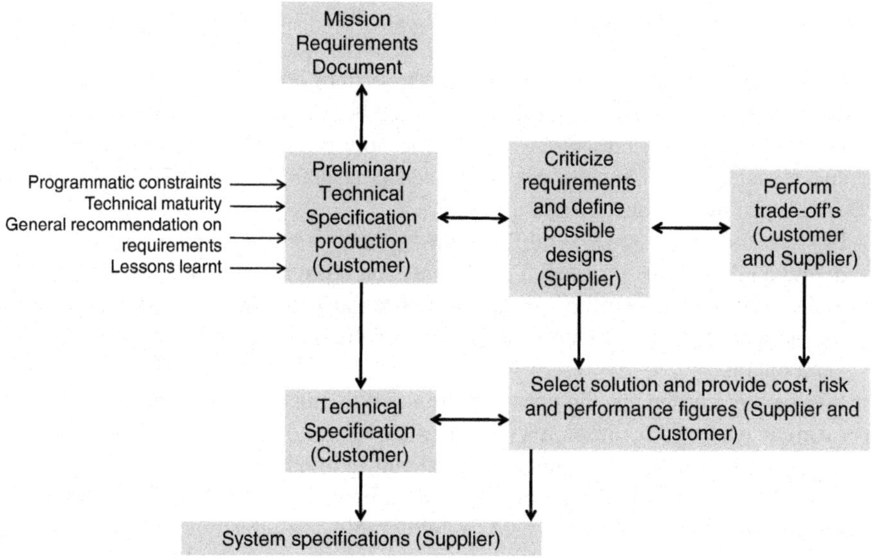

Figure 3.2. The Path to Derive the Fundamental System Specifications as per ECSS (Technical Requirement Specification, ECSS-E-10-Part 6C, ESA 2009).

implementation. The technical specification is the firm basis for providing credibility to the performance, risk, and cost estimations needed for a well-informed decision on whether or not to approve a fully funded project. When the project is nearing completion, the technical specification becomes the document of reference for the acceptance of the implemented system and for determining if the supplier has fulfilled his obligations and can be fully paid. All these reasons make it essential for the technical specification to be precise and utterly unambiguous.

The technical specification should contain general information related both to the context of the document itself and to the context of the project as a whole. The former should include items such as: the owner, distribution, management rules, scope, references, and definitions of terms and abbreviations. Information related to the project's context should include: a summary of mission aims as put forth in the mission requirements document, programmatic framework and programmatic constraints, and details on different aspects of the system or life cycle.

The most important component of the system requirements document is the inventory of technical requirements it contains. These include:

- System functional requirements. These define what tasks or actions the system must perform in order to fulfill the consumer's aims.
- Requirements on the observables. These define the observations provided by the system in terms of quality, quantity, speed, and mode of delivery.
- Instrument requirements. These include all those directly related to the characteristics and performances of the instrumentation carried by the satellite.
- Satellite technical and design requirements. These specify the satellite's components, including both specific requirements derived from the mission's aims and

general ones governed by the currently accepted methods of the different engineering disciplines.

- Operational requirements. Those related to system operability, such as utilization profiles, levels of autonomy, and interactions between satellite and ground.
- User interface requirements. These define how products are requested from the space system and how the data produced is transformed into mission products distributed to the consumers.
- Product assurance, and assembly integration and verification requirements. These define quality levels demanded by the mission's physical implementation including reliability or availability levels, required standards of quality, and specifications for the manufacture, assembly, and testing of the satellite, and other space system components.
- Interface requirements. On occasion the satellite must interface with elements not forming part of the mission's principal design and implementation effort; for example, with a specific launcher or existing ground station. These existing elements will impose their own interface requirements on the rest of the end-to-end system. In other cases, the mission might employ an already existing instrument or one being designed in a different programmatic framework. A detailed Interface Requirement Document will be needed to ensure the instrument's integration into the satellite's design.

The level of detail in the system requirements document shall be higher for requirements applying to more innovative elements or components (for example, requirements related to a new instrument), while requirements applying to conservative components that are likely to be very similar for similar missions do not need to be defined to a very high level of detail. This is the case of design requirements on well-proven subsystems like electrical power or satellite structures. At any rate, the level of definition must be high enough to allow a reliable estimation of cost, schedule, and programmatic risks for the binding agreement between the supplier and the customer.

3.2.1.4. Industrial satellite requirement specification

The first-level supplier – the prime contractor – will produce a specification in response to the system requirements document produced by the customer. The structure of this specification will be similar to the structure of the customer's system requirements document though much more detailed, and each requirement detailed in the original document will generate several derived requirements in the document produced by the first-level supplier. The latter will be written with the proposed design in mind and the resulting design will imply specific sets of requirements and allocation of functions regarding the subsystems of the satellite being specified.

This specification will focus on the satellite and is likely to consider elements such as the ground stations or the launcher as beyond the supplier's control and external to his effort. In line with this approach the specification will treat these external elements as having interface requirements needing to be proposed, discussed, and agreed upon, but will not impose any requirements on their internal functions. In some missions, some or all the instruments to be carried by the satellite also fall outside the responsibility of the satellite's prime contractor and will be treated as external elements specified only through their interface requirements.

The next step is a thorough review of the Satellite Requirements Specification by the customer and the first-level supplier with the aim of reaching an agreement on the viability of the mission which demonstrates confidence that it can be implemented through a design with a reasonable level of risk, within the imposed cost and scheduling constraints and that it would fulfill the consumer's requirements.

3.2.1.5. System support specifications

The supplier will also produce supplementary specifications as needed, covering in detail aspects such as thermomechanical or electrical and electronics design guidelines. Specifications concerning specific areas that apply to the entire satellite project are known as "System Support Specifications." Taking into account the importance of the verification of test approach and test methods in determining the credibility of the mission's cost and performance estimations, defining test requirements is another crucial step.

3.2.1.6. Interface requirements specifications

As mentioned above, certain elements interfacing with the satellite fall outside the main supplier's purview. Each one of these elements will be defined by an Interface Requirements Specification, provided by the customer to the supplier, which the supplier will respond to with documents specifying the interfaces from the satellite's point of view. As also mentioned earlier, interfaces with the launcher and ground segment must be specified, and sometimes with the instruments as well, or with any other satellite unit that might be furnished directly by the customer to the supplier.

3.2.1.7. Specification production, project phases, and project milestones

A project's phases and their corresponding milestones have been described in Figure 1.6. The mission statement is normally produced before the start of Phase 0. The first draft of the mission requirements document is produced at the end of Phase 0. The first draft of the system requirements as a preliminary technical specification is produced at the end of the Phase 0 and is updated and improved during the first half of Phase A. The System Requirements Specification progressively becomes a technical specification as Phase A advances. During the second half of Phase A the number of flexible requirements is greatly reduced and the end of Phase A should coincide with a system requirements document that has definitively become a technical specification and that is totally free of ambiguities. The Preliminary Requirements Review at the end of Phase A must produce a well-consolidated system requirements document as this is the only way of providing credible cost, risk, and schedule estimations, often crucial at that stage. The industrial satellite requirements specifications, the Satellite Support Specification, and the interface requirements specification shall be agreed to during the System Requirements Review that takes place a few months after the beginning of Phase B.

3.2.2. Requirements types

Requirements can be divided along different lines:

- Depending on the perspective of the requirements, they can be expressed in science or in engineering terms.

- Requirements can be divided into rigid and flexible groups. As explained in the previous section, flexible requirements are crucial to reaching a convergence among consumer requirements, sponsor constraints, and supplier design, while rigid requirements are needed for providing credibility to the performance and cost estimations.
- Depending on their aims, requirements can also be divided into three major categories: functional, performance, or design requirements.

These three categories will be discussed below.

3.2.2.1. Science versus engineering requirements

Requirements can be expressed in either scientific or engineering terms. Consumer wishes regarding quality are expressed in terms of observables, such as ocean circulation accuracy by m/s of ocean current, wind recovery accuracy by m/s of wind. The mission requirements document will contain many requirements – described from the consumer's perspective – in observable "science related" terms. The instrument requirements such as signal-to-noise ratio or observational bandwidth will be expressed in engineering figures meaningful to instrument designers. The observable-related quantities will have to be translated into instrument-related quantities and this entails the establishment of the observing concept.

3.2.2.2. Rigid versus flexible requirements

To ensure that the requirements do not drive the mission outside its cost or risk boundaries, they must be flexible; so it is advisable that the highest-level requirements always be expressed in terms of ranges instead of as single, fixed values. It is possible to classify the performance levels indicated by their consumers into three categories:

- Threshold, the limit at which the observations become ineffectual and are of no use for the targeted application
- Breakthrough, the value at which a significant improvement in the application is achieved
- Objective, the observation's optimal limit

The availability of key mission requirements in terms of threshold, breakthrough, and objective values will provide the required flexibility during the earliest stages of the design, allowing mutual adjustments until a balanced set of requirements for all the main mission elements is achieved, and which the design can respond to by fitting within the cost and risk guidelines. This is the moment when the preliminary technical specification turns into the technical specification and the prior flexibility can disappear. Even after this convergence process it might still be necessary to indicate that some requirements are not yet totally mature, in which case indications such as "requirement to be confirmed" or "to be defined" can be employed. These markers can be eliminated as the design process, and hence the evolution of the requirements advances.

3.2.2.3. Functional, performance, and design requirements

In general, all requirements can be classified into three categories: functional, performance, and design. Functional requirements, which can be numerical or nonnumerical, answer the question: **what must the system do?** Performance requirements answer the question: **how much?** And they are always expressed numerically. Design requirements answer the question: **how?** And can be numerical or nonnumerical. The latter are not normally related to a specific mission's needs but rather to accepted procedures of the different engineering disciplines involved in the system's design and implementation.

3.2.3. Requirements for technical specifications

Once the technical specification stage has been reached, all the requirements shall become fixed and will share a defined set of features. These must be:

- Clear: describing the required performance, and expressed in quantifiable terms as far as possible.
- Verifiable: in order to provide adequate credibility to the cost value, and should be written with a view of verifying them in a practical way.
- Justified: in terms of the way the requirements are needed for the fulfillment of mission's aims. At every level the customer must specify the minimum requirements necessary for their suppliers to be able to reach the project's objectives. Minimization of requirements is a key factor in minimizing supplier costs and maximizing supplier responsiveness.
- Configured and traceable: all requirements shall be properly identified and subject to configuration management so backward and forward traceability can identify what lower-level requirements they generate and what higher-level requirements they derive from.
- Unambiguous: so that all requirements are utterly unequivocal.
- Unique: each technical requirement shall appear only once.
- Identifiable: each requirement shall be identified with a unique alphanumeric reference.
- Singular: each unique alphanumeric reference shall apply to a single, separately stated requirement.
- Self-contained: so that each requirement is complete and self-contained, requiring no supplementary information to understand it.
- Consistent: so that no two requirements are in conflict with each other.

The need for verification not only entails differentiating between requirements that must be verified by analysis versus those to be verified by testing, but also the choice or development of the actual verification methods. This will be treated in detail in Section 3.5.

3.3. Requirements engineering

The responsibility for system implementation is clearly in the hands of the supplier, as is that of system design, though the latter is usually shared with the system customer. Nevertheless, the responsibility for system-level specifications is completely in the customer's hands, acting as the consumer's architect. As we are examining the process of satellite design from the architect's perspective, requirements engineering is given particular emphasis. Proper requirement apportioning and engineering are the architect's most effective means for helping to reach an efficient definition of the mission.

Requirement engineering must ensure the:

- Proper interpretation of the consumer's needs
- Production, discussion, and consolidation of the actual requirements
- Creation, control, and maintenance of a coherent set of system and lower-level specifications
- Full traceability of the requirements within the set of specifications, until the final verification closeout, ensuring that each requirement is referred to the adjacent upper-level requirement that generated it

The purpose of requirements engineering is to provide designers with all the necessary data to accurately specify an approach that can satisfy the perceived needs. Requirements engineering can be considered to have been satisfactorily completed when all lower levels have accepted their statement of requirements and their consequences have been sufficiently evaluated. Customers should agree that the product defined by these requirements will fulfill their needs, and the means for demonstrating that the product will indeed comply with them has been correctly identified. The primary output of the requirement engineering process is a detailed system requirements document agreed upon by the customer and supplier, supported by sufficient supplier-provided documentation demonstrating the feasibility and acceptability of all critical areas of the proposed design solution. At each level, requirements must be:

- Captured
- Analyzed, critically assessed, and validated
- Apportioned and allocated
- Maintained

The capture of requirements requires understanding the consumer's expectations and entails discussions with the highest-level supplier to arrive at a complete understanding of the implications of the requirements on the final design. The capture process is supported by a functional analysis for defining the functional elements and apportioning the corresponding functional requirements to each element and as well as any interactions between them. This combined process of discussions between consumers and suppliers and functional analysis will capture all the requirements directly related to the needs of the particular mission. These "mission-specific requirements" will be supplemented by requirements likely to be repeated from one mission to the next, which deal with accepted practices within the different engineering disciplines. The mission's specific programmatic framework, which is defined by the sponsor, not the consumer, will define levels of tolerance to failure; hence the product assurance, redundancy, and reliability requirements. The sponsor may also impose requirements such as the use of a specific launcher. In any case, any new requirements must be submitted to review in line with the already-mentioned guidelines to ensure that unnecessary requirements for the execution of the mission are not included.

The resulting specification shall be analyzed with an eye to identifying requirements that might have a large impact on cost or risk while providing only a marginal improvement in performance. This identification of potentially expensive requirements shall be applied to the different stages of the system during its entire life cycle, from implementation to disposal, and a critical review of all requirements is crucial for deciding if and in what way they are applicable to the mission under consideration. This will produce a validated list of necessary requirements that are internally consistent.

Higher-level system requirements shall be apportioned to the different subsystems, and the resulting derived requirements shall be included in the subsystems' specifications. Each higher-level requirement is likely to be apportioned between several lower-level requirements. This apportioning may be:

- Functional, in which a higher-level function is broken down into several lower-level ones
- Numerical, in which a higher-level performance figure is apportioned between all the error sources that will contribute to that higher-level figure, providing one specified figure for each error source

These allocations shall be applied in parallel with the functional analysis and with the progressively more detailed design process. Requirement apportioning determines the level of difficulty of the design and implementation task for each individual lower-level contributor to the higher-level overall figure that is being apportioned. Requirement apportioning is a powerful tool for ensuring a balanced design that avoids situations in which the requirements of one lower-level contributor are more demanding and difficult to fulfill than another's, without correspondingly higher funds or time allocation. **A fair allocation of requirements is one of the customer's most effective instruments for optimizing the system**. This is further discussed in Section 5.5.2 for functional apportioning and Section 5.5.4 for numerical allocation.

During the process of creating specifications it is a general rule that the supplier at each level may pass a requirement from a higher level down to a lower-tier supplier. Any flowed-down requirements then become the source of the lower-level-derived requirements in the appropriate specifications. It is the responsibility of the higher-level supplier to ensure consistency and compatibility between requirements at their own and lower levels. It is also the responsibility of the highest-level supplier – that means, of the industrial prime contractor responsible of the whole system – to establish a clear "specification tree," a document that shows the relationships between all the specifications for the space system, from the highest to the lowest levels, for each segment of the system. Each high-level requirement will produce a cascade of lower level, more detailed requirements that will drive different parts of the space mission. Figure 3.3 depicts the cascade of requirements starting from the fundamental classes of needs of the consumer: data quality, data quantity, and the way of delivery of the data. The meaning of the terms used is explained in detail in Chapters 8 and 9, but most of the terms are self-explanatory and adequate to illustrate the "cascade of requirements" resulting from the apportioning of functions and budgets.

Requirement maintenance consists of the processing and documenting of all requirements at each level, as well as the recording of the entire cascade of requirements derivations. Any change in requirements must be processed and

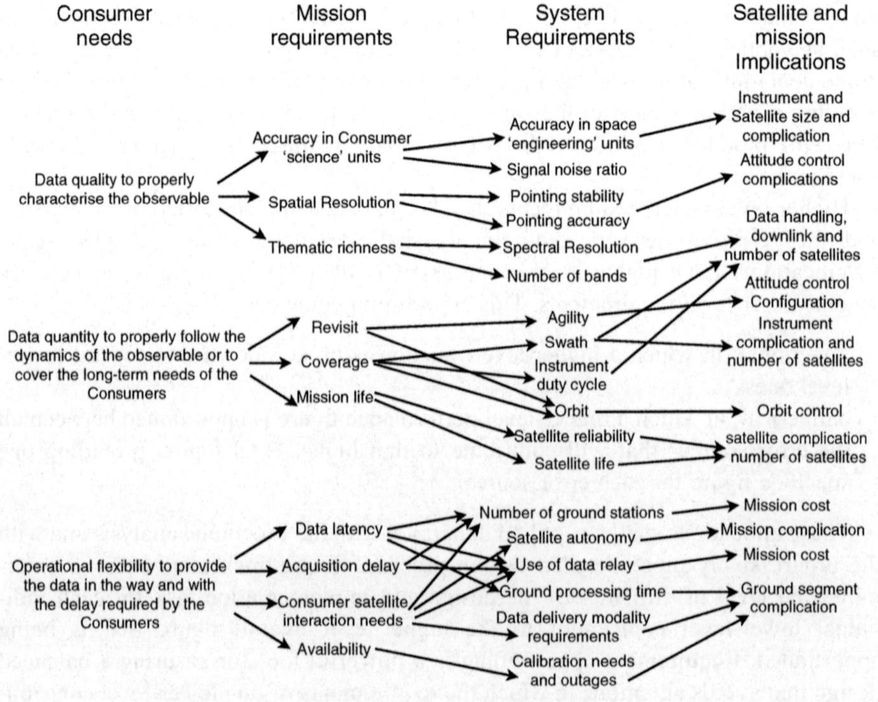

Figure 3.3. Example of Cascade of Higher- to Lower-Level Requirements and Resulting Implications on the Satellite and the Mission.

recorded in order to guarantee full visibility and traceability of each requirement at all levels: from the highest and most comprehensive to the lowest, most detailed ones. Computerized tools are available to help carry out this task.

3.4. Value engineering

The aim of value engineering is not to verify if a mission fulfills its requirement but to determine – numerically – how well the mission fulfills the consumer's needs. Requirements are established to ensure that a mission fulfills its aims; the corollary is that a mission that fulfills all its requirements will fulfill its aims. Near the end of the mission implementation process, documents known as verification control documents or conformance records provide a list of all the requirements, confirm that they have been implemented, and provide information on how they have been fulfilled. Requirement fulfillment only provides yes/no testimony, since as far as system validation is concerned, no apparent difference in requirements is relevant beyond whether they have been fulfilled or not fulfilled. Nevertheless, throughout the design process, and especially during the early stages of mission implementation, figures must be provided to help determine the relative value of alternative designs under consideration for fulfilling the needs of the consumer. The system's accomplishments must be quantitatively measured; that is, numerical measurements of the system's effectiveness must be established in order to ensure that the consumer's aims for the mission are being met. Value engineering consists in the measurement of system effectiveness. The primary aim of value engineering is to quantify in numerical terms how well the system meets the mission's objectives.

3.4.1. The different values of requirements

As said in the previous paragraph specifications are a collection of requirements. To fulfill his contractual duties the supplier must demonstrate compliance with all requirements; from this viewpoint all requirements are equal. Nevertheless, some requirements are clearly more important than others:

- Some requirements are absolutely fundamental for the mission to be of any value to the consumer; for example, missions to study weak astronomical events located far away in deep space have to be sensitive enough to detect them and accurate enough to characterize them properly. Clearly the requirements related to accuracy and sensitivity would be fundamental for such missions.
- Some requirements have major impacts in many aspects of the design; the need to fly a satellite at low altitude over the Earth's surface will impact not only almost every subsystem of the satellite but also the overall data flow and the operations approach.
- Some requirements are both things at the same time: fundamental for the consumer and impacting many aspects of the design.

Identifying the most important requirements is not difficult, since they derive directly from fundamental consumer needs and because they become the main and most difficult to fulfill from the earliest stages of mission's definition. Design decisions connected with the fulfillment of these requirements generate the longest discussions and entail the most difficult decision processes. To support the system engineering process, the employment of mission effectiveness figures directed toward the fulfillment of these key requirements is crucial. Each requirement is always unique and addresses a single performance such as the minimum angular resolution of the images obtained by the instrument. Nevertheless, mission effectiveness figures normally integrate several key requirements into a single value that should be especially meaningful to the consumer, in line with keeping the mission focused on the latter. For example, a good mission effectiveness value for a mission studying global vegetation would be the ability to gather high-accuracy estimations of vegetation dry mass in Kg/Ha. This mission effectiveness parameter would integrate many mission requirements, including several key requirements regarding the instrument and ground processing. In some cases, on the other hand, a single key requirement might also be a reasonable mission effectiveness parameter. For instance, the weakest recoverable signal of a particular mission is likely to be a specific requirement indicating the kind of minimum signal capable of being distinguished by the instrument from among the signal noise, while also being clearly important and meaningful to the consumer.

3.4.2. System effectiveness metrics

The availability of numerical estimations of mission effectiveness permits:

- The comparison of alternative concepts and their ranking in terms of the relative satisfaction of consumer needs
- The analysis of any mission concept from the perspective of effectiveness versus costs or risks, allowing the optimal design points for that concept to be identified

For comparisons to be meaningful and to facilitate decision making, a mission cost value must be calculated in addition to the mission effectiveness metric. Cost engineering is crucial to the comparative evaluation of alternative concepts, for correctly choosing the best design point, and for ensuring a well-documented decision on whether or not the mission will become a fully funded project. **Mission effectiveness and mission cost analysis are the two sides of the process needed to reconcile mission requirements with programmatic constraints,** and this book will dedicate ample space to comment on them, and to point out their strengths and weaknesses. Cost will be discussed in Section 4.8 and mission effectiveness in Section 5.6.

Figure 3.4 expands on the roles of the actors depicted in Figure 1.2. This expansion provides the flow of information among the actors during the architectural definition phases. In line with the account given in Section 3.2.1, the figure includes two arrows indicating the expression of needs and requirements; the "mission needs" arrow represents the consumer's requirements as provided to the customer in the mission requirements document, while the "system requirements" arrow represents the customer's requirements as provided to the supplier in the system requirements document. The mission effectiveness models and the cost estimation models enable the assessment and reconciliation of the legitimate – and contradictory – needs of consumer and sponsor.

The formulation of a mathematical model epitomizing system effectiveness is also a positive step because it forces a systematic and detailed analysis of all the design features or natural events that will play a role in the mission's performance. This kind of analysis also produces good understanding of the different factors influencing the fulfillment of requirements, how these factors can be broken down into lower-level factors, and how all these factors interact. Moreover, a mathematical model can assist

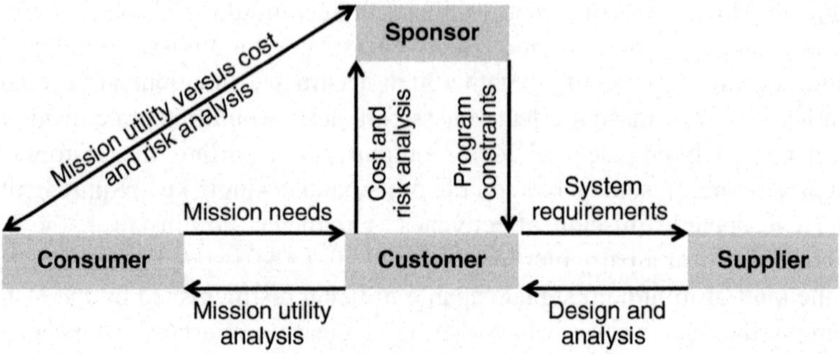

Figure 3.4. Value and Cost Engineering as Tools of the Customer/Engineer to Reconcile the Needs of the Consumer and of the Sponsor.

with the aim of guaranteeing that the identification of necessary requirements has been suitably exhaustive and that all essential points pertaining to the correct fulfillment of the mission have been accounted for.

Analyzing the mission requirements document, Section 3.2.1 divided the fundamental needs of the consumers into three groups. Mission effectiveness figures can also be divided into the three same groups:

1. What to deliver? Output quality figures, which determine to what extent the mission products are able to provide high-accuracy information about the sources, parameters being measured, or targets of observation. High-accuracy information allows the consumer to properly identify and study the relevant observable items.
2. How much to deliver? Output quantity figures, which determine to what extent the mission produces large amounts of data, providing fast coverage of large areas of interest in a short time, and for how long the mission can provide this service.
3. How to deliver? Responsiveness figures, which determine how fast the mission can react to the users' potentially variable observational needs, or how long it takes to acquire, process, and deliver the requested data to consumers.

Output quality figures for system effectiveness must be true end-to-end values, encompassing the behavior of all the elements involved in delivering the final mission products. This will require collaboration among experts in diverse engineering and scientific fields; for example, determination of ocean currents from satellite measurements calls for contributions from ionosphere specialists. Apart from this need for multidisciplinary collaboration, the essence of output quality determination is the creation of good instrument performance models. This will be discussed in more detail in Section 8.7.3. True end-to-end quality figures are expressed in the relevant observational units; thus, ocean currents recovery accuracy would be expressed in m/s. Instrument requirements, however, are expressed in instrument units: swath, bands, noise-to-signal ratio, etc. Analysis of different instrument performance ranges can determine how relaxing or tightening instrument requirement affects the overall performance sensitivity. This is an obligatory step in a project's early stages for identifying optimal alternatives.

Compared with the complication and wide scope of output quality figures, output quantity figures are simple to produce and understand and can be generated with the sole aid of space engineers. Satellite life and instrument swath are parameters of this kind. The simplicity of these parameters, however, does not keep them from having dramatic impact on the whole space system in cases where it is necessary to improve them. To increase an instrument's field of view, for instance, entails increasing its size, its data output, its power consumption, and its weight, producing a cascade of effects on the satellite and the system as a whole. Extensive analysis will have to be undertaken to ensure that this change in the field of view is efficient or even feasible. Clearly, changes in satellite life would also have a major impact on numerous satellite components. Satellite life depends on the redundancy approach, reliability levels, in-orbit degradation of the materials used, and the amount of consumable items such as fuel.

Some responsiveness figures, such as data delivery delays, are straightforward numerical values. Other measures of responsiveness, for example, the possibility of consumers to modify their observational needs at will, are not directly numeric and require interaction between consumer and customer to reach a clear understanding of what the real needs are. Demanding responsiveness requirements significantly influence satellite operations, the design of the overall data flow architecture, which must take into account data flow from the consumer to the satellite operator, thence to the satellite, to information acquisition by the satellite, and back to the ground segment, encompassing data processing and final delivery to the consumer. This will be studied in Sections 8.4.5 and 8.4.6 and again in Chapters 11 and 12.

Three classes of system effectiveness, with their concomitant examples, have so far been examined, yet Figure 3.4 assumes implicitly a "single" mission utility value that could be compared against cost. Theoretically the existence of a single well-defined value permits the definition of optimal design points. Nevertheless, obtaining a trustworthy single value for mission utility or system effectiveness is not a realistic goal, due to the difficulty of integrating the diverse aspects of mission effectiveness into a single overall value. This issue, involving ways to produce optimal system designs, will be discussed further and expanded in Section 5.6.

3.5. Requirements and verifiability

Section 3.2.3 explained how all requirements need to be verifiable, and that merely indicating whether verification would be accomplished by analysis or testing was not enough, and that it is necessary to have in mind practical ways of verification. Most requirements on the majority of missions can be verified by means of well-established procedures, and indicating the verification method is sufficient for understanding how the requirement will be verified and how complicated and expensive it will be. This is not always possible, however, and some cases will require intensive efforts to determine if and how the requirements can be verified.

Ideally, each item should be tested in an environment identical or very similar to conditions in space and demonstrate that it is able to function as required. Nevertheless, for requirements that are not easily tested, current space sciences analysis can provide an excellent level of credibility by employing a wise combination of partial tests, analysis aimed at an overall view, and tests devised to validate the analysis. Testing is expensive, and there is a clear tendency toward reducing the number and fidelity levels of tests, despite the obvious increase of risk this entails.

It is possible that a new mission will demand requirements that are difficult or even impossible to verify through testing. This can occur for tests associated with two principal circumstances:

- Tests related to the measurement of subtle physical effects that might be perturbed by the Earth environment, especially by the Earth's gravity
- Tests of high-level items such as an entire instrument, a whole satellite, or the complete end-to-end command and data flow network, where available test facilities are too small or would be unable to adequately reproduce the real-life environment for the entire high-level item

A good early understanding of how to properly verify the requirements in these two types of situations will be fundamental for the mission and must be addressed within the scope of the architectural definition phases, but the verification approach can be complex and difficult to design, cost estimates can be tricky to calculate, and establishing if verification is even feasible is not always easy. This is another example of the intrinsic contradictions in high-level system design, represented in the V-scheme of Figure 1.5: making the most important design decisions, the highest-level ones, earliest, when the level of definition is the poorest and when the resources available are the smallest; while actual verification of those most important design decisions will be performed toward the end of the mission's implementation.

In missions with special verifiability problems, early support from verification specialists is crucial. This is especially true when the physical effects to be verified are so small that ground testing would be inadequate and the level of confidence provided by supporting analysis is weak. Missions whose aim is to research the fundamental physics of gravity are of this nature, as are missions that must accurately maintain the relative positions of several satellites flying in formation, or those deploying large and flexible structures. In some cases in which a requirement is impossible to verify, technology demonstration missions are sometimes necessary in order to verify techniques in-orbit that are impossible to verify on the ground and are fundamental to the eventual success of an ambitious, expensive, and complex mission.

It is the duty of the customer acting as architect to present the sponsor and consumer with alternative scenarios in which the likely risks and costs are presented clearly, in which difficult or impossible-to-verify requirements are properly indicated, and practical ways ahead are provided. Then it is up to the sponsor to give the go-ahead for providing the funds to proceed and to the consumer to accept or reject the distinct scenarios presented. The emphasis on addressing the verification aspects of a project in the early definition phases is inevitably constrained by the limited available information. Therefore, the system engineer must always be thinking ahead toward the later phases, to ensure that he is not responsible for introducing requirements at any level that will prove difficult, costly, or even impossible to verify. This difficult task can be summarized as follows:

- If the design of a component, equipment, or subsystem introduces a significant adverse influence upon another part of the system, carefully consider the alternative solutions so as to avoid such impacts.
- If the adopted design at any level cannot be readily analyzed without having to make assumptions that cannot be justified, consider the alternative solutions that permit the analysis to proceed with confidence.
- If the adopted design at any level cannot be readily tested without unacceptable risk and/or cost, consider the alternative solutions that permit the test to proceed.
- If the adopted design at any level invokes requirements that cannot be verified, change the design.

Hence, verification must be considered throughout all the project's phases and not dismissed as something to be done later. If a system invokes requirements that cannot be verified, it should not be built.

Chapter 4

Constraints and Design

Keywords Constraints, External Environment, STEP analysis, Cost Analysis, Risk Analysis, Schedule Analysis, Technical Readiness Levels, Development Approach, Model Philosophy, Space Project Aim

Systems are designed according to specifications and are also verified against them, but there are many factors affecting and limiting the design that are not requirements or contained in a specification: they are the constraints.

This chapter will study how constraints influence the design of space systems. This chapter starts by analyzing the politics and the policy around space projects. It then continues by analyzing space projects according to type, size, and acceptable level of risk. The chapter continues with analysis of cost, risk, and scheduling. The chapter ends with a discussion on management trends and their impact on design. This chapter is not only about engineering procedures and management policy but also about politics.

4.1. Requirements versus constraints

This text will divide the factors influencing the material and programmatic design of a space system into two types: requirements and constraints:

- Requirement: anything that is included in a formal specification and that will be verified during the implementation process. It is internal to the project, explicit, and, during the architectural design phases of a project, can be negotiated and perhaps modified. The consumer is normally the source of requirements.
- Constraint: anything that limits or affects the system design or implementation. It can be implicit or explicit and can be external to the project. Frequently a constraint cannot be modified or negotiated. The sponsor is normally the source of constraints.

Constraints could include such things as the need for the project to be dedicated to a specific purpose, being below a given cost, completed within a given timeframe, or being launched using a specific launcher. Political, "hidden" constraints can also strongly drive the design. Examples of this could include a political desire for national prestige or for international cooperation that would favor, or even force specific mission design features or suppliers. Going back to the recurrent theme of correct design by balancing

M.A. Aguirre, *Introduction to Space Systems: Design and Synthesis*, Space Technology Library 27,
DOI 10.1007/978-1-4614-3758-1_4, © Springer Science+Business Media New York 2013

conflicting interests, during the design process, constraints will push against requirements until a balance is reached. In the original proposal of the Hubble Space Telescope the diameter for the mirror was 3 m. Under pressure by the sponsor to reduce costs, the customer proposed to reduce it to 1.8 m, which would have made the telescope compatible with well-mastered manufacturing processes. The consumer objected strongly, arguing that an optical device that small would not represent an improvement over already operating ground-based telescopes. The final diameter of the Hubble telescope was 2.4 m, which was the compromise between all the conflicting needs.[1]

4.2. The external environment of a space project

Any space project sits inside a larger environment. From a thermodynamic point of view, space design is an open process that interacts with its social, political, technical, and economic environment and that can adjust its behavior and goals as an answer to these external influences. The requirements as established by the consumer generate an internal control loop strategy for the designer trying to fulfill the mission aims. Nevertheless, there is actually a second, external[2] feedback driven by the constraints. The internal loop sets the objectives and monitors progress toward them through internal interfaces. The external loop adjusts the mission goals based on external feedback (Figure 4.1).

The reaction of the external environment could force the designer to change the objectives set by the mission requirements. This double loop image is in line with the analysis done in Section 1.2, which considers the space architectural definition process as an interaction between consumers, sponsor, customer, and supplier. The constraints forced by the external environment -- cost, schedule, risk, and political pressure – will force this readjustment of requirements and design during the architectural definition phase. The case of the Hubble telescope above is a good example of this.

4.2.1. STEP analysis

One classical approach for analyzing the external environment of any endeavor is the STEP analysis.[3] This analysis studies the social, technical, economic, and political environment around the endeavor. A good understanding of these factors around a

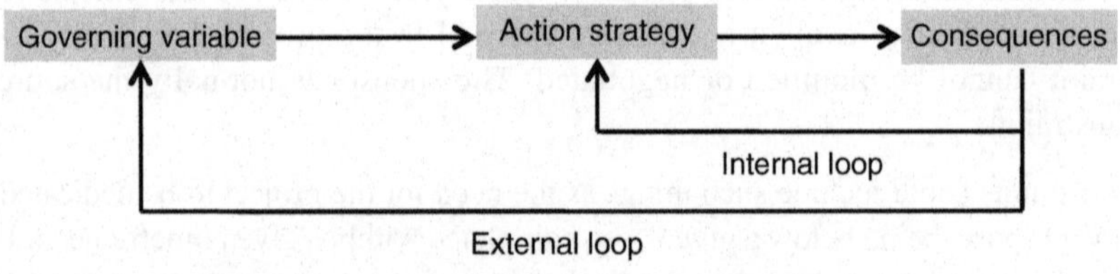

Figure 4.1. Double Loop Learning Strategy in an Open System.

[1]The Space Telescope, Steven Smith, Committees in Congress, Congressional Quarterly Press 1984.
[2]Strategy Changes and Defensive Routines, Argyris Putman McLain, Pitman 1985.
[3]The Challenge of the External Environment, U.K. Open University 1992.

space project will always allow the production of a design with a higher possibility of being approved and becoming a reality.

4.2.1.1. Social environment

The changing social perception of a national space program will influence which space system could or could not become a reality. This social perception can change in different directions[4] for the different fields of space research: it could improve or degrade for Earth observation with respect to astronomy or manned versus robotic research. In the beginning of the space age, social appreciation for space research was very high. This high appreciation was tolerant of failures; six Ranger flights to the Moon failed before the seventh was successful. The general perception is that this positive attitude does not presently exist. On the other hand, the social support of space research has been high and rather constant since the beginning of the space age. This positive perception has been reinforced by the role of Earth observation as a powerful tool for weather prediction and environmental protection.

4.2.1.2. Technical environment

Space design is fundamentally a technical activity and the technical environment cannot be considered "external." During the design phases, technology readiness analysis will characterize the problem and risk mitigation approaches, and specific technical developments will be recommended to reduce the technical risk. This is the classical risk analysis of all space design and will be studied in Section 4.9.

4.2.1.3. Economic environment

Public or private, the feasibility and optimal design for a space system will depend on the economic environment. Most space endeavors are government funded; the status of public finances, which depends on the economic environment, will discourage or encourage the implementation of more or less ambitious space projects. ESA's budget has had a long history of peaks and valleys, driven by the economic environment. The general economic mood will also influence the implementation of private space projects. The positive economic environment and the economic optimism of the mid-1990s made possible the implementation of ambitious private telecommunication programs like Iridium and GlobalStar.

4.2.1.4. Political environment

The overwhelming majority of space projects are done with government money. It is a legitimate concern for political representatives to care about the use of public funds. This makes the approval and implementation of any space project a political process that will be subjected to changes forced by the political environment. NASA's Mission to Planet Earth's financial envelope dropped from $17 billion to $8 billion in 1992. This dramatic reduction coincided with a positive attitude toward the study of the Earth as demonstrated by NASA's decision, that same year, to create two co-equal offices within the agency: one for planetary science and astrophysics and the other for Earth observation. This is a clear illustration of the permanent tension between the

---

[4]Evolving Public Perceptions of Spaceflight in American Culture, R. Launius, NASA, Acta Astronautica 53, Pergamon Elsevier 2003.

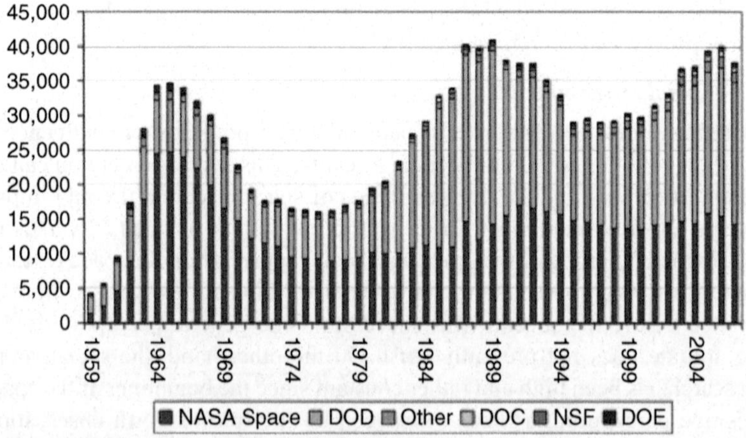

Figure 4.2. Historical USA Space Budget at 2004 Fiscal Year $ (The USA Government Space Budget, Paul Shawcross, OMB, October 2008).

unlimited desires of the consumer and the limited resources of the sponsor, which will force mutual adjustment between requirements, constraints and the design answering both of them. The package deal that created ESA in 1973,[5] included the authorization to start three projects: Ariane, Spacelab, and Marots. Each one of these projects was sponsored by a major European nation and it was a fundamentally political decision. Figure 4.2 provides the funding oscillations experienced by the US space budget. NASA and the Department of Defense (DOD) dominate the expenditure and there is a continuously increasing DOD percentage of participation.

In the USA, the yearly budgetary process through Congress makes the political environment relevant during all the phases of the mission implementation. In Europe, where multiyear funding is the norm, the political environment is most important during the process of approval of the project.

4.2.2. Forecasting and scenario analysis

This text identifies two types of forecasting:

- Short- to medium-term, limited scope forecasting
- Wide-scope strategic forecasting

Limited scope forecasting deals with the evolution of specific areas, which are principally in the technical and scientific domains. This limited scope provides forecasting with a high level of predictability. These predictions will be based on on-going and foreseen technical research on the specific fields under consideration. They will be used for the definition of the mission risk and for the establishment of development activities needed for the mission implementation. This type of forecasting is compulsory during the definition phases of any space project, and the data package supporting the decision on full project implementation always includes them.

[5]ESA Achievements Publication BR-250, ESA 2005.

Long-term "strategic" forecasting predicts wide-field and long-term evolution of technical or nontechnical areas. This type of forecasting is done to support large space projects and long-lasting space programs, but is not normally performed for average single-satellite space projects.

Long-term wide-field forecasting has been demonstrated to be very difficult: in 1995, the Teal group predicted that by 2005 we would have launched 1,000 communication satellites.[6] This prediction, done by competent people with adequate resources, did not become a reality. Past predictions about the rapid development of space industrialization or commercial Earth observation have also failed. Nevertheless, strategic forecasting is necessary to provide the overall framework for any large size or multiactor space endeavor, for example, the evolution of operational Earth observation as a whole. Taking into account past experience, the forecasting should always be done with caution and a large dosage of modesty.

Scenario analysis[7] is the process of analyzing possible future events by considering alternative possible environmental developments and process outcomes: the varying scenarios. The analysis is designed to provide improved decision making by allowing more complete consideration of environmental evolution, process outcomes, and their implications. Scenarios do not try to forecast the future but to establish the impact of alternative futures. This will provide a more thorough understanding, allowing the establishment of robust strategies that minimize risk whatever the future evolution. As in the case of strategic forecasting, scenario analysis is also especially useful when designing large or complex systems or systems of systems.

4.3. History of selected past space endeavors

Political evolution has always had a strong impact in space design. Three examples will be provided in the following two sections.

4.3.1. Private versus public communications and Earth observation

Private space communication endeavors have existed from the beginning of the space age. Hughes had private designs for communication satellites in GEO as early as 1959. Telstar was a successful private venture of AT&T already in 1962 but the political mood of the time discouraged private space projects and the American Congress created the monopoly Comsat just 1 month after Telstar was launched. Nevertheless, this situation of national monopoly did not last long; by 1974, Comsat's monopoly was broken by the U.S. Federal Communication Commission, and cable operators entered the market. Hughes' private design, HS-376, in Ku-band appeared afterwards, and this led to the development of private satellite communications in the USA. In the international field the monopoly of Intelsat lasted much longer. Intelsat

[6] Aviation Week and Space Technology, Mc-Graw Hill, 16 January 1995.

[7] The Challenge of the External Environment, U.K. Open University 1992.

was created in 1964 and was not privatized until 2001. Nevertheless, PanAmSat had successfully broken the monopoly of Intelsat by the mid-1980s. In Europe, which has always been more public-property-oriented than the USA, Eutelsat was created in 1983 and remained as an intergovernmental organization until its privatization in 2001, but the first private European satellite communication operator was SES, later SES-Astra, created in 1985 and which is still in business. During certain periods, the political mood has favored public space communication, but whenever privatization was allowed, private industry was able to provide successful space satellite communication programs.

At the beginning of the space age, the capability to observe the Earth was exclusively in the hands of the military sector. In 1965, the American Department of the Interior planned to develop a satellite for civilian mapping and geology but the program was not approved. Finally, NASA launched the Earth Resource Technology Satellite ERTS-1 in 1972. This satellite, which was later called Landsat 1, became the first Earth observation civil satellite. This started an era of public and open provisioning of satellite data for Earth observation. In 1979, the Carter administration declared the program "Operational" and shifted responsibility from NASA to the National Oceanographic and Atmospheric Administration (NOAA) in the U.S. Department of Commerce. In 1984, under the Reagan administration, Congress passed a law privatizing Landsat, which at the time had already launched four Landsat satellites. This created the private company EOSAT. EOSAT, to recover its investment, increased its prices. This generated a 77 % reduction in the use of Landsat data from 1984 to 1990. The business generated was unable to keep Landsat viable. Furthermore, the public Spot-1 satellite was launched in 1986 and undermined Landsat's market. In 1992, Landsat was taken over by the government but a procedure for licensing of private Earth observation data was established. This has generated a moderately successful commercial Earth observation sector, mainly in the area of high resolution imaging, but the provision of global images is still solidly in the hands of the government in the USA as well as the rest of the world. In contrast to commercial satellite communication, and despite strong political support, the development of private Earth observation has been slow and patchy.[8]

4.3.2. Apollo

By 1960,[9] NASA's long-term plans called for a manned circumlunar flight by 1970. This decision was motivated by political considerations. The failed invasion of the Bay of Pigs in Cuba had taken place the month before, and the Soviet Union had launched Yuri Gagarin into orbit just a few days before the Bay of Pigs debacle. Political leaders were asking USA to "catch up" with the apparent space technology gap. President Kennedy chose a lunar landing. With this decision Kennedy spurred NASA's subsequent growth during the next few years.

---

[8] Mixed Signals, a History of the Landsat Program and U.S. Commercial Remote Sensing Industry, Michael Belligan, Michigan State University 2005.

[9] Can Democracies Fly in Space, W. D. Kay, Praeger 1995.

At the time, three possible means of sending astronauts to the Moon were under discussion: direct ascent, rendezvous in low Earth orbit, and the rendezvous in lunar orbit. The short timeframe established by Kennedy favored the simpler Moon rendezvous. The direct ascent would have needed a launcher much larger than the Saturn V. This launcher, called Nova, was on the drawing board at NASA Marshall Spaceflight Center. The low Earth orbit rendezvous would have required several Saturn launches to assemble near the Earth orbit the components necessary to land on and return from the Moon. The lunar rendezvous was the only alternative that promised to land on the Moon by 1970 as Kennedy wished, and which required just a single Saturn V per flight. On the other hand, the selected mission architecture was highly specialized and did not allow the development of a range of general-purpose elements as the low Earth orbit rendezvous would have allowed. During the early 1960s, the increasing budgetary support was maintained even when confronted with a large series of technical failures: NASA had to wait until 1965 to have a "reasonable" success rate larger than 90 %, but in the politically favorable environment of the early 1960s the budget of NASA went from $462 million in 1960 to $5,016 million in 1964. Nevertheless, the budget remained flat from 1965 onwards and already began experiencing declines by 1968. At the time it had to fight for funds with the Vietnam War and with the "Great Society" social program of L. B. Johnson. Once the lunar landing had demonstrated the superiority of Capitalism over Communism, which was the real aim of the sponsor, the top-level political support stopped and NASA had to learn to survive with flat or decreasing budgets.

4.4. The programmatic framework as constraint

Very often, space systems being designed are located within a larger-sized programmatic framework; they are part of a space program. This space program would have been created with a specific objective and it would have established specific constraints on the projects that are part of the program. ESA's Core Earth Explorers Programme[10] rules establish that the missions proposed for implementation shall address broad Earth observation scientific problems, that they will be selected after peer review, and that their predicted cost shall be below a specific amount. NASA's New Millennium[11] was created in 1995 to test advanced technology in space flight. Missions in that program shall demonstrate advanced technologies that will enable space science missions with significant leaps in technical capabilities. Any design exercise for a possible new ESA Explorer or NASA New Millennium mission should respect the rules of these programs.

Programmatic constraints are clearly stated, known beforehand by all and, most of the time, are not negotiable. These conditions provide compulsory guidelines for mission cost, satellite size, allowable level of risk, mission objectives, mission implementations, and many other aspects. These constraints have to be acknowledged and respected by the mission designer.

---

[10] The New Candidates for ESA Earth Explorers Core Missions, A. Tobias, Acta Astronautica 52, Pergamon Elsevier 2003.

[11] The New Millennium Program Architecture and Access to Space, L. Herrel, IEEE Aerospace Conference, IEEE 2005.

4.5. Types of projects by project aims

Taking into account the fundamental aim of the project, space projects can be divided into four classes:

- Capabilities demonstration mission
- Technical demonstration mission
- Advancement of sciences mission
- Operational mission

Each one of them has different key mission drivers and is driven by different mission protagonists.

4.5.1. Capabilities demonstration

These are missions where the overriding need is to demonstrate that the organization leading the mission has the capability to implement it successfully. The motivation is fundamentally political and is driven by the sponsor. The Apollo project, the race to the Moon, was fundamentally of this nature. It was necessary to demonstrate that America was able to put a crew on the surface of the Moon before the Soviets did. At a most modest scale, a medium-sized country can decide to put a satellite into orbit as a demonstration of its technological capabilities. The most important characteristic of these missions is that, recalling the definitions of roles outlined in Section 1.2, they are not consumer driven but sponsor driven. This situation is perfectly compatible with that of the mission fulfilling valuable scientific or operational aims: Apollo provided important information on the origin of the Earth-Moon system but the scientific output is rarely important as the origin of the mission. Capability demonstration programs normally carry strong political support; nevertheless, they also work within cost limits and it is the duty of the engineer to always provide minimum cost solutions. These projects, which often carry a heavy load of national prestige, have to be designed with a very high probability of success.

As a summary it can be said that these missions shall be designed so that risk of failure is minimized, while cost is kept within bounds, and that high-quality scientific output or a long mission life are desired and welcome but not compulsory.

4.5.2. Technical demonstration

Technical demonstration missions want to verify in orbit a technical development related either to a specific observing technique or to satellite platform technology. The above-mentioned, New Millennium, is a good example of this kind of mission. As for capability demonstration missions, technical demonstration missions can provide interesting mission output with outstanding scientific value, but this is not their primary aim. . The New Millennium mission EO-1 validated a multispectral camera, which was a significant improvement over the previous Landsat 7 ETM+ and in the process of validating the camera it produced a large number of Earth images. To make these images accessible to the interested public the “EO-1 Extended Mission” has been created. This Extended Mission supplements the Baseline Mission, which deals with the original aim of technology demonstration and science validation.[12] Recalling again Section 1.2, technical demonstration missions are customer and supplier driven but the consumer plays an important supporting role because he will be the last user of the information provided by the mission and of the technical advancements demonstrated. These missions allow the development of risky technology, and being just capability demonstrations, the mission shall be kept within strict cost limits. To compensate this double difficulty, these projects are failure tolerant. This is the domain where high-risk high-reward solutions can be demonstrated. As in 4.5.1, to have a long life is not a primary characteristic of this type of mission.

These missions are risk tolerant, cost limited, and can have a short life but the mission life and the quality of the mission products should be enough to clearly validate the technology under consideration.

4.5.3. Advancement of science

The aim of an advancement of science mission is to solve a well-defined scientific problem. To do so, it has to produce data of a specified quantity and quality. The above-mentioned ESA's Earth Explorers Program missions are examples of advancement of science. Advancement of science missions is normally initiated by a group of scientists who propose the mission and establish its objectives; that is, they are consumer driven. These scientists often federate themselves and act as "mission owners." To be able to solve the scientific problems it addresses, the mission often requires the use of advanced technology, which also implies a high level of technical risk. In order not to waste the heavy human investment required for the preparation of the project, these missions have to be designed to be successful. This explains why, in case of failure, the mission is often repeated: ESA's science mission Cluster was lost with the explosion of the first Ariane 5, but the satellites were rebuilt, successfully relaunched and the mission ended up being a total success. Advancement of science missions are placed within a programmatic framework that limits the amount of money available for its implementation. If fulfilling the desires of the consumer is incompatible with the sponsor's ability to pay the mission will not fly. Advancement of science is the typical mission where it is necessary to balance conflicting desires in order to come up with a design whose risk, performance, and price are acceptable to all the actors. In the past, the provision of the instruments and the processing of the data produced by the mission were separate from the satellite design, and were in the hands of the mission owners or "Principal Investigators" that originally proposed the mission. They provided the instruments, processed the data, and published the discoveries made possible by the mission. Today, there is a clear trend to design the instruments and the ground processing and distribution of the data as just another element of the space mission.

Advancement of science missions requires pushing state-of-the-art technology, but to ensure mission success within the allocated cost envelope, this push should be kept within bounds. In cases where the technical advancements required for the mission

---

[12] http://eo1.gsfc.nasa.gov/

are too high, it would be advisable to start with a precursor technical demonstration mission, moving to the full-fledged advancement of science mission only after the successful in-orbit demonstration of the precursor. Mission life should be long enough to allow solving the problem under consideration. Missions with a longer life than that specified may be accepted as a bonus but very long life would not normally be a key mission driver. These missions must be designed to be successful and use advanced technology, but they are cost limited. They need to fulfill the specified data quality and mission life.

4.5.4. Operational

Operational missions answer specific "permanent" societal needs of clear desirability. These needs are extremely varied: communications, weather prediction, climatic change and environmental monitoring, and military intelligence. The missions underwritten by NOAA in the USA and Eumetsat in Europe are all operational missions. An example of an operational mission is Sentinel-3,[13] which will provide information about the status and evolution of the oceans and land surfaces. These missions are also consumer driven. The fundamental driver for these missions is the long-term and reliable provision of data of consistent and acceptable quality. These missions do not require pushing state-of-the-art science, though the observational techniques necessary to deliver the data required by the consumers might include moderately high levels of new techniques that would be accompanied by an associated level of risk. Operational mission data is normally distributed openly and publicly, is quality controlled, and has specific delivery delay requirements. In operational missions, the instrument and data processing are integral parts of the mission and are treated with the same set of rules as the other elements. In operational missions, as in science missions, the desires of the consumer must be limited by the sponsor's ability to pay. In the case of operational missions the cost situation is made even more difficult by the need to design a system able to provide the data with a very high level of availability and for a long, ideally indefinite, period of time. To ensure long-term data delivery, operational missions require a series of satellites and heavy investment in ground processing and distributing infrastructure. These missions have to be designed to be successful, but could use the multiplicity of satellites to reduce the reliability requirements of any one of the satellites. Mission life is an important asset, both for the overall mission and in order to reduce the number of satellites needed.

In summary, these missions should be designed to be successful, but this is compatible with a moderate level of reliability in individual satellites. They use moderately advanced technology and have moderately demanding requirements for data quality. Long individual satellite and overall mission life is a strong requirement. They are expensive and so cost minimization is of special importance to them.

[13]GMES Sentinel 3; A Long Term Monitoring of Ocean and Land to Support Sustainable Development, Y. Baillion, J. Juillet, M. Aguirre, B. Berruti, IAC-07-B.1.2.04 2006.

4.6. Type of projects by projects criticality

As already discussed in the previous chapter, not all the projects have the same level of tolerance to failure and projects can be classified by their criticality. ECSS[14] has established several categories depending on the acceptability of risk; NASA has an identical classification.[15] On the basis of the criticality level allocated to the project, requirements for its management will be relaxed or tightened. NASA NPR 8705.4[16] divides its missions into the following classes (Table 4.1):

- For class A, all practical measures to achieve mission success shall be taken and the highest quality assurance standards shall apply.
- For class B, stringent assurance standards shall apply but minor compromises, still maintaining a low risk to mission success, can be applied.
- For class C, a medium level of risk of failure may be accepted and reduced quality assurance standards are permitted.
- For class D, a medium to high risk of failure is permitted and minimal quality assurance standards are permitted.

Depending on the classification, there is a different attitude with respect to: single-point failures, model philosophy, qualification and acceptance testing, electronic and electronic parts, reviews, software development, satellite telemetry coverage, and the level of production of formal quality assurance documentation. The final level of project management requirements will depend not only on the criticality of class A to D but also on the mission cost.

Table 4.1. NASA Projects Criticality Classification.

	Class A	Class B	Class C	Class D
Priority	Highest	High	Medium	Low
Complexity	Very high	High to medium	Medium to low	Low
Life	More than 5 years	2–5 years	Less than 2 years	Less than 2 years
Cost	High	High to medium	Medium to low	Low
Alternative reflight opportunities	No alternatives	Few or no alternatives	Some or few	Significant reflight opportunity

[14] Space Project Management, Policy and Principles, European Cooperation for Space Standardization, ECSS-M-00, ESA 1996.

[15] NASA Procedural Requirement NPR 8705.4, Risk Classification for NASA Payloads, Office of Safety and Mission Assurance NMI 8010.1A.

[16] Risk Classification for NASA Payloads. NPR 8705.4 Appendix A and B, NASA 2004.

4.7. Types of projects by project size

As just discussed above, NASA recognizes that projects with the same level of criticality but of different sizes would require different management requirements. The larger a project is, the more numerous its interfaces and the longer the communication channels; therefore, the larger a project is, the more formalized procedures for information flow it needs. On the other hand, small satellites lead naturally toward simplified implementation approaches. This can be the result of a formal relaxation of management requirements, or can be produced naturally by the small size of the team involved in the satellite implementation. A small size provides short and easy communication chains that allow solving complex multidisciplinary technical problems easily, by informal discussion between experts. In large projects, multidisciplinary problem solving requires the use of formal procedures that define the roles and responsibilities of the different stakeholders in the problem, and that must be followed to arrive at a solution.

In many applications, a given set of requirements can be fulfilled by several small satellites or by a single large one. A constellation of small satellites, each one carrying a small narrow-field-of-view camera, could provide the same performance as a large satellite carrying a larger, more costly camera with a wide field-of-view. The implementation through a constellation of small satellites has three benefits:

- It allows reaping the benefits of short communication chains mentioned above.
- It allows a progressive, step-by-step, robust mission implementation. This allows the use of lower reliability design solutions, and a lower level of management requirements, while still ensuring a high level of overall mission availability.
- It raises the prospect of extending the mission, if desired, at modest cost by the launch of replacement or spare spacecraft.

On the other hand many laws of physics do not scale well with size, and there are missions that need to be large. There are limits to what small missions can provide. The advantages and limitations of small missions as "alternatives" to normal missions will be discussed in detail in Chapter 13.

Whatever the management approach and whatever the size, all space missions are unique and should be designed and managed to be successful. This is especially true with advancement of science missions, as defined in Section 4.5.3, because they are normally implemented using a single satellite.

4.8. Cost

We have introduced cost engineering in Section 2.6.5. In the present section we will analyze how to deal with cost as a design constraint, how cost drives the design, and how to prepare a cost estimate that is good enough to allow a well-informed decision for project go-ahead.

In the previous headings of this chapter we analyzed the different types of projects and how they have different risk and reward sensitivities. Nevertheless, the first thing to be said is that the obligation to "design to cost" applies to all projects, not just those projects that have a "design to cost" label. If a project is part of a program with absolute cost guidelines where the cost must be below a given threshold, the design will become an exercise in providing the best possible performances within the established program cost boundaries. In cases where absolute explicit cost boundaries do not exist, the sponsor will still have a range of expenditures that he is able to tolerate. It is the responsibility of the designer to know this range and to accommodate the design to be cost compliant. In general the designer will use bottom-up "parametric" cost estimation tools, as part of internal trade-offs to arrive at the selected baseline. To provide the necessary level of quality, the preferred alternative, or alternatives, shall be evaluated in more detail using a more complex but more reliable bottom-up cost estimation method. It is especially important to provide a cost estimation that is comprehensive and where it is clear what is and what is not included. The cost of the implementation of a mission is not only the cost of the satellite; a proper estimation of the cost of new space systems shall be comprehensive and include all the nonsatellite acquisition costs:

- Customer's internal costs
- Launch service
- Satellite operations
- Technology predevelopment
- Data utilization

Almost all space projects are implemented through the interaction of the customer (the space agency) and the supplier (the space industry). The space supplier will provide the industrial cost, but the total cost to the sponsor will include the cost for maintenance of the customer team. These costs represent around 20 % of the cost of the satellite.

The selection of a launcher for a mission is done from a small set because the number of high reliability launchers is limited and because the launcher selection might be further reduced by a possible constraint to buy the launcher from a specific source. Nevertheless, accommodating the satellite in a smaller, cheaper launcher while still fulfilling the system requirements is perhaps the best way to reduce overall costs without reducing mission performance. Launchers can also be shared, but to share a launch will introduce supplementary programmatic rigidities or problems with the possible orbit. Depending on the selection, launches can vary from 10 % to 30 % of the cost of the satellite.

Satellite operations include all the activities needed to keep the satellite in orbit working according to the mission's needs. It is a long-term activity, which requires the development of control hardware and software and the services provided by highly paid experts. Special activities like launch and early orbit and commissioning phase need heavy teams to inspect and validate the satellite and its equipment while in orbit. All these costs need to be included in a comprehensive cost analysis. Operations costs are around 10 % of the cost of the satellite. Operations costs have fixed and variable components. The fixed component will be an investment cost independent of the mission duration, while the variable cost will be a function of the mission duration. There is an increasing tendency toward the reduction of operations costs through using satellite autonomy; on the other hand, clever operators have been able to save missions that apparently were totally lost. For many missions a single recovered satellite can compensate the cost associated with the maintenance of a large and experienced team kept on stand-by.

Data utilization includes: the costs associated with the process of converting the data produced by the satellite into products of increasing levels of sophistication and the cost of distributing these products to the final users. This cost entry is especially important in operational missions. On the other hand, in the case of small science missions with a "principal investigator" this cost will be hidden as an overhead of the institution doing the research with the data produced by the satellite. As in the case of operational costs, it has a fixed and a variable part. Data utilization costs can go from 5 % to 20 % of the satellite cost. The highest percentage will be for operational missions, which require normally complex data distribution systems. A mission without a proper allocation for data utilization will still be in orbit and it will demonstrate the capability of the organization that put it there, but it will not provide the scientific results that should have been its final output.

The most important element for taking into account nonsatellite costs is transparency. It must be very clear what is and what is not included, and why. Some programs could include outside the allocated budget such elements as the launchers, data utilization, the instruments, or the platform. If these are the rules of the program, the designer must follow them. Nevertheless, a correct design shall take into account the overall cost to the sponsor. The existence of these off-budget costs can produce unbalanced designs. A typical case is to use "project-free" large launchers or large standard platforms to put small payloads into orbit that would have been compatible with much smaller and cheaper overall mission concepts. Another aspect to be taken into account is that the addition of all the nonsatellite costs is quite close to that of the satellite. While there is much emphasis on reducing the cost of the satellite, a proportionate effort needs to be given to reducing the cost of the nonsatellite parts of the system. This is especially clear if we consider that activities like mission operations and data utilization, being ground based, appear to be far more appropriate for meeting ambitious cost reduction targets through the employment of modern approaches like in-orbit autonomy, in-orbit and on-ground artificial intelligence, and advanced data processing and distribution methods. On the other hand, large-scale data utilization subsystems for operational missions have an unfortunate tendency to be more costly and be delayed with respect to the corresponding space segment: for example, NASA's Mission to Planet Earth data distribution system, Envisat's ground segment of ESA, or the MetOp ground segment of Eumetsat. It might be the case that to design a good ground data distribution network is much more difficult than the space designer assumes. The early involvement of experts in satellite operations and in data processing and distribution subsystems is compulsory to provide a reasonable estimation of the overall cost of a space project.

4.8.1. Top-down cost estimation

The top-down cost estimation approach provides the cost of a satellite, which is normally the largest cost element of the overall space system, and is commonly treated as a function of the satellite's dry mass and its level of technology innovation.

Top-down cost models are obtained by creating databases of the cost of earlier, similar satellites and deriving correlations between cost, mass, and innovation level.

Figure 4.3 provides an example of the application of this method. The CCOSTAT model is fed with data from a number of projects. The mass and the cost are known numerical data. Innovation is reduced to a numerical parameter that is a function of the inheritance, or novelty, of the satellite with respect to the previous missions that have been used to create the cost model database. The quality of the model will

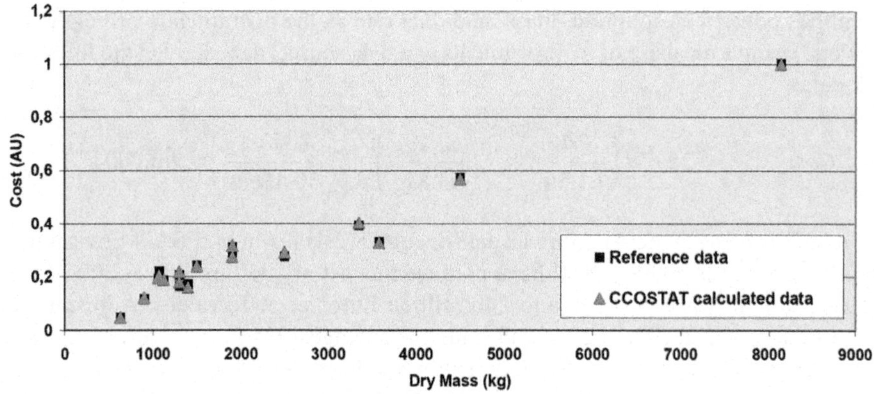

Figure 4.3. Parametric Cost Estimation for Low Earth Orbit Earth Observation Satellites (Cost Estimations of Sentinel-3, SEN3-ASP-TN-144-2, Thales Alenia Space, 2007).

depend of the abundance, applicability, and reliability of previous examples that are used for developing the procedures used to transform the elusive concept of innovation into a set of clear and unambiguous numerical values for the new satellite into consideration. The reliability of this type of estimation depends upon the previous knowledge used in the creation of the models and of the availability of a large database of past satellites with similar characteristics to the new satellite under consideration. The term "similar characteristics" implies:

- Similar technical requirements and their difficulty
- Similar project organization
- Similar risk management (Section 4.9.1)
- Similar technology development approaches (Section 4.9.3)
- Similar model philosophy (Section 4.9.4)
- And similar schedule targets (Section 4.10)

The creation of high-quality databases and top-down cost estimation models is a complex and expensive undertaking, and private companies are reluctant to reveal the models they use to make estimates. Public institutions are more open and a list of cost models used by NASA[17] can be found but as of August 2009 no public information was available on top-down cost models used by ESA. Instruments constitute the novel element of all missions, and there are many classes of instruments with widely different levels of complexity, making the cost estimations of instruments an especially difficult task. NASA employs a cost estimation model based on Principal Component Analysis. The idea behind this type of analysis is to identify the instrument's most important set of independent parameters. Principal Component Analysis is used in many scientific areas, from human genetic history to vibrations, where it is essential to identify the main factors behind a system's complex behavior. NASA

[17] http://cost

identifies power consumption, mass, and data rate as the instrument's principle cost drivers. Using a database of 37 instruments as a data source, they derived the following formula:

$$Cost_{payload} = \$25.6M \left(\frac{P(W)}{61.5W}\right)^{0.32} \cdot \left(\frac{m(kg)}{53.8kg}\right)^{0.26} \cdot \left(\frac{R_b(kbps)}{40.4kbps}\right)^{0.11} \text{(in 2004 \$)}$$

ASAR is a Synthetic Aperture Radar (Section 8.2.4) flown by ESA's Envisat mission. The formula above would have predicted a cost of \$332 million at 2004 dollar values. In reality the cost came to 240 million Euros at 2001 values. Adjusting for inflation between 2001 and 2004 and using the average 2004 Euro to dollar exchange rate of 1.25, the price of ASAR predicted by the formula was accurate to just a 4 % error. This is a remarkable example of top-down cost estimation's power to provide reasonable estimates of complex space items.

Wertz and Larson dedicate an entire chapter of their book on space mission analysis and design[18] to cost modeling, and they provide a selection of top-down cost estimations that can be used to derive preliminary cost estimations. The book also includes cost estimation algorithms for the estimation of "transversal" tasks such as ground segment or assembly integration verification and testing.

Top-down cost estimations can be used with a reasonable level of confidence especially when used to ascertain relative comparisons of cost estimates among a mission's distinct design alternatives. Top-down cost estimations permit comparisons such as that depicted in Figure 3.4 of the previous chapter to be performed. Figure 3.4 indicated the link between a mission's utility and its cost. The existence of numerical mission effectiveness figures and mission cost estimates enables the proper trade-off of mission implementations, which in turn allows for the convergence between the conflicting drivers of cost containment and performance excellence.

4.8.2. Bottom-up cost estimation

Bottom-up cost estimation proceeds by dividing all the work to be done in the definition, implementation, and operation of a space system into packages and estimating the price of each one of them. The finer the division of the work, the larger the number of work packages and the more comprehensive the final cost estimation. The estimation of each one of the packages is done by consultation with the help of experts with experience in the cost of similar packages done in the past. The following procedures can be used:

- Analogy to similar equipment/subsystem/system-level costs, taking into account the amount of new development required
- System activities estimated and validated by expert-opinion
- Expert judgment from technical specialists in combination with cost references, in case the amount of new development is extensive
- Expert judgment from technical specialists only, if references are not available

[18] Space Mission Analysis and Design Space, Wertz and Larson, Kluwer.

It is also possible to supplement a bottom-up cost estimation with some top-down parametric values; for example:

- Parametric cost models for equipment and subsystems with a low level of innovation
- System-level parametric cost models, based on observed trends for relevant references

The final cost is the addition of all the individual costs. To create a detailed bottom-up cost estimation requires a large amount of effort and it cannot be repeated many times during the process of definition of a space mission. It is also evident that to be able to provide a high-quality prediction of the cost of the mission it is necessary to have a fairly detailed level of design, including a rather detailed approach to the development and to the test philosophy; test is one of the main cost drivers. Detailed bottom-up cost estimations are normally produced by the supplier and delivered to the customer when the definition phases of the possible new mission are finished. The bottom-up cost estimation will be discussed between the supplier and the customer and will be finally provided to the sponsor to provide credibility to the cost estimations.

Bottom-up cost estimations are management documents and used mainly for management purposes. When the supplier provides it, he commits himself not only for an overall cost but also for a detailed decomposition. For this reason, the supplier will deliver the cost estimation only when reasonably sure of the credibility of the estimation.

4.8.3. The risk of cost estimations

Cost estimations – either top-down or bottom-up – carry a risk of being wrong. It will be possible to quantify this risk by defining the cost estimation not as a single value but as a probability distribution with its mean and its variance. There are methods that estimate that probability distribution and it would be possible to determine cost estimations that would ensure a given level of confidence that the final cost will not exceed the determined cost. The probability distribution will depend on:

- The innovation of the mission under consideration. This takes into account the cost growth caused by unseen complexities that will be revealed as the design matures into more details.
- The accuracy and quality of the parametric cost models themselves that depend not only on their intrinsic quality but also on the applicability of the model to the specific context of the new cost estimate.
- Project management decisions to account for cost risks induced by potentially negative events, as well as potentially positive opportunities, that may occur or not and that are under the direct responsibility of the project manager.
- External events to account for cost risks or opportunities that originate from external influences outside the direct control and responsibility of the project manager.

It is possible to combine all the factors above by mathematical procedures – for example, Monte Carlo simulation – and derive a final total probabilistic project cost. It is necessary to determine the percentage of probability that is going to determine the level of financial reserves that the space project will use to deal with unexpected events during the project's implementation. The value corresponding to 70 % confidence level is often used to determine the cost reserve for the project.

4.8.4. Single satellite versus multiple satellites cost

Most of the time satellites are single units with a reasonably high level of novelty. Even in many cases where the new satellite uses standard components, they are combined in new ways; the new satellite is always a new design. This will imply that the cost of a second – recurrent – model will be much cheaper than for the design, manufacturing, and assembly of the first model. A good rule of thumb is that the first recurrent unit of something very new can be 0.5 times the first "prototype" unit. A second factor is that the building of subsequent units becomes cheaper as the number of units increases. This is called the learning factor. There are rules based on previous experience with series production that provide guidelines for the estimation of the cost of a series of identical satellites.

4.9. Risk constraints

The term risk denotes the degree of variability in the outcome or result of a particular action. The risk management focus should be in the avoidance of undesired outcomes such as technical failures or cost overruns. In this section we are going to study how risk management in conjunction with mission aims drive the physical and programmatic design. In the context of space projects, risk has two dimensions: the likelihood of the outcome and the severity of the outcome. It is not an accident that we speak of risk management and not of risk engineering. To accept or not to accept a given level of risk is finally a management decision, and the most that can be requested from engineers is to document the case as thoroughly and in as unbiased a manner as possible, so that the manager can take the final decision. The present book focuses on the design work previous to a formal decision on project go-ahead, before a project manager was appointed. In these mission architecture definition phases, the architect shall provide the sponsor with a fair depiction of the risks of the mission and the developmental approach chosen to minimize them. With this information, the sponsor will approve or reject the mission.

4.9.1. Qualitative risk management

Space projects are risky: space deals with a low number of satellites – most of the time a single satellite – and the first time something is verified in the real situation is in space where there are few opportunities for software changes and almost no opportunity for hardware changes. The management of risks includes:

- The systematic identification, evaluation, and classification of all risks causes and consequences
- The systematic definition, implementation, control, and verification of actions appropriate for the elimination of risks or their reduction to an acceptable level

Due consideration of risk is necessary from the earliest stages of mission design, and it is necessary to provide a comprehensive and unbiased estimation of mission risk to allow a well-informed project approval or cancelation. Risk management, albeit having a strong science/techniques component, is a management activity, and managers are responsible for managing the risks and for their planning and implementation, in a cost-effective way. The project manager must make appropriate arrangements to prevent those risks from arising, to prevent their consequences from materializing, and to ensure the necessary reporting to allow appropriate overall supervision. The estimation of risk is based on the assessment of all risk from the point of view of: likelihood of an event occurrence and severity of the consequences of the event occurring. The severity of consequence is divided into:

- High, if the event would impede the achievement of the commitments formally made with respect to the consumer or sponsor for a program or project
- Medium, when the event would affect the achievement of the committed performance or planned schedule or costs, typically taking months to restore the situation or requiring funding beyond plans within the approved margins
- Low, when the event would, at the most, affect the planned distribution of performance, schedule, or cost, typically taking a few weeks to restore the situation and possibly requiring redeployment of project resources

The likelihood of occurrence is divided into:

- High, when the event is very likely to occur, and plans or controls are not available or cannot be put in place to prevent the risk or the consequences from materializing
- Medium, when the event is likely to occur, or current plans or controls are not expected to prevent the risk or the consequences from materializing, and substantial additional actions would be required
- Low, when the event could occur, but is not likely to, or current plans or controls are expected to be sufficient to prevent this type of event or its consequences from materializing

Following the rules of composition between likelihood and severity used in Table 4.2, it is possible to directly classify each risk as: critical (dark gray), major (middle gray), and significant (light gray).

Table 4.2. Level of Risk Criticality.

Likelihood of occurrence	High	Major	Critical	Critical
	Significant	Significant	Major	Critical
	Low	Significant	Significant	Major
		Low	Medium	High
		Severity of Consequence		

Risks have to be characterized, analyzed, and mitigated and the status of the risk shall be tracked throughout the project. Experts in the different fields will identify the risks associated with the different fields. These technical risks will be associated with the novelty and the difficulty of the system under consideration and that is dependent on the ambition of the targets sought by the consumers. Once identified the risks are listed and documented following formal procedures.

The management of risk is a specialized domain but the responsibility of risk management is in the hands of the project manager. Before the beginning of the project a responsibility of the architect is to be aware of the risks of the mission, mainly of the risks associated with the level of technical maturity of the technologies considered for the mission. Technology maturity levels and technology development will be covered in detail in Section 4.9.3.

- The management of the risks shall be in accordance with the rating of the risk, namely: critical, major, or significant.
- Significant risks are resolved through risk avoidance or risk reduction and control actions. Typical risk avoidance is via design change. Typical risk reduction and control could include: detection by analysis, direct monitoring or inspection, and early application of preplanned responses to limit the extent of the consequences. In some cases, for example, excessive cost of risk-reduction measures, the significant persons in charge – the project or program manager – may accept significant risks.
- Major risk shall be resolved primarily through risk avoidance actions or by risk transfer. There may be actions to modify baseline configurations, plans or costs, or to set-up alternative operating ways and means. Acceptance of major risks shall normally be avoided unless they are unavoidable and in that case they must be agreement at high management level.
- Critical risks shall be resolved primarily through risk avoidance or by risk transfer. Potential avoidance measures may involve substantial redeployment of resources and organization or renegotiating agreements or requesting exceptions with external parties, for example, with the consumer, sponsor, or supplier, or by modifying commitments, including their funding. Acceptance of critical risks shall be avoided unless they are unavoidable and then agreement at the highest management level is essential.

A risk can be accepted if it is considered to be sufficiently low in terms of its likelihood of occurrence and the severity of its consequences. To limit the severity of the consequences of the risk protective actions are taken, for example, shielding to protect against radiation damage or dampening for high levels of vibration. The likelihood can be reduced through preventive actions, for example, undertaking a predevelopment activity or adding redundancy.

Risk management can be applied at all phases of a project and at all levels. Note that risk management is an end-to-end process, and not only for the satellite, since if the ground segment of a space system does not work then nothing works. Envisat and MetOp and MSG are examples where managing risks with the ground segment implementation proved to be an important activity. Therefore, ground segment risks are to be treated as seriously as space segment risks, especially for an operational system. Note that the principal aim of a space system is not to demonstrate that we can fly a satellite, but to produce a specific information stream and this needs ground processing.

Some risk analysis is done without a formal risk management framework, which often implies that a large project team already exists, for example, the phase A satellite development plan is not a formal risk management document but a document to manage risk.

Risk shall be identified, characterized, and mitigated. The risks are identified and characterized by the experts in the different technical fields where the risks lay. Once they are identified, they are recorded and formally characterized in tables and templates, assigning to each one the appropriate class of risk. The mitigation of risk is applied throughout the project. Milestones and external reviews mitigate risk, and so does technical development both prior to, within, or parallel to, the project schedule. Risk is also mitigated by the use of contingency management reserves on cost or schedule. Formal techniques include the establishment of critical items and issues list plus all the traditional cost and schedule control systems.

4.9.2. Quantitative risk management

Numerical probabilistic risk analysis can characterize risk. The full most comprehensive analysis of this type will include modeling a comprehensive set of scenarios leading to all undesired end states including loss of crew – if applicable – or danger for the public. In this most comprehensive approach all the basic causes are modeled, including: hardware failures, human errors, process errors, and others. A simpler approach focuses in a limited set of events, most of the times in mission failure and analyzes only specific types of failure mechanisms – like probabilistic failure analysis of parts – that are specially adequate for numerical treatment. This is the well-established branch of availability and reliability calculations. Mission specifications establish the availability of the service provided by the space mission. Satellite specifications establish the probability of a satellite to be still operating correctly after a number of years. Product assurance specialists (Section 2.6.2) perform this branch of quantitative risk management.

The present section will concentrate in numerical techniques different from availability and reliability. There are techniques such as probabilistic risk analysis, probabilistic network schedules, or Monte Carlo models that are able to provide probabilistic assessment of the risk involved in the estimation of parameters like: performance, cost, or schedule. At NASA this type of comprehensive probabilistic risk assessments (PRA) are conducted for new programs in situations of high criticality.[19] These situations include:

- Human space flight.
- Special issues of public safety – for instance when flying nuclear payload as done in Ulysses (Section 9.6.2).
- High visibility programs with high schedule criticality. This would be the case for ambitious planetary missions requiring a very specific launch date.

[19] Quantitative Risk Analysis at NASA, William Vesely, NASA.

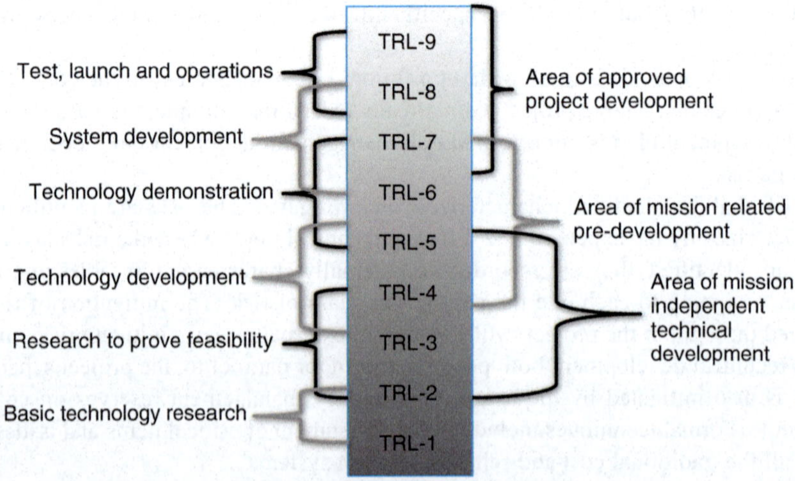

Figure 4.4. Technology Readiness Levels and Mission Implementation.

ESA that has had consistently a much smaller emphasis in manned space does not perform systematic PRA different from "traditional" availability and reliability analysis. Quantitative risk assessment requires establishing probability density functions. This will be studied later in Section 5.8 that will analyze tools for numerical support to decision making.

4.9.3. Technical readiness and technical development

Technology readiness levels are used extensively to define the level of maturity of a technology. Increasing levels means increasing maturity. They are shown in Figure 4.4, and four development environments are defined.

The following steps are defined[20]:

- TRL-1. Basic principles observed and reported
- TRL-2. Technology concept and/or application formulated
- TRL-3. Analytical and experimental critical function and/or characterization proof of concept
- TRL-4. Component and/or breadboard validation in laboratory environment
- TRL-5. Component and/or breadboard validation in relevant environment
- TRL-6. System/subsystem model or prototype demonstration in a relevant environment
- TRL-7. System prototype demonstration in a space environment
- TRL-8. Actual system completed and flight qualified through test or demonstration
- TRL-9. Actual system flight proven through successful space mission

[20] Technology Readiness Levels Handbook for Space Applications, Artemis Innovation Management Solutions 2008.

Technology levels TRL1 to TRL 3 are associated with basic research and they are done without a specific implementation in mind. Technology levels TRL3 to 5 already point toward developing usable implementations, but they still aim toward a range of applications. TRL 5, 6, and 7 are the logical technology levels to be considered during the early stages of designing a new space mission but before the new project team is fully implemented. TRL 6 to 9 are normally performed within the framework of an already approved mission. The conclusion is that to keep the level of risk within reasonable limits, it is crucial to have performed at least the first four levels to include the technology in the design of a new system, and to level 6 or higher for its implementation.[21]

In some cases a possible new space mission will require the development of a very special observing technique. In this case all the developmental effort, including the basic research, will be done with a final aim in mind. Nevertheless, this development will still be rather "detailed mission design independent" because the reachable performances for the technology could not yet be predictable and because it would be too early to foresee the physical implementation that the technology under development would require.

When a new mission is proposed it already has a preexisting body of technical research available that it can identify and use. During the definition stages of the new mission, the status of the technology will be studied and this will allow identification of areas that are already well developed and areas where there are gaps. The normally long time needed to define and to approve or reject a new space mission offers the opportunity for steering the previously existing development in a direction that is useful for the new mission under consideration. The amount of money available before the project is fully approved will be moderate; but this period will allow difficult technology to mature, wrong paths to be explored and discarded, and the best alternatives to be found. Often the development of technologies needs time and an already approved program cannot tolerate the delays that would be introduced by failure in the development of a key technology. A new space project should avoid proposing the use of a development that requires a new technology with a low TRL level. If this were absolutely necessary to fulfilling consumer requirements, these factors should be clearly indicated and discussed before the project is approved. In an extreme case, and for very challenging technologies, it would be necessary to start with a technology demonstration mission before the related advancement of a science mission could be undertaken.

Technology readiness levels are not static quantities that once obtained will stay forever. There are two factors that will reduce the status of technical maturity of a technology:

- The status of a technology changes when used in a new environment. A device successfully flown in a given type of environment is TRL 9 only when flying again in a similar environment. It will not be possible to use a technology demonstrated with a short flight around the Earth and to claim that it is flight qualified for a

[21] Technology Readiness Level Schedule Risk and Slippage in Spacecraft Design, Gregory Dubos, J Saleh, Robert Braun, Georgia Institute of Technology, Journal of Spacecraft and Rockets Vol 45 No 4 July–August 2008.

mission to a far-away planet. It will be necessary to perform dedicated development to gain confidence on the behavior of the technology in the new environment.
- Know-how and technological maturity can be lost, persons can retire, or special manufacturing materials or procedures can be discontinued. This is a very real possibility in the space field that works normally with short series production and with infrequent contracts. The problem of obsolescence is specially acute in the area of electronic components because the average lifespan of a electronic technology is normally much shorter than the time necessary for the implementation and operation of a space mission.

4.9.4. Developmental approach and model philosophy

Before authorizing a new space project, and to be able to provide a clear cost and risk estimation, it is essential to establish a developmental approach including:

- Predevelopment of critical technologies, to be performed before full project implementation begins
- Developmental approach and model philosophy to be applied during project implementation

Predevelopment work expenditure is normally small with respect to the overall cost of the project, but, as mentioned above, it provides time for technology development. The project development work will be based on the model philosophy chosen. Different models of the satellite and its assemblies will be built. These models will have objectives progressively more ambitious until the final model will be used for flight:

- Development and breadboard models, which are use to understand and improve the design of the items under test.
- Engineering models are used to characterize the functional behavior of the items under test. They shall be functional representatives but they do not need to be fully physically representative of the final item to be flown.
- Qualification models shall be functional and physically representative of the flight hardware. Qualification model shall be able to provide the in-orbit performances when tested in an environment fully representative of the in-orbit situation. They are used to formally validate – that means "qualify" – the adequacy of the item for the intended purpose in the intended environment.
- Flight model. This is the model that will actually fly.

The model philosophy will be a function of the novelty of the satellite, of the risk acceptance-avoidance of the project, and of the cost and schedule constraints. The establishment of a developmental approach and its associated model philosophy will provide credibility to the cost, schedule, and risk estimations of the future space project, and it is fundamental for a well-informed decision on project go-ahead to be made.

In the past a very deliberate approach toward engineering development with a correspondingly "heavy" approach to model philosophy was the norm. Today, it is the exception. ESA's mission Ulysses, which was designed in the mid-1970s, had three models: an engineering model, a qualification model, and a flight model; and the qualification model was fully representative of the flight hardware. The building of these models was totally sequential, so that the lessons learned from one model were implemented in the next one. This produced a long and costly implementation, but the risk was minimized. Present cost and schedule pressures and the evolution of technical factors, principally computers and computer-based simulations, have led to much more streamlined approaches being used. Modern analytical tools such as thermal, structural, and electromagnetic interference, allow a very high level of confidence on predicting the behavior of satellites, and "virtual models" allow the use of test configurations which mix real physical hardware and computer-residing simulations. This allows test benches that provide progressive confidence in the components being designed and allows minimizing the work prior to building the flight model. With this approach the flight model will be the first "physical" model where all flight-standard satellite components will come together, and it will also be used as the "formal" qualification model. This is the "proto-flight" approach where the model to be flown is at the same time its own prototype and goes through qualification tests before flying.

The selection of the model philosophy is very closely related to the risk management process and the technology readiness status on a program. Deliberate development approaches reduce risk but increase cost and schedule. The selection of the development approach and of the resulting model philosophy will be the result of balancing four conflicting forces (Figure 4.5):

- The programmatic constraints (schedule, funds, …) established by the sponsor
- The development status and resulting technical readiness levels, heritage, and novelty of the mission under consideration

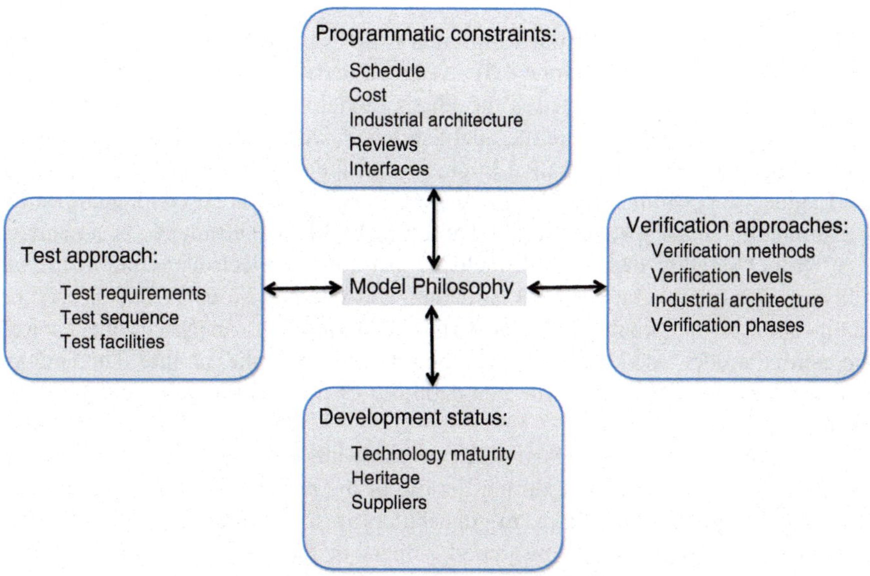

Figure 4.5. Model Philosophy Definition Process (System Engineering, ECCS -E- 10, European Cooperation for Space Standardization, ESA 1996).

- The overall verification strategy of the mission that will depend of the requirements on the mission
- The test approach that will depend on factors like test requirements, intrinsic test difficulties, and test sequencing; for example, the possibility of doing – or not – tests in parallel

The problem of the adequacy of the analysis and simulations is how representative they are of the flight hardware and its environmental conditions. As stated above, the improving fidelity of computer-based simulations provides increasing confidence in their efficacy. Furthermore, this problem also applies to testing: ground tests of space hardware, for example, vibration, thermal, or large deployable functional testing, are nothing but "simulation" of the real space conditions. From this perspective, both analysis and tests can both be closer or further away from the real in-orbit situation. The conclusion is that the verification approach, by testing, by analysis, or by a combination of both, shall always be critically analyzed for adequate representation of the space conditions. This representational adequacy is especially important because highly representative tests are sophisticated and sophisticated tests are expensive. To reduce the sophistication of verification, large margins can be used, but large margins imply overdesigning and overdesigning is also expensive. Again, the need for balance appears in our struggle to arrive at an optimal design. A supplementary lesson of this discussion is that sophisticated tests, being cost drivers, should be analyzed with special care in order that a credible cost estimation for a space project can be made.

It is common for sophisticated testing to provide test results that require yet another layer of analysis and interpretation before the results can be applied to the flight conditions. This is because in these sophisticated tests, it is not possible to exactly simulate the flight environment. It is only possible to approximate them.

The recently approved Sentinel-3 mission is an example of a modern streamlined developmental approach (Figure 4.6). At satellite level, the only full "physical" model that is going to be tested is the Flight Model/Protoflight Model (FM/PFM), which will be used for full satellite qualification as well as for flying. Equipment is going to be tested in representative Engineering Models (EM) or highly representative Engineering Qualification Models (EQM). These models are used for qualification at instrument – or subsystem – levels. These EM/EQMs will ultimately be assembled in a "virtual satellite EM," which includes instruments, electronic equipment, and software, all assembled together. A Structural and Thermal Model (STM) is used for qualification of the structure and for early validation of the analytical, mechanical, and thermal models used to predict the environment within the satellite. The Suitcase is a dedicated model to study the compatibility of the end-to-end communications chains.

The employment of combined software-hardware environments will allow the use of models that will evolve during the life of the project: from the early phases' mission performance simulators, to contributors to the qualification, to operations training tools, and finally to on-ground helpers in the case of in-orbit failure of equipment.

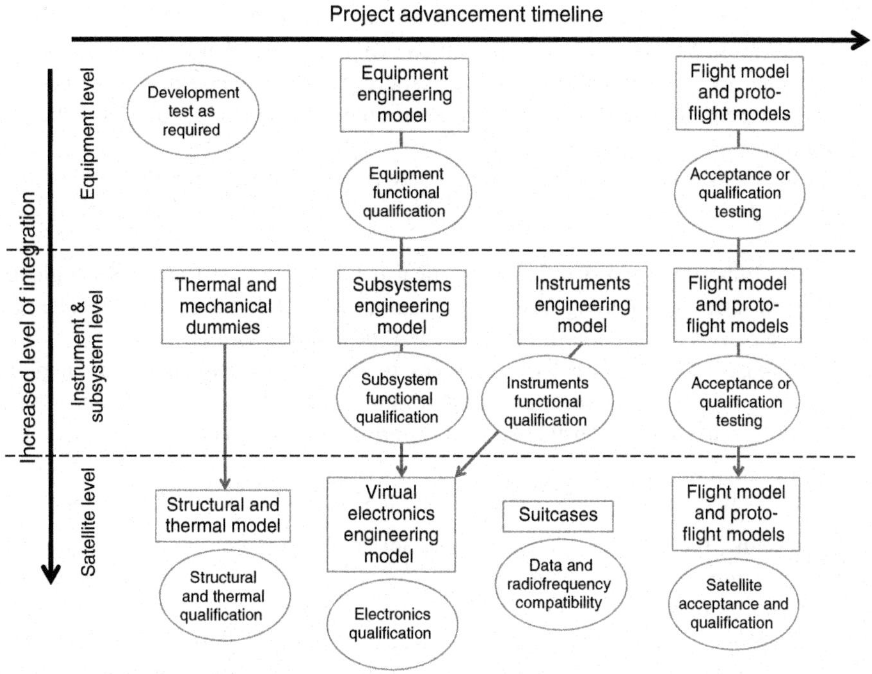

Figure 4.6. Model Philosophy of a Modern Earth Observation Operational Satellite: ESA's Sentinel-3 (Sentinel-3 Quotation Executive Summary, Thales Alenia Space, 2007).

4.10. Schedule constraints

Any space endeavor is a set of activities that are connected in time. The number and complication of the activities, the need to perform these activities in series or the possibility of doing them in parallel, and the level and quality of resources available for the execution of the work will directly influence and contribute to the derivation of the schedule of the whole endeavor and of each of their parts.

Complex projects require a series of activities, some of which must be performed sequentially and others that can be performed in parallel with other activities. This series of sequential and parallel activities was first modeled as a network for project management in 1957, and is known as the Critical Path Method (CPM). CPM uses fixed time estimates for each activity in a deterministic approach. It enables managers to determine how the combination of parallel and sequential activities can combine to define the shortest possible schedule; that is, the longest sequence of activities that can obligatorily be executed in series. The Program Evaluation and Review Technique (PERT), on the other hand, allows for randomness in the completion times of activities.

To determine the expected duration of any endeavor it is necessary:

- To divide the whole endeavor into individual activities with a clear beginning and end.

- To allocate preliminary resources to the different activities.
- To determine the necessary duration of each activity in consultation with experts. Durations are a function of the level of resources – human and material – available for the execution of each activity. The duration figures can be fixed or probabilistic profiles.
- To establish logical links of precedence between the beginning and the end of each activity; for example, the manufacturing of an item cannot start before it has been designed.
- To determine the longest duration "critical path" sequence of activities logically linked in series and running from the beginning to the end of the endeavor. The sum of these activities will determine the overall program duration.
- To increase the assignment of resources to the activities in the "critical path" sequence in order to reduce the overall duration, and to look for an optimal allocation of resources allowing the shortest duration in accordance with the allowable resources.
- To update the allocation of resources and provide final schedule estimates.

That means, the schedule of a project is not an independent parameter. The time needed for project implementation will derive directly from the development approach and model philosophy, which will be a function of the level of novelty of the design proposed and of the failure-tolerance of the project. The desire to compress schedules without giving due regard to model philosophy, failure-tolerance, and the level of novelty should be resisted. The model philosophy of space hardware development is based on learning lessons from the early models, which can be implemented in the later ones; therefore, the level of allowed overlapping between models is limited. The only way to answer strong schedule constraints is to design a project that uses mature technologies, thereby allowing a streamlined model philosophy and a simplified review approach. Communication satellites, which have a very low level of novelty and a correspondingly simplified model philosophy, can be implemented in 2 years or less. This is not the case with most satellites that require development of new technology pushing the state of the art.

The architectural definition phases of all satellites will include schedule estimations for the duration of the full program implementation. As stated above, the schedule will have a critical path that will be provided by the need to sequentially design, build, test at component level, assemble, and finally test at satellite level. The overall schedule will include bar charts for the overall satellite and for the most important elements of the mission, such as the instruments. The schedule has to be realistic and must provide enough time to allocate to potential failures in the test program. This schedule prediction, which should be firmly based on the overall development philosophy, will complement the cost information as fundamental data for the mission approval or cancelation by the sponsor.

4.11. Management trends as constraints

Management trends are management fashions that can constrain the design. The constraints can come as an explicit directive, for example, to provide design features in the satellite that allow in-orbit servicing by astronauts, or as an implicit mood, for example, situations where management encourages or discourages high levels of risk. Examples of trends include in-orbit servicing by astronauts, large standard platforms, and small satellites. Almost all the trends that we have seen since the beginning of the space age were based on quite sound reasoning, and they were adequate in some contexts. On the other hand, almost all of them were overadvertised and in real life they have all demonstrated limitations.

In the 1970s the USA decided to rely principally on the use of the Space Shuttle for low Earth orbit missions, which was eventually proved to be an unwise decision, and meant that a large number of such missions would incur the costs of using a man-rated system when this was unnecessary. This trend became an explicit space systems design directive when in 1978 Jimmy Carter ordered the expendable launch vehicles to be phased out in favor of the Space Shuttle. A good example of the type of mission design that this directive produced was Shuttle-6, which carried and released in low Earth orbit a 2,200 kg TDRSS-1 satellite. This satellite used an IUS upper stage to reach its intended geostationary orbit from the Shuttle's orbit. This approach put many satellites successfully in orbit, but the original predictions regarding Shuttle turnaround times and costs proved to be far too optimistic. These predictions claimed that the Space Shuttle would be capable of putting payloads in low Earth orbit for $400 per pound, but the actual cost was closer to $5,000–10,000.[22] The original aim was to launch 26 shuttle flights per year, but the maximum number has been just six.

The 1980s were the time of in-orbit servicing by astronauts, made possible by the use of the Space Shuttle as the launch vehicle of choice. The Challenger crew serviced Solarmax in space in April 1984. The satellite was designed for in-orbit servicing but the astronauts also performed successful unforeseen repairs on the Coronagraph using duct tape and pliers. Here we have a clear example of a mission saved by in-orbit servicing. The use of astronauts for the servicing of the Hubble Space Telescope rescued the mission from failure, but the requirements of astronaut servicing entailed an orbit that was inadequate for an astronomy mission. This is another example of the contradictory impact of trends. The different influences acting on the design of the Hubble mission design will be discussed further in Section 7.2.

The 1980s[23] were also a period of extensive use of standard platforms as tools of cost reduction. In the 1980s NASA decided to reduce the costs of deep space missions by standardization in two types of buses: Mariner Mark II for outer solar system missions and Observers for the inner solar system. Cassini was the only successful Mariner Mark II mission. The 2,500 kg Mars Observer, which failed to reach Mars, was the only Observer mission. On the other hand, the use of standard platforms in the series of NOAA operational meteorological satellites and later, with a different model in NASA-NOAA's Terra, Aqua, and Aura, led to cost reductions as well as successful performances.

The 1990s were the years of "faster, cheaper, better." This trend came with a higher tolerance to failure and the desire for an intense rethinking about how things

[22] Restoring US Launch Capacity, Robert Zirkle, "Space National Programs and International Cooperation" Boulder Colorado, Westview Press 1989.

[23] Space System Failures, Harland and Lorenz, Praxis.

were done and how they could be done better. This renewal helped to reinvigorate NASA after a string of high visibility failures in the early 1990s which included the discovery of the Hubble aberration in 1990, the failure in the deployment of the high-gain Galileo antenna in 1992, and the crisis in GOES that in 1992 was three years behind schedule and over twice the planned cost. The actual implementation of the "faster implementation cheaper" philosophy included successes, like Clementine, and failures, like Mars Climate Observer and Polar Lander. Again, the actual implementation provided both lights and shadows. It was clear that the missions trying to be at the same time fast and cheap while having ambitious objectives and being objectively complex had a very large probability of failure.[24] This topic is retaken in Section 13.2.3 and specifically in Figure 13.1.

The current decade appears not to display any overwhelming trend, albeit small satellites and the use of nanotechnology could be considered the fashionable trends of recent years. Small satellites can solve many problems, but the goal should never be to design a small satellite, but rather the smallest satellite that fulfills the mission's requirements. Nanotechnology is also quite appropriate for some applications, mainly for deep space missions, but many of the laws of physics do not scale well with size. If experience of the past should be our guide for the future, the trend toward small missions shall be welcomed as a salutary measure for cost reduction, but we should treat it with a healthy dosage of skepticism: it will not be applicable in every case.

All the trends above featured advantages and disadvantages and they all included successes and failures on account of it. The problem has been more the trend's over-promotion, than the trend in itself. Engineering has its own form of "rough justice": either the hardware works or it does not. If it works, the design is successful; if it does not the design is unsuccessful and there is no level of oratory, persuasive argumentation or orders from headquarters that can open a jammed deployable antenna. If a management trend becomes a management directive, the designer shall take it as just another explicit system constraint and include it in the basic design. It is also the designer's duty to point out the implications of that trend in the design, to indicate where and how it is adequate, and inform the decision maker of those implications. It will ultimately be the decision maker, usually the sponsor, who decides whether or not to implement the trend.

---

[24] When Is a Satellite Mission Too Fast and Too Cheap? David Bearden, MAPLD International Conference 2001.

Chapter 5

System Design as a Synchronic Process

Keywords Space System Elements, Space System Architecture, Cost Constrained Design, Risk Constrained Design, Schedule Constrained Design, Functional Analysis, Trade-off, Functional Apportioning, Budget Apportioning, Concurrent Design Facility, Dependability, Mission Effectiveness, Safety Margins, Design Heuristics

The word synchronic in the title of this chapter is defined as "the analysis of something as it exists at a particular moment in time." This term was first applied in linguistics to refer to the analysis of a language from the perspective of all its rules, features, and interactions at each moment in time. Parallel to the synchronic perspective, linguistics also introduced the term "diachronic" to refer to the study of how languages change with time. As in linguistics, the process of designing a space mission can be analyzed from either a synchronic or diachronic viewpoint. The present chapter will discuss typical mission designs from a synchronic perspective, while the following one will take a diachronic approach.

Analyzing the design process from a synchronic standpoint involves the ways in which requirements, constraints, systems breakdown, analysis, accepted design procedures, heuristics, and design decisions interact with one another. Our discussion in this chapter will begin by recalling the physical elements making up an end-to-end space system. The second section analyzes space system specification and design and how they relate to the principal aim of this book, as defined in Section 1.1, to produce an "architecture" for the space system. This is followed by sections describing how various constraints – those of cost, risk, and scheduling – influence system design and how for each mission the varying degrees of emphasis given to each of the constraints influences the design. The next section deals with the different types of requirements and how they influence the design. The fifth section describes the specific tools used for design: performance analysis, functional analysis, trade-offs, breakdowns, functional apportioning, budget apportioning, and how concurrent design facilities can help mission design. A final subsection of that section discusses dependability. The following section looks at the design from an end-to-end perspective, with subsections examining measurement methods for mission effectiveness and their limitations, the problems associated with numerical and probabilistic approaches to design optimization and on the impact of safety margins, and the mistakes and errors regarding the actual performances of space systems. The chapter ends with an analysis of nonnumerical aids to design: heuristic processes and the need for developing sensitivity in seeking for a balance between conflicting needs.

M.A. Aguirre, *Introduction to Space Systems: Design and Synthesis*, Space Technology Library 27,
DOI 10.1007/978-1-4614-3758-1_5,

5.1. Space system elements

In Figure 1.3 the design of space systems was described as a recursive process where the highest-level elements are defined first, followed progressively thereafter by lower-level components. The first step in any design is to define the elements resulting from a meaningful first-level decomposition of the end-to-end system. Wertz-Larson[1] uses a division into architectural elements, including:

- The subject or event being observed, hereafter called the observable
- Payload or instruments acquiring information about the observable
- Spacecraft bus providing support to the instruments
- Launcher
- Proposed orbit
- Ground segment
- Communications architecture
- Operational concept

ECSS,[2] on the other hand, uses a two-level division into physical segments:

- Launch segment, divided into launcher and launch facilities
- Ground segment, divided into mission control center, communications network, ground stations, payload data management facility, and ground support equipment
- Space segment, divided in spacecraft bus and instruments

Wertz-Larson's functional approach is the best suited for the process of defining a mission's overall architecture, and this text will employ a division nearly identical to that of Wertz-Larson; the only difference being the division of the ground segment into two different elements as in the ECSS approach: mission control and instrument data processing. This supplementary division is selected because these two parts of the ground segment are driven by different design criteria and do not interact strongly with one another. The resulting division into elements followed here are (Figure 5.1):

- Observable: the events we want to observe. The observable under study is the root of the mission. It will be seen in later chapters that the observable directly shapes preferred operational methods as well as data usage.
- Orbit: the satellite's path. The orbit is selected from the rather limited set of possibilities allowed by the laws of Newton and required by the observable.
- Launcher: the device used to inject the satellite into space. Launchers are usually selected from a limited set of possibilities: first, because reliable and reasonably priced launchers are rare, and second, because programmatic constraints often forbid or favor launchers from specific national sources.
- Instruments: the devices that obtain the information about the events to be observed. Given that each mission is aimed at very specific observational needs, instruments are generally designed specifically for each mission.
- Platform: the device that supports the instruments. The platform helps instruments to fulfill their functions and sends and receives information from the satellite to the

---

[1]Space Mission Analysis and Design, Wertz-Larson, Microcosm Presm, Kluwer 1999.

[2]Policy and Principles, ECSS-E-00A, ESA 1999.

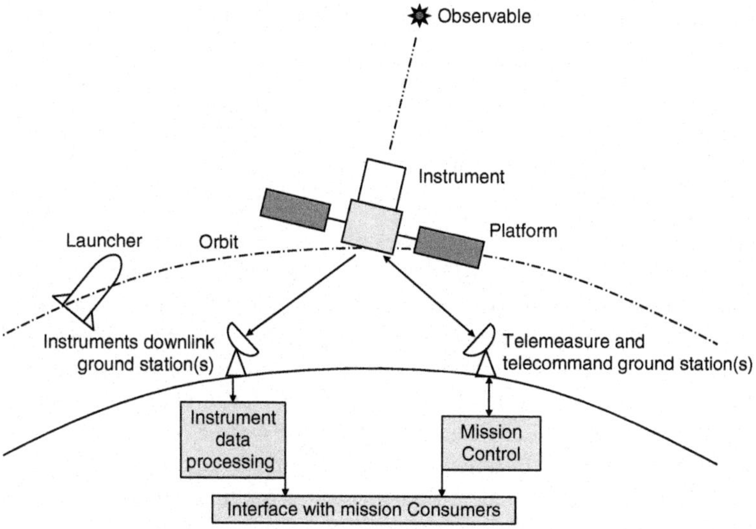

Figure 5.1. Highest-Level Space Mission Elements.

ground. Depending on the circumstances, platforms are specifically designed for each mission, or adapted from a previous design.

- The satellite that is made up of the platform and the instruments.
- Telemeasure and telecommand ground stations and mission control centers. The operational ground stations send commands to and from the satellite, and receive information related to the status of the platform and instruments. The mission control center monitors the information provided by the satellite about its own status and prepares and sends commands to control the satellite. The operation's ground elements are normally based on an existing infrastructure with adaptations as needed. For any needs not covered by existing infrastructures, new elements are designed and built.
- Instrument downlink ground stations and instrument data processing and distribution centers. The instrument ground stations receive the data produced by the instruments about the observable. The Instrument Data Processing Center processes and distributes this information to the consumers. Existing infrastructures are employed as far as possible, though many or even the majority of components for this element will require mission-specific systems.
- Both the instrument and the mission control centers will interface with the consumers receiving their requests and delivering the mission products.

It should be recalled that some missions require more than one satellite. The need for multiple satellites might arise from a need to provide long-term services by flying successive satellites, or in order to split demanding requirements between several satellites flying simultaneously, for instance, missions requiring large amounts of data or a high frequency of observation.

The definition of any space system must start from a broad definition of each of these elements and the interfaces between them. Once this has been established, a detailed definition of each element can be developed.

5.2. System specification, system design, and system architect

Before receiving definitive authorization for a fully funded project, the process of mission design involves:

- The customer, in most cases a space agency
- The industrial supplier, in most case a space engineering company

In principle, the customer establishes the requirements and the supplier will produce – and later implement – the design; nevertheless, functions allocation, the corresponding system decomposition, and budget allocations are all part of the iterative process of system specification and system design. The process of establishing requirements never entails specifying concrete physical design parameters, but a specification can perfectly well specify functional or even physical decompositions of the system. This effectively combines the processes of both system design and system specification. The consumer, for example, might require the rapid delivery of data once it has been produced. This generates a requirement for acceptable data latency, but the customer could well specify the use of a data relay satellite, or encourage or discourage specific locations for ground stations. This perfectly legitimate kind of requirement stipulates an allocation of functions and strongly influences the design. By specifying the approach he prefers, the customer has a strong hand in shaping the design provided by the supplier. On the other hand, the specification must never demand design features, and the customer's involvement in the detailed design is forcefully limited.

Both the customer and supplier are involved in the design process and before project approval both will need to maintain an architect's broad perspective in order to allow a progressive top-down definition of the system. As the system becomes progressively more defined the design team will separate into those with an architect's perspective and those focusing on detailed design. The mission architect can be defined as collectively the group of people defining the mission and who at the same time are keeping a generalist perspective over the mission. This concurs with the definition of system architectural design offered in Section 2.2 as consisting essentially of a special perspective over system engineering and system design. It is evident that any person making design decisions having multiple or major impacts on a mission is performing the duties of a system architect, because these kinds of decisions can only be made from a generalist perspective. This is true for early design decisions in areas related to instruments, mission orbit, or operational approach.

It shall be noted that no new architectural work is necessary in implementing a system that is not innovative, in which most components are standard. Architectural design has been defined as the work prior to a well-informed decision to go-ahead with a fully funded project, and projects with low novelty levels can be initiated with little previous work. It can be said that **only unique systems need to be the subject of the architectural design process**, but nearly all space systems are in fact unique. On the other hand, many apparently noninnovative space systems always need some type of adaptation when applied. In some cases, an existing satellite is used in a different environment, requires redesign of obsolete hardware, or is forced to use a different launcher or data flow, for example, a new data relay satellite. In this case some level of new architectural design will be needed.

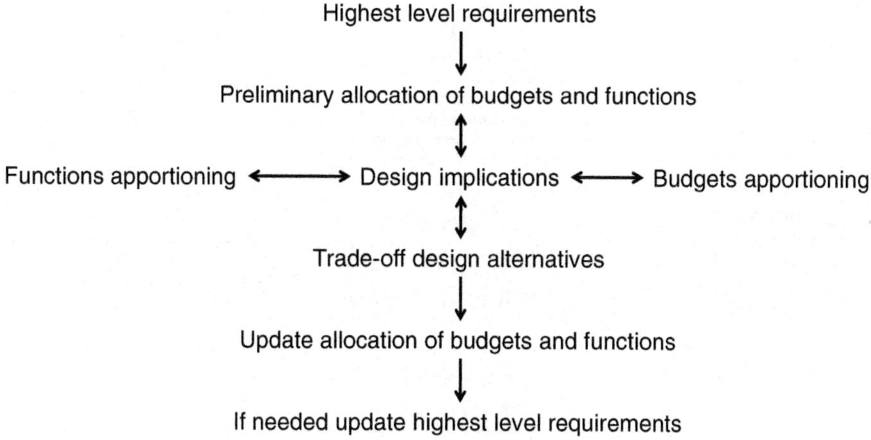

Figure 5.2. Requirements Establishment and Design Implications.

Design decisions always entail trade-offs between alternatives. The supplier offers design alternatives, trades-off between them, and chooses a winning approach according to previously established selection criteria. Still, the customer always reviews and discusses the supplier's choice of approaches. In the highest-level trade-offs—such as orbit selection or the attitude control concept – the customer is directly involved with the trade-off process, but trade-offs related to detailed design are left almost completely in the hands of the supplier. The relationship between all these activities – requirements establishment, design, apportioning, and trade-offs – are summarized in Figure 5.2, which shows how agreements reached on design alternatives will often entail the updating of allocations and even of highest-level requirements. As we have indicated repeatedly, functional design, physical design, and requirements establishment are all tightly interrelated before project approval.

The process of defining and implementing the system includes four kinds of activities:

- Scoping and certification, which both form part of the system's specification domain and are therefore principally the customer's responsibility
- Partitioning and aggregation, which form part of the system design domain and are therefore principally the supplier's responsibility

Scoping is a series of procedures for selecting and rejecting approaches defining the mission, defining requirements and constraints, and deciding what is inside and outside the system. Scoping is the art of defining the system's boundaries and establishing its requirements and constraints. Proper scoping requires a clear aim. The consumer will provide the aim in mission statements or a mission requirements document (Section 3.2.1). At the same time, the sponsor provides constraints that limit the choice of acceptable mission definitions; for example, demanding that the mission be designed to cost or that failure cannot be tolerated (Figure 5.3).

The certification or verification of a system offers a guarantee to both the paying sponsor and the nonpaying consumer that the system is fit for use. The customer's role is to first define a system that can be unequivocally certified and then eventually

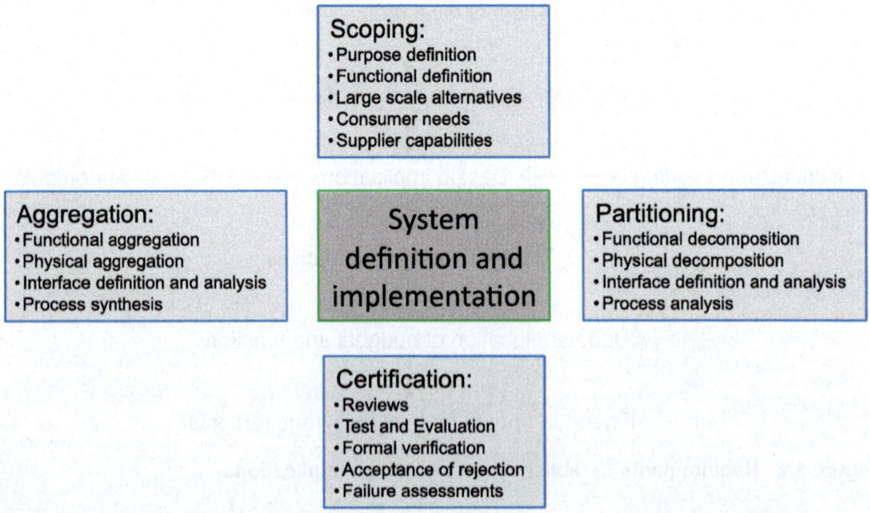

Figure 5.3. The Four Aspects of System Definition (The Art of Systems Architecting, Maier Rechtin, CRC 2000).

run the certification. Effective certification, that is, qualification or acceptance, implies the ensuing payment of the agreed-upon price to the supplier. This is the reason it must be completely unambiguous. To provide credibility to the cost and performance estimations, the architectural definition process requires defining a system that can be clearly verified and establishing unambiguous requirements that can be properly tested. This is reflected in the heuristic sentence: **Define how a requirement is to be verified at the same time it is established**. The need for verifiability was recalled in Section 3.5.

Aggregation and partitioning are the grouping and separation of related solutions and problems. They are two faces of the same coin: proper system design. This entails maintaining a balance between the analytical/partition activities – such as functional and physical decomposition – and those addressing synthesis/aggregation – such as the analysis of end-to-end performances. A balance in the allocation of functions and resources to the different system elements and components is also crucial.

System aggregation and partitioning is more art than science. It is synthesis oriented, not analytical, and science is nearly always analytical. Establishing a scientific system aggregation and partition method or tool kit is a difficult task, because the process is unstructured and eclectic; heuristic but also artistic. The whole second part of this book starting in Chapter 7 and finishing in Chapter 12 is an attempt to develop the sensitivity necessary to properly perform the aggregation and partition of a space system.

5.3. Design against constraints

Section 4.4 introduced the concept of a programmatic framework as the set of rules established by the sponsor for the implementation of potential new space projects. Sections 4.5 and 4.6 studied different types of space missions and classified them according to their aims or level of criticality. All these sections revolve around the need to establish a rational balance between the future mission's level of tolerance toward: cost overruns, failure risks, and final performance. The establishment of these different levels of tolerance is not a decision of the architectural definition team but of the special programmatic guidelines of the project, which in certain cases will just have broad guidelines, while in others they will establish specific and compulsory plans on:

- The number of, and the workload associated with the performance of, formal reviews (Section 6.4)
- Definition of the degree of documentation to be formally delivered by the supplier to the customer
- Definition of the developmental logic and number of components and satellite models to be manufactured and tested before the final flight model (Section 4.9.4)
- Reduction or increase in the level of product assurance efforts
- Number and fidelity – hence cost – of tests
- Establishment of the supplier's level of dependence with respect to the customer and establishment of the customer's level of involvement with the implementation process

The set of all these decisions can be called the mission's "programmatic guidelines." Most of the measures indicated above are applied during the mission's implementation phases, not the architectural definition phases; that is, the architectural definition team's work directly related to programmatic directives will be modest. Nevertheless, the programmatic directives' impact on optimal mission design is crucial because the design must be coherent with them.

The following sections will analyze the impact of cost risks and schedule on system design.

5.3.1. Cost

There is a heuristic system design maxim that claims, "**Customers will accept delays but will tolerate cost overruns only with extreme difficulty**." This means that cost is the most important constraint.

Cost containment is at the heart of all proper engineering, and a slightly tongue-in-cheek definition of engineering could be "accomplish with one euro or dollar what any fool could do with ten." According to this viewpoint, each proper system definition decision must be made with cost containment in mind; hence, design for cost minimization would be the essence of all proper design, not a mere component of it. Nonetheless, it is both possible and practical to analyze cost reduction measures and their impact on system design as a separate issue, and this will be the aim of this section. Cost reductions can be sought in two different areas:

- Measures in the way the mission is designed and organized could reduce costs; these alternative approaches are discussed in Chapter 13.
- Proper mission definition effort can reduce cost. The aim of the present section is to discuss this.

In cases where the sponsor requires that the mission be strictly design to cost, the designer must play with performance and risk in order to satisfy the cost constraint while keeping the risk of failure moderate and providing performance levels still acceptable to the consumer. Reducing performance will directly reduce costs, and a design-to-cost mission is one whose performance is inevitably bound by the cost limit. Excessive reductions in performance can also be avoided by increasing risk; for example, by eliminating redundancy, using lower-quality electronic components, or by reducing the frequency of communications between the satellite and the mission control center. Any of these measures will reduce costs without reducing performance, but the first one risks ending the mission in case any key nonredundant component fails. The second measure would reduce the satellite's reliability by the use of low-quality components, while the third measure will increase the interval between any in-orbit failure and the possibility of ground intervention to remedy it.

One always-valid way to simultaneously reduce costs and risk is to devote a proportionally large amount of the budget to the mission architectural definition phases. This investment will be repaid with much larger cost reductions during the project implementation phases. The following quotation from NASA's system handbook[3] is relevant here:

> Typically, by the time the preferred system's overall concept is selected, between 50 and 70 percent of the system's life-cycle cost has been "locked in." By the time a preliminary system design is selected, this figure may be as high as 90 percent. This presents a major dilemma to the system engineer, who must lead this selection process. Just at the time when decisions are most critical, the state of information about the alternatives is least certain. Uncertainty about costs is a fact of systems engineering. This suggests that efforts to acquire better information about the life-cycle cost of each alternative early in the project life cycle potentially has very high payoffs.

Two important messages are conveyed here:

- **When the fully funded project begins, almost all the important decisions influencing the final cost have already been made.**
- **Early heavy efforts regarding system definition will lead to optimal design solutions that will reduce final implementation costs.**

So mission architectural definition phases must be long, detailed, and well funded if costs are to be reduced. This apparent contradiction between expending a bit more at the beginning of the design process to avoid expending much more at the end was discussed in Section 1.1, and is supported by the data in Figure 1.1.

Another approach to cost reduction that also avoids risk increase is to produce robust designs that reduce them simultaneously. The word "robust" is used here to indicate designs that are simple, well proven and fit for purpose; for example, simple-shaped satellites without deployable solar arrays provide both high reliability – reducing the risk of failure – and lower costs. The cubic configuration of the highly effective, low-cost SSTL satellite shown in Figure 11.10 is an example of this. Another example, also from SSTL, is the use of old-fashioned, low-performance but cheap and well-proven processors for satellite data handling (see Section 11.4.4). Robust simple designs have a snowball effect, providing cost reduction potential throughout the system. A simple satellite, for instance, minimizes the need for interventions by the ground segment for control and monitoring of the satellite.

---

[3]NASA System Engineering Handbook, SP-610S, NASA 1995.

A third possibility for reducing costs without lessening performance is to improve the flow of information among the supplier, customer, and consumer. Concurrent design, that is, designs in which all the technical and scientific expertise interact together in a single place, also enables experts to understand the problems faced by one another. This mutual interaction permits a better understanding of what requirements can be changed or relaxed with a minimum impact on consumer satisfaction, and those that cannot be relaxed without damaging the mission's essence. The easy flow of information, on the other hand, could be in contradiction with necessary formal methods of project control. Nevertheless, during a mission's architectural definition phases the customer and the supplier have yet to establish large project teams, the volume of documentation generated is still moderate, and it is possible for everyone involved to be aware of everyone else's activities.

5.3.2. Risk

Risk refers to the possibility that something that threatens the mission's integrity could occur, such as component failure, major mission performance degradation, or even cancelation of the mission itself. Risk was thoroughly examined in Chapter 4: Section 4.9 discussed risk constraints and techniques for risk management; Section 4.9.1 introduced qualitative procedures for classifying and handling risk; Section 4.9.2 introduced quantitative risk management; Section 4.9.3analyzed the concept of technical readiness levels; and Section 4.9.4analyzed development approaches and model philosophy. The present section will discuss the impact on design of risk acceptance or avoidance.

In the early stages of the space age, risk-averse missions entailed above-average funding and a comfortable schedule. The additional time and funds allow a conservative model philosophy, a large number of detailed project reviews, and comprehensive realistic testing. A comfortable schedule also facilitated staggering the development stages so that the manufacturing of the flight model does not start until all the components and elements have been thoroughly validated. This relaxed approach to program development and funding is relatively slow and is now out of fashion. Section 4.9.4compared the highly conservative three-model approach (development, qualification, and flight) of the ESA's 1970s Ulysses mission, with the model-and-a-half approach (avionics test bench and proto-flight) favored today. On the other hand, during the late 1990s NASA's excessive push in the direction of "faster cheaper better" led to several major mission failures. There is a permanent tension between reducing cost and increasing risk but thorough (hence slow and well funded) development of critical technologies before full project go-ahead always improves the technology readiness level and it is extremely effective at reducing risks to the mission.

One issue deserving mention is that most methods for risk characterization and management are qualitative, not quantitative. The formal techniques for a project's criticality classification (see Table 4.1 of Section 4.6); the classification of risks as significant, major, and critical (depicted in Table 4.2); or the classification of technology readiness levels from 1 to 9 (Figure 4.4) are qualitative. This makes risk estimation methods apparently less rigorous than numerical cost or performance estimations. Nonetheless, the qualitative definition of a mission as risk-averse or as risk-tolerant provides clear guidelines for the system's design, both technical and programmatic.

The definition of qualitative risk guidelines will also act as a guide for the establishment of numerical availability or reliability requirements.

Operational missions (Section 4.5.4) fulfill extremely important social needs – such as weather prediction – and have low levels of tolerance to failure. These missions use relatively conservative technology and normally employ multiple satellites, allowing the use of in-orbit or on-ground spares in case of failure. In general, the implementation of a mission using several small satellites instead of a single large one will reduce risk, and this approach does not inevitably increase costs by unacceptable amounts. Technology demonstration missions, by contrast, are risk-tolerant. They are also cost constrained, so the mission definition can perfectly well select low-cost but high-risk solutions. At any rate, any serious mission definition effort must assume a plausible evolution of the readiness of the technologies required by the mission, and the proposed mission concept must not demand dramatic and unlikely technical breakthroughs in order to be viable.

5.3.3. Schedule

Section 4.10 analyzed space mission schedule determination. We saw there that the duration of any endeavor is strictly determined by the tasks to be performed and by the resources available for executing them. Also, the heuristic maxim at the beginning of Section 5.3.1 pointed out that schedule delays are generally less critical than cost overruns. It can be deduced from all this that efforts to ensure schedule compliance during the definition phase will be proportionally smaller than the corresponding efforts to minimize cost or risk. Nevertheless, there are situations in which schedule fulfillment is critical, for example:

- Operational missions that provide vital public services, which must be delivered without interruption.
- Some classes of advancement-of-science missions can only be performed at specific times. The aim of ESA's Giotto mission was to study Halley's comet and its launch was timed to rendezvous with the comet. A strict respect for the schedule was crucial for successfully intercepting the comet; otherwise the mission would have become useless.

It is vital for schedule-critical missions to have a high level of technical readiness from the very beginning of the project. The use of components featuring a high level of technical readiness will reduce schedules because they increase the likelihood of a smooth project implementation as well as permitting the simplification of the model philosophy and a decrease in development and qualification tests, while still maintaining reasonable risk levels for the overall mission implementation. In Giotto's case, the satellite bus's design was closely based on ESA's Geos satellite, which had proven its effectiveness before development of Giotto had begun. Nevertheless, the Geos satellite needed certain modifications and developments in order to cope with the dusty and hostile environment around comet Halley. As we saw for designs needing to satisfy cost constraints, the use of well-funded mission definition phases will ensure that all technologies are mature and that the implementation phases will unfold smoothly and without delays.

5.4. Design against requirements

Design is fundamentally design against requirements and the second part of this book, Chapters 7, 8, 9, 10, 11, and 12, will discuss in detail how to design against the requirements of the distinct parts of a space system. So the present section will just focus on how requirements impact design in general, and how design responds to, and interacts with, them.

The European ECSS[4] collection of standards provides a sequential view of the process:

1. Needs are researched
2. The requirements associated with these needs are established
3. A system design fulfilling the requirements is defined
4. The system is implemented

This sequential view is also implicit in the traditional names for the design phases recalled in Section 1.5. Strictly following this scheme, requirements are treated as fully fixed. Nevertheless, as discussed in Chapter 3, the consumer's wishes cannot be transformed into firm requirements until both the consumer and sponsor understand the risks and costs involved in their implementation, until the sponsor has agreed to finance the project, and until the supplier has provided a reasonable design with a well-substantiated cost estimation. Therefore, during the early phases of mission definition, design and requirements interact until they mutually accommodate each other, reaching a balance between needs, means, and feasibility, between requirements, constraints, and design.

As the consumer is at the origin of the mission, the most important requirements will be connected to the fulfillment of the consumer's needs. Nearly all the requirements directly related to the three kinds of system effectiveness metrics – data quality, data quantity, and mission responsiveness (see Section 3.4.2) – will be among the most important and will strongly shape the mission. This is not to say that requirements not directly related to consumer needs are unimportant. In a mission with stringent and unavoidable limits on satellite mass, all the design requirements related to structural design and mass margins would be important; nevertheless, even in situations like this, the consumer's needs will drive the mission, because satellite mass will have been determined by the need to include all the instruments required by the consumer, and if the consumer agrees to eliminate one or more instruments, the mass problem will automatically disappear. It is obvious that reducing performance, that is, relaxing requirements, will reduce both costs and risks. The mission architectural definition team must identify what requirements can be relaxed; in order to significantly reduce costs and risks while keeping the consumer essentially satisfied. One clear way to accommodate design and requirements is to express requirements not as a single value but as range of possible performances: threshold, breakthrough, and goal (Section 3.2.2). This expression of needs in terms of ranges is quite normal in the very early stages of system definition and facilitates the study of alternative designs with increasing levels of performances and with their correspondingly higher

---

[4]Project Phasing and Planning, ECSS-M-30, ESA 1996.

levels of cost and risk. Section 3.4.1 pointed out that although all requirements should be fulfilled they are not all equal and that some of them are more important than others. It was also mentioned that identifying the most important requirements is not difficult because the whole process of mission definition revolves around them. A good heuristic rule is to consider the requirements' importance from the perspective of the Pareto Principle. The Italian sociologist Vilfredo Pareto pointed out that in many situations 80 % of the outcome derives from 20 % of the causes. Applied to requirements, the principle could be reformulated to provide two heuristic maxims; the first is in line with the previous statement that not all the requirements are equal:

- **80 % of the effort employed to satisfy requirements will be absorbed by just 20 % of the requirements.**

The second applies to each individual numerical requirement and could be formulated thus:

- **80 % of the total effort will be needed to fulfill the last 20 % of the required performance.**

N. Augustine[5] has produced a slightly different formulation: **"The last 10 % of the performance sought generates 1/3 of the cost and 2/3 of the problems."**

Heuristic maxims are not laws of physics, but in this case their thrust is clear: design efforts should focus on the few areas of absolute interest to the consumer and on vigorous discussions with the customer about possible requirements relaxations that would yield a cheaper and less risky system. Nevertheless, the mission's ultimate value is measured in relation to the consumer's needs, and the final decision on requirements relaxation shall be in the consumer's hands.

5.5. Tools for design

5.5.1. Analysis and design

Analysis is a Greek work whose original meaning referred to untying a knot. It implies the rational examination of something by dividing it into its individual parts and understanding how they interact. Analysis includes any structured approach to rational problem characterization and solving. ECSS defines[6] analysis as "a method, which entails performing a theoretical or empirical evaluation by accepted analytical techniques. The selected techniques may typically include systematics, statistics, qualitative design analysis, modeling and computer simulation." Hence, analysis includes not only mathematical quantitative methods but also rationally systematic qualitative approaches to understanding problems. Analysis is the basic tool available to system design because during the architectural definition phases test results will be rare.

From the architectural perspective, analysis can be divided in two parts:

- System analyses
- Engineering disciplines analyses

---

[5]Augustine's Law, Law Norman Augustine, AIAA 1983.

[6]Glossary of Terms, ECSS-P 001, ESA 2004.

System analysis includes all inquiries aimed at supporting decision makers on global courses of action. System analysis examines relevant goals and the tools needed to accomplish them, as well as comparing the cost, performances, and risks of the available options for reaching them. Consequently, system analysis includes all of the cost, performance, and risk analyses discussed in previous sections. From the start, the stated aim of this book has been that of keeping a global perspective, focusing on the higher-level aspects of mission definition. So this and the following sections of this chapter will concentrate on qualitative and quantitative system analysis approaches: trade-offs, functional analysis, budgets apportioning, use of safety margins, heuristic processes, and end-to-end mission effectiveness metrics. Highest-level decision making demands high-quality support, as detailed, realistic, and quantitative as possible. In some cases, when the structure of the model and uncertainties allow, a numerical, analytical solution will be feasible. This will be studied in Section 5.8.

To support the system definition process, the use of quantitative and qualitative analytical methods from the different engineering disciplines will be crucial. These methods include quantitative "mathematical" procedures such as orbital, structural, and thermal analyses, as well as a qualitative rational comparison of alternatives; for example, comparing liquid or solid lubricants for use in space mechanisms. Also in line with our stated global perspective, these procedures will not be covered in detail in this book. Although Chapters 7, 8, 9, 10, 11, and 12 will systematically examine all relevant areas of space design, describing the numerous design decisions to be made and how they interact and the analytical methods used in the different engineering disciplines will be touched on only briefly.

Any system design, even the most preliminary, must include a demonstration of its feasibility, supported by an analysis that specifies the dimensions of key design parameters and demonstrates that it fulfills the mission's key requirements. Nevertheless, a single-point design supported by an analysis demonstrating the fulfillment of requirements by no means guarantees that the design is optimal, and it is always necessary to support system design with alternatives including an analysis of the parametric sensitivity of the system's performances to changes in the most important requirements or design features. One design point is insufficient; it is necessary to research many alternatives.

Analyses performed during the architectural definition phases previous to project approval will have different levels of complexity. The analysis supporting the design's earliest stages is always brief and has limited reach, while those undertaken during later stages will be progressively more thorough. The need to begin with a simple feasibility assessment is valid for all levels of system design from the highest and more abstract to the more detailed and concrete. Initially, the general feasibility of the highest-level overall architecture design must be established by means of a very simple calculation and heuristic processes. A second design loop will permit the much more detailed trade studies to be made, supported by end-to-end mission effectiveness calculations, – mainly instrument-related – as well as a mission effectiveness parametric sensitivity analysis for the overall architecture. The first set of basic feasibility studies of the mission's less innovative and risky elements will sustain the second design loop. The final set of documents allowing a well-informed decision to be made regarding authorization of the fully funded project must provide credible cost, risk, and performance estimations. In order for these to be credible, it is necessary to provide more than a simple preliminary or overall design. The necessary degree of detail will be different for each design component, and must be in line with the levels of novelty and risk associated with each component:

- Standard components possessing a high technology readiness level require a low level of detail.
- Components influencing the fulfillment of important mission requirements or needing special modifications or developments must be designed and documented to a fairly detailed design level.
- Components playing a fundamental role on the fulfillment of key mission requirements (principally mission products quality) or that need extensive new developments must be designed and documented to a very detailed level, as is often the case with instruments. The design must be supported by comprehensive analysis and as well as actual manufacturing and testing of selected particularly risky components.

Space system design documentation must also be end-to-end and should include not only the design of the satellite but also the design of the ground segment. The ground segment design should include a functional explanation of how to operate the system and how to interface with the consumers: how their requests are to be received and how instrument data will be processed and distributed. The architecture must also include mathematical models providing end-to-end mission effectiveness figures.

5.5.2. Functional analysis and functional decomposition

Function is the intended effect of a system and functional analysis is the process of identifying, describing, and relating all the functions that a system must perform in order to fulfill its objectives. Functional analysis supports the followings tasks during the space mission definition process:

- Ascertainment all the requirements to be fulfilled.
- Identification of system effectiveness measures.
- Identification of mission reference scenarios regarding nominal and degraded modes of operation and study alternative operational approaches.
- Assessment of requirements concerning the environments to be encountered throughout the mission.
- Identification via trade-off studies of the alternatives able to meet the system's objectives.
- Providing insights for the builders of mission effectiveness models.

This list includes almost all the highest-level activities required in the earliest phases of mission definition. As we can see, most of the work done at this stage is actually functional analysis.

Functional analysis starts from system requirements and derives the functions, which must be performed to satisfy all the highest-level functional requirements. Each function shall be described in terms of inputs, outputs, and interfaces. As always with space mission definition processes, functional analysis is performed following a top-down hierarchical decomposition of functions, which permits the derivation of lower-level functions from the higher-level ones, producing function trees. This process in turn permits deducing sets of lower-level requirements from each higher-level functional requirement. At the top level a mission statement or mission requirements document shall provide the functions. The functional analysis will be the source for the first functional specification (Section 3.2.1) that is a key milestone in the definition of any space system. This process of system decomposition will create a parallel set of functional blocks that will define a functional concept and will drive associated physical concepts.

The most common functional analysis technique is the Functional Flow Block Diagram. These diagrams describe the network of events leading to the discharge of a function. The flow diagram represents the logical sequence of what is happening, depicting inputs and outputs, tasks, and relationships. What they do not show are durations, which means that flow diagrams are not a good tool for understanding the fulfillment of time-critical requirements. This is accomplished by timeline analysis, which includes durations and is used to analyze the necessary requirements for time-critical functions. Another functional analysis technique is that of N2 diagrams, which are used to develop data interfaces in software-intensive systems, but they are not normally used in defining satellites. They employ square matrixes of N rows by N columns that register the information flows between functions in row i with functions of column j, or mark a zero if there is no interface.

As said above, most of the work in the early stages of space system definition is actually functional analysis, but this does not means that "functional analysis" documents must be produced by strictly following one or another of the techniques described in the previous paragraphs. In point of fact, the functions of a space system and the elements performing these functions are very similar from one project to the next and there is normally no wide scope for considering dramatically different system approaches. Indeed, the analysis undertaken in Section 5.1 is a high-level functional analysis that can be applied to almost any space system, and the division into elements shown in Figure 5.1 illustrates the fundamental functional decomposition of nearly all space systems.

5.5.3. Trade-offs and design

A trade-off is a rational approach helping to select alternative solutions. Applied to design, the purpose of the trade-off is to ensure that the optimal design is selected. In principle, a trade-off requires the following steps:

1. Understand and make explicit the requirements and system effectiveness metrics (Section 3.4.2).
2. Devise design alternatives able to meet the requirements.
3. Ideally, determine a mission-effectiveness function, which combines all the effectiveness metrics into a single value. This function shall be a mathematically useful and objective one.
4. Include uncertainties by determining probabilities and determine sponsor risk acceptance/aversion profile.
5. Evaluate the design alternatives in terms of their mission effectiveness.
6. Evaluate these alternatives in terms of cost and risk.

7. Rank the alternatives according to a selection rule that shall include the benefit and cost and that should take into account probabilities on nonnominal outcomes and the risk aversion profile of the sponsor.
8. Select one or several best alternative according to the selection rules.
9. Continue the design work only with the selected alternative(s).

The ideal situation would be when all the steps from 1 to 9 could be implemented. This would permit a fully numerical handling of decision making able to determine not only the best alternative but also the best performance/cost point for each alternative. The steps indicated above are summarized in the following figure.

This purely numerical scenario is an ideal that should be strived for but that in practice is rarely possible. The limitations of numerical reliability calculations will be discussed in Section 5.5.6, and the limitations of mission effectiveness models will be reviewed in Section 5.6.2. Risk analysis – as discussed in Section 4.9 – is fundamentally quantitative, and an entirely numerical approach to trade-off selection could provide a false sense of security, frequently concealing more than it reveals. It is often better to make subjective but explicit and legitimate value judgments than to buttress the selection with the apparent support of mathematical solidity. **Apparently unbiased numerical methods often hide highly subjective choices within their algorithms**!

One design heuristic maxim claims that "**the eye is a good architect**" that is, that design solutions pleasant to the eye are good design solutions. Given the dangers associated with strictly numeric trade-off procedures, the importance of simplicity and even esthetics as a supplementary aid to trade-offs should not be underestimated. It is important to develop an intuitive understanding of overall design solutions that embody an optimal use of resources and a balanced allocation of functions. As already said in Section 5.4 the aim of the whole second half of this book from Chapters 7, 8, 9, 10, 11, and 12 is just to develop this intuitive understanding (Figure 5.4).

It is also necessary to make decision at many different levels and in many different situations, and the degree of the trade-offs complexity and formalism will be different for each case. The following levels can be identified:

- Lowest-level trade-offs, concerning specific issues of detailed design such as the use or not of a special kind of solar cell. The corresponding technical specialists usually perform these trade-offs.
- Medium-level engineering trade-offs, which drive many lower-level design decisions. For instance, the choice of orbit altitude or selection of satellite's configuration. The supplier and customer architectural definition teams perform these trade-offs, always maintaining a generalist perspective.
- Highest-level trade-offs to select the optimal mission architecture that use end-to-end mission effectiveness figures, and that incorporate parallel assessments of cost and risk. These trade-offs are also performed by the supplier and customer architectural definition teams, but the consumer must be kept informed and involved in helping to define end-to-end mission effectiveness algorithms, for example, and in contributing to the selection of the most promising alternatives.

According to the different types of trade-offs, there are different procedures:

- Direct reasoned selection between compared alternatives
- The Analytical Hierarchy Process (AHP) or the more complicated but powerful Multi-Attribute Utility Theory (MAUT)

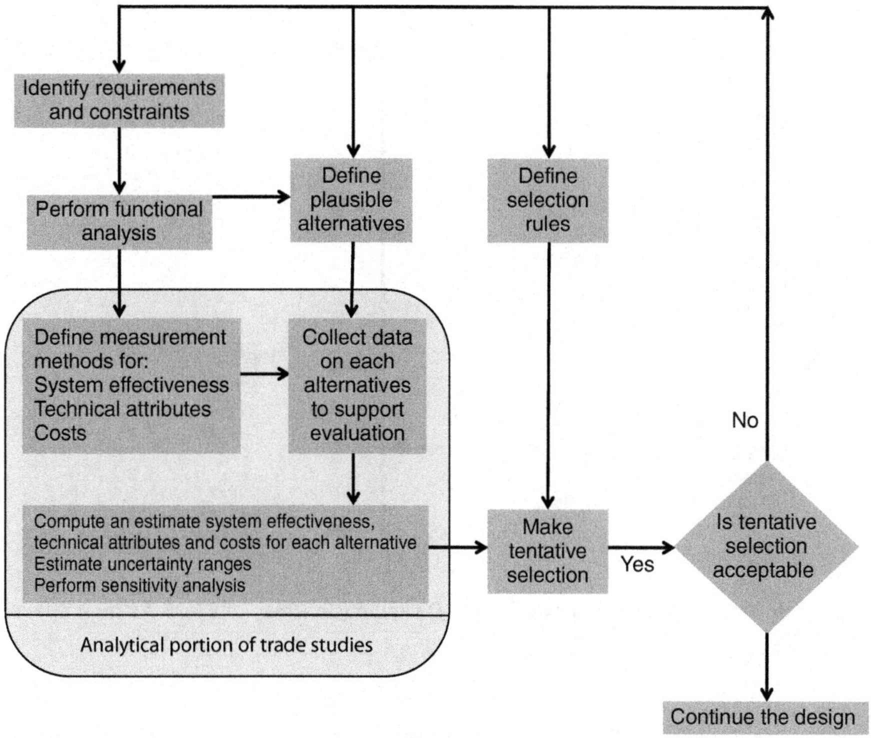

Figure 5.4. Trade-Off Flow (NASA System Engineering Handbook, SP-610S, NASA 1995).

- The use of fully numerical approaches based on mission effectiveness metrics as described in the numbered list (1–9) at the beginning of this section
- Heuristic processes and insight that is commonly related to past experiences

The methods related to the first two bullets are going to be discussed in the following paragraphs. Mission effectiveness will be discussed in Sections 5.6.1 and 5.6.2 and heuristic processes and insight in Section 5.7.

Direct comparison (Figure 5.5) is adequate for selecting between well-determined alternatives on issues of detailed design. The figure below provides an example of a trade-off between different alternatives for radio-frequency power generation. The trade subject in the example is to determine the best source of radio-frequency power generation for an active microwave instrument (Section 8.2.4). The figure also establishes a set of reasonable trade alternatives and trade criteria. The trade options are discussed and compared from the viewpoint of the trade criteria, using mainly or exclusively qualitative reasoning for the selection. In areas of detailed design such as this, the opinion of a single, experienced person is enough to decide upon an optimal solution. If the person happens to be the designer assigned to design and implement the component under review, then his opinion should be given special weight since he is ultimately responsible for the item's delivery and should enjoy the final say on an issue for which he is accountable.

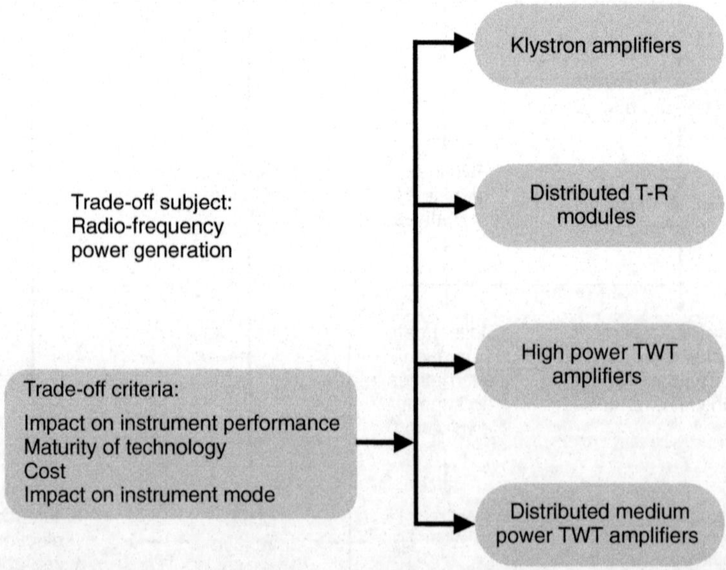

Figure 5.5. Trade-Off for Low-Level Design Decision (Courtesy of EADS Astrium Space).

AHP is a decision-making technique in which a figure of merit is determined for each of several alternatives through a series of pair-wise comparisons. AHP is normally performed in six steps:

1. Create a set of evaluation objectives: mass, technical risk, science-return, and many others.
2. Determine the relative importance of the objectives by pair-wise comparisons between them.
3. Establish a series of weights for the different objectives; highest for the most important evaluation objective, to the lowest for the least important.
4. Instruct evaluators to make separate pair-wise comparisons of the alternatives with respect to each evaluation attribute.
5. Produce a ranking for each alternative by addition of the evaluation marks multiplied by the weights allocated to each evaluation criteria.
6. Discuss the result with the evaluators and iterate the process to arrive at a consensus ranking.

This method can be used for all types of design decisions.

MAUT uses utility theory to create a mathematical model by quantifying the desirability of the alternatives under consideration. It is necessary for scenarios where—unlike AHP above – a numerical treatment of uncertainty or risk is necessary. The end result is a function, which represents the preferences of the evaluators with respect to a number of design attributes. The function creation is based on mathematical axioms regarding the way individuals will behave when confronted with uncertainty. It is based on establishing the evaluators' preferences by determining their level of indifference to the best outcome with probability p and worst outcome with probability $1-p$ when compared with an intermediate value that could be obtained with absolute certainty. This will allow the creation of a utility function for the particular attribute of the form.

$$U(x) = ae^{bx} + C$$

Different functions will be created for the different attributes. This will allow combining them using weighting factors to be agreed. This will create a multiattribute utility function. The alternative to be selected shall be the alternative providing the highest multiattribute utility value.

Independently of the trade-off mechanisms, selection rules for determining the optimal outcome must be established. The criteria are numerous and will vary according to whether qualitative or quantitative effectiveness or ranking values are applied to the different alternatives. When there is a lack of numerical values for mission effectiveness, it is possible to create a relative ranking by employing the AHP or MAUT procedures described above, after which a selection can be made of the highest-ranked alternative or the best alternative that fulfills some supplementary constraint, such as being below a given cost boundary or above a particular numerical quality performance that is considered compulsory.

5.5.4. Budget allocation engineering

The specification will include many high-level numerical requirements that apply to the space system or to the satellite; for example, the total mass of the satellite, the accuracy of data recovered by the instruments, or the actual pointing of the instruments carried by the satellite. The actual performances of the system with respect to each one of these high-level numerical requirements are a function of many factors distributed throughout the system, for example:

- The overall mass of the satellite will be a function of the mass of each component it carries.
- The instrument's accuracy will depend on factors such as aperture size, detector quality, thermal control efficiency, the calibration approach, the quality of on-ground processing to eliminate spurious signals, the use of external validation data for comparison, etc.
- Knowledge about the pointing of the instruments will depend on the mechanical and thermal stability of the instruments and satellite structure, the satellite's attitude determination and control, and the end-to-end system's capability, or lack thereof, to use on-ground processing to determine the actual pointing by comparing the figures provided by the satellite with well-characterized references over the Earth or celestial sphere.
- The total number of images produced by an optical Earth observation mission will depend on factors such as the presence of adequate solar illumination, the absence of clouds, the time necessary for pointing and repointing the instruments, the time needed for calibration, the time needed to ensure that the satellite pointing direction is stable enough for proper image acquisition, and many others.

In each case it would be necessary to determine how many lower-level sources can affect the higher-level numerical performance, and once that all the causes are identified, it is necessary to apportion the overall high-level numerical performance between all the relevant contributors. In this way the single high-level figure will derive a multiplicity of low-level numerical requirements, one for each contributing cause. **This apportioning process is called budget allocation and it is one of the most powerful tools to optimize and to drive the system design process**, because different apportioning will allocate different values to different sources and this will make easy, difficult, or just impossible different design alternatives. For example, the chosen allocation of mass figures for the structural or thermal subsystems of the satellite will make possible or impossible specific structural or thermal concepts, it will determine the risk and cost associated with their implementation and delivery, and it will be possible to make harder the fulfillment of one or the other subsystem by just changing the allocated figures. Budget allocation is a very powerful tool, not only to drive the design but also to drive the level of risk and cost associated to all the lower-level components involved in the particular high-level budget. Relationship between budget allocation and design has already been depicted in Figure 5.2. The budget allocation is fundamentally a process to establish requirements; that means, it is a responsibility of the customer.

A possible heuristic maxim on budget apportioning would be: **to allocate as much of the budget as necessary to each error source in order to balance the resources available for fulfilling the requirement with the difficulty involved in its fulfillment**. This means that the implementation difficulties of all suppliers must be balanced. A fair allocation of resources that is fair for each and every supplier will be also optimal for the system as a whole; that is, it will produce the specified highest-level performances while reducing overall cost and risk. It is the duty of the customer to identify if the allocation of requirements is unfair for a given supplier. **The space agency shall be proactive to avoid the prime industrial contractor allocating budgets in a way that is detrimental to some subcontractor because an unfair allocation will be finally detrimental for the whole project**.

Another aspect to bear in mind concerning the relationship between budget allocation and design is that all budget allocation is undertaken with some design in mind, albeit preliminary. For instance, the allocation of the overall mass budget of a satellite is simultaneously top-down and bottom-up, entailing a preliminary top-down allocation based on previous experience, followed by bottom-up estimations once mass estimates derived from actual component designs start becoming available. As a consequence of this interrelation between design and budget allocation, any major change in the satellite concept (a change in satellite configuration, for example) will require changes in budget allocations to the different subsystems of the satellite.

Budget engineering must always include the ability to accurately calculate the effects of lower-level design decisions on the higher-level requirements. This ability to calculate upward is crucial for establishing the entire system's actual performance once the performances of all the contributing causes have been determined. This means that there is a descending path for correct budget apportioning and an ascending one for overall performance evaluation. The ability to control both paths should be maintained at all levels of system decomposition, from the most general to the most detailed. This is a supplementary example of the principle that during the definition phases design and specification interact until they reach a common accommodation.

Different error budgets will have different error decomposition and summation rules. The simplest rule is that of direct addition; for example, the mass of a satellite is the direct sum of the masses of all its individual components.

Many error sources are probabilistic. In that case the addition rules for different error sources will be derived from the combination of the mathematical expectations of these error sources. Expectation is defined as the integrated product of the probability of an outcome multiplied by the value of that outcome. In the case of uncorrelated random errors (electronic noise, for instance), the errors will add "root mean square." In this case two errors with a standard deviation (the square root of the expectation) of 1 will combine to produce a total error with a standard deviation of 1.41.

In some cases, determining the correct apportioning requires the performance of specific or complex analysis. For example, if an instrument can only produce images of the ground when there are no clouds and when the Sun is located high on the horizon, it would be necessary to determine the cloud provability for the time of day when the Sun is high enough to snap the image. An especially complex case of error apportioning is the allocation of overall pointing requirements for the satellite. This will be analyzed in Section 9.8.2.

5.5.5. Concurrent engineering

Concurrent engineering defines a space system with the help of multidisciplinary teams composed of all the relevant technical experts, who interact both directly and simultaneously. This simultaneous interaction permits the evaluation of the downstream and upstream implications of any design decision, and facilitates rapid verification of its feasibility and adequacy, not just from the viewpoint of the engineering specialty that made the design decision, but from the perspective of its implications for other engineering domains as well. Concurrent engineering generally requires the use of an automated data-exchange environment so the different technical experts can communicate with one another; but the essence of the approach is direct communication, allowing experts to freely discuss the implications of each design option. Concurrent engineering not only involves engineers specializing in areas such as satellite thermal control or data handling, but also specialists in areas of engineering support such as reliability or cost estimation. Furthermore, the best use of concurrent engineering requires the involvement of experts on observational techniques to be used and on the instruments to be carried by the satellite. All space missions are designed around the instruments, and the availability of specific instrument experts is crucial for the optimal accommodation of the instruments with regard to the rest of the satellite and the mission as a whole. The presence of consumers is generally not required as part of the process of concurrent engineering. However, their presence, at least at key moments, will allow a better understanding of the difficulties implementing designs that satisfy the requested requirements. Their presence could prompt the rephrasing or relaxing of some requirements, making them easier to fulfill while maintaining perfectly acceptable performances for the mission.

The essence of concurrent engineering lies in the swift transmission of information among all those involved in the mission's definition, allowing each one to understand the issues faced by all the others. This is another example of the observation made at the end of Section 5.3.1 that one of the few things capable of reducing both cost and risk simultaneously is the easy flow of information during a mission's architectural definition phases. This is one of those truly rare occasions when "faster-cheaper-better" can actually be realized.

5.5.6. Dependability

Dependability is a general term embracing the system availability and its three governing factors: reliability, maintainability, and serviceability. Availability is defined as the ability of a system to perform its required function; the availability figure is defined as the system's total operating time divided by total time. Reliability is defined as the numerical probability of failure occurrence. Serviceability is the capability of acting on a system to prevent or correct failures. Satellites cannot be physically maintained or serviced in situ, but ground segments can perform scheduled maintenance tasks, software updating, for instance, or service tasks such as correcting recoverable failures, activating redundant processes, or implementing work-around procedures. This kind of servicing can make missions that have suffered apparently unsolvable problems operable again.

The definition of numerical figures for acceptable levels of availability and reliability is part of the overall task of system specification, which is the responsibility of the customer.

5.5.6.1. Numerical reliability calculations

Reliability has been defined as the numerical probability of something unpleasant occurring. The relative frequency with which particular events occur is indicated by a probability distribution, which defines the failure rate. If the failure rate were constant and if failure occurred randomly, which are reasonable assumptions and widely employed, then reliability would be a negative exponential function curve that decreases with time (Figure 5.6).

Numerical analysis for the evaluation of availability and reliability is well established. These methods are based on adjudicating failure rate figures to components. This adjudication of figures works quite well with electrical and electronic components that are used in large numbers – so the analysis is statistically representative – and operating in well-characterized environments. Nevertheless, this approach is only moderately successful for dealing with the real-life behavior of mechanical components used in small numbers and in widely different environments. Furthermore, it cannot deal with events like design or assembly mistakes, or errors of operation. This demonstrates that in order to design a properly functioning space system its reliability must be studied, but that it is unwise to focus exclusively on its reliability figures. **Reliability figures are meaningful but they cannot be taken literally**. Experience has demonstrated that well-built satellites last much longer than reliability figures appear to indicate. Examples about that will be provided at the end of this section.

Reliability figures can be improved by:

- Using high-quality/high-price space-qualified components
- Using redundant "duplicate" components that can take over if one member of the set malfunctions

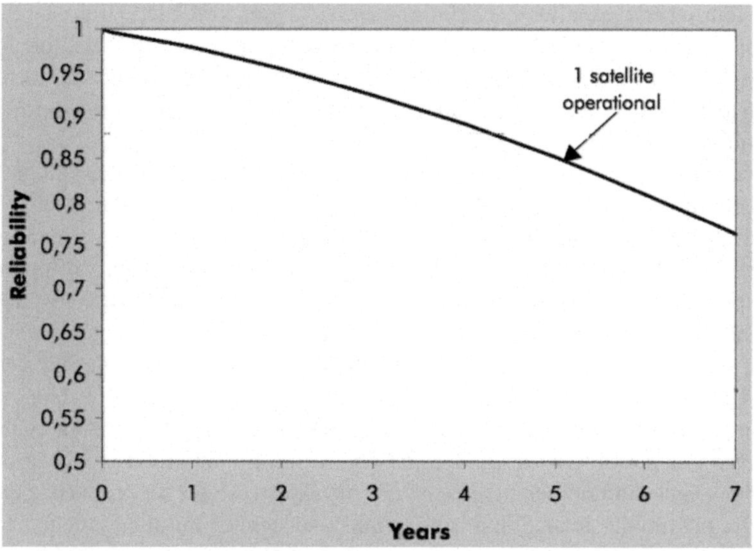

Figure 5.6. Reliability Reduction with Time for One Satellite with a Specified Reliability of 0.75 After 7 Years of Operation.

- Protecting electronic components from radiation by shielding them with metal
- De-rating, or the use of high-capacity components to perform low-capacity functions; the lower stress of use will increase the component's reliability with respect its nominal usage

An extreme case of redundancy is to double the number of satellites. In the early years of the space age NASA frequently launched more than one satellite with the declared intention to ensuring that at least one was functioning properly. NASA's 1975 Viking mission to Mars employed two satellites with this idea in mind, but it turned out that both worked flawlessly. This approach is still extensively used for operational missions with demanding availability requirements, for example, meteorological missions or GPS constellation.

As mentioned above, reliability analysis works poorly outside the sphere of electronics and does not reflect how satellites function in real life. Nonetheless, numerical reliability analysis is an absolutely necessary tool for proper system definition, though the results are better understood when interpreted as more comparative and qualitative than as literal and quantitative. Reliability analysis is compulsory to compare alternative implementations, to balance the reliability figures of each satellite's distinct parts, and to achieve reasonable levels of redundancy for the different parts of a satellite or system.

5.5.6.2. Redundancy

Redundancy is provided by the availability in the satellite of two or more identical items able to fulfill the same functions. If one assembly fails the satellite will use a redundant assembly to continue operating. **Redundancy protects against uncorrelated random failures. Redundancy does not protect against an underlying common cause, unknowns, or mistakes**. In 1982, flying over Indonesia, a commercial aircraft lost its four jet engines all at once. From a classical – uncorrelated failure – reliability point of view that would have been absolutely impossible; nevertheless, the underlying cause was that the satellite was flying through a cloud of volcanic dust dense enough to extinguish the combustion in the four engines. The fourfold failure was due to an unforeseen underlying common cause.

So far the discussion has centered on "normal" redundancy, which increases the dependability of a system by duplicating identical components. Functional redundancy is an alternative approach in which the system responds to component failures not by switching to the still-functioning duplicate component, but by operational work-around and the use of components in ways that were not originally intended. The main communication antenna of NASA's Galileo mission to Jupiter failed to deploy, and a work-around was accomplished by software fixes that further compressed the science data and images. The compressed images were then returned through the low-gain antenna, although at a severely reduced data rate due to the low-gain antenna's performances limitations. This work-around greatly reduced the data throughput and the number of images the mission was able to produce, but transformed a potentially total failure into a fairly successful mission. Adequate safety margins combined with ingenuity have saved more than one mission that might have ended prematurely in total failure.

5.5.6.3. Heuristics

The above discussion has already provided heuristic maxims, like: "**Reliability figures do not work well with mechanical components**" or "**Reliability figures are meaningful but they cannot be taken literally.**" It is a generally accepted space systems design heuristic rule that **satellites shall avoid having single-point failures** that result in the system's complete failure. But the concept of single-point failure also does not apply well to the area of mechanical design; to the point where it can be said "**Every satellite will have mechanical characteristics that can become single-point failures.**" The best way to provide highly reliable mechanical design is to have a good understanding of the satellite's environment, of proper design analysis and testing, and the use of robust design and safety margins to provide supplementary confidence.

Unknown environments are especially hazardous with respect to classical "redundancy" approaches for ensuring system availability because such an environment is perilous for both nominal and redundant functional paths. **Redundancy does not protect against unknown environments**.

Another way to provide improved reliability is by implementing functional redundancy or implementing the system using several satellites, allowing for a progressive "gentle" degradation if individual satellites were to fail. **Satellite constellations provide good availability** because the functions they provide remain available (albeit with reduced performance) even after the complete failure of one or more elements of the constellation. The use of a constellation of multiple satellites is common in operational missions that – being necessary to fulfill routinely important social needs – require the highest level of availability.

5.5.6.4. Mission life and reliability

A mission life reference duration that can be used as a guideline for the mission's design must be established. Reliability figures decrease with time, but beyond reliability, mission life is also relevant to the design of items with a limited life; the amount of fuel needed to perform orbit maintenance, for example, or the progressive degradation in the efficiency of solar cells affecting the satellite's power capacity. These factors also limit the satellite's operational life. Figure 5.6 presented reliability as a function of mission life. This figure employs typical space system design values, and due to the increasingly downward tangent of the negative exponential reliability curve, it would be unlikely for a satellite to survive much beyond the specified time. In real life, on the other hand, well-designed and well-built satellites survive longer than specified and predicted by reliability calculations. The examples are many:

- ESA's COS-B mission was designed for 1 year of life with 2 years of consumables, but lasted 7 years, until all its consumables were used up.
- IUE, launched in 1978 with a design life of 3 years, lasted an incredible 18.6 years.
- The OTS communications satellite was designed to last 3 years, but its transponders continued working for 12 years.
- Soho had a designed life of 3 years and still operates 14 years later.

A good idea of the difference between predicted and actual satellite life is provided by the in-orbit performance of the Meteosat series of operational meteorology satellites:

Satellite	Launch	End
1	11/77	11/79 (imager failure)
2	6/81	11/91
3	6/88	11/95
4	3/89	11/95 (imager failure)
5	3/91	Still in service in 2005
6	11/93	Still in service in 2005
7	9/97	Still in service in 2005

Meteosat 1, 2, and 3 satellites were designed for 3 years of life, while satellites 4, 5, 6, and 7 were designed to last 5 years, using conventional reliability calculations. The actual lifespan of all the satellites but the first was consistently longer than their calculated values. It is evident that conventional reliability calculations underestimate the lifespan of well-designed and built satellites. This is due to the existence of both explicit and hidden safety margins. The impact of safety margins will be discussed in Section 5.6.3 below.

5.6. Design and mission performances

Space systems engineering texts provide a useful simplified view of the multidimensional design space that is normally compressed to a cost versus performance curve as depicted in Figure 5.7 below. A system providing no performance will still have

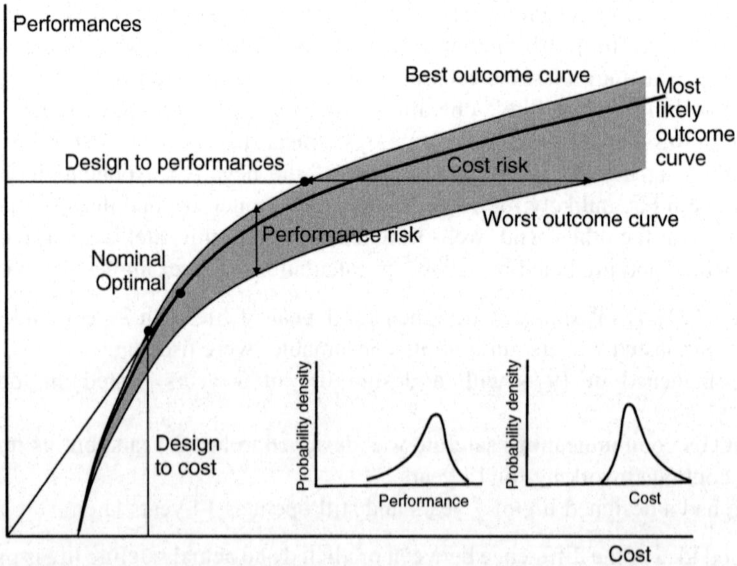

Figure 5.7. Performance Versus Cost Curve and Possible Design Strategies.

some cost. Starting from that point the system performances will increase fast for small increases in cost, and as the design moves to higher and higher performances, the cost increases associated with each performance improvement will be larger and larger because the technology needed to implement the system will become more and more sophisticated. The shaded area represents a probabilistic performance inherent to the design. There is an intermediate most likely performance curve framed by upper/optimistic and lower/pessimistic limits, which provide an indication related to a given set of probabilities of performance. The most likely curve is not centered because outcomes with a performance much higher than the "most likely" are less probable than outcomes with a performance much lower than it. That means the performance probability distribution is asymmetric. Costs larger than the most likely are very likely but performances better than the most likely are rare. In the other hand, designs always include safety margins. These safety margins will ensure that the "most likely" performance is larger – sometimes much larger – than the specified performance as stated in the formal specification of the mission. This will be discussed in detail in Section 5.6.3. As indicated in Figure 5.7 there are multiple possible design points:

- In the majority of cases, the aim is to provide the least expensive design that is good enough to satisfy all the requirements. This is designing to the specified performances, an approach that entails cost risks.
- When there is some leeway on performances and cost, the ideal is clearly the system that provides the best relation between them: the optimal performance per unit cost.
- When the cost is absolutely firm "designing to cost" will provide the best possible performance within the given cost constraint. Design to cost entails performance risk.

- In extreme cases it is advisable to aim for the lowest possible cost that would still produce a meaningful mission – meaningful for those with the authority to approve the project. This might be the case for technology demonstration missions.

Performances and cost are probabilistic, not deterministic. Quantifying these uncertainties involves producing performance and cost probability density functions. Nevertheless, it must not be believed that the curve above represents a solid mathematical reality, as would a probabilistic analysis of pointing errors or random acceleration. Space system performance is not a linear one-dimensional parameter and cost and risk will depend on many factors, which will be difficult to analyze with rigorous mathematics alone. Disraeli is quoted as saying. "Sober people use lampposts for illumination and drunks use them for support." **Complex numerical calculations are excellent ways to gain insight and provide illumination with regard to design alternatives but mathematics alone cannot be used to decide the optimal design selection**. The limitations of numerical effectiveness parameters will be discussed in detail in Section 5.6.2.

5.6.1. Mission effectiveness metrics

Section 3.4.2 introduced mission effectiveness metrics. The present section will expand on mission effectiveness and their use as tools for design optimization. Figure 5.8 summarizes the most common mission effectiveness parameters.

As indicated in item 3 of the numbered list at the beginning of Section 5.5.3, when comparing alternative designs, ideally there would be a single mission effectiveness parameter providing an integrated estimate of the mission's final value for the consumer. Figure 5.7 of the present chapter implicitly assumes that this is possible. In that case, it would also become possible to perform all nine steps of the numbered list of Section 5.5.3, and the design trade-off would become purely numerical.

As said in Sections 3.2.1 and 3.4.2, the fundamental needs of the consumers and the corresponding System Effectiveness Metrics will be in terms of:

- What to deliver? Those are the parameters of the "Output Quality" column of Figure 5.8.
- How much to deliver? Those are the parameters of the columns "Output Quantity" and "Life" of Figure 5.8
- How to deliver? Those are the parameters of the columns "Availability" and "Responsiveness" of Figure 5.8

It is always possible to assign – in consultancy with the consumer – numerical weights to the different numerical effectiveness parameters listed in Figure 5.8. This would provide a truly integrated, transparent, and mutually agreed-upon overall value of system effectiveness that would allow mission designers to define the optimal system design, as implied by Figure 5.7. It is also possible to use qualitative or relative effectiveness metrics that would provide a ranking of the alternatives under consideration, though relative ranking cannot be used to compare different alternatives from the perspective of mission effectiveness per unit of cost. In real life, the limitations of numerical approaches to overall mission effectiveness are severe. The next section will analyze these limitations.

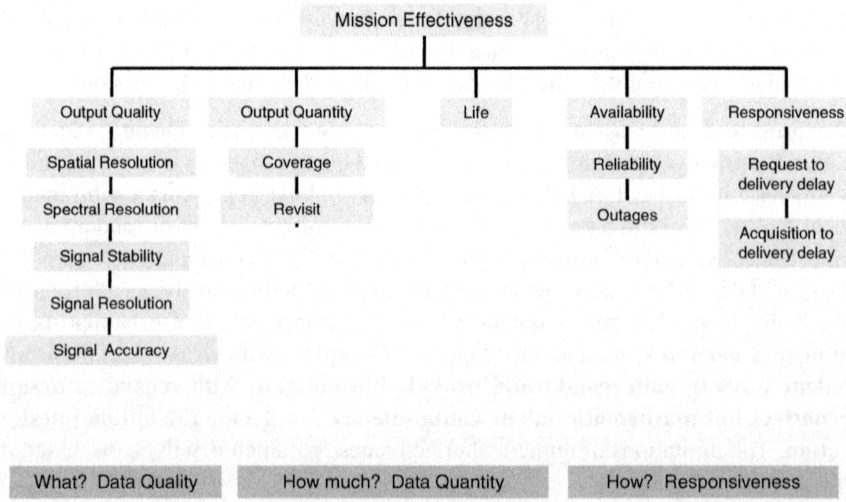

Figure 5.8. Common Mission Effectiveness Parameters. The Parameters Are Divided in "What" "How" and "How Much" Categories.

5.6.2. Effectiveness metrics limitations

Optimal designs require the use of models that permit quantifying measures of effectiveness and cost. Models have their limitations, however, and obtaining good figures of overall system effectiveness is often difficult. Practical effectiveness models only feature some of the many possible key requirements of special importance to the consumer. Models supporting the different steps necessary to providing a true system-level effectiveness figure may not have been properly integrated. Furthermore, true performance scales are not linear. Even if we focus on the easiest and most quantifiable parameters, for example, image resolution, doubling the resolution does not imply doubling the usefulness of the mission for the consumers. Sometimes a small increase in performances produces a very large increase in mission effectiveness because the small increase reveals a totally new set of observables, opening up new areas of application and research to the consumers.

Figure 5.8 presented different types of mission effectiveness metrics. Some of these types of metrics are numerical and can be obtained using algorithms based on physical and mathematical laws. The weakest astronomical signal that an orbiting telescope is able to detect, the time the telescope needs to complete a comprehensive survey of the heavens, or the time required for the telescope to react to a specific observation request are all unambiguous and objective numerical values. Nevertheless the dependency of the mission effectiveness on these parameters is not linear. A telescope not sensitive enough to detect the events of interest to the consumers will be insufficient even if its sensitivity is quite close but still below the required one. A system intended to survey the entire sky will not provide 90 % satisfaction if it only surveys 90 % of the sky. It can, however, still produce good science, and consumers would prefer it to no mission at all. Establishing sets of values for threshold, breakthrough, and target criteria, as discussed in Section 3.2.2, can diminish the problem, but the existence of real discontinuities affecting consumer satisfaction makes it clear that a nice, monotonically increasing performance curve, as depicted in Figure 5.6, is always a simplification of reality.

Models always include assumptions about the real world and they always leave something out. They are only able to produce accurate results when addressing quantifiable parameters in which the mathematical methods used for the derivation of the figures is well understood, and that is not normally the case when dealing with the highest-level system definition issues. There is a significant difference between the really important questions in cost-effectiveness discussions and the questions that can be truly treatable from a mathematical model perspective. That being the case, the model will produce insights but will not be able to provide definitive answers to the most important questions on its own. On the other hand, numerical models that "only" provide insights are far from useless; indeed the resulting insights are fundamental tools for system definition, and system designers must interpret model results and convey their interpretations to the consumer and sponsor in a way that captures the essence of the real problem to be answered: identifying the best way for consumers to obtain what they want while satisfying the limitations established by the sponsor.

Unlike overall mission performance, overall mission cost is an unambiguous parameter and the best estimation of how much the future system is going to cost conveys a clear and powerful message. Nevertheless, cost estimations are notoriously unreliable. Bottom-up (Section 4.8.2) cost estimations depend on the judgments of the expert providing the figure; that is, they are subjective calculations influenced by the expert's degree of optimism. Establishing a bottom-up cost estimation is a long and complex process that requires ample time and a detailed breakdown of the job, with its associated well-defined design. Missions are approved well before the system's definition is detailed enough to allow an item-by-item bottom-up estimation of costs. Bottom-up price models also lack the flexibility required for estimating widely different alternative concepts or distinct options within a given concept. The flexibility provided by top-down cost estimations (Section 4.8.1) makes them far more useful for determining an optimal design. Even so, a top-down cost model is never more than an observed correlation between cost and mass. Therefore, to provide credible cost estimations the top-down model shall be based upon cost data of older systems reasonably similar to the new one. The similitude shall be not only technical but also programmatic. The cost of a radically new system or being developed in a radically new programmatic framework is very difficult to predict with top-down models.

System difficulty or complexity is another subjective parameter, correlation does not equal causation, and the past is frequently a poor guide to the future. As in the case of numerical system effectiveness metrics, the top-down cost model only provides an insight into future costs, but as with the former, the insight is valuable in itself, and the cost figures thus derived are the best tool for comparing alternative designs.

We can conclude therefore, that **providing numerical values for overall system effectiveness should be considered as a support for sound reasoning, not as a substitute for it.**

5.6.3. Safety margins, mistakes, and errors

Among all the factors influencing the process of space mission definition, there are two that greatly influence the real-life performance of space systems, albeit in opposite ways: safety margins and design errors.

Any system needs to fulfill the consumers' requirements in a consistent way, within a particular environment and during a specified period and with an extremely high level of confidence. To accomplish all this, engineers use safety margins. For example, any thermal design includes explicit numerical temperature values that are added as margins to the worst temperature range that the design will actually face. These margins will ensure that the resulting design will be able to cope with a broader temperature range than the worst range expected during the life of the system. Margins are used in every area of design, and engineering is thoroughly steeped in a philosophy of prudence. All systems possess not only explicit safety margins employed as part of the design process, but hidden ones as well. On the other hand, due to this hidden safety margin, an average low Earth orbiting satellite's thermal control subsystem will perform better over time than predicted by the relevant calculations. For perfectly known conditions, safety margins can be greatly reduced; but space engineering is a field in which elements of novelty are common, and where the conditions that the satellite must withstand are seldom perfectly known. Both explicit and hidden safety margins provide robustness to space systems, enabling them to survive and operate correctly when confronted with unexpected situations or conditions. Coping with the unexpected is the norm in space missions whose aim is opening new domains for science. The designer must expect the unexpected, and comfortable safety margins offer the best tool for doing so.

A by-product of safety margins is that the most probable performance of a well-implemented system will be superior, often far superior, to the performance requested by the formal specification (Figure 5.9). Section 3.2 defined specifications as legally binding documents delimiting the minimum performance parameters to be fulfilled in order for the customer to accept and pay for the work undertaken. The contractual relationship between the customer and supplier centers on the specified value, but the estimation of the mission's performances must include a probabilistic evaluation of the performance, differentiating between most probable performances and guaranteed performances. The consumer shall be informed of both so that he acquires a real understanding of what is likely to occur. ESA's ISO mission (Section 7.2) studied the cold universe and required cooling the instrument with liquid helium. The liquid cooled the instrument through evaporation and was lost to space so that the mission ended when all the liquid Helium was used up. The satellite carried enough liquid Helium to guarantee a specified mission life of 18 months, but in the event, the Helium and the mission lasted 28 months. These valuable 10 additional months were the result of ample safety margins and conservative analytical methods. It is logical to make the consumer aware not only of the guarantied specified performance but also of the – better – most probable one. This "most probable" performance will give the customer a more realistic understanding of the mission products' real quality enabling him to plan for the mission's most likely duration.

Safety margins improve a system's most likely performance. Numerically explicit safety margins can be used to produce probabilistic distributions of performances that

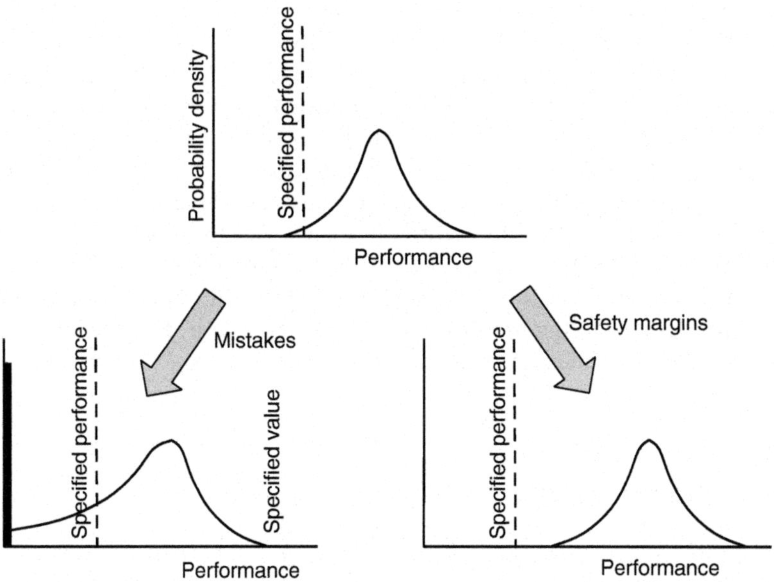

Figure 5.9. Impacts of Mistakes and Safety Margins in Performance Probability Density Functions. Mistakes Increase the Probability of Very Poor Performances and Generate a Zero Performance "Mission Lost" Peak. Safety Margins Improve the Actual Performance.

take them into account. Hidden safety margins, on the other hand, are not explicitly stated anywhere and cannot be easily treated numerically, even though they are absolutely real.

Space mission definition and implementation are complex processes and human error is almost inevitable. Space research opens new paths to knowledge, and it is not surprising that space systems encounter situations and conditions that were unforeseen during the design phase. Errors and unknowns have the opposite effects from those of safety margins. Noncatastrophic errors or unknowns will move the entire density function downward and will also raise the probability of poor performance. Large errors – large enough to result in mission failure – will provide a mission performance of zero. In order to take into account this quite credible total failure probability, the real probability density function will have a secondary peak on zero performance. This means that real-life performance density functions are bimodal. They have a peak near the expected most likely average and another near zero performance. The second peak must be high enough to represent the finite probability of having zero performance.

Ulysses (Section 8.3.5) is a good example of mistakes, errors, and safety margins compensating for each other to finally produce a successful mission. Ulysses deployed a 7 m long boom to study magnetic fields and electrical waves around the Sun. Just after launch the Sun heated the boom, pumping energy onto it, and making it oscillate. This was completely unexpected. The boom displayed a nonlineal stiffness behavior – it was very stiff against high forces but less stiff against low forces – and this had not been properly tested or analyzed, which was a mistake. Nevertheless, and due to the use of safety margins, when withstanding strong forces the boom was much stiffer and stronger than originally specified. The boom was finally able to withstand the oscillations originated by the Sun, and as the satellite's attitude control was also overdesigned, it was able to eliminate these oscillations as soon as the satellite's orbit moved it away from Earth, thus decreasing the Sun's intensity, a few days after the beginning of the mission. Just after the phenomena was discovered, it was possible to study in detail the true behavior of the boom under the Sun's influence and controlled by the satellite's attitude control system. It was concluded that the behavior of the boom, as excited by the Sun-boom geometry and controlled by the satellite's attitude control system, was such that the oscillations would not occur again during the duration of the mission. This turned out to be true, and Ulysses and its boom functioned flawlessly for more than 19 years until the mission finally ended.

5.7. Nonnumerical support to decision making

All the design and programmatic decisions imply risk. To reduce this risk the decisions shall be supported by nonnumerical and numerical rational and scientific procedures. This section provides a summary on nonnumerical procedures that have been discussed several times in the previous sections. This section will provide a summary on nonnumerical procedures and the next one will do the same for numerical procedures. Space system engineering is rooted in science, and science is rooted in numbers. This chapter emphasized the importance of a nonnumerical "more art than science" perspective regarding design. Section 5.2 is based on nonnumerical reasoning. Heuristics, of course, provide rules of thumb and educated guesses that help in the process of space system definition. Heuristics represent for strategies using readily accessible, though loosely applicable, information to control problem solving in human beings and machines. Heuristics are obtained by trial and error; that is, they derive from the experience of the person formulating the heuristics. Quite a number of heuristic maxims or tenets concerning mission definition have been provided in this chapter; they come from a variety of sources, and a few are original to this text. Heuristics are especially valuable in situations where strict mathematical analysis can be misleading, and during the earliest stages of a project because during these early phases many important decisions must be made based on very little information. Programmatic constraints, which will ultimately be derived from political or policy considerations, will also introduce unavoidable nonnumerical considerations into the process of system definition. The Apollo program only trained astronauts as Moon geologists for the very last flights, proving that the Apollo's primary aim was to demonstrate that it was possible for the USA to land on the Moon before the Soviet Union, not to study the geology of the Moon. Political constraints are real and the designer shall design the system taking them into account. As said in Section 5.5.3 many design decisions at all levels are taken performed by specialist after a purely rational nonnumerical approach. Direct reasoning (Figure 5.5) and The Analytical Hierarch Process are nonnumerical procedures to help in the selection of optimal design alternatives.

5.8. Numerical support to decision making

Deterministic and probabilistic numerical tools can be used to support decision making.

5.8.1. Deterministic approaches

In most situations it will be possible to model the behavior of a system using deterministic mathematical formulations. This will allow the derivation of numerical parameters that can be compared and used to select the best design alternative. Highest-level design decision must be made with the aid of numerical system effectiveness figures. Figure 5.7 analyzed the concepts of mission performance and mission cost and how they can be used to define different types of optimal alternatives. It is possible to evaluate a telescope's image quality as a function of the aperture, and it is also possible to use parametric cost estimates (Section 4.8.1) to derive the cost of a mission as a function of the aperture of the telescope as well. This provides a performance versus cost function that permits comparing alternative mission concepts.

Specialists at all levels use numerical methods to compare alternative designs and to help select the one that best fulfills the multiple requirements of the mission. For example, structural specialist will use finite element analysis to compare alternative structural concepts in order to find the lightest design compatible with specified strength or stiffness levels or the strongest concept compatible with specified mass requirements.

5.8.2. Nonprobabilistic numerical approaches in situation of uncertainty

The outcome of strategies or decisions might actually come to fruition in an environment different than that originally predicted. The occurrence or nonoccurrence of future events could change what is considered to be the optimal strategy.

The assignment of probabilities to possible future events is a difficult task and in some cases may be impossible. Nevertheless, it is still possible to take into account the possible existence of these future events and compare their impact on the selection of optimal strategies, even without making any assumptions about the relative probability of these events. There are a number of criteria that can be applied to defining different types of optimal strategies.

The Laplace criterion proposes that when comparing the merits of several possible decisions, the strategy with the best arithmetically determined average performance shall be chosen. This criterion assumes that all possible futures are equally probable, that a straight average provides a true measure of value, and that the strategy with the optimal value shall be chosen.

Section 4.9 provided an analysis of risk and Section 5.3.2 analyzes design and risk. Space projects are expensive and unique and space engineering is risk averse. A conservative "maximin" strategy recommends selecting the decision with the maximum – the least bad – worst outcome. In case of a low level of risk aversion the "maximax" approach recommends to select the decision that maximizes the best outcome. Both maximin and maximax are extreme alternatives; most space scientists prefer an intermediate approach. The Hurwicz criterion assumes a compromise figure α $0<\alpha<1$. When $\alpha=0$ the decision is Maximin and when $\alpha=1$ it is Maximax. Once α is selected, the rule requires the computation of:

$$\text{Max}_i \left\{ \alpha[\max E_{ij}] + (1-a)[\min E_{ij}] \right\}$$

Where E_{ij} is the performance of the ith decision when the jth future occurs.

5.8.3. Probabilistic approaches

In many situations it will be possible to assign probability distributions to the likelihood of the various possible futures affecting each decision. This permits creating probability density functions for each decision. Figure 5.7 provided possible probability density functions for a design to performance or a design to cost decision. In this case the tools of probabilistic analysis would be used to support decision making.

One possibility is to select the alternative with the best expectation. Expectation was defined already in Section 5.5.4 as the integrated product of the probability of an outcome multiplied by the value of that outcome. Recalling Figure 5.7, if designing to performances and in cases having a cost probability density function, it will be possible to compare the cost risk expectations for different design alternatives and select the alternative that minimizes the cost expectation. The same approach could be used when comparing several design alternatives in a “design to cost” situation. The expectations of the different performance probability density functions can be calculated and the design alternative with the best performance expectation can then be selected.

Chapter 6

System Definition as a Diachronic Process

Keywords Project Phases, Project Reviews, Milestones, Technical Development

The word diachronic means "concerned with the way in which something has evolved over time." It is often opposed to synchronic. The previous chapter analyzed space system definition as a synchronic process; the present chapter will analyze it as a diachronic process. Together they provide a comprehensive overview of the system definition process.

As indicated in Section 1.1, the aim of this book is to analyze the proper definition of a space system up to the point where the design is detailed enough to allow an accurate estimation of the system's effectiveness, cost, and risk, enabling a well-informed decision for or against a fully funded project. This implies that the design progresses step-by-step over time, as a diachronic process with increasing levels of definition. The aim of the present chapter is to describe the necessary steps to reach this goal.

The chapter begins by pointing out that the system definition process is at the same time recurrent and linear, and it continues with two complementary sections: the first one describing the recurrent procedure that is used to advance the definition, followed by an analysis of system definition as it advances linearly through the three standard phases: 0, A, and B1, which normally occur before project's formal approval. Next there is a discussion of the milestones that enable the system definition to advance from phase to phase, and a comparison of the logic of a specific "real life" system definition with the descriptions provided in the previous sections. The chapter winds up with a section describing necessary developments in technical maturity and science, which are parallel to the main definition process but that are not normally part of it.

6.1. The system definition process as recurrent and linear

Section 1.4 provided a dual, and to a degree, contradictory view of the design process: Figure 1.3 portrays it as recursive and cyclical, while Figures 1.4 and 1.5 depict it as linear and progressive. The truth is that though the definition process is indeed recursive, each cycle implies an increasingly better level of definition; that is, there is a progressive advancement from cycle to cycle.

Section 1.5 introduced the project phases as defined by the commonly used standards. Of these phases – from 0 to F – those normally included in the system definition process are Phases 0, A, B, and C.

M.A. Aguirre, *Introduction to Space Systems: Design and Synthesis*, Space Technology Library 27, DOI 10.1007/978-1-4614-3758-1_6, © Springer Science+Business Media New York 2013

The first three phases defined in Section 1.5 where:

- Phase 0. Establishment of mission aims
- Phase A. Feasibility
- Phase B. Preliminary definition
- Phase C. Detailed definition

The process, however, is never as neatly sequential as implied by these phase designations, where there is a continuous flow from the establishment of needs, to the establishment of requirements formalizing those needs, and then a design fulfilling the requirements. Actually, Phase 0 already involves a fair amount of requirements establishment, and even some design, while Phase A requires a fairly detailed system design (detailed enough to provide good cost and risk estimates, allowing a well-informed project approval decision). It is far better to view the process as recursive, consisting of several steps with a similar structure and aims but with an increasing level of detail, and correspondence among user needs, requirements on system elements, cost, risk, and design. In real life, the working logic and aims of Phases 0 and A are similar, as are the results, confirming the view that the system definition process is recursive. Successive iterations result in greater understanding and can reveal critical requirements and design drivers.

Nevertheless, the process advances and the first "formal" set of mission requirements establishing the consumer's demands on the customer is typically produced at the end of Phase 0, while the first formal set of system requirements that clearly and unambiguously establish the customer's demands on the supplier comes at the end of the Phase A. The end of Phase 0 might also allow rough cost estimations, a preliminary analysis of mission cost and risk and early technology development plans – prior to full project go-ahead – in order to reduce the project's future overall risk. Still, in order to reach a level of design detailed enough for credible cost and risk estimations, the second design iteration provided by Phase A must be performed. This is because it is impossible to produce a credible cost estimation until a firm set of requirements and a subsequent design are established, and that will not happen until the end of Phase A or B1, depending on how the work is organized.

6.2. System definition as a recursive process

It is unavoidable to describe the diachronic process of space system design without starting from the steps defined by J. Wertz in his standard text, *Space Missions Analysis and Design*[1]. He defines the following steps:

- Define objectives:

  1. Define broad objectives and constraints.
  2. Estimate qualitative and quantitative mission needs and requirements.

[1] Space Missions Analysis and Design, James Wertz and Wiley Larson, SLT, Kluwer 1999.

- Characterize the mission:

  3. Define alternative mission operation concepts.
  4. Identify alternative mission concepts.
  5. Identify system design drivers.
  6. Characterize, i.e., design each proposed mission including all its mission elements to the necessary degree of detail.

- Evaluate:

  7. Identify critical requirements, i.e., identify the most important performance figures that are driving the entire design.
  8. Evaluate utility and cost of each alternative mission concept.
  9. Define the baseline.
  10. Formalize or update requirements.

Step 1 of the process frequently occurs prior to the involvement of a mission architect; with the consumer producing a mission statement that describes the basic aims of the mission, its uses, and frequently also recommends an observing technique, including a specific instrument technology. This approach is driven by users/consumers. In other cases it might be a new observing technique that spurs a search for possible applications; this second approach is technology driven. Whatever the approach, as soon as a general statement of objectives is produced, the mission architect will proceed to analyze the mission. The process of defining broad objectives was introduced in Section 3.2.1 and will be further analyzed in Section 8.1 when discussing observation techniques and observables. Chapter 4, especially Sections 4.4, 4.5, 4.6, and 4.7, analyzed sponsor-generated constraints and how they apply to different types of projects. An early analysis of the programmatic constraints by the architect is also necessary to ensure that the mission concept can be achieved within the budget, schedule, and risk limits established by the sponsor.

Step 2 embraces the tasks necessary for creating the first set of formalized qualitative and quantitative mission requirements and will create a progressively more defined mission requirements document delineating the mission performances from the consumer's viewpoint, which will eventually become the source of the system requirements document delineating mission performances from the customer's perspective. This process was described in Section 3.2.1. The first formal "Mission Requirements Document" is normally issued at the end of Phase 0, when the first loop of the design process has been completed. This document contains all the quantitative and qualitative requirements crucial to the consumer, as well as some key numerical requirements regarding the instruments of special interest to the consumer for ensuring the high quality of the data to be produced.

Step 3 includes the establishment of broad guidelines: for data production and delivery; for the communication architecture; for tasking, scheduling, and control; and for the overall mission timeline. These broad guidelines must be produced in collaboration with the consumer because the nature of the observables will determine how they will be observed. Observables will also shape the mission operation concept in terms of data type, data flow, and data speed, and the kinds of interactions that will be necessary among the orbiting instrument, ground operations, and consumers.

These factors will shape how the mission will operate and will establish criteria for the interfaces between the satellite, ground control, and consumers.

In Step 4, mission characterization continues with the definition of alternative mission concepts including designs for a mission concept, payload, satellite bus, launcher, orbit, ground system and communications architecture, as well as a concept for mission operations architecture, in line with the operations concepts outlined in Step 3. So concepts must be provided for all the elements described in Figure 5.1, Section 5.1, which together produce a space mission. This step will establish the mission's highest-level architecture and will define the interrelations between its elements. In the main, the definition of alternative mission concepts entails taking functional design decisions, i.e., deciding which elements perform which functions and how they will be performed.

Step 5 identifies system design drivers. Drivers are the design features that will drive the mission in terms of performance and cost/risk. Designing a space system requires many design decisions; nevertheless, not all design decisions are of equal importance and key performances and allied design decisions that are crucial to mission performances must be identified.

Step 6 expands on Step 5, providing designs for the system elements. The design's degree of detail depends on the specific stage of the mission's definition. In Phase 0, this step will focus on the most important and innovative aspects of the mission, which nearly always involve the instruments. During Phase A, this step requires a fairly detailed design of all the mission elements, with a very detailed design of the mission's most innovative elements in order to provide credible estimations for overall mission cost, schedule, and risk and to provide good quality numerical estimations of mission effectiveness.

In Step 7, the broad mission drivers identified will be transformed into critical requirements that will determine in turn other lower-level design decisions as well as lower-level requirements. The process of establishing the alternative architectures also allows the identification of these critical requirements. Note that critical requirements (for example, a telescope's aperture size) are related to important system effectiveness parameters that will be used to judge the value of the mission from the consumer's point of view. Section 3.4 discussed system effectiveness. System effectiveness will be further analyzed in Section 8.5 when discussing end-to-end mission performance simulation.

Step 8 quantifies mission effectiveness and the cost of the alternative mission concepts identified in Step 4, permitting a comparison of the different concepts. Mission effectiveness must be measured with respect to the most important requirements identified as mission drivers, which are likely to be the most important in satisfying consumer needs as well as the most demanding (as far as implementation difficulty, cost, and risk are concerned) when creating designs to meet them. Mission evaluation entails estimations of cost, risk, and scheduling. Relevant methods were described in the corresponding sections of Chapter 4.

Step 9 involves performing trade-offs and choosing the best design option. The choice must balance the performances derived from the system's effectiveness with the programmatic constraints established by the sponsor. The best solution will depend not only on fulfilling the specified mission performances but on the programmatic environment as well.

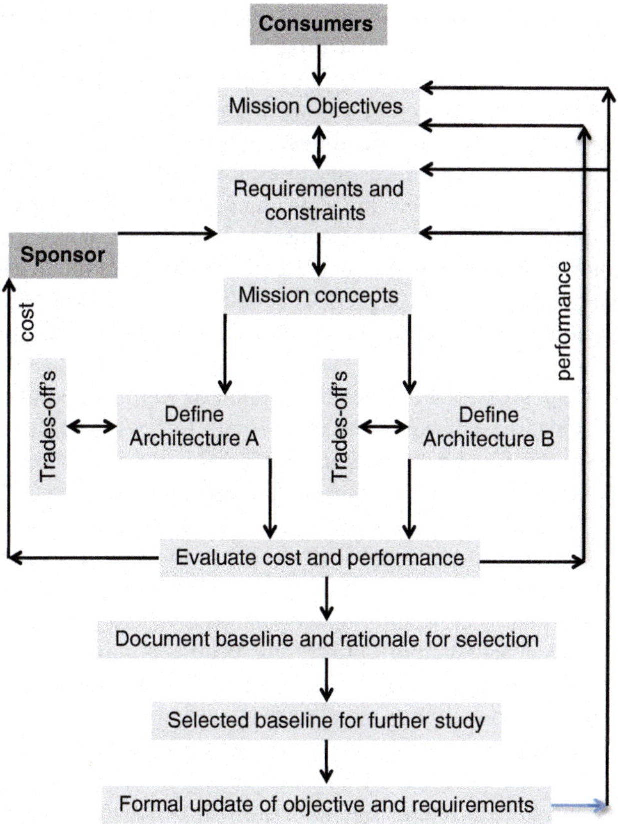

Figure 6.1. Cyclical Space System Concepts Exploration and Update of Mission Objectives and Requirements.

Step 10, in line with the recurrent nature of the procedure described in this section, refers back to Steps 1 and 2. Step 10 will update the higher-level requirements and also formalize the lower-level requirements resulting from the design decisions. This step enables the original mission and system requirements documents to be updated and expanded, providing a higher level of certainty.

Figure 6.1 provides a visual representation of these steps, their interrelations and how they interact with consumer requirements and sponsor constraints. As indicated in Figure 6.1, this process is understood as cyclical and it must be repeated until the proposed mission is finally approved or is canceled. Analysis performed at each step may cause reconsideration of high-level synthesis decisions and trigger another episode of synthesis and analysis.

6.3. System definition as a linear process

Project phases were introduced in Section 1.5; the present section will expand on the material presented there and, in line with the aims of this book, will focus on describing the phases prior to a fully funded project go-ahead: Phase 0, Phase A, and in part

Phase B1. The process of system definition involves manipulation of models, and as the design advances the models become progressively more concrete.

Prior to project go-ahead, available funds are moderate and the teams involved are coordinated by the person acting as system architect. The architect's role is not restricted solely to high-level considerations. Architects also look at specific components and domain details when necessary to establish feasibility and determine performances that are important to the consumer. The overall process is one of high-level structuring and synthesis, frequently based on nonnumerical heuristics and insight followed by rational and numerical analysis of all the necessary details. The second part of this book – Chapters 7, 8, 9, 10, and 12 – will attempt to foster this nonnumerical insight by presenting and analyzing numerous examples of space system design.

The present section offers a detailed description of well-defined phases with outputs and reviews clearly delineated and in line with ECSS standards, and compares them with NASA's definition of phases. In general, ESA uses the same names as NASA, but ESA's phases are denser and contain more output, even to the extent that a portion of ESA's Phase A is found in NASA's Phase B. In real life, however, the contents of each phase can be slightly different than the references provided by the two agencies. When the actual contents of Phase A are lighter, Phase B1 should be included as the last stage before full project go-ahead. Where Phase A is proportionally denser, a well-substantiated full project go-ahead is possible at the end of Phase A. Whatever the case, it is important to understand the need for a sufficiently detailed level of mission definition at the moment of the decision on full project go-ahead. If a good level of definition is achieved the project can proceed, independent of the name given to the phase in which it was achieved.

In any case, and to help reduce mission risk, it is better to wait until project go-ahead and to accomplish enough preparatory work to solve as many problems as possible. As Figure 1.1 showed, missions in whom the definition phases were well funded did not experience large cost overruns during their implementation.

6.3.1. Phase 0

ECSS[2] defines Phase 0 as identifying the overall needs, and establishing a preliminary concept and a set of overall requirements for a mission. The main activities during this phase are:

- Analyze requirements and mission constraints: physical, operational, and programmatic aspects.
- Analyze basic feasibility of the observing technique.
- Analyze alternative architectures.
- Propose a preliminary architecture and basic design of the observing system with emphasis on the critical aspects.
- Produce mission requirements.
- Provide a rough cost estimate and rough schedule estimate.

[2] Project phasing and Planning, ECSS M 30, ESA 1996.

In Phase 0, the consumer or customer shall evaluate the nominal limit and external environmental conditions in which the system will undertake its mission, and shall also establish the expected performance. The customer must also define and characterize the functions of the system; identify and evaluate different possible system concepts and critical, innovative aspects; and identify the environment for possible project implementation. The expected output of this phase includes:

- Mission requirements document
- System preliminary functional specification
- System concepts, concepts trade-off, and supporting analysis
- Project framework including target budget
- Technology risk and status

Altogether, Phase 0 is a trial-run of the mission definition process as described in Figure 6.1. It provides a first opportunity for adjusting the consumer's performance needs whilst remaining within the cost constraints and the sponsor's appetite for risk. The resulting mission concept or concepts are still tentative, but provide reference points for subsequent work. The last task of Phase 0 is to identify developments essential to advancing the technology readiness level of mission-critical items, a crucial step for reducing mission risk; this will be covered in Section 6.4.1.

NASA[3] indicates the main aim of Phase 0 as producing a broad spectrum of ideas and alternatives for the mission, from which new programs and projects can be selected. According to NASA's view, Phase 0 shall include not only interaction with the consumer to produce the mission's preliminary objectives and justification, but also a possible operation concept; mission architecture; and cost, schedule, and risk estimates. To accomplish all this, it is necessary to provide a preliminary mission design with an especially detailed focus on the most novel and critical elements of the system. This approach emphasizes the role of Phase 0 as a source of ideas that may or may not become reality. It is not only possible but advisable to initiate many more Phases 0s than missions actually expected to become reality, because the cost of this phase is comparatively moderate and makes possible the natural selection of the most promising mission concepts.

A mission design with an acceptable level of detail for a Phase 0 could in theory be accomplished and properly documented in a concurrent design facility in a period of about 1 or 2 months. Nevertheless, Phase 0's mission aims are not yet clearly defined and they entail constant interaction among consumer, customer, and supplier to enable mutual accommodation of design requirements and constraints, which means that a Phase 0 is not merely a preliminary design stage responding to a set of requirements; it requires enough time to carry out all the necessary interactions, and will actually take from 6 months to 1 year to complete.

6.3.2. Phase A

ECSS defines Phase A as defining the mission's feasibility. This phase establishes the system's functional requirements and provides a solid system definition along with

[3] NASA System Engineering Handbook, SP-610S, NASA 1995.

performance and risk estimates. All the mission areas that are risky, novel, or connected with the fundamental performance requirements must be studied and defined in detail. This phase must:

- Analyze and further refine the already existing mission and system requirement documents.
- Produce and compare alternative architectures and designs for all the elements of the system.
- Propose a baseline design: mission, platform, payload, and ground segment functional block operations concept.
- Provide analytical support to the system definitions under consideration to demonstrate that they are compatible with system requirements.
- Perform trade-offs at all levels in order to choose a baseline definition.
- Allocate requirement to system elements and also derive requirements for the internal and external interfaces.
- Provide the preliminary design for all the elements of the end-to-end system and analyze and design all the critical elements with the necessary level of detail to properly quantify them with a view to demonstrate their technical and economic feasibility and adequacy.

This repeats the activity performed in Phase 0 but with a higher degree of definition and progressively higher level of certainty. This demands additional interaction between the consumer and customer to reach an agreement on requirements not made firm during the previous phase. Customer-supplier interaction will be especially intense in this phase because they have to produce a firm system design baseline. The proposed baseline shall be analyzed not only to ensure the mission's basic feasibility but to produce credible estimations of performance cost and risk.

So Phase A generates a firm set of system requirements, an observational concept, and an end-to-end system definition and estimation of performances. The outputs resulting from this phase are:

- Mission design definition and supporting analysis
- Developmental approach for the implementation of the mission: product tree, major development risks and mitigation approach, model philosophy, and if needed, organization for the future project
- Cost and risk estimations
- End-to-end mission performances
- Consolidated system functional specification including interface requirements between mean mission elements

Phase A must demonstrate that a feasible and well-substantiated design for the system exists. It is not necessary to define in detail every single item of the system, but it must be detailed enough to ensure the credibility of both the system performance and the cost and risk estimates. This obliges all the novel components of the system to be designed and analyzed to a fairly high level of detail. The need for all critical components to be defined with a high degree of detail is in line with our basic definition of system architecture as a dossier that enables a well-informed decision to be made on project go-ahead.

At the end of Phase A, the project goes through a number of reviews to verify the requirements to be fulfilled, the standard of technical definition, and the system's risk and development status, which ultimately lead to the project's approval or rejection. Project reviews will be studied in the following section.

During Phase A the verification activities are focused on the assessment of the general project requirements and the definition of the development and verification approach, including the Model Philosophy and the associated Hardware Matrix. The results are discussed at the Preliminary Requirement Review (PRR) in order to assess the feasibility of the development and verification program.

Also during Phase A, it is necessary to perform the appropriate tasks for reducing the mission risks identified in Phase 0 and which should be running in parallel with Phase A. These activities (described further in Section 6.5.1) include design, manufacturing, and testing of the critical items under review.

NASA also considers Phase A as a structured version of the previous phase that should include: the development of top-down requirements and their corresponding requirements evaluation metrics and criteria; the identification of alternative operations concepts, system constraints, and system boundaries; and the provision of alternative design concepts with supporting feasibility and risk studies, cost and schedule estimates, and advanced technology developmental needs and plans. In NASA's scheme, Phase A should determine the feasibility and desirability of a suggested new major system, and its compatibility with strategic plans.

6.3.3. Phase B1

Phase B1 may or may not be performed before the project receives formal approval. Phase B1 builds on the requirements outlined in the system requirements document produced in Phase A and fixes the apportioning of functions and performance (and hence interfaces) to the system's elements. This phase also refines the design work done during Phase A by producing a fairly detailed analysis of all the system's parts, including the less innovative and less risky ones. The decision to include or skip Phase B1 will be made by the sponsor. It is advisable to include this phase in larger or more advanced systems calling for more detailed analysis of all the system's aspects and their interrelations, in order to reduce the risks associated with project approval.

ECSS calls Phase B the "definition phase." According to this approach, Phase B continues and deepens the system definition created in Phase A. Phase B1 provides a higher-level definition, while Phase B2 refines the definition to the most detailed level. The tasks associated with Phase B1 are:

- Based on the requirements formulated in Phase A, derive and negotiate requirements for the subsystems and requisite lower-level components.
- Freeze basic mission elements: launcher, instrument concept, satellite platform, operational approach, and data processing.
- Analyze alternative design implementation for subsystems and components that were not analyzed in Phase A.
- Select subsystem designs.
- Support the selected designs by analysis, verifying the fulfillment of requirements.

- Freeze interface specifications between main system elements.
- Produce preliminary specifications for elements and subsystems.
- If the project is already fully approved, produce management plans and supporting documentation.
- Provide a preliminary list of mission products and their characteristics.
- Refine end-to-end mission simulation.
- Provide a preliminary design for mission operations.
- Provide preliminary designs for ground data processing and consumer-to-mission operations interactions.
- Perform early developmental tasks in critical areas as needed, including design manufacturing and testing of technologies crucial to mission success.
- Assessment of manufacturing, assembly, verification, and test approaches and associated costs.
- Procurement of long-lead items; that is, items requiring long waiting periods before being delivered.

The outputs of the Phase B1 will be:

- Detailed mission design definition and supporting analysis
- Assessment of performance versus performance requirements
- Major plans: management, development, assembly integration and verification, product assurance and safety
- Controlled system configuration
- Updated cost and risk estimations
- Updated end-to-end mission performance
- System-wide technical specification and support specifications including frozen definition of interface requirements between the main system elements

In the area of system design the transition from Phase A to Phase B1 is smooth because the same design approach is employed for lower-level subsystems and components; nevertheless, the design of the system during Phase B1 will always require the involvement of a larger number of technical experts than in Phase A.

In the area of project management, Phase B1 requires the production of more detailed documentation on project organization and cost and risk estimation. Phase B1 also calls for a fairly detailed description of the test approach, including detailed definitions of complex tests that will become cost and schedule drivers. The documentation produced by the supplier, however, will be of a preliminary nature and not legally binding until the project is formally approved and commenced.

If the project has already been approved and fully funded by the end of Phase A, Phase B1 will represent a major break with the previous phases because a project team must be created with separate support, engineering, and management roles and the establishment of a hierarchical structure among its members. The formal launch of the project will also make the supplier's cost and schedule commitments legally binding.

NASA's B phase is definitely lighter than ECSS's corresponding phase of the same name. For NASA, the aim of Phase B is to establish the project's baseline, which includes a formal flow-down of the highest-level performance requirements to a complete set of system and subsystem specifications for space and ground elements. The baseline consists of technical and business aspects of the project. In the technical part of the baseline these include: requirements and specifications, designs, verification, and operations plans. In the business portion, schedules, cost, and management plans are produced. The establishment of a baseline implies the implementation of the configuration management procedures.[4] During the Phase B, NASA performs a "nonadvocacy" review, in order to judge the soundness of the project being considered. This review provides confidence prior to formal authorization. This means that in NASA's scheme, Phase B comes before project approval.

6.4. Mission milestones and reviews

Milestones are management events enabling the system definition to proceed step-by-step until full project go-ahead is received. As each milestone is successfully passed, funds for the next step are allocated, until final approval results in the allocation of funds for the entire system's implementation. The allocation of funds and the authority to proceed is in the hands of the sponsor. Two situations are possible:

- The sponsor delegates approval responsibility to the consumer.
- The sponsor makes the decision himself.

In the first case, some form of scientific "peer review team" analyzes different missions and ranks them in terms of relative merit, approving one or more for continuation. In this case, the sponsor will have previously provided guidelines on the maximum cost allowable per mission and on the number of missions allowed to move forward. In the second case, the sponsor undertakes an analysis of the potential performances, risks, and costs of each mission and either authorizes or rejects them. The customer shall provide all the documentation necessary for enabling a well-informed decision as well as advise or clarification where needed, though approval is the exclusive decision of the sponsor. The number of mission approval milestones varies from project to project and the sponsor, in consultation with the consumer, defines their exact number. The final milestone will be the authorization for project go-ahead. In order to proceed with project implementation, it must be demonstrated that the system satisfies the consumer's requirements within the constraints of the sponsor. Questions such as the following must be addressed before formal approval is conceded:

- Are the allocations contained in the system specifications sufficient to meet mission objectives?
- Are the evaluation criteria established and realistic?
- Are measures of effectiveness established and realistic?
- Are cost estimates established and realistic?
- Has a system verification concept been identified?
- Have appropriate plans to support projected system development milestones being initiated?
- Have critical technology development issues been identified along with approaches to their solution?

---

[4]NASA System Engineering Handbook, SP-610S, NASA 1995.

Mission reviews are examinations of the technical status of a system at a particular point in time. Their primary purpose is to provide a comprehensive and independent assessment of the technical progress being achieved. Independence is provided by the participation in the review of external experts not involved in the mission's definition process. Every review involves three parties:

- Independent technical experts from the customer's organization not directly involved in the mission
- The customer's mission team
- The supplier

Reviews are events in which the customer evaluates the work performed by the supplier. Relations between them are regulated by contractual agreements that clearly determine the quality and quantity of work to be provided and how it is to be delivered. The contractual relationship between customer and supplier also specifies the nature and number of the reviews to be performed. Two teams represent the customer in the review: the customer team directly involved in day-to-day work with the supplier, and the customer's team of independent technical experts. All customer representatives review the work produced by the supplier, but in addition to this, the independent technical experts evaluate the work performed by their colleagues involved in the day-to-day interactions with the supplier.

The customer divides his review into groups dealing with the main areas of technical expertise: instruments, satellite platform, operations, quality assurance, management, and others. The number of groups varies according to the complexity of the mission. To ensure the review's independence the number and authority of customer representatives who are members of the project team is balanced with those who are independent. Above these groups there is a review board that discusses issues of special importance that could not be agreed upon at group-level, and that decides on actions to be taken regarding these issues as well as providing a final judgment on the status of the system implementation. The majority of board members remain outside the project. The supplier also organizes teams to confer with the customer's technical experts, while the supplier's project management group will also discuss with the review board any issues left unresolved by the technical experts. In principle, neither the consumer nor the sponsor is directly involved in the review process. Nevertheless, both are informed of the outcome, which becomes part of the documentation accompanying the mission milestone process. During the earliest phases of mission definition, when requirements are not yet firmly established, intense, albeit informal, involvement of the consumer is crucial to the review process. Consumer awareness of the relevant technical discussions will permit a fuller understanding of the difficulties involved in satisfying the requested mission performances, which directly impacts final agreement on mission requirements.

Reviews are a long and expensive process requiring many hours of analysis and discussion resulting in actions and recommendations to be implemented by the customer and supplier. But they also provide a unique opportunity to produce a detailed independent evaluation of the system's status. The presence of the customer's independent reviewers is especially valuable for identifying hidden problems, but also makes the reviews long and cumbersome. This is why there is a tendency to streamline the review process, significantly reducing the effort, cost, and time involved.

This will be further examined in Chapter 13 when discussing cheaper mission implementation methods.

6.4.1. Review procedure

Reviews are performed according to formal procedures proposed by the customer. These procedures define the review's objectives, the participants, their affiliations and responsibilities, the designation of specialized groups, the list of documents under review, the overall schedule, planned meetings, and the procedures and tools for the creation and processing of review item discrepancies (RID). RIDs are items for discussion referring to points contained in the documents under review and that the customer review team wants to discuss with the supplier. RIDs are produced and discussed until there is agreement on how to proceed between customer and supplier. Besides the processed RIDs, the different review groups will produce documents evaluating the quality of the work performed as well as suggestions or recommendations on how to proceed. These documents, along with any unresolved RIDs are presented to a customer review board with decision-making authority for all outstanding questions. This board will also provide final resolutions on the review's outcome, and will issue final directives and recommendations on how to proceed.

6.4.2. Reviews during the mission definition stages

Figure 1.5, which is inspired by ECSS[5], identifies the following reviews during the mission definition phases:

- Mission Definition Review (MDR) at the end of the Phase 0
- Preliminary Requirements Review (PRR) at the end of the Phase A
- System Requirements Review (SRR) at the end of Phase B1

The next paragraphs will describe their contents in some detail.

6.4.2.1. Phase 0

Phase 0 ends with an MDR to confirm the mission requirements and to assess the basic feasibility of the proposed mission concept, with special emphasis on the observing techniques. The MDR has a number of features that makes it different from the other reviews:

- Its main aim is to confirm the mission requirements agreed to between the consumer and the customer; therefore, the consumer's participation is crucial.
- In Phase 0 the customer and supplier teams involved in the work are very small.
- The system's specifications are still tentative so reviewers cannot resolve RID "discrepancies" between the supplier's concept and the system's provisional requirements.

With these factors in mind, the MDR is never a formal review adhering to a formal procedure, but more a set of discussions between representatives of the consumer,

[5] Organization and Conduct of Reviews, ECSS-M-30-01, ESA 1999.

customer, and supplier directly involved in the mission definition, along with certain customer experts external to the project, who are principally specialists in the areas of risk assessment, cost estimation, and observation techniques. They provide an independent assessment of the proposed concepts. The aim of these discussions is to agree on a set of mission requirements giving rise to a mission concept in line with the cost and risk guidelines provided by the sponsor.

6.4.2.2. Phase A

Phase A ends with a PRR. This provides a sound concept baseline and freezes the system requirements document containing the highest-level system requirements. The aims of this review are:

- To evaluate the results of Phase A.
- Evaluate the validity of the concept proposed in response to the preliminary system requirements document.
- Release the system requirements document (SRD).
- Verify downward the consistency of the higher-level SRD requirements with the allocations of functions and performances to lower-level system elements.
- Derive preliminary interface requirements between system elements from the allocation of functions and performances.
- Provide detailed risk, cost, schedule design, and development approaches for the project.

The PRR has to demonstrate that the proposed preliminary design satisfies the agreed-upon requirements; i.e., that the proposed concepts are feasible, that the expected performances are in line with user needs, and that the cost and schedule are in line with programmatic constraints. The PRR can well be a formal review in line with the approach and procedures described in the previous sections. Two situations are possible:

- The decision on project go-ahead will be made immediately after the PRR.
- The decision on project go-ahead will be delayed until the SRR.

If the project go-ahead decision is taken just after the PRR, its outcome must be detailed enough to provide clear and credible estimations on feasibility, adequacy of the proposed concept, risk, cost, and expected final performances, in order to allow a well-substantiated decision on the project's go-ahead. Among other things, this means that all the SRD's requirements should be clear and unambiguous and that the allocation of functions and performances to the various mission elements should be quite near to final status, albeit still preliminary and not yet fixed frozen.

If the project go-ahead is delayed until the SRR, the PRR may or may not be formal. Formalizing the PRR will provide a supplementary verification of the adequacy of the proposed concept and is strongly recommended if the system being defined is especially large, complex, or risky. For small and low-cost systems with higher tolerance to failure, and in cases where the SRR follows soon after the PRR, it is possible to perform the PRR as an informal evaluation avoiding the costs of two formal reviews – PRR and SRR – being undertaken with only a brief period between them.

6.4.2.3. Phase B1

Phase B1 ends with a system requirements review. The SRR's aims are the same as the PRR's, but expanded and more detailed. The aims of this review are:

- Examine the compatibility of the system concept with the allocation of higher requirements to lower-level subsystems.
- Review the proposed concept, which includes a fairly detailed definition of all the elements, and evaluate the analysis supporting the concept's suitability.
- Review the proposed framework for initiating the project: plans and procedures, necessary resources, cost and schedule estimations, possible industrial organization, recommended early development activities, etc.
- Examine the compatibility of the system interface specifications with system components.
- Freeze the system requirements document and the resulting interface between the system elements.

The output of Phase B1 includes a detailed definition of all the mission's critical components and a preliminary definition of all the noncritical ones. This means that experts from all the engineering disciplines must participate. Unlike Phase A, Phase B1 includes participation by experts in "transversal" engineering support disciplines (Section 2.6) like assembly and test engineering and quality assurance. When not done after the PRR at the end of Phase A, the outcome of the SRR will enable final confirmation of the program and the decision to start development. When the SRR is the last review before full project approval, it is always a formal review providing an independent evaluation of the work done during the project's architectural definition phases.

6.5. Parallel developments

In order to undertake a new space program with an acceptable degree of risk, an optimized system is not sufficient. Also essential are:

- An adequate level of maturity of the technologies to be employed by the mission.
- A thorough understanding of how the observations obtained will be translated into useful information for the consumers.

Both of these aspects require specific developments to be performed parallel to the main design effort. These developments are strongly linked to the main definition effort (Phases 0, A, and B1), while normally remaining independent of it. This section will focus on these two aspects of the overall process.

The amount of time, money, and effort invested in these two facets must accord with the risk avoidance and funding guidelines provided by the sponsor. Missions with a high degree of responsibility, in which failure is prohibited, require a more intense level of effort on both technical development and an understanding of what we might call the mission's scientific background.

6.5.1. Technical maturity improvement

As Section 6.3.1 indicated, assessment of the "technology risk and status" is one of Phase 0's outputs. Technology maturity levels were examined in Section 4.9.3; from level 1, the least advanced, to the most mature, when the technology has been demonstrated in orbit. Figure 4.4 also indicated the development areas associated with the different technology readiness levels. Research, which involves Levels 1 and 2, entails confirming basic feasibility and identifying possible applications and is normally performed prior to the beginning of the mission definition process. The most advanced development levels, 7 through 9, are costly and are only meaningful if performed in the framework of a fully fledged project. The intermediate levels, from 3 to 6, are best performed during the mission's definition phases; Phase 0, Phase A, and B1.

The importance of technology development is a function of the type of project involved (see Section 4.5). Technology demonstration missions (Section 4.5.2) are risk tolerant and can reasonably start at Level 4 or even 3. Advancement of science or operational missions, on the other hand do not tolerate high levels of failure and are usually much more expensive. Hence, both kinds of missions demand that all the technology flown be defined to a degree that minimizes the uncertainty of cost, performance, and risk. This means that specific development plans, to be executed in parallel to the Phases 0 and A, must be established for technology that is fundamental to the mission and whose technical maturity level is between 3 and 5 with the target to advance all the necessary technologies to level 5 or 6 before the formal approval of the project.

In the early stages of technology development the allocation of adequate time for trial and error is crucial. This ample provision of time can be compatible with moderate expenditure. The mission definition period from Phase 0 to Phase B1 is indeed richer in time than in funds, and is the right milieu for advancing the technology readiness level enough to reduce project risk. Already approved missions enjoy ample funds, but a development failure in a key technology will have a major impact since it could delay implementation of the entire project.

6.5.2. Scientific understanding advancement

The proper design of any space mission demands an adequate understanding of the entire mission, including the science behind its application, that is, how the data produced by the satellite will be transformed into information useful to the consumer.

A proper understanding of these aspects will help justify the mission and enable a more reliable estimation of its performances. Well-defined procedures for translating the orbiting instrument's output into high-quality useable information, combined with suitably developed science, will always facilitate project approval.

The two main types of activities in this area are:

- General advancement of the scientific background that is going to be advanced through the implementation of the future space mission
- Development of specific algorithms that can be used to provide a true end-to-end model of mission performance

Any truly end-to-end model will require the integration of many different modules, produced by experts in various scientific fields. Section 8.5 and Figure 8.19 provide good examples of these models' complexity; to determine final performances of a radar altimeter satellite providing information on ocean circulation the required modules include: radar engineering, ionosphere science, precision orbit determination, and atmospheric dynamics, in addition to help from the ocean dynamics specialists and oceanographers.

These end-to-end models are crucial for gauging the output of the images with respect to costs. This will enable not only comparison of final mission performances with costs, but also between different design alternatives, allowing selection of the best possible solutions. Given that they are important tools for the design process as a whole, these models should be produced as early as possible.

As is the case for technology maturity levels, these activities are carried out in parallel and in close coordination with the main thrust of mission design. This coordination is even closer for activities involved with the production of end-to-end mission effectiveness tools because they are directly linked to mission design work.

Chapter 7

Introduction to the Design Domains

Keywords Design Interactions, Design Domains, System Breakdown, Astronomical Observatory Satellite, System of Systems

This is the first chapter of the second part of the book. It examines the interrelations between the components and elements of a space system and how they interact together during the design process. To study these interactions, the overall design space will be divided into domains. Each design domain is defined as a specific area where multiple design decisions interact strongly with one another, while they interact only weakly with design decisions external to the domain. The main tool for the study of these interactions is the comparison and analysis of past design examples. To provide an introductory perspective on the highest-level design decisions, the chapter analyzes the evolution over time of a very specific type of mission: the "Astronomical Observatory" mission. A number of past astronomical observatory missions will be presented and their highest-level design decisions – orbit, overall data flow, attitude, overall satellite configuration, operations – will be analyzed and compared.

This chapter will also review several general topics that are better discussed before the design domains themselves are studied in detail in the following chapters. These topics are related to supra-satellite design decisions: single versus multi-satellite missions and system of systems.

7.1. Design interactions and design domains

As mentioned in Chapter 1, we will not describe in detail the contents of the different space engineering disciplines, but concentrate instead on the interactions between all the engineering design decisions and how they drive the system towards the final design. In order to organize the study of these interactions, to make them more manageable and understandable, the overall design space will be divided into the following domains:

1. The observables and instruments domain
2. The orbit and attitude domain
3. The satellite configuration and mechanical domain
4. The operations data flow and satellite electronics domain
5. The instruments output data flow domain

M.A. Aguirre, *Introduction to Space Systems: Design and Synthesis*, Space Technology Library 27,
DOI 10.1007/978-1-4614-3758-1_7,

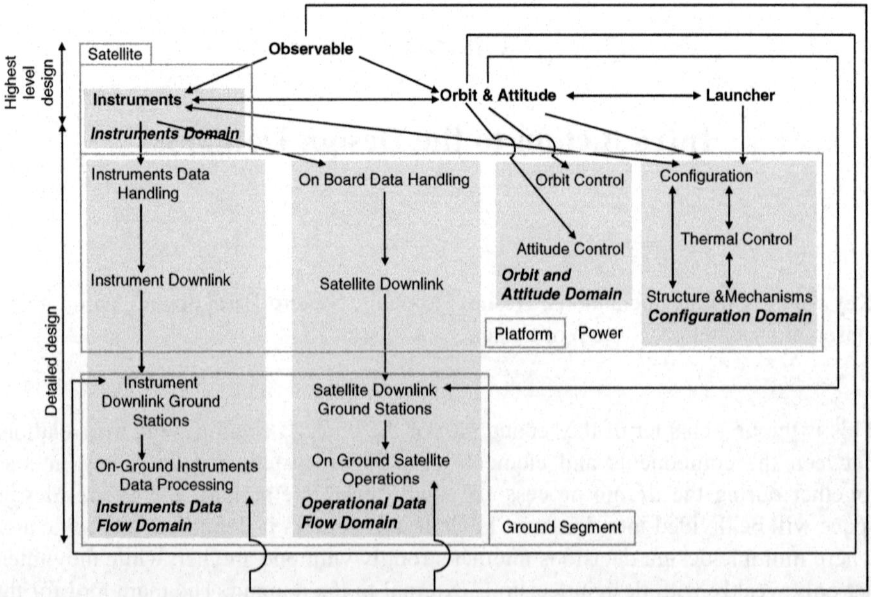

Figure 7.1. Top-Down Requirements Flow, Resulting Design Interactions and Resulting "Natural" Design Domains.

The top-down derivation of requirements and the most important interactions between space system elements and components, and the resulting division into "natural" domains can be seen in Figure 7.1.

Observables are at the top of the design process. Observables drive the instruments or payload concept and the manner in which data shall be acquired and distributed. The instrument and the observables will determine the selection of the satellite orbit and attitude. The nature of the observable will also drive the data acquisition and delivery needs of the consumer, which in turn will drive the operations and instrument data utilization approaches. At the same time, the instrument, the orbit, and the attitude will strongly interact with each other until they fit together. The orbit and the amount of data produced by the instrument will determine the frequency of the space-to-ground contacts. The orbit will drive the selection of the launcher, whose performances will act as a boundary for the orbit and for the satellite's size and shape. Once we have a highest-level design where all the elements fit together, we can proceed with a second, more detailed design loop, which will produce the detailed architecture. This process has been discussed in Sections 6.2 and 6.3 and it is also depicted in Figure 7.1 above.

The instruments will be at the root of this second design loop. The instruments themselves will require progressively better-defined designs, which will provide the instrument performance and interface features. The instruments data output will drive the design of the satellite on-board data processing and of the on-board dedicated telecommunication equipment that will download the information to ground. In parallel, the already performed definition of the basic instrument data utilization approach will push upward and contribute to the definition of the number of ground stations, the on-board instrument data storage, and the instrument data downlink speed.

The needs of the instruments will also drive the overall satellite on-board data handling: directly through their specific needs, and indirectly through their impact on the electrical architecture of the platform. At the same time, the already established satellite operational approach will push upward and help in the definition of the on-board satellite monitoring and control architecture and of the satellite bidirectional operational communications hardware.

The need to accommodate the instruments and provide their required fields of view will also fundamentally define the satellite's physical configuration. The instruments will also drive indirectly the configuration through the previous decisions on satellite orbit and attitude, which define the geometry around the satellite. The configuration will also be constrained by the launcher through the limitations that it will impose on satellite mass, center of gravity location, and volume in the launch configuration. An independent configuration driver will be the need for electrical power of the satellite, which will drive the size and location of the solar array.

The orbit and attitude requirements will drive the design of the corresponding orbit and attitude sensors and actuators and of the strategy for attitude and orbit control. The attitude and orbit control components will also have usually less demanding design interactions with the satellite configuration and with the on-board data handling.

The design of the electrical power provision components can be done rather independently from the other design elements. The exception is the size, and configuration of the solar array that must be considered in the configuration domain, for both its stowed-for-launch and in-orbit configurations.

At the end of this second design loop a comprehensive design for the satellite will be available. To check for consistency, it will be necessary to revisit the fitting of all the design decisions previously taken.

The interrelationships depicted in the Figure 7.1 and discussed above demonstrated that to divide the design process into the five domains of the figure is indeed a logical way to introduce order in the design process.

In the following sections, this text is going to introduce the different design domains. Each domain will be revisited and expanded in the following chapters.

7.1.1. The observables and instruments domain

Space missions acquire information over natural or man-made physical objects. These objects are the observables. The observable, the things the mission wants to know about, is at the root of the mission and it drives the design of all the mission elements.

Sometimes the data on the observable can be acquired using widely different techniques, for example, the atmospheric wind can be observed using active scatterometers (as in MetOp[1]), passive polarimetric microwave radiometers (as in Coriolis/

[1] http://www.esa.int/esaLP/LPmetop.html

WindSat[2]), or active Doppler LIDAR (as in Aeolus[3]). In some cases the selection of the optimal observing technique is in the hands of the space system designer but quite often the selection is previous to the beginning of the design process. In these cases the consumer or sponsor already provides the selection as a system constraint. This is normally the case for advancements of science missions; to provide an instrument concept, which appears in principle adequate for the task, will give more credibility to the mission concept. On the other hand, in the case of operational missions that are normally addressing issues where the science is settled, the designer could be entrusted with the responsibility of studying alternative observing techniques to determine the optimal one.

As already said in Section 5.6.1 and Figure 5.8, the consumer shall identify his needs about the observable in terms of:

- Data quality
- Data quantity
- System responsiveness

Hipparcos,[4] launched in 1989, was an ESA mission that allowed determining the position of the stars with an accuracy of 0.001 arcsec, the position of 120,000 stars was cataloged, the desired mode of data acquisition was systematic, being a continuous survey of the sky, and the data was sent to Earth as it was produced, where it was processed until the data received was sufficient to produce an overall map of the sky: the star catalog that was the output of the mission. Therefore, the mode of operation of the mission and utilization of the data was fairly constant and automatic during the whole duration of the mission.

The consumer needs, being related exclusively to the observable, would be normally independent of the observing technique, but they would have been established aware of what is possible and very often with a specific observing technique in mind. Three situations are possible:

- The observable can be studied by measuring in situ characteristics of the space around the orbital path: magnetic or electric field, atmospheric density, or gravity gradient.
- The observable is going to be "remotely sensed" but it is located over the surface of the body that the satellite is orbiting.
- The observable is going to be remotely sensed and it is far away in the deep space.

Each one of these three situations will produce very different interrelations between the observable and the orbit. This is going to be discussed in the next section.

The features of the remote sensing instruments will depend on the wavelength and intensity of the radiation being measured. In the particular case of Hipparcos, the instrument was a telescope that allowed the determination of angular distances between stars, by combining over a common focal plane the images of stars located with quite different angular positions.

---

[2]http://directory.eoportal.org/pres_CoriolisWindSat.html

[3]http://www.esa.int/esaLP/LPadmaeolus.html

[4]Hipparcos in ESA Achievements, SP-250.

7.1.2. The orbit and attitude domain

In the case of observables measured in situ, the requirements for the orbit are clear: the orbital path shall travel through all the areas of interest to the consumer. For these kinds of situations the satellite's attitude is somewhat independent of the observable, and it can be selected according to other criteria. For example, the attitude of the Ulysses mission (Section 8.3.5 and 9.6.2), which monitors the electric and magnetic fields around the Sun, was driven by the need to point the satellite's high-gain communication antenna toward Earth.

In cases where the observable will be "remotely sensed" but is located above the surface of the body that the satellite is orbiting, the selection of the orbit will automatically determine:

- The distance to the observable
- The areas that can be observed from the satellite's orbit

This is the case of all remote sensing Earth observation satellites, and generates a very strong interaction between orbit selection and data quality and quantity. The distance to the observable, that is, the orbit altitude, will drive the level of quality, spatial resolution, and radiometric accuracy of the information. The areas under the orbit as the satellite proceeds along its orbital path will determine the coverage provided by the instruments. In these kinds of missions the orbit altitude and the instrument characteristics need mutual adjustment to reach a reasonable compromise.

In the case the payload's field of view of the satellite would be pointing toward Earth, as the satellite orbits the Earth, the resulting pointing attitude will rotate with respect to the inertial space at the speed of one turn per orbital period. In cases where systematic coverage is needed, the satellite will be continuously pointing down towards the Earth. In cases where the mission requires swift and repeated coverage of specific areas, the attitude of the satellite should change, in order to allow the instrument to be directed for as long as possible at the desired targets; this means that the satellite must be agile.

In astronomical missions, the observable is at a much greater distance than the body being orbited. In this case, the orbit selected must offer uninterrupted visibility of the celestial sphere where the observable is located. If the mission involves phases where a large sector of deep space is surveyed or mapped, followed by a phase of pointing toward a number of preselected celestial objects, the spacecraft's attitude control must cater for these quite different sets of requirements. Particular attention must be paid throughout the mission to avoid pointing sensitive instrument detectors at the Sun or Moon that could damage the sensors. In astronomical missions, the attitude of the satellite will be determined by the need to point in the direction being observed, while maintaining an adequate field of view for communications with Earth and for receiving electric power from the Sun. This generates a strong link between the orbit and attitude domain and the configuration domain. It is important to keep in mind that as soon as the orbit is selected, the resulting space environment, for example, atmospheric drag or radiation levels, has also been determined.

The selection of the mission's orbit and attitude is one of the earliest and most fundamental decisions of the mission design process. As can be seen in Figure 7.1, it is also the design decision richest in interfaces. The selection of the orbit is chiefly driven by the observables and by the instrument, but is also strongly allied with launcher selection, with the satellite's configuration, and with the number and location of ground stations, both operational and instrument downlink ground stations.

7.1.3. The satellite configuration domain

The physical shape of the satellite is its configuration. The configuration is strongly determined by two requirements:

- The need to provide in orbit the right observing geometry for all of the satellite's components: instruments, antennae, thermal radiators, and solar arrays
- The need to provide structural support during assembly and launch

This drives two configurations: a stowed configuration on-ground and during launch, and a deployed configuration in orbit.

In orbit the instruments will require not only an adequate line of sight to the observed targets, but also a proper view of cold space for heat dissipation. Electromagnetic emitters and receivers either for satellite communications or detection of observables also potentially interact with each other, so it is often necessary to place them as far away as possible. The position and orientation of the solar array must also be adequate for acquiring the necessary amount of solar energy in all phases of the mission. Governed by the need of electrical power, the satellite's solar array is frequently very large, which in turn requires specific deployable mechanisms. The volume allocated to the satellite by the launcher strongly limits the satellite's configuration. When stowed inside the launcher the satellite's configuration is driven by the need for adequate strength, stiffness, and compactness, providing the required dynamic clearance from the inside of the launcher's payload fairing. When deployed, the configuration is driven by the need for open fields of view and separation or isolation. The size of the satellite's main body is frequently smaller than the necessary distance of separation. In that case it is essential to provide specialized deployable structures that provide the necessary separation. To cope with the demanding mechanical environment generated by the launcher, the satellite's configuration must include a structural skeleton, which provides strength and stiffness during launch. Satellites are normally attached to the tip of the launcher; therefore, the satellite on the launcher will be in effect a cantilevered beam. The on-ground environment also influences the configuration design; assembly, verification, testing, and transportation generate their own environments that the satellite structure must be able to withstand, including transportation by land, sea, or air. The configuration of a particular satellite is also sometimes strongly influenced by programmatic constraints, such as the desire to retain modularity of payload and platform, any need to adopt a preexisting platform, or multiple launch constraints, which may impose accommodation restrictions. Other technical issues are the need to limit longitudinal and lateral center of gravity position at launch, which in turn drives internal equipment accommodation; access for installing and removal of units at any stage of on-ground assembly; platform sensor and antenna fields of view, thruster accommodation and thrust vectors; and deployed stiffness and nonnominal thermal design scenarios. The location and field of view for communication antennae is also a design driver.

Figure 7.2. GOESS NASA/NOAA Geostationary Meteorological Operational Satellite Configuration (Wikipedia Commons. Image ID:spac0255, NOAA In Space Collection).

The satellite attitude strongly impacts the configuration concept. It is instructive to compare the configuration of two satellites with the same orbit and purpose, but with different attitude control approaches, for example, the three axes-stabilized American GOES and the rotation-stabilized European MSG. The configuration of these satellites is totally different. GOES and MSG are operational meteorology Earth observation missions and both have geostationary orbits. As can be seen in Figure 7.2, GOES is a cube with six faces. In the inertial space, it rotates once per orbit around the axis, perpendicular to the orbit:

- The Nadir face, which points continuously toward Earth, is where the Earth observing instrument should be located. It must provide space for the payloads, sounder, and imager, and for the communication antennae, S-, and L-bands, also pointing towards Earth.
- The two faces parallel to the orbital plane. They are perpendicular to the Nadir direction, and to the solar array rotation axis. Both faces avoid direct illumination by the Sun. One accommodates the solar array drive mechanism, and the other the radiators required for cooling the instruments as well as the satellite as a whole.
- Another face accommodates the interface with the launcher.

MSG rotates at 100 rpm around the axis perpendicular to the orbit. This provides a strong gyroscopic inertia for this direction, and good pointing performance. Due to the rotation, there is no Nadir direction and the satellite has a cylindrical symmetry with top, bottom, and sides:

- As the spacecraft spins, its sides scan the plane perpendicular to the rotation axis, which roughly coincides with the plane where the Earth and Sun are located. The instruments and the solar cells are located on these sides and they sequentially scan the Earth and the Sun. This allows the instrument to observe the Earth and the cells to acquire solar energy from the Sun.

Figure 7.3. MSG ESA/Eumetsat Geostationary Meteorological Operational Satellite Configuration (http://esamultimedia.esa.int/images/meteosat/15_O.jpg).

- The axis perpendicular to the orbit, which is also the satellites' axis of rotation, defines the top and bottom faces. Both faces avoid direct illumination by the Sun. One accommodates the communication antennae with the Earth, and the other the radiators required for cooling the instruments as well as the satellite as a whole. The communication antennae number and location are designed and controlled to compensate for the rotation of the satellite keeping always the satellite sending the information toward the Earth (Figure 7.3).

7.1.4. The satellite operations data flow domain

This domain deals with the functions that the end-to-end space system must perform in order to operate correctly, and with the technologies and devices used to fulfill these functions. This domain includes both the in-orbit and on-ground elements. At satellite level, this domain is driven by the nature of the observable, by the instrument used to observe it, and by the electronic architecture of the instrument and the platform. At ground segment level this domain is driven by the instrument and platform operations concept and the resulting operational requirements, which are also driven by the nature of the observables.

The elements that are included in this domain are the traditional satellite on-board data handling and electric power components, the satellite-to-ground, and ground-to-satellite, on-board communications subsystems, and all the ground-located satellite control elements. The satellite operations ground stations and any data relay satellite is also included in this domain. This domain is strongly tied to the instrument data

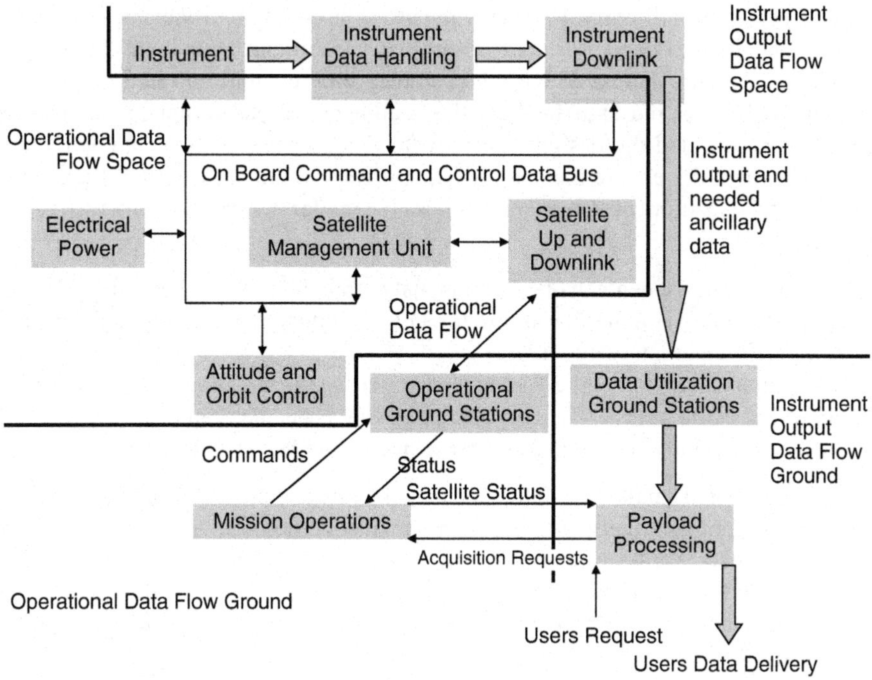

Figure 7.4. Functional Blocks and End-to-End Data Flow for a Space Mission with Indication of the Separation Between Domains and Between Ground and Satellite. The Thick Arrows Indicate Large Data Flows.

flow domain; that will be presented in the next section, but is better considered separately because:

- They have different key drivers. The operational data flow is driven by the need for the operation of the instruments: routine or interactive. The instrument data flow is driven by the requirements of data quantity and speed of delivery.
- The operational data flow is bidirectional between ground and satellite. The instrument output data flow is unidirectional from satellite to ground.
- The operational data flow involves the complex handling of moderate amounts of data. The instrument output involves the simpler handling of large amounts of data.
- In Earth observation missions, albeit not in space sciences missions, both data flows are implemented using different space and ground hardware, and in many cases work at different speeds and are specified at different frequencies.
- The processing of the data produced by the instruments and the satellite operations require different types of skills and are usually performed by different institutions.

The interactions between both domains can be seen in Figure 7.4. On-ground, the most important interaction takes place when the payload processing center provides observational requests to mission operations. In the opposite direction, mission operations provide information on the status of the instrument that might be necessary for the appropriate processing of the instrument output. In orbit, the satellite management unit will control the instruments as it controls all the satellite's electronic components. Despite the many in-orbit and on-ground components included in this design domain, compared with the complex interrelations and interfaces of the previously considered design domains, the satellite operations design domain is more straight forward, and the design decisions flow top-down without many complications. The main drivers are the operational approach to the mission, and the characteristics of the instrument. There are two possible modes of operation: interactive and systematic.

- Systematic instruments work continuously without requiring frequent ground operations intervention. This allows for a highly autonomous operational data flow with rare human intervention.
- Interactive, observatory-like, instruments requiring the handling of user requests to observe specific targets. The resulting observing schedules can change frequently and even suddenly. This requires intense interaction between space and ground.

The operational data flow will include the on-board production of the data required for monitoring of the satellite's status. In nominal conditions the satellite shall be able to autonomously handle most events. In nonnominal situations, the satellite shall communicate the event to ground, and depending on the nature of the nonnominal event, the satellite shall either take autonomous action or wait for ground commands.

Once the operational approach has been established, it becomes necessary to adjust the frequency of contacts with the operational ground stations. This selection will also be driven by the selection of the mission orbit. Satellite components, like the on-board data handling computer or the avionics buses, are directly determined by the need to monitor and control the satellite and its instruments. Once the operational needs of the instrument and the platform have been established, electronic monitoring and control hardware requirements must be derived, and designs for all these elements produced.

7.1.5. The instrument output data flow domain

This domain deals with the functions that the end-to-end space system must perform in order to: receive data acquisition requests (if needed), process the acquired data, and deliver them to the consumers. As in Section 7.1.4 above, this domain includes both in-orbit and on-ground elements. As in Section 7.1.4, the primary driver for this domain is the observable and instrument needs. The principal design drivers will be the amount of data produced by the instrument, and whether or not there is a need to provide the data to the users within a specified delivery time.

Included in this domain are all the satellite's in-orbit components handling instrument data output: processing, storage, modulating, and downlinking. The on-ground reception stations, payload processing centers, data distribution networks, and user-interface facilities also form part of this domain.

As in the case of the operational data flow, the requirements derivation process and the resulting design are quite straightforward: the quantity of data produced by the instruments and the speed-of-delivery requirements, combined with the satellite orbit, will determine the number and location of the ground stations – or data relay satellites if available. Typically, the satellite will downlink the on-board instrument data every time one of the ground stations becomes visible. When the speed of delivery is of paramount importance the use of data relay satellites might be necessary. The instrument data throughput and the longest period without ground station contact will determine the on-board data storage needs. Depending on the nature of the data, data compression hardware might or might not be used. As a rule, advancement of sciences mission data should not be compressed, but operational missions can tolerate data compression. Depending on mission requirements, special data processing may be necessary, for example, signal scrambling for security. Finally, the necessary data downlink speed and the space-to-ground link budget will shape the satellite data downlink hardware. Once ground stations have received the data, it becomes necessary to determine the main processing steps needed for delivery of the different data products required by the users. Some users might prefer relatively unprocessed data, or to use their own processing schemes, while other users might prefer highly processed data. The processed data must be delivered to the user on time and with a guaranteed level of quality. This requires the design of adequate on-ground installations to act as an interface with the consumers. The processing of the data often requires interfaces to external sources of auxiliary data and also the use of data produced by other space missions.

7.2. The astronomical observatory missions as an example of space system design

The flow down of the requirements and the interrelation between domains has been discussed in Section 7.1. The upper part of Figure 7.1 shows that the observable, the instruments, the observational approach, the configuration, the orbit, and the attitude are strongly interrelated and that it is necessary to arrive to a global architectural design before it was possible to continue with more detailed design work. This section will expand and explain this first set of design interactions through the use of commented examples. This section will compare the overall architecture of a number of missions, launched at different times but all of the same type. They are all "Astronomical Observatory" missions. In these missions a telescope is launched, and astronomers request specific observations. After they have been performed, the observations are downlinked to the ground, processed, and distributed to the astronomers that requested them. Such a mission requires:

- High-quality instruments, which normally means large aperture telescopes
- Extended and uninterrupted periods of observation of the targets
- An observation direction unobstructed by the Earth, Moon, or Sun
- Fast and easy access by the customer (astronomers), to observing times and swift delivery of the observational results
- Fast and flexible, ideally interactive, communication from ground to satellite and from satellite to ground
- Carefully specified spacecraft pointing accuracy, stability, and control
- Good thermal and thermoelastic design
- Long-lasting mission life

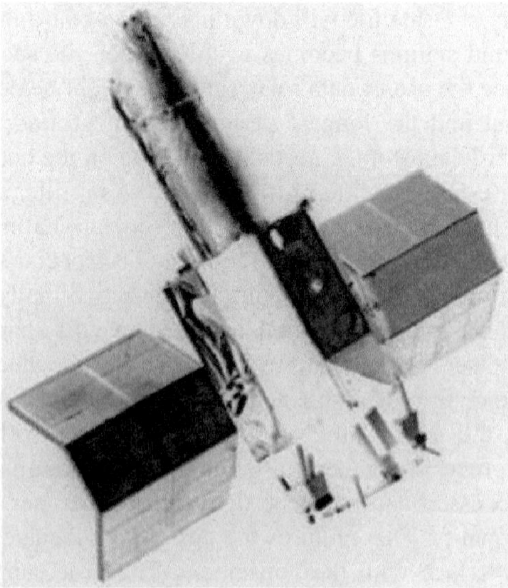

Figure 7.5. IUE Satellite (http://www.esa.int/images/iue_2_l.jpg).

To illustrate this evolution, the following missions have been selected: IUE, ISO, Hubble, XMM-Newton, Herschel, and JWST. They encompass an extensive time period, from the IUE in the 1970s to 2014 when JWST will be launched. The description of the missions will highlight factors affecting the highest-level design decisions: orbit selection, basic configuration, mission life, and data flow to and from the satellites.

7.2.1. Mission descriptions

7.2.1.1. International ultraviolet explorer (IUE)

IUE[5] was a collaborative NASA/ESA project to perform ultraviolet astronomy with a 45 cm aperture telescope. It was launched by a Delta 2914 rocket in 1978 and had a mission life of 19 years. It was injected in a geosynchronous orbit over the Atlantic with 32,050 km of perigee and 52,254 km of apogee and it returned 100,000 ultraviolet spectra from 9,000 celestial bodies. It was the first satellite operated flexibly to allow astronomers to request fast reaction observations to targets of special interest, with provision of the results made available in less than 1 h after the request (Figure 7.5).

It had two centers for receiving requests and for provision of observations: NASA Goddard near Washington DC and at Villafranca near Madrid. Both centers also controlled the satellite and received and processed the data produced by the telescope. Due to the chosen orbit period and location, the satellite was under continuous ground

[5] ESA Achievement, BR-250, ESA 2005.

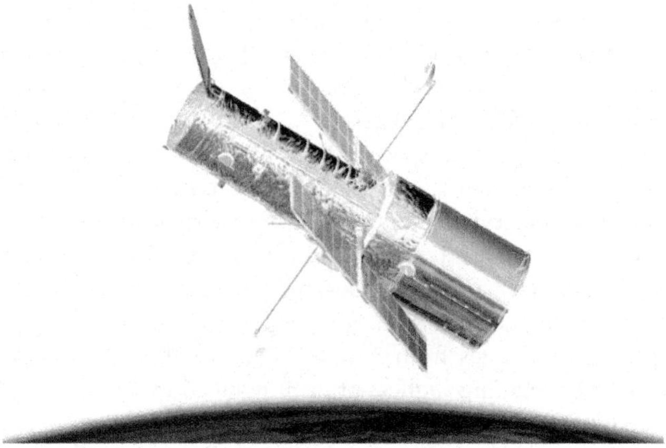

Figure 7.6. Hubble Space Telescope.

control: 8 h a day from Villafranca and 16 h a day from Goddard. The data was downlinked at speeds up to 40 kb/s. With its 19 years of life, IUE has become one of the longer-lasting satellites ever.

7.2.1.2. Hubble space telescope

The Hubble space telescope is a NASA mission with some ESA collaboration, for high-quality, general-purpose astronomy from space including: ultraviolet, visible, and infrared radiation access. It carries a high aperture telescope of 2.4 m of diameter. It was launched by the space shuttle in 1990 to an orbit altitude of 600 km, and an inclination of 28°. It was designed for in-orbit servicing by astronauts. This allowed upgrading of instruments, solar arrays, and electronics. It remains in orbit and has been serviced five times by astronauts, including in-orbit repair of optics, renewal of the solar array, and updating of electronics (Figure 7.6).

The Space Telescope Science Institute (STSCI), located in Baltimore, is responsible for the scientific operation of the telescope and the delivery of data products to the astronomers. The daily satellite monitoring and the management of the in-orbit satellite and instrument operations are in the hands of NASA Goddard, which provides this service 24 h per day.[6] Astronomers can apply for time on the telescope, calls for proposals are issued annually, and proposals are either selected or rejected, as the demand for time is between six to nine times that available. It is also possible to request special "target of opportunity" periods, when the observation of important transient events requires a swift reaction.

The visibility of the celestial sphere from Hubble's orbit is strongly hampered by the Sun and the Earth. To prevent the Sun entering the telescope, there is a pointing avoidance angle of 45° half-cone around the Sun. The telescope's low-Earth orbit also means that the Earth obscures almost half the celestial sphere. There is a

---

[6] http://hubble.esa.int/ and from http://www.stsci.edu/hst/ and http://hubble.nasa.gov/index.php

so-called continuous viewing zone, at roughly 90° to the plane of Hubble's orbit, in which targets are not occulted for long periods. Due to the inclination of the orbit and the resulting precession, the location of this zone moves slowly around the celestial sphere with a period of 8 weeks.

Hubble data is stored on the spacecraft. When launched, storage facilities consisted of reel-to-reel tape recorders, but these were replaced by solid-state data storage facilities during two servicing missions. From the onboard storage facilities, data is transferred to ground via the Tracking and Data Relay Satellite System (TDRSS[7]), located in the Geostationary orbit and designed so that satellites in low-Earth orbit can communicate with their mission control facilities during 85 % of the orbit. TDRSS can provide data rates from 2 to 48 Mb/s, but this system has to be shared with other missions, including highest-priority manned missions. The data is transmitted from the GEO located TDRSS satellites to its ground station in White Sands New Mexico and then to the Goddard Space Flight Center that links it to the STSCI in Baltimore for processing and distribution to the users.

7.2.1.3. Infrared space observatory (ISO)

ISO[8] was an ESA mission performing infrared astronomy in the 2.5-240 μm range and employing a telescope with a 60 cm aperture. It was launched by an Ariane 44-P in 1998 to a geosynchronous orbit with 1,038 km of perigee and 70,578 km of apogee. It had a design life of 18 months—limited by the depletion speed of the 2,100 l of He at 1.8 K used to cool the telescope. The actual duration of the He supply in orbit was 28 months. It was able to provide good quality observation when located above the Earth's radiation belt, which occurred during 17 of its 24-h orbit. It was operated as an astronomical observatory with reception of requests and delivery of the requested observations, and had two ground centers that controlled the satellite and interfaced with the consumers: the primary one located at Villafranca, near Madrid provided visibility during 12 of the 17 theoretically available hours for observation, while the secondary center, located at Goldstone California, allowed observation during the remaining 5 h. During its lifetime the satellite performed 26,000 individual observations. The satellite's data rate was 33 kb/s from which 24 kb/s where available for instrument output data and the rest for the monitoring of the satellite (Figure 7.7).

7.2.1.4. XMM-Newton

XMM-Newton is an ESA mission whose aim is to perform astronomical observation in the X-ray part of the spectrum in the energy range between 20 nm and 0.1 nm. It has three grazing incidence telescopes, each one with 0.7 m of aperture. Each telescope has its mirror module – made up of 58 concentric mirrors – and its focal plane. It was launched in 1999 using an Ariane 5 rocket. The satellite was put into a highly elliptic orbit with 7,365 km of perigee and 113,774 km of apogee, an inclination of 38.9° and with an orbital period of 48 h. The mission was still operating correctly in 2009. The satellite has no on-board storage memory and observations are down

[7] www.spacecomm.nasa.gov/spacecomm/programs/tdrss/default.cfm

[8] ESA Achievements, BR-250, ESA 2005.

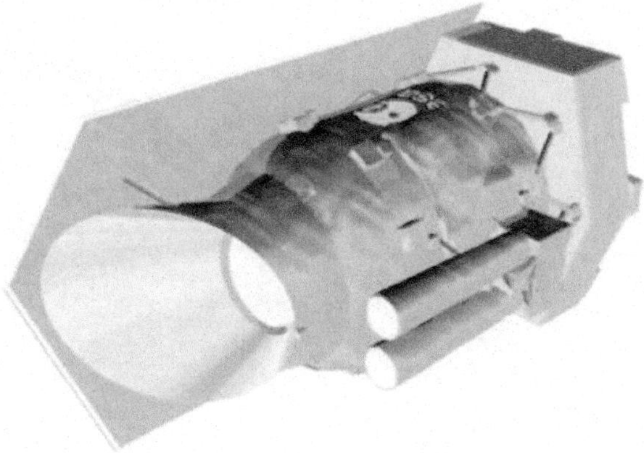

Figure 7.7. ISO Configuration.

linked at 70 kb/s in real time to stations at Perth (Australia), Kourou (French Guiana), and Santiago (Chili). Observations are possible for up to 40 h of the 48 h of the orbit period when the satellite is above Earth's radiation belts.

Newton time for observation is divided in four categories:

- Open time that is accessible by regular open announcement of opportunity to the worldwide scientific community.
- There is "discretionary time" – about 5 % of the overall time – that the XMM-Newton Project Scientist can grant discretionary. Within this discretionary time there is "target of opportunity" time for observations, which could not have been foreseen at the time of the deadline of the announcement of opportunity.
- There is a calibration time – also about 5 % of the overall time – for calibration ad health monitoring and control of the satellite and instrument.[9]

Normally XMM is operated in a preplanned manner. Thus all observations must be specified in full detail well in advance by the proposers (Figure 7.8).

7.2.1.5. Herschel

Herschel[10] is an ESA mission whose aim is to perform astronomical observations for the first time in the far infrared range between 100 and 600 μm. It has a large aperture telescope with a diameter of 3.5 m. It was launched in May 2009 using an Ariane 5 rocket. The launch was shared with another ESA mission: the Planck satellite. Both satellites were put into a halo orbit around the L2 Lagrangian point. The orbit is heliocentric, and located 1.2–1.8 million km from Earth. The anticipated mission life is 3 years. As in the case of ISO, the instrument life is limited by the depletion rate of the 2,160 l of He at 1.65 K used for cooling. The satellite stores the information produced by the instrument in a 25 Gbit on-board solid-state memory (Figure 7.9).

[9] XMM Newton Announcement of Opportunity Policies and Procedures, N. Schotel, ESA 2005.

[10] ESA Achievements, BR-250, ESA 2005.

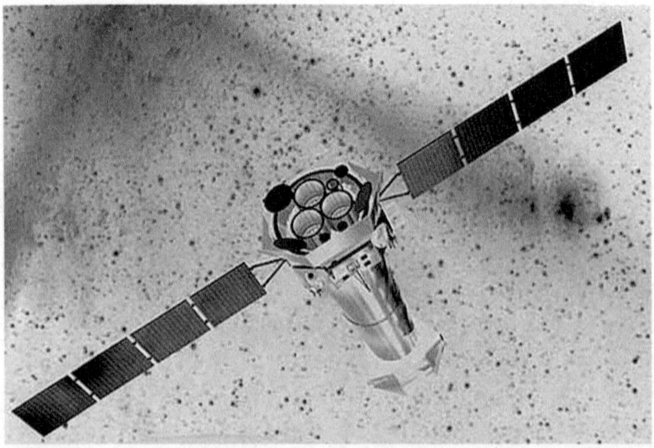

Figure 7.8. XMM-Newton Configuration (ESA achievements BR-250, ESA 2005).

Figure 7.9. Herschel Configuration (http://esamultimedia.esa.int/images/spcs/herschel/herschel_8_high.jpg).

The stored information is downlinked at 1.5 Mb/s to a ground station in Perth, Australia, for up to 3 h per day. The satellite is operated from Darmstadt, Germany, and the data is distributed from Villafranca, near Madrid. NASA will run a parallel distribution center located at Caltech to serve US astronomers.

7.2.1.6. The James Webb Space Telescope (JWST)

JWST is a NASA mission with ESA participation. It will be, like the Hubble Space Telescope, for which it is the replacement instrument, a general-purpose astronomical telescope, but designed to specialize in observing the infrared spectral region. Its aperture diameter will be 6.5 m. An Ariane 5 ECA will launch it in 2014 into a heliocentric halo orbit around the Lagrangian L2 point. Its distance to the Earth will vary from 1.2 to 1.8 million km with a periodicity of 6 months. The satellite is designed

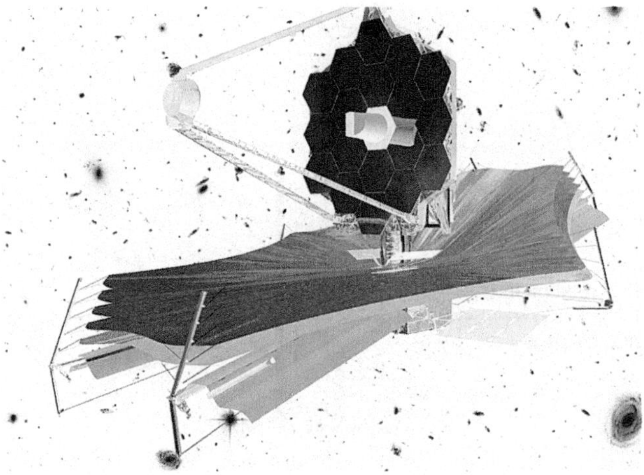

Figure 7.10. JWST Configuration (From James Webb Space Telescope (JWST) Observatory Architecture and Performance, John Nella, Charles Atkinson, AIAA-2004-5986, American Institute of Aeronautics and Astronautics Space Conference 2004).

to provide a mission life of more than 10 years. This long life is made possible by the use of a sophisticated Sun-shield system. Its design and objectives make the use of liquid He for cooling unnecessary (Figure 7.10).

The[11] Jet Propulsion Laboratory (JPL) will provide communications between the mission flight operating center and the JWST, using the 34 m diameter antennae of the Deep Space Network (DSN) located in California, Spain, and Australia. The DSN will support a command uplink rate in S band of 16 kb/s, a real-time S band telemetry downlink rate of 40 kb/s, and Ka band high-rate telemetry downlinked at 28 Mbps that is able to downlink up to 232 Gbits of recorded telemetry data in a single 4-h contact per day. The satellite operations and data utilization approach will be identical to Hubble's. The Space Telescope Sciences Institute in Maryland will act as interface with the consumers and will receive and process the data gathered by the DSN.

The JWST's configuration is dominated by a large sun shield intended to block radiation from the Sun as well as from the Earth and Moon. This ensures that the telescope operates protected from the radiation of the Sun and thereby keeps its sensitive optics cold enough to meet its mission objectives.

7.2.2. Comparison of the missions

These missions have many things in common, and there are similarities in their implementation. Nevertheless, many mission elements are different. Below, we will analyze the similarities and differences between them, and extract lessons from the analysis.

[11] http://www.stsci.edu/jwst/externaldocs

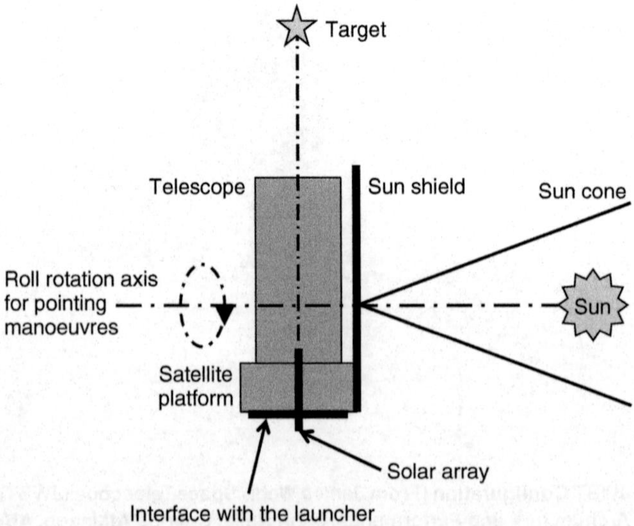

Figure 7.11. Configuration of "Telescope" Satellites and Cone Where the Sun Will Be Always Located.

7.2.2.1. Instrument and configuration

All the physical configurations are driven by the need to accommodate the instrument: the telescope. It requires support during launch, as well as for pointing it once in orbit. The configuration must be able to point at targets while making sure to avoid the Sun and the Earth. The configuration must also allow the solar array to point toward the Sun, and the communication antennae to point in the right direction toward Earth. The configuration must be as compact as possible to allow moderate inertia and high rigidity for fast maneuverability between targets and to allow for easy storage inside the volume provided by the launcher.

In all these missions the telescope sits right on top of the satellite platform. This provides all the satellites with a cylindrical symmetry during launch that fits nicely in the cylindrical space provided by the launcher and offers a high capacity for withstanding loads from all sources down the telescope, to the platform and to the interface with the launcher. In orbit and to protect the telescope from solar light entering directly into the telescope perturbations, it will be also necessary to establish an exclusion area defined by the highest Sun angles that the telescope can tolerate. To cope with these Sun angle limitations the satellite is forced to point the telescope to the targets located on the celestial sphere by rolling around the Sun direction. That means, in-orbit, the Sun will be always inside a cone defined by the satellite-Sun direction (Figure 7.11).

In all cases but one – the JWST – the sightline of the telescope is along the longitudinal axis of the launcher. This is by far the simplest way to accommodate the telescope in the launcher, but the aperture of the telescope is limited by the diameter of the launcher's payload fairing envelope. To get around this limitation, the sightline of JWST's telescope is perpendicular to the longitudinal axis of the launcher. This permits the 6.5 m diameter of the aperture to fit within the 4.5 m diameter provided by Ariane 5, but a series of complex deployment mechanisms are necessary to attain the operational satellite configuration. The structure of JWST can be seen also in Figure 10.18.

Telescopes that work in the infrared: ISO, Herschel, and JWST, need special protection from the heat and light generated by the Sun or Earth. In these cases, either the solar arrays act also as Sun shields – as for ISO, Herschel – or dedicated deployable sun shields are needed, as is the case for JWST. Telescopes working with more energetic radiation, like Hubble, IUE, and XMM-Newton, do not have this limitation; they do not need a Sun-shield and can use conventional solar arrays that are deployed from the satellite platform.

7.2.2.2. Orbits

The observables of these types of missions are extremely remote, so the distance to the observable is essentially the same for all orbits, although Hubble is sometimes tasked with solar system imaging. Therefore, the orbit will be determined by the visibility of the observables by the telescope, the avoidance of the Sun, and the flexibility of the communications with Earth. The ideal orbit is an orbit far away from Earth and with a predictable geometry with respect to the Sun. This orbit allows days and even weeks of observing time in a specific direction of space unmolested by the Earth or the Sun. It seems that the ideal orbit would be a heliocentric orbit around a Lagrangian point. JWST and Herschel indeed fly around the Lagrangian L2 point. In this orbit the Sun and the Earth are in roughly the same direction, and measures taken to protect the telescope from the visibility of one will serve as well for the other. This location also serves to avoid the thermal transient environment experienced by spacecraft in low-Earth orbits. Nonetheless, this orbit needs large launchers to reach the orbit, as well as powerful communication antennae to send and receive information from L2, which is up to 1.8 million km distant from Earth. This is the reason why older missions did not employ this orbit.

Geosynchronous orbits are also good because they provide prolonged contact with the same ground station and permit close to a day of observation without serious interference by the Earth or Sun, while allowing the employment of the large selection of launchers that are optimized for communications missions on geostationary orbits. This was the orbit chosen by IUE and ISO. IUE's orbit allowed continuous operations because its path passed directly above Earth's radiation belts. ISO and XMM, on the other hand, had highly elliptic orbits that forced the payload to switch off during several hours per day – when the mission was inside Earth's radiation belt. In the case of ISO this alternative was chosen to optimize the total mission duration, which was limited by the amount of He coolant on-board: it was more efficient to have a larger amount of He in the highly elliptical orbit than a smaller amount in an orbit that would permit continuous operation of the telescope but would have required a larger amount of fuel. In the case of XMM, the limitation imposed by the launcher also made it advisable to choose a highly elliptical orbit, which interrupts observations for just a few hours every 2 days, than to circularize the orbit to permit completely uninterrupted instrument use. Final orbit selection is always influenced by numerous factors, including: maximizing the time available for uninterrupted observations of a specific target, long mission lifetime, long instrument operating duty cycle, satellite operational flexibility, and launch constraints.[12]

The trend of the 1980s in favor of in-orbit servicing and shuttle use forced the Hubble to choose an orbit accessible to the space shuttle. This resulted in several very positive consequences. Astronaut servicing enabled repairs to be made to the telescope as well as for the upgrading of the solar array, the processing electronics, gyros, thermal insulators, battery, and the mass memory. Nevertheless, the chosen orbit comes with some strings attached; because it adds the limitations inherent in avoiding Earth, which at such a low orbit, hides almost one half of the celestial sphere, to the unavoidable limitations of avoiding the Sun. The problem of contact with Earth was solved by the use of the real-time TDRSS communication system. However, Hubble shares the TDRSS system with the manned International Space Station, which has priority over Hubble, limiting the operational flexibility that the use of TDRSS would have promised. Hubble's final lifespan will be close to 20 years, which is testimony to the advantages of in-orbit servicing allowing its sensors to take advantage of significant improvements in technology since the original detectors were designed. Nevertheless, IUE has survived 18 years of operations, even though it was designed for just 5 years of life, and the next generation of observatory satellites is designed to last for 10 years or more. Hence, a properly designed satellite of this type is likely to have a very long actual life.

7.2.2.3. Attitude

The telescope shall be able to point to the different targets that could be located anywhere in the celestial sphere. As far as satellite in-orbit attitude is concerned, the satellites are all inertially stabilized and must be operated by pointing the whole satellite in the direction of the target to be observed. In all cases the communication antenna is located on the satellite platform and points toward Earth. The Earth's direction changes as the satellite flies along its orbit. To decouple the pointing of the antenna and the pointing of the telescope the communication antenna has to be pointable.

The pointing of the telescope's opening toward the Sun must be avoided and the satellite can only maneuver by rolling around this direction. The Sun's direction rotates in the inertial space once per year and in order for observations to be performed, it is necessary to wait until the Sun is on the platform side of the Sun-shield and the target on the telescope side of the sun-shield. In the case of JWST, the Sun-shield protects the instrument when its line of sight, and thus the satellite attitude, is located within a ±25° belt perpendicular to the direction of the Sun[13] (Figure 7.12). This defines a 50° functional line-of-sight sector between 85° and 135° from the satellite-Sun direction. The region of the sky near the celestial north and south poles is continuously accessible. The medium latitudes are accessible during a long single

---

[12] XMM Overview, H. Barre, ESA Bulleting Issue 100, ESA 1999.

[13] James Webb Space Telescope (JWST) Observatory Architecture and Performance, John Nella, Charles Atkinson, AIAA-2004-5986, American Institute of Aeronautics and Astronautics Space Conference 2004.

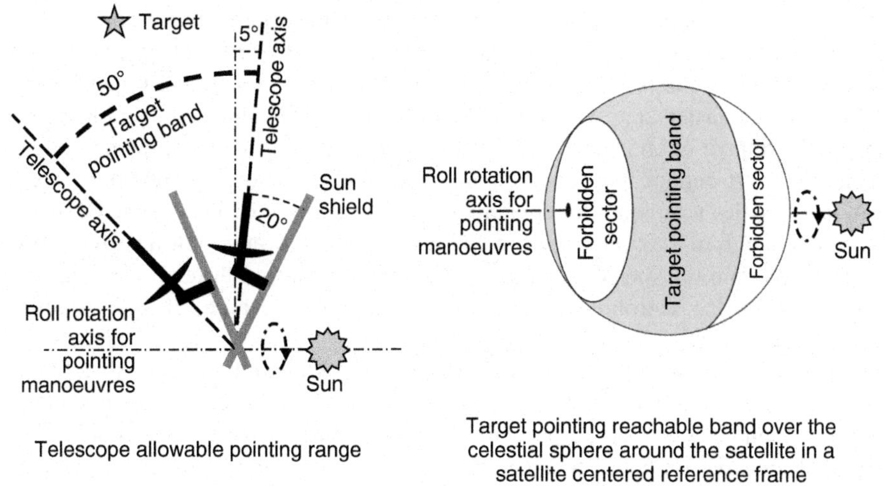

Figure 7.12. JWST Area of Observation Limited by Sun and by the Satellite Configuration.

season of 197 days, and are unreachable during the rest of the year. Regions that are near the ecliptic plane are reachable during two yearly seasons of 51 days, separated by two seasons of 131 days when they are inaccessible. As explained in Section 7.2, opportunities for observation of the different requested targets are allocated by a mission observation plan elaborated on the ground that has to take into account the seasonal opportunities provided by the Sun-satellite-observable geometry.

7.2.2.4. Operations and utilization

It might have seemed beforehand that rapid and direct access to the observatory by the consumer would be essential; however, this is not the case, for two reasons:

- In real life the available observation time is limited and the observation requests far exceed the available time. The establishment of a "peer review" process acting as gatekeeper for the telescope and adjudicating the available time to the worthiest requests is compulsory.
- The complex observing geometry, which is a function of the seasons, and the need to balance the different observation requests make it necessary to create a long-term "seasonal" observing plan to optimize the satellite's use while respecting the satellite's visibility constraints. It is not possible always to observe anywhere.

From time to time mission coordinators offer an "Announcement of Opportunities," through which the users community can convey their wishes. After a filtering process, a number of observational initiatives commensurate with the available time are selected and are scheduled for the appropriate period according to their requirements. This results in a long-term satellite observation plan, which is established well in advance of the actual observations. This plan consumes a large chunk of the available observation time. At the same time, all missions include a "fast track" contingency plan, in case an unexpected and important astronomical event should require interrupting the normal flow of observations to point the satellite toward new targets. The response time of this faster cycle is normally of the order of several days to weeks, including the time needed for the gatekeepers to review the request, albeit using a much faster approval process. In July 2009, Hubble was commanded to image the impact of an asteroid on Jupiter within hours. In general, the operational scenario for astronomical observation missions is complex, and the creation of observation schedules is an important planning task for the mission. These schedules must be rigid enough to maximize the time allocated for observation but flexible enough to accommodate unexpected events. Scheduling is achieved through close interaction between the astronomers and the engineers in charge of the satellite's operations.

As we can see, uninterrupted communication with the satellite is desirable, though not compulsory. Older missions, such as IUE and ISO, did indeed maintain continuous communication with Earth while their instruments were operational. They required no large on-board memory for storing instrument-produced data because data was downlinked as it was produced. This afforded the missions a large operational flexibility. IUE was actually operational for 24 h a day, while ISO was operational only during the period of 17 h a day when the instrument was above the Van Allen radiation belts. Both missions employed two ground stations, one in America and the other in Europe.

Hubble, which was using the TDRSS data communication network, featured a totally different overall data flow that would have allowed a continuous and real-time data dialog with the satellite, but the need to share the TDRSS network with other users – like the International Space Station – and the need to review and agree on observations, produced a system that is identical to all the others as regards the attending and processing of astronomers' requests. During periods when the TDRSS is not available for Hubble's use, the satellite uses its on-board memory to store data. Astronauts upgraded the telescope's memory from tape-based to the much more flexible solid-state variety. We have seen that the availability of large capacity on-board mass memories permits the reduction of expensive and long contacts with ground stations. The future JWST and Herschel will be in contact with ground stations for just a few hours a day. This will reduce the use of ground stations and the mission's overall cost while still allowing enough flexibility for satellite operations while providing almost uninterrupted instrument operations. The technical advancement offered by on-board mass memory has produced a next generation of missions that do not need continuous contact with Earth.

7.2.2.5. Satellite life and mission life

The need for astronomical observatories is continuous. Ideally a constant flow of missions would be able to provide uninterrupted observing capabilities in the different parts of the spectrum. However, this has not been the case. Still, astronomers have enjoyed a continuous flow of observatory missions that have provided a quite reasonable long-term coverage of needs with increasingly better instruments. This increase in quality can be clearly seen by comparing ISO's 0.6 m of aperture with the 3.3 m of Herschel, or Hubble's 2 m with the 6.5 m of JWST. It can be argued that all these missions are actually a system of systems providing complementary observing capabilities to advance our understanding of astronomy. This "unplanned" continuity is a result of the considerable current operational life – much longer than designed for – of all the missions listed here, coupled with the political pressure applied by consumers on sponsors to ensure new projects are approved. Nonetheless, astronomers' political muscle has been inadequate to the task of obtaining a long-term formal commitment to continuity on the part of politicians, though they are getting closer to that goal.

ISO had a limited life due to the reliance upon the finite supply of He used as cooler. JWST, which observes similar phenomena, uses sophisticated deployment mechanisms to allow a much longer mission life. The advantages of the longer life are paid for with a complicated configuration that will be discussed in detail in Section 10.3.3.

It is worth emphasizing that the actual life of these missions exceeded all expectations. Both IUE and Hubble will end up having lasted for nearly 20 years.

7.2.3. Observatory mission highest-level design interactions

The previous two sections analyzed the fundamental design decisions of several astronomical observatory missions. This section will supplement that analysis by providing an introduction to how higher-level requirements and design decisions interrelate with one another. This introduction will not be exhaustive, but will concentrates on how design and specifications relate to image quality and quantity, satellite size, and instrument pointing needs. These interrelations are especially complex and provide a good illustration of the multiple implications of design decisions.

Figure 7.13 depicts the cascade of interrelations and design decisions generated by requirements regarding data quality and quantity. Mission astronomers will require that the data to be produced be of a specific quality and quantity. These terms are studied closely in Sections 8.4.2 and 8.4.4, respectively. Here it suffices to say that spatial resolution is the smallest distance between two image features distinguished by an instrument, while radiometric resolution is the smallest difference in signal intensity measured by the instrument.

Good spatial and radiometric resolution depends on correctly sensing weak signals arriving from their limited place of origin. In order to properly detect signals an instrument with a large signal collecting area is needed. This will determine instrument size, which in turn will influence the satellite's overall size and mass. The design implications of instrument and satellite size were thoroughly discussed in previous sections. To avoid image smearing during the integration period needed for acquiring the image, good spatial resolution demands that the instrument's line of sight be highly stable. This requires a very precise command of the satellite's attitude, providing accurate line-of-sight control while avoiding mechanical perturbations. Achieving high pointing stability becomes even more difficult in large, heavy, and flexible satellites. The satellite mechanisms, and the attitude actuators themselves, introduce mechanical pointing perturbations that alter the instrument's pointing stability. High-resolution astronomical observatories require a strict control of internal mechanical perturbations. This could entail the use of active or passive damping techniques, further complicating the design. It is crucial that both the attitude control system and any internal sources of mechanical perturbations be specified so that

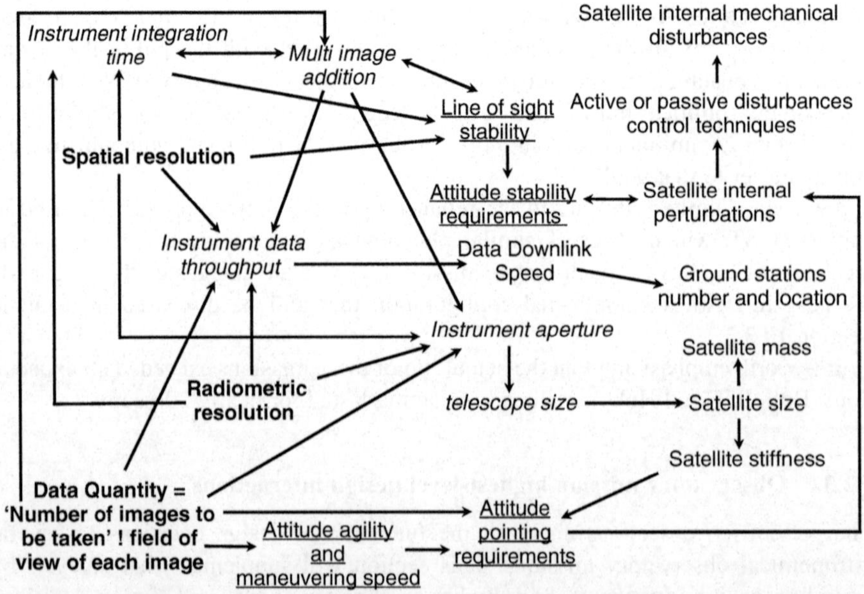

Figure 7.13. Interrelation of Requirements. Consumer's Needs (*Bold*), Derived Requirements on: the System (*Regular*), Instrument (*Italic*), and Satellite Attitude (*Underline*).

instrument pointing and pointing stability are compatible with the instrument's resolution.

In order to reduce the problems described in the previous paragraphs, it is possible to provide strong integrated signals by increasing the total integration time through the addition of several images of shorter duration. This will relax the instrument size and line-of-sight stability requirements. Nevertheless, it will increase the instrument's data throughput and the time needed to produce each final observation. This integration time increase will reduce the number of observations produced by the mission, with the corresponding reduction in the mission's value for the astronomers/consumers.

Consumers will always prefer the acquisition of as many observations as possible, and each one with a large field of view. Large fields of view require large and complex instruments. The acquisition of a large number of images demands a high level of agility to enable fast repointing from one target to another. This in turn requires a powerful attitude control system and a stiff satellite. A satellite's stiffness and agility are in direct contradiction to instrument aperture and satellite size. The use of powerful attitude control actuators is also at variance with the need for smooth and careful attitude control described above. Satellites producing large amounts of data need fast and powerful data downlink subsystems and/or a large number of judiciously located ground stations to provide extended periods of contact with the satellite. The proper consideration of pointing perturbations is an area requiring an architectural perspective. Pointing perturbations will be examined in detail in Section 9.8. The paragraphs above are just an introduction and the interactions discussed in this section will be retaken and expanded several times in the Chapters 7, 8, 9, 10, 11, and 12.

7.3. Multi-satellite design aspects

There are design decisions regarding the number of satellites required to fulfill a mission's objectives as the same requirements can be fulfilled by one, several, or many satellites. Examples are:

- One communication satellite in a geostationary orbit can provide continuous communication services to the users in a given region on Earth. A constellation of satellites in low-Earth orbit, like Iridium, (Section 9.6.3 and 10.3.4) can be designed to provide continuous communication over the whole Earth.
- A small number of satellites – or just one – carrying instruments with large fields of view will provide similar observing capability than a large number of satellites with correspondingly smaller fields of view.
- A larger number of cheaper satellites with lower reliability and shorter mission life will provide similar length of service than a smaller number of more expensive satellites with higher reliability and longer individual satellite lives.

In some cases – like the geostationary example provided above – the selection of the number of satellites is related to the selection of the orbit but in the other two cases mentioned, the constellation of satellites will travel along very similar or identical orbits.

Satellites collectively providing a service shall have controlled orbital positions; that means, they shall be a constellation; Section 9.11 will concentrate in the design and maintenance of constellations. The present section will study the design drivers that will favor using a larger or smaller number of satellites to provide a similar set of performances.

7.3.1. Data quantity and quality versus number of satellites

If the mission requirements can be fulfilled with a single satellite, the mission is commonly implemented using a single satellite. Nevertheless, employing several satellites to provide the same service has a number of advantages:

- It permits smaller, more robust, and simpler instruments and satellites and a lower development risk for the mission. Small robust satellites, using proven technologies, can be at the same time cheap and reliable and can be accommodated in lower-cost launchers. Robust satellites have been already discussed in Section 5.3.1 and will be discussed again in Section 13.2.4 when discussing low-cost satellites.
- The Earth's surface is curved and instruments observing regions located far away from the satellite's ground track will see those areas strongly foreshortened or distorted. The use of several satellites each with a small swath instead of a single one with a large swath will allow observation with minimum image distortion.
- Each satellite of the constellation will sequentially downlink the information stored on-board as they pass over the ground station. This allows a longer and more efficient – lower speed and longer duration – downlink of the data produced by the missions.

- Multiple satellites reduce risk and make missions more robust with respect to failure: the failure of one satellite does not imply the end of the mission but only a degradation of performances.

For the same set of programmatic constraints and for the same style of project organization, splitting the capabilities of the instrumentation among more than one satellite – providing that mission requirements allow this—is more likely to imply an increase rather than a decrease in total mission costs. In the other hand, to provide the same function with multiple satellites would allow the use of mature technology that can be implemented using lean management techniques (Section 13.2.3 and 13.2.4) and the result could be an overall cost reduction.

Reducing the size of an instrument reduces its aperture. When an instrument's aperture is reduced, not just the quantity but also the quality of the data is reduced. Using "synthetic aperture" techniques where several instruments are combined may solve this; nevertheless, synthetic aperture methods are not always applicable and they entail demanding requirements for controlling the relative positioning of each individual instrument. For many applications it will not be possible to provide the same mission data quality with many small instruments instead of a single large one.

As we can see, there are many factors that influence the decision to deploy one or many satellites; the advantages and disadvantages of one or the other implementation are finely balanced, and must be carefully weighed for each particular mission. Recent history offers examples in favor of all three forms of implementation: single, dual, and multiple satellite missions:

- Most missions that can be implemented as a single satellite are implanted that way. All the missions described previously in Section 7.2 are implemented using a single satellite.
- RapidEye[14] consists of five satellites, each with one small camera providing a swath of 70 km and 6.5 m of resolution. Each satellite is simple enough to weigh just 150 kg, but combined they provide a very high data production rate and a powerful mission performance.
- Sentinel-3's (Section 8.3.3 and 9.6.1) initial designs considered both one and two satellite options. The final decision was for two satellites. This case will be studied in detail below.

Sentinel-3's objective is to provide operational information about Earth's land and ocean environment. Ocean observation is made difficult by the Sun's high reflectance off the sea, which negatively impacts image quality when glare from the Sun prevents instrument observation. Consumers demanded a mission able to provide full glare-free coverage in less than 3 days. For ocean observation, Sentinel-3 features an instrument consisting of several individual cameras. Each camera provides a field of view of approximately 12°, and they can be arranged in different configurations. The problem was how to arrange the individual cameras to fulfill the 3-day revisit requirement

[14] The RapidEye Spacecraft, Dr George Tyc, Dc Manfred Krischke et al., 4th IAA Symposium on Small Satellites for Earth Observation 2003.

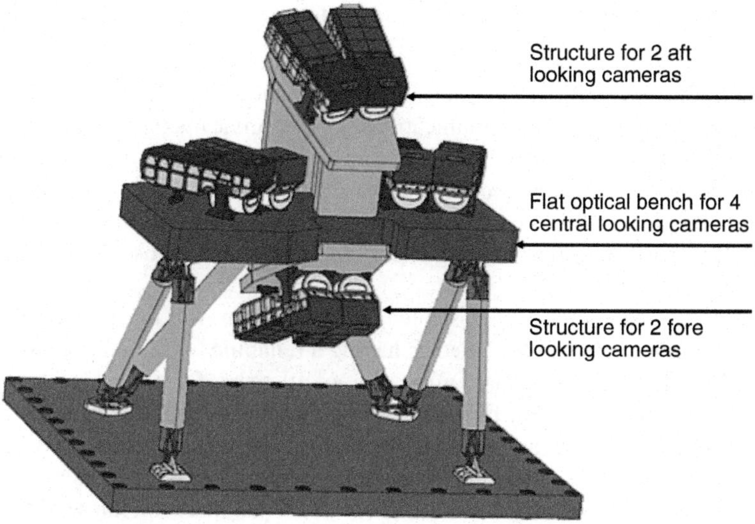

Figure 7.14. Sentinel-3 OLCI Cameras Are Assembled Fanned over the Optical Bench to Provide the Overall Swath Whatever the Sun Illuminating Conditions (Courtesy Thales Alenia). This Alternative Was Actually not Selected for the Instrument.

in the most efficient way. During the definition phase of this mission two alternatives were studied:

- In the first, six cameras were placed over a flat optical bench. In this case, two satellites were needed to fulfill the 3 days coverage requirement because Sun glare was polluting observations provided by some of the cameras.
- In the second alternative, two pairs of two cameras pointing 30° ahead and 30° behind the Nadir direction substituted for the two cameras at the center. In this case, one satellite was able to fulfill the coverage requirement because either the cameras pointing ahead or those pointed behind were free of Sun glint.

The use of two pairs of two cameras provided an effective and fast Earth covering single satellite mission. Nevertheless, this required the use of a much larger, heavier, and complex instrument, since the flat optical bench of the six-cameras solution was substituted by a cross-like three-dimensional optical bench carrying eight cameras. The change increased the instrument length from 0.85 to 1.45 m and also complicated the installation of the calibration devices and cooling system. On the other hand, the instrument allowed the mission to fulfill the ocean coverage performance in the 3 days required by the consumers. One satellite carrying eight cameras was cheaper than two satellites carrying six cameras each. Nevertheless, the final mission implementation favored the simple instrument solution, and two satellites were necessary to fulfill the 3 days requirement. The complex eight-camera arrangement was judged too risky and bulky. Furthermore, the simpler solution finally chosen allowed a staggered implementation of the mission by launching one satellite after another, thus ensuring a robust mission design (Figure 7.14).

7.3.2. Mission life versus number of satellites

As we mentioned in Section 4.5, operational missions need to last for as long as possible because operational missions fulfill permanent societal needs, for instance environment monitoring or communications. To determine operational life it is necessary to decide how the mission life is going to be shared between individual satellites. Overall mission life must be a clearly defined requirement. It must be determined if the requested mission life is compatible with a single satellite, though there are limits to what duration is realistic, and long operational missions are normally implemented as a series of satellites.

Figure 7.15 examines the possible implementation for the above-mentioned Sentinel-3. Each satellite is specified as having a reliability of 0.8 after a mission life of 7 years, and of being deorbited after a life of 10 years. Each individual satellite's reliability curve declines over time as its reliability deteriorates. This means launching replacement satellites (as indicated by the arrow on the horizontal axis) when the probability of a satellite continuing to operate correctly becomes too low. There are three curves for the availability of at least one, two, and three satellites. For 20 years of life (240 months), and six satellites launched during this period, the availability of at least two functional satellites being in orbit is always above 0.8, and is higher than 0.9 during most of the mission, which was the original requirement of the system.

In real life, well-built satellites frequently have lives well above the guarantied values provided by reliability calculations. In Section 5.5.6, this book noted that Meteosat 5, 6, and 7, which were designed for an operational life of 5 years, were still functional in 2005, having performed for 14, 12, and 8 years, respectively. Nevertheless, the availability calculation of Figure 7.15 provides a good overview of possibilities, and sheds light on alternative scenarios. The combination of cost constraints and the well-established long life enjoyed by satellites ends by advising the

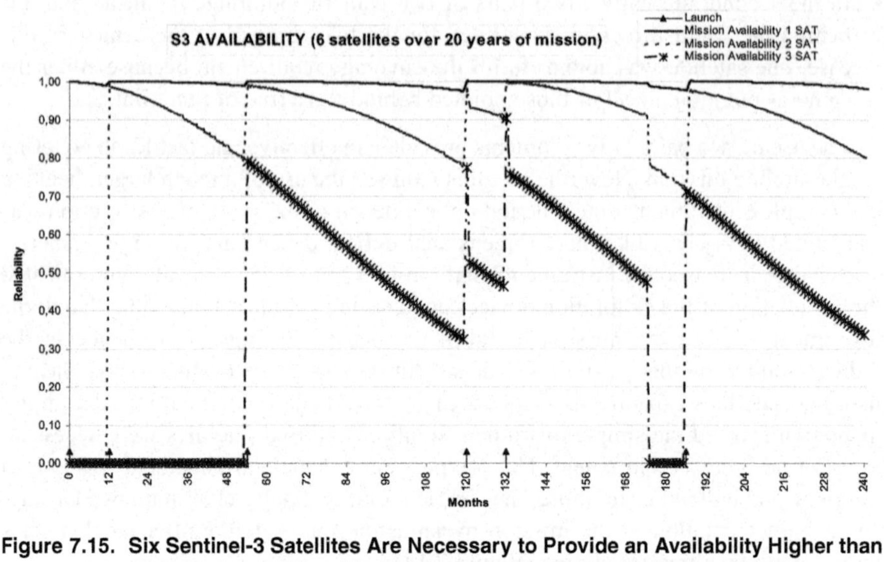

Figure 7.15. Six Sentinel-3 Satellites Are Necessary to Provide an Availability Higher than 80 % During the Specified 20 Years of Life (Courtesy Thales Alenia).

use of a smaller number of satellites than the number actually indicated by availability analysis. Meteosat Second Generation, whose first satellite was launched successfully in 2002, will employ three or four satellites to cover the mission's nominal operational life up to the year 2018, and just three Sentinel-3 satellites are foreseen in order to provide a system with an operational life of 15 years.

7.4. Systems of systems

The system of systems approach to analysis and design has been used in defense and information technology for some time, and it was recently introduced in space system design for two different purposes:

- The integration and interoperability of existing space assets with each other and with ground-based systems, so that together they can provide services that they are unable to provide separately. This is more oriented toward the provision of new services with existing space assets.
- To move from an asset-oriented system design to a service-oriented design, focusing on the requirements from the end-user. This is more oriented to the design and provision of new services with a mixture of new and existing elements.

The system of systems approach is especially relevant for dealing with the delivery of services related to areas such as civil and military security, disaster management, and environment monitoring. Systems of systems approaches offer:

- *Operational independence*: each constituting system can operate in isolation. When operated in isolation each constituting system is able to fulfill a specific set of requirements that were the original reasons for creating it.
- *Managerial independence*: each constituting system could be designed and implemented in isolation and could also be actually managed and operated in isolation.
- *Evolutionary development*: the system of systems, consisting of independent systems, does not appear fully formed. It is not implemented in a single project but made up of existing and new systems that can evolve, disappear, or reappear at any time.
- *Emergent behavior*: the system of systems is able to provide services that none of the independent systems in isolation is able to provide.
- Geographic distribution so that the components can interchange only information.

The methodology for the analysis and design of system of systems is based on the use of formal architectural frameworks (Section 2.2.2) and formal system description languages. The aim is to define the environment in a rigorous way that would allow identifying the problems, mainly those involving information interchange, that would be necessary for providing the required services. The system of systems is implemented using a combination of new and existing ground and space systems. A proper analysis would be able to provide not only a static view, but an evolving one as well, for indicating how the systems inside the system of systems will evolve, and what measures will be necessary to ensure a mid- to long-term delivery of the specified services.

The overall Earth observing capability presently in space is one example of a system of systems as officially recognized by the concept of GEOSS (Global Earth Observation System of Systems).[15] Understanding the Earth: its weather, climate, oceans, atmosphere, water, land, geodynamics, natural resources, ecosystems, and natural and human-induced hazards is crucial, and all the Earth observation satellites in orbit contribute to these tasks. Observations of the Earth constitute critical input for advancing this understanding. The ultimate vision is to realize a system of systems that allows decisions related to the environment to be well informed by coordinated, comprehensive, and sustained Earth observations. GEOSS is made up of all existing independent Earth observation satellites. The overall aim is for each of these missions' administrators to accept and implement interoperability arrangements centered on technical specifications for collecting, processing, storing, and disseminating shared data and products by employing agreed-upon standards and data sharing arrangements. This interoperability would be focused on interfaces in order to minimize any impact on the affected space missions.

The SSTL DMC constellation was made up at the beginning of 2010 of six satellites built and operated by Surrey Satellite (SSTL) and with the following resolutions and ownerships:

Satellite	Owned by	Imager resolution
Deimos-1	Deimos Imaging SA	22 m
UK-DMC2	SSTL	22 m
Beijing-1	BLMIT	32 m, 4 m PAN
UK-DMC	BNSC	32 m
NigeriaSat-1	NASRDA (Nigeria)	32 m
AISAT-1	CNTS (Algeria)	32 m

Each member of the consortium has an independent satellite imaging capability, which under the consortium agreement is made available to support common objectives. All members have signed a nonlegally binding agreement to make available up to 5 % of satellite imaging capacity. Such an original arrangement allows for quite an unusual operational approach to the analysis of data. Each satellite of the constellation provides images of Earth in three bands with a resolution of 32 m, and several boast panchromatic instruments with resolution up to 4 m. Acting together as a constellation; these satellites are able to observe any part of the Earth with a mean delay of 1 day. It is possible to reschedule these satellites in 24 h (depending on the resources available), and the time from imaging to downlink ranges from 10 min to 24 h depending on the location of the targets and the ground stations. The time from data reception to product release varies from 4 h (not orthorectified) to 48 h (orthorectified using reference control points in the ground to produce subpixel accuracy). All the satellites together have capabilities that they do not have separately and DMC can be also considered a system of systems.

[15] GEOSS 10 Years Implementation Plan, The Group in Earth Observation (GEO), ESA BR-240, ESA 2005.

Chapter 8

The Observables and Instruments Domain

Keywords Instruments Engineering, Space Instruments Types, Optical Instruments, Microwave instruments, Observational Requirements

This design domain was introduced in Section 7.1.1. This chapter starts with an analysis of the relationship between an instrument and its observables. It continues with a brief introduction to the types of payloads that are carried by satellites. It continues by examining five space missions:

- ESA's Earth observation satellite Aeolus, which determines the speed of the wind in the atmosphere using a Doppler wind laser as the active element of an optical instrument.
- The NASA-ESA astronomy satellite JWST, which uses an exceptionally large telescope to observe the relatively cold early universe. It carries a passive optical instrument.
- ESA's mission Sentinel-3, which carries – among other devices – a radar altimetry instrument, which measures sea levels, permitting derivation of ocean currents. The radar altimeter is an active microwave instrument.
- The ISRO-CNES tropical water-cycle mission Megha-Tropiques, which carries two passive microwave radiometers to study water vapor, cloud properties, and precipitation.
- ESA-NASA Sun magnetosphere observing mission Ulysses, which carries a set of magnetic and electric field sensors. These instruments are in situ measuring instruments and cannot carry out remote sensing.

These five missions encompass the five main categories of instruments used by satellites. They provide a good overview of the kinds of problems associated with the implementation of different instruments.

The chapter continues with the analysis of the observational needs as design drivers and what type of features instruments should have to produce good data. The next section of the chapter covers the end-to-end performances, providing examples of how complex it is to determine the "final" performance of the instruments and of the space mission carrying them.

The correct allocation of functions is one of the two fundamental mechanisms for achieving an optimal design; the other is the correct apportioning of budgets. The next four chapters, dedicated to analyzing the different design domains, will examine in detail the allocation of functions and budgets in the design process of each domain.

M.A. Aguirre, *Introduction to Space Systems: Design and Synthesis*, Space Technology Library 27,
DOI 10.1007/978-1-4614-3758-1_8,

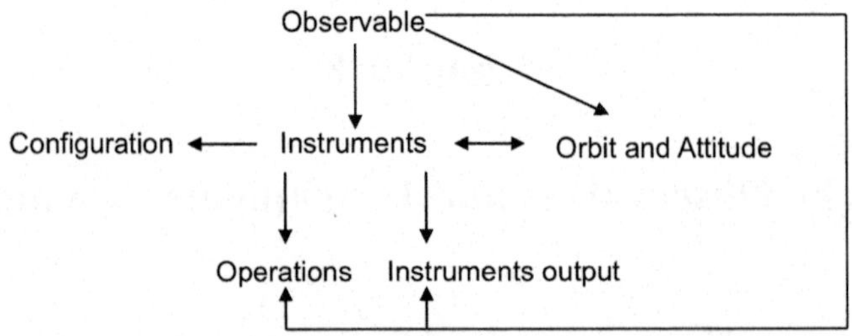

Figure 8.1. External Interrelations of the Observables and Instruments Domain.

This chapter will analyze this in two sections: one dedicated to the allocation of functions in an instrument and the other to the allocation of budgets.

Figure 7.1 of the previous chapter presented the requirements flow-down process. The figure's upper part depicted the highest-level decisions made early in the design process, most of which are related to observables. Figure 8.1 focuses in these decisions.

The purpose of space missions is to acquire information on the objects or processes under observation. In order to acquire that information, space missions use instruments; that is, the observables and the observing instruments represent the foundations of the mission and will shape the space system's design from beginning to end. All design decisions involve compromises and bidirectional interactions; nevertheless, instrument requirements – being so integral to the mission's fundamental aims – usually govern the design of all the other system components. The exception is when the observable coupled with instrument requirements unavoidably push the mission beyond the fundamental programmatic constraints of cost, risk, and scheduling. This kind of situation would entail either a redesign in conjunction with reduced mission objectives, or even cancelation of the mission as a whole.

8.1. Observables and instrument selection

In principle, a mission's consumer will be interested in receiving the data they require for studying the target object and/or processes of interest, and not in the observing technique employed. An oceanographer would be interested in receiving a certain quality and quantity of ocean circulation data, not in whether the data was gathered by radar altimetry, thermal infrared sensing, Doppler radar techniques, or by buoys measuring the ocean currents in situ. It could be argued that a system designer starting from consumer needs should offer a choice of alternative observing techniques and should attempt to come up with the most efficient technique for fulfilling those needs. Nevertheless, this is not usually the case and as pointed out in Section 7.1.1, the choice of observing technique is most often made by the mission's consumer prior to the start of the design process; hence, the choice is presented to the mission's designer as a design constraint.

In the interrelation between observables and observing techniques, two situations are possible:

- New observing techniques enabling breakthroughs regarding the quality or quantity of data gathered from observables. This is more common on advancement-of-science missions (see Section 4.5.3).
- Mature situations where more than one well-proven observing technique is able to provide data on the observable. This is more common in operational missions (see Section 4.5.4).

In the promotion of new space missions involving new observing techniques, scientists occupied with the observable and scientists favoring the new observing technique always act as a team. Before the architect begins considering how to implement any potential new mission design, both the observables and the observing-techniques scientists should have produced a viable preliminary observing concept. ESA's mission Aeolus studies winds by measuring the Doppler shift generated by wind-carried particles on a laser signal emitted by and reflected back to the satellite. Early support for the mission came from atmospheric and laser scientists. Atmospheric scientists promoted Aeolus because it featured a scientific breakthrough – the obtention of wind profiles in cloudless air – while laser scientists supported it for its potential value to the development of their research field. Thus, the space mission designer was faced with designing a mission using a currently active optical technique and was not starting from a totally clean slate.

In the case of operational missions, the observing technologies are normally mature and the mission designer has more freedom to select the instrument concept from within a set of possible observing techniques. Returning to the example provided in Section 7.1.1, atmospheric winds can be observed using active scatterometers (as in MetOp[1]), passive polarimetric microwave radiometers (as in Coriolis/WindSat[2]), or active Doppler LIDAR (as in Aeolus[3]). Ocean surface winds can also be obtained as a by-product from radar altimetry and high-altitude winds assessed by measuring cloud displacements from a temporal series of images. The consumer could provide their requirements regarding the observable without any reference to the observing technique, allowing the designer to study a number of options for the most cost-efficient observing technique and mission implementation which best meet the project's goals.

Nevertheless, consumers usually know the strengths and weaknesses of existing instruments and will always take that knowledge into account when expressing their needs. Operational agencies such as NOAA or Eumetsat normally federate consumers and produce the highest-level documentation concerning observables, as well as requirement documents based on existing types of observing techniques. The result is, once again, that the mission designer will not be starting from the observable but rather from one or several possible instrument concepts.

In the rare cases where the architect needs to perform instrument trades, he will take his cue from the mission's fundamental aims and will study all available instrument alternatives. Each instrument alternative will focus on observing a different

---

[1] http://www.esa.int/esaLP/LPmetop.html.

[2] http://directory.eoportal.org/pres_CoriolisWindSat.html.

[3] http://www.esa.int/esaLP/LPadmaeolus.html.

aspect of the events under observation, and will offer different advantages and disadvantages as regards mission performance and development costs and risks. These different instrument alternatives might also entail distinct basic architectures for the mission. Finally, the optimal alternative will be chosen based on its balance of advantages and disadvantages vis-à-vis the mission aims. This kind of comparison among widely differing options is a difficult and prolonged process. Properly performed, it involves a thorough assessment of needs, the apportioning of the highest-level requirements to the elements of the system, designs fulfilling the apportioned requirements, the establishing of alternative instrument concepts detailed enough to allow an estimation of end-to-end performances, and estimations of costs and risks for each alternative. These procedures are far from simple and cannot be frequently repeated.

8.2. Elements and components involved in the observables and instruments domain

Chapters 9, 10, 11, and 12 begin with an overview of the elements and components involved in the design domains to be discussed, even though the book's purpose, as stated in Chapter 1, is to "describe and analyze space systems not from the point of view of their elements but from the point of view of the interrelations between them." Chapter 1 also indicated "It is expected that the reader has already learned about space system technologies from the many excellent texts already available covering these subjects." In accordance with these statements, the aim of this and the following overviews is not to present in detail the science and technologies related to each domain, nor even to provide primers of them, but rather to recall the functions performed by the components of the domain and to serve as an introduction to an analysis of their interrelations.

As in the previous section, we will examine the components of this domain by classifying them according to instrument type, not by type of observable. Satellite payloads can be divided into six types:

- Remote sensing, passive optical instruments
- Remote sensing, active optical instruments
- Remote sensing, passive microwave instruments
- Remote sensing, active microwave instruments
- In situ sensing instruments
- Communications payloads

The first four categories follow the standard division of remote sensing space instruments engineering. From the architect's viewpoint, this is also a sensible division because each one of these classes has different implementation problems. For example, passive microwave instruments are not driven by solar illumination, whereas passive optical instruments are. These four types of instruments detect electromagnetic radiation coming from the observable being studied. In the case of passive instruments this radiation will be either radiation naturally emitted by the source or solar radiation reflected by it. Active instruments produce and emit their own radiation that is scattered by the event and detected back by the instruments. In situ measuring instruments are exceptionally diverse. They include magnetic, electric, and gravity field sensors, as well as devices such as mass spectrometers, able to detect the chemical composition around the satellite. Communications payloads are carried not only by communications satellites but also by all satellites that always need to communicate with ground.

8.2.1. Passive optical

Passive optical instruments detect the intensity of the optical radiation arriving from a specific direction of space; hence, they provide spatial (direction) and radiometric (intensity, frequency) information. Depending on the frequency they are registering, these instruments can detect gamma or X rays, ultraviolet, visible, or infrared radiation. The instrument is able to detect one, several, or many frequency bands. The instrument can detect radiation produced directly by the observable or, in the case of the Earth or other planets illuminated by the Sun, solar radiation reflected by the observable.

Instruments that detect electromagnetic radiation have been traditionally classified as imagers, spectrometers, and radiometers. Imagers gather mainly spatial information, spectrometers mainly spectral data, and radiometers radiation intensity data. Nevertheless, most modern instruments feature all three: good radiometric quality, good spectral resolution, and good spatial resolution. However, one of the three aspects always functions as the main design driver. Panchromatic imagers for high-resolution Earth observation, for example, are driven mainly by spatial resolution.

Passive optical instruments include the following components (Figure 8.2):

- Telescopes collecting energy arriving from the region of space being observed
- Components such as filters or spectrometers that disperse incoming radiation into frequency bands in accordance with observational requirements
- Focal planes where radiation is detected and transformed into a signal
- Pointing and scanning components for orienting telescopes toward areas of interest

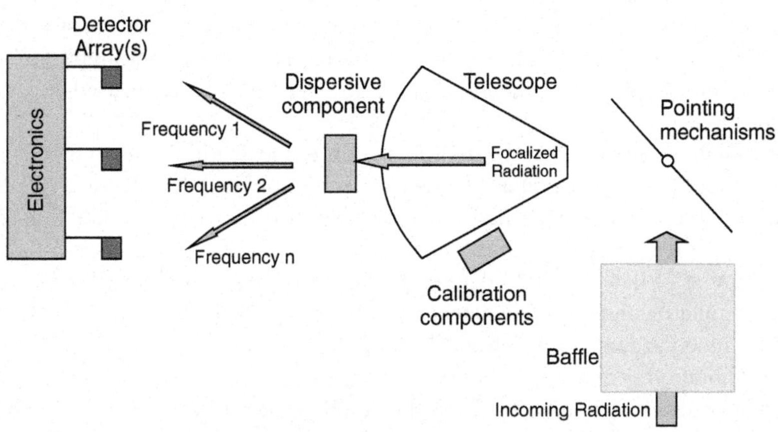

Figure 8.2. Passive Optical Components Functional Block Diagram.

- Components to calibrate the instrument's radiometric response
- Baffles that protect the entrance to the instrument from radiation not coming from the observation target

Passive optical instruments also include the necessary mechanical and thermal components as well as the electronics required for processing of detected signals and for the control of the instrument.

Focusing telescopes feature refractive lenses or reflective mirrors that concentrate the radiation reaching them from the observable. The number of groups of optical elements can also be used to classify telescopes: Schmidt telescopes use one set of optical elements groups, Cassegrain two, and three mirror anastigmatic (TMA) three. Instruments with a larger number of groups of optical elements are better quality but are also more complex. Nonfocusing telescopes using masks to determine the direction of incoming radiation are used for examining higher-energy X-rays or cosmic rays.

Pointing mirrors, normally controlled by pointing mechanisms, are required for pointing the instrument or its entrance optics toward the observable. The area of interest for a particular observation may be much larger than the telescope's field-of-view, in which case mechanisms for scanning the entire area can be employed. Sometimes the satellite as a whole is facing in the optimal direction; in that case pointing mechanisms will be unnecessary (Figure 8.3). At any rate, the telescope's scanning speed must be fast enough to completely cover the area of interest within the required time.

In some cases the protection of the telescope against unwanted light sources requires the use of large, complex, and often deployable structures called baffles that conspicuously drive the configuration of the satellite. Figure 8.6, provides a good example of a large baffle. Figures 7.5, 7.6, and 7.7 also depict baffles of different sizes. Nevertheless, Herschel (Figure 7.9) and JWST (Figure 7.10) do not use cylindrical baffles but large flat screens to avoid light from unwanted sources – mainly the Sun – entering or affecting the telescope. In many cases the external baffles are large and introduce complex accommodation design issues for the satellite.

Most passive optical instruments acquire information in more than one frequency band, so it is necessary to include optical components capable of separating the incoming radiation into as many bands as required, and each band must cover the required bandwidth. Filters are introduced to separate the incoming light into a few to a moderate number of frequency bands and spectrometers are used to provide many bands with fine spectral resolutions.

The radiation will be detected by a sensor array located in the instrument's focal plane and made up of individual detectors. Each individual detector will provide a signal proportional to the radiation intensity coming from a specific direction and in a specific frequency band. The instantaneous field of view of the instrument divided by the number of detectors will determine the angular resolution of the observation and the angular resolution of the detector multiplied by the distance to the observable will determine the spatial resolution.

The calibration component provides radiation of a well-known intensity and spectral character that can be used to verify the output of the detectors with that carefully controlled input.

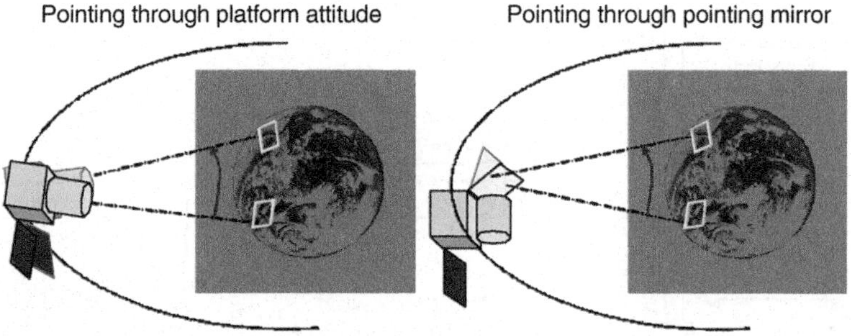

Figure 8.3. Instrument Pointing Alternatives: Satellite Versus Mechanisms (Courtesy of EADS Astrium).

The instrument also requires electronics to read and digitize the signal produced by the detectors, and electronics for instrument data processing and instrument control.

It is necessary to generate an "image cube" with three dimensions: along the image, across the image, and along its spectral content. Nevertheless, the detector array will consist of one of the following: a single detector, a one-dimensional line of detectors, or a two-dimensional matrix of detectors. To derive the three dimensions of the image cube from the zero, one, or two dimensions of the sensor array requires the use of supplementary tools to provide the missing dimensions. These tools are:

- The movement of the satellite along the orbit providing an along-track swath
- Scanning pointing mechanisms, providing an across-track swath
- Dispersive components and multiple detector arrays providing the spectral content

In "whiskbroom arrangements" one or a very small number of detectors are mechanically scanned to provide the across-track swath. One single detector with adequate temporal response should be able to provide an image if the scanning speed is fast enough to allow the across-track swath to be scanned in the time it takes the satellite to travel one pixel of along-track spatial resolution. Using more than one pixel will proportionally reduce the required scanning speed. In all cases the velocity of the satellite along the orbit provides the along-track swath. The frequency dimension will be provided by the use of dispersive components sending the different bands to different sensor arrays, one per frequency band of interest (Figure 8.4).

In "push broom" arrangements, the focal plane detector array will have a long dimension possessing as many detectors as needed to cover the entire swath required. As with the whiskbroom arrangement, the along-track scanning will be provided, by the movement of the satellite along its orbit. This approach requires complex detector arrays, though scanning mechanisms are not needed. As before, the frequency dimension will be provided by the use of dispersive components sending the different bands to distinct long linear arrays, one per frequency band of interest. A matrix array consisting of a number of rows and columns can employ each row for a different frequency. Dispersive components or filters will split the radiation so that the necessary radiation frequencies illuminate the different rows. Matrix arrays can also be used to improve the radiometric performance. As the satellite moves, the different rows can

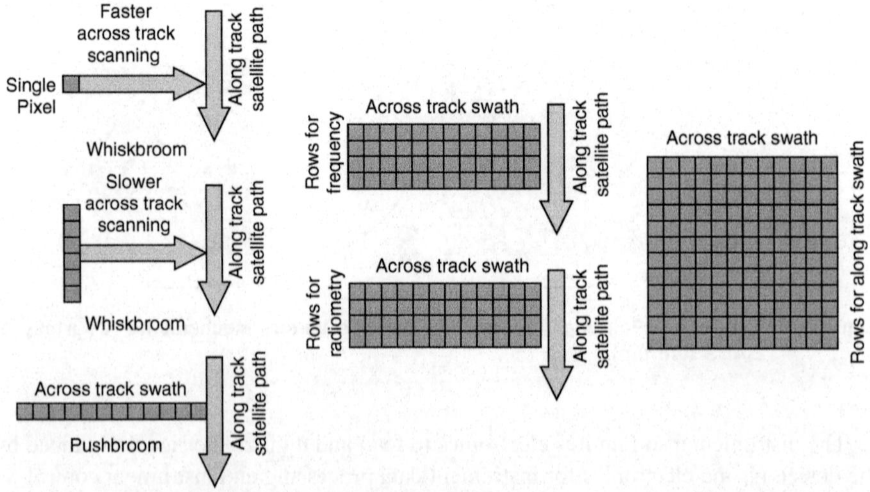

Figure 8.4. Single and Multielement Whiskbroom, Single Band Pushbroom, and Step-and-Stare Sensing Schemes Used for Frequency and One-Dimensional or Two-Dimensional Spatial Coverage.

be dedicated to observing the same ground spatial resolution cell at the same frequency. This will increase the actual observing time and allow a better radiometric performance.

In step-and-stare arrangements, a square two-dimensional detector array will provide spatial information in the along- and across-track directions. Once the image is acquired, the satellite, or appropriate mechanisms, will move the line of sight to the next instantaneous field of view until the total area of interest is covered.

The instrument design will be shaped by the following set of requirements:

- The spatial resolution will determine the telescope's aperture diameter and focal length. This drives the size of the instrument and of the whole satellite.
- The spectral resolution will define the nature of the dispersive component: filters or spectrometer.
- The number of spectral bands will drive the dispersive components and the detector array(s) concepts.
- High radiometric accuracy requires high radiation collection capabilities: large telescope apertures and long integration times by the detectors. Higher radiometric accuracy will also require the use of sophisticated calibration components.
- The time specified for providing synoptic views of the area of interest will determine the instrument swath, which in turn will influence the number of pixels of the detector array. Larger swaths entail large aperture optics, high data throughput, and complex data processing electronics.

The telescope concept, the focal plane concept, and the scanning concept must also be determined. For its part, the instrument concept and size will govern the satellite's interface requirements, namely: instrument mass and volume, visibility and pointing needs, power and energy consumption, instrument data throughput, and thermal interfaces. In some cases the instrument will also make special demands on the satellite's design, such as whether the whole satellite will be agile or if the payload requires special thermal cooling. The instrument's visibility needs will also be a strong influence on the selection of the satellite's configuration, orbit, and attitude.

The design of optical instruments – optical concept, spectrometer concept, focal plane and detector array concepts, aperture size, and focal length – is a highly specialized task and all design decisions related to them are left in the hands of specialists. These specialists will also provide architects with the designs for the required mechanisms, structure and thermal control concepts, and information on the risks associated with the designs and ways to reduce them in the preliminary design stages. The development of large detector arrays is an especially long and complex process.

Most instrument-related design decisions encompass fundamental system-level implications, and the architect must provide relevant feedback to specialists, especially in those cases where design decisions might have detrimental impacts on other components of the system. Accordingly, architects must always be prepared to arrive at reasonable compromises regarding basic mission parameters such as instrument aperture, orbit attitude, satellite configuration, and overall system data throughput.

The instrument aperture and the resulting overall instrument size will require repeated interaction between specialists and the architect. In some cases it will be impossible to avoid the use of optical components of dimensions incompatible with their accommodation in the launcher. This will require the use of deployable optics or even synthetic aperture optics. The deployable optics of JWST (Figures 7.10 and 8.7) is a good example of this approach.

Decisions regarding the instrument pointing concept and the allocation of pointing functions – pointing by the instrument or pointing by the satellite – will also demand frequent interaction between the different specialists and the architect.

Instruments requiring high swath, high resolution, and the recovery of a large number of frequency bands will produce large amounts of data. This data throughput will drive the downlink data flow architecture. The architect will need to interact with specialists and the consumers to arrive at compromises concerning instrument performance, instrument design complexity, in-orbit or on-ground data processing, and data downlink architecture.

Instrument thermal control and thermal stability needs and their corresponding design solutions also have system-level implications that require interaction between the architect and specialists.

8.2.2. Active optical

Active optical instruments direct optical wavelength radiation toward a target under study and detect the subsequent backscattered radiation. Lasers are used as the source of the emitted radiation from active optical instruments because laser light is coherent and possesses a well-defined wavelength.

There are different types of Laser-based active optical instruments:

- Laser Doppler emits laser pulses that are backscattered by dust or molecules in the atmosphere. The return signal is analyzed to detect the Doppler shift produced by the velocity of the dust and the molecules, permitting the derivation of atmospheric wind data.

- Backscatter Lidars send laser pulses and measure the intensity of the backscattered signals. This allows the derivation of cloud properties or dust density profiles.
- Laser rangefinders measure the delay of the backscatter signal to determine the distance from the instrument to the target. This distance can be used to detect parameters like treetop height or ice thickness.
- Lasers can also be used to detect chemical composition of the atmosphere by detecting fine absorption or emission lines associated with specific chemical species.

Both active and passive optical instruments feature the same components, and the previous section's account is applicable, except that active instruments also require a laser source and laser emitting optics. Active optical instrument can operate with one or several wavelengths. If several wavelengths are used different laser sources must be provided, as well as the necessary dispersive components to select the frequency bands of interest, along with as many detector arrays as there are frequency bands of interest.

The fundamental design decisions affecting active optical instruments are governed by the need to receive a sufficient quantity of backscattered radiation to detect the parameters under observation with the necessary accuracy. This can be accomplished by increasing either the laser emitter's power or the collecting capability (aperture size) of the receiving telescope. High-power lasers require lots of electrical power and will produce large amounts of heat that needs to be dissipated from the laser source. To produce high-powered, highly reliable and long-lasting lasers is not easy, and extended active optical missions require high levels of redundancy in the number of laser sources in order to offset their eventual failure in orbit. Increasing the receiving telescope's aperture will reduce the laser source's necessary power, but large telescopes are heavy and difficult to accommodate, driving up the size of the entire satellite and increasing mission costs.

It must be decided if the instrument will have a constant or a variable line of sight. With a single line of sight, the instrument can only provide data along a line defined by the satellite's flight direction and attitude. If the line of sight changes rapidly enough, the instrument will be able to provide a swath. This increases the mission's yield but complicates the concept, because each cell covered by the swath requires its own laser pulse and because the laser pulse's line of sight and the pointing of the receiving telescope needs to change quickly enough to allow coverage of the entire swath. These are burdensome requirements that greatly complicate the design.

Active optical instruments require excellent dimensional stability in both the instrument and the satellite as a whole. This demands the use of high-performance thermal control and first-rate high dimensional stability structures. Dimensional stability is made more complicated by the heat generated by the laser source. The dissipation of this heat requires the use of special cooling devices.

The provision, or not, of swath is a basic trade-off that requires the balancing of the consumers' desire for large quantities of good quality data and the very serious difficulties of providing an active optical instrument able to provide it. The final decision will depend on how the architect balances the tensions between consumer needs and the sponsor's willingness to pay the consequent costs.

As in the case of passive optical instruments, the actual design remains in the hands of specialists. The architect shall take the lead on issues related to the trade-off between laser power and telescope aperture size. Large instrument size has a major impact on a satellite's mass, configuration, and cost. High laser power leads to a short mission life and increased mission risk. The architect must balance the costs associated with large instrument apertures and the risk associated with high laser energies. Orbit altitude is another area where the architect shall take the lead. Flying at low altitudes reduces distance to the target, and permits the use of lower laser power and reduced telescope size. Nevertheless, low-altitude orbits require larger amounts of fuel for orbit maintenance and more frequent sources of pointing disturbance from thruster firings. The architect must balance instrument complexity with the consequences for the orbit control and attitude control system.

8.2.3. Passive microwave

Passive microwave instruments measure the intensity of the natural microwave radiation produced by the events under observation. Passive microwave instruments are independent of the Sun's illumination, and clouds are fairly transparent to them. This independence from light and atmospheric conditions is a major advantage of this kind of instrument. Like passive optical instruments, passive microwave instruments can also detect spatial information, spectral information, radiometric information, or any combination of them, but in general the principle design determinant is the radiometric quality, that is, the accurate measurement of radiation intensity.

Passive microwave instruments, as with passive optical instruments, are divided into categories according to their frequency range. Passive microwave radiometers from 1 GHz to 150 GHz are common. Observation at 1.4 GHz provides soil moisture and sea-surface temperatures; wind over the ocean can be detected at 10 and 18 GHz; rain can be detected in the ranges around 18.7 GHz; water vapor at 23.8 GHz; liquid water in clouds at 36.5 GHz; and ice in clouds at 137 GHz. Passive microwave instruments can also provide temperature sounding and atmospheric humidity. Temperature can be sounded in the 60–63 GHz range and humidity in the 50–57 GHz range.

Microwave instruments can be either single pixel, providing information solely over the point in the antenna's line of sight, or imagers providing images over a given swath. For the latter the antenna must scan to provide the desired coverage.

A passive microwave radiometer's basic detection chain includes: the collecting antenna, a low-noise amplifier to amplify the arriving signal, a filter to select the frequency of interest, and the detector. Heterodyne receivers also include an oscillator that produces an intermediate frequency that is combined with the incoming radiation and later amplified and detected. Passive microwave radiometers also require calibration devices, control and processing electronics, structure, baffling, thermal control, and scanning mechanisms. Instruments designed to study the lower frequencies of the microwave range are proportionally larger and use different focusing and detecting technologies than instruments studying the higher end of the range. Microwave radiometers require an adequate calibration of the instrument. This is normally accomplished by providing a calibrated source of microwave radiation and a view of cold deep space as a second reference. Taking into account microwave radiation's longer wavelengths, the spatial resolution will be limited by the aperture of the instrument. The spatial resolution of microwave instruments is much coarser than the resolution of optical instruments. Earth observation optical instruments have resolutions from the kilometer range to below 1 m, while Earth observation microwave instruments have resolutions in the 1 km to tenths of a kilometer range. Microwave instruments use scanning mechanisms with rotating reflective surfaces. If high spatial resolution is required the rotating surface must be large and the perturbations generated by the rotation of the large reflecting surface must be minimized by using counter-rotating or balancing mechanisms.

As in the case of optical instruments, the most important decision is to define the aperture of the receiving antennas, which will determine the instrument's size and will be driven by the required spatial resolution. The number of bands to be observed will depend on the nature of the events being studied and on the consumer's requirements. The receiving horns, filters, and detectors for the different frequencies must be accommodated, and the detection approach, direct or heterodyne, must be decided upon.

Due to the proportionally low resolution of microwave instruments, their data throughput is moderate, hence the resulting requirements on data processing and data downlink are also modest and the resulting data handling designs are easy to implement. On the other hand, microwave instruments can be large, making their mechanical and thermal design complex.

As in all the previous cases, the design of the instrument is the responsibility of instrument specialists, but scanning microwave instruments also call for the early involvement of structural, mechanical, and thermal control specialists to solve the mechanical and thermal problems that will arise as a consequence of its large mass and size, by its rotating elements and by its thermal and dimensional stability demands. The architect should mediate the interaction among all these specialists. Furthermore, any unbalance in the rotating components will produce forces and torques that will alter the pointing of the satellite; this will inevitably involve interactions among the instrument and attitude control specialists that will also need mediating by the architect.

A supplementary problem requiring the architect's attention is the presence of potential artificial sources of radiation that can pollute the radiation emitted by the source being observed. Ground-based radars and communication antennas can emit radiation that the instrument receives mixed with the signal produced by the source. International regulations establish which bands are protected for Earth observation instruments detection. The consumer, the architect, and the instrument designer must be aware of this regulatory environment. The satellite itself will also be carrying communication antennas or radars that are sources of microwave radiation. These antennas must be designed to avoid contamination of the signal being measured by the microwave radiometer. This means the use of emitting frequencies that are well separated from the sensitive receiving frequencies. To further reduce the problem, the emitting and receiving antennas must be physically separated. This may have a severe impact on the satellite's configuration and also requires the architect's attention.

8.2.4. Active microwave

As in the case of active optical instruments, active microwave instruments emit microwave radiation to illuminate the area being observed. The advantage of active microwave instruments is that they produce their own illumination of the target,

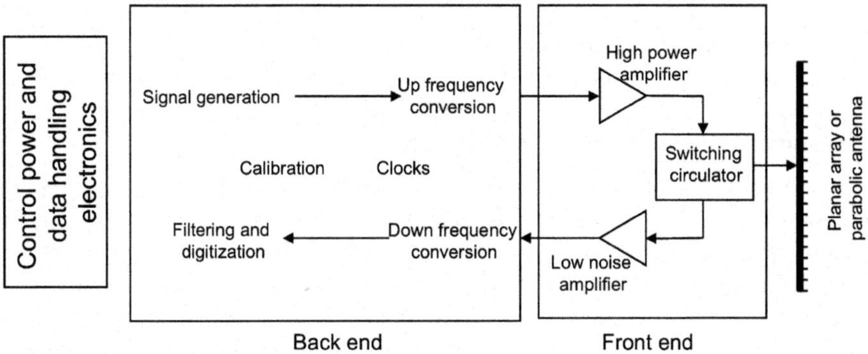

Figure 8.5. SAR and RA Functional Block Diagram.

allowing observation independent of lighting and climatic conditions, including the absence of sunlight or the presence of clouds. Active microwave instruments have been used widely for Earth observation.

Imaging radars send radar pulses that are reflected back and transformed into images of the observed target. Taking into account the wavelength of the radiation and the distance between a satellite and its target, the normal instrument aperture needed for any reasonable spatial resolution would be impossibly large. To solve this problem satellites use Synthetic Aperture Radars (SARs), which provide spatial resolutions comparable to the resolutions provided by optical instruments. SARs, which are side-looking radars, use sophisticated post-processing of the received signal to emulate a receiving antenna with an aperture equal to the distance traveled by the satellite while it illuminates each observed point. SARs are powerful imaging instruments routinely providing up to meter-level resolutions with antennas in the order of up to tens of square meters at orbit altitudes of 600–800 km and with swaths in the order of tenths to hundreds of kilometers.

Scatterometers provide information on the texture of the surface being observed by studying the characteristics of the backscatter radiation. Scatterometers are used to determine wind speeds over the water, since the texture of the air-water interface changes with wind speed. They can also be used to detect ice or snowmelt.

Radar altimeters are Nadir-looking radars that emit pulses and determine the distance to targets by measuring the time it takes for the signal to return to the instrument. Radar altimeters are used to study the water levels of oceans, rivers, and lakes.

Cloud and rain radars are able to determine the density of clouds and the intensity of precipitation by analyzing radar signals backscattered by clouds particles, raindrops, or snowflakes. Radar altimeters and cloud and rain radars are normally swathless "sounders," though they can be made to provide moderate swaths, of the order of many tenths of kilometers.

Active microwave instruments can be divided (Figure 8.5) into the following components, which are going to be described in the following paragraph:

- Front end, which includes the antenna, either parabolic or planar, and also other elements used for emission and reception of the radar signal
- Back end, which includes all the necessary electronic for the operation of the radar
- Instrument control and data processing electronics

The front end of the instrument produces and receives the high-frequency microwave radiation. The antenna receives and emits the microwave radiation. Synthetic Aperture Radar (SAR) antennas are either parabolic dishes or flat arrays and are normally large, measuring tens of square meters; cloud, rain, and snow radars are of intermediate size, between a few and ten square meters; and radar altimeters are small, with sizes around 1 m^2. Instruments working at longer wavelengths are proportionally larger but emit smaller amounts of radiating energy, thus simplifying the electrical power demand and thermal dissipation configurations. Large antennae require complex stowage and deployment mechanisms to solve the contradiction between the necessary small volume accommodation inside the launcher and the required large area configuration in orbit. The mechanical and thermal design of large deployable antennae is a complex specialized engineering domain that shall be taken into consideration from the earliest stages of the system definition. Flat array antennae have a large number of similar radiating elements organized and distributed so that they form the desired antenna beam. Parabolic antennae concentrate the emitted or received signal in the focus of the parabola. Active antennae have located on their surface a very large number of small electronic emitting and receiving modules. The behavior of these modules can be reconfigured in a multiplicity of ways allowing the antenna to provide different radar modes with different swaths and resolutions. Active antennae have a variety of radio frequency power architectures, such as fully or partially distributed radio frequency power, or powered by row or columns of active elements. Nonactive antennas have a centralized front end, and the signal is distributed to the parabolic antennas or to the flat array radiating elements.

The back end is the part of the radar that operates at lower frequencies and that performs the detection and digitization of the backscatter radiation. The back end also generates the signal to be emitted, which will be modulated and amplified in the front end. The back-end functions normally include: the up- and down-frequency conversion, the calibration hardware, and time references. The instrument will also include all the necessary electronics for power production and delivery, data handling, and instrument control.

Low-frequency L- or P-band SAR's possess very large antennas and require complex deployment mechanisms and sophisticated structural design. Higher-frequency radars such as C-X, or Ka-band call for smaller but still fairly large – usually deployable – antennas. Higher-frequency SAR's generally provide demanding spatial resolutions and feature comparatively smaller antennas as well as proportionally high levels of energy emission, which require complex power generation and heat dissipation systems.

Active microwave instruments can emit and receive in single or multiple polarizations in one or multiple bands and are able to provide interferometric products. SARs employing active antennas are flexible and able to operate in different modes; some with more swath and lower resolutions and others with enhanced resolution but smaller swath. A scheme must be designed for the selection of the different modes to be used. Active antennae are also able to point in different directions and provide fast scan within large ranges. Nonactive antennae can only change pointing by slowly changing the attitude of the satellite.

The choice of frequency or frequencies and the polarizations to be used depends directly on the application: ocean and ice observation favors C-band, vegetation L-, or P-bands, near optical quality imaging favors the X-band, and land classification requires multiple frequencies and polarizations. Therefore, the choice of frequencies and polarizations to be used is not normally a design decision, but a requirement that the designer must fulfill.

Active microwave instrument design is a complex process involving the definition of many strongly interrelated parameters including: the radar pulse repetition rate; the pulse bandwidth; the allocation of time windows for pulse transmission and reception; pulse energy, antenna width and length; the different combinations of swath and resolution; and the maximum and minimum incidence angle of the microwave beam over the ground. The specialists must perform multiple iterations in order to produce a design capable of fulfilling mission requirements. These iterations are integral to the instrument design process.

The choice of the antenna concept and shape (parabolic or flat array, and if flat array, active versus nonactive) are critical design decisions. Active microwave instruments use electronic scanning to provide swath widths and their placement on the ground, so scanning mechanisms are unnecessary. The large size of many SAR antennas, however, requires complex deployment mechanisms demanding careful design. The data throughput, power consumption, and heat dissipation of SARs are usually very high, and become even more severe for high-frequency high-resolution SARs. High-resolution SARs cannot operate continuously because it would be impossible to downlink the enormous quantity of data produced, and because the antenna and back-end electronics would overheat. The amount of time allocated for operation must be established by balancing the consumer's desire for large data production, the regions of interest to be imaged, and remain compatible with the various platform subsystem design limitations.

Because SAR antennas are large and can involve complex deployable structures that strongly influence the stowed and deployed configuration of the satellite, a SAR will dominate the payload-platform interfaces, power and energy generation and management, satellite operating and safe mode attitudes, and system data circulation, even if there are other instruments on the same spacecraft. SARs also require the formulation of operational strategies for providing optimal coverage or revisits. All these aspects have implications for the overall mission design and demand intense interaction among instrument specialists, mission architects, and all other mission design specialists.

Radar altimeters and the other active, nonimaging, microwave instruments are generally smaller, produce smaller amounts of data, use moderate amounts of power, and feature simple interfaces with their host satellite. In these cases, specialists will provide the architect with the overall instrument concept and a well-defined set of interfaces.

In orbit, SARs emit electromagnetic radiation that could interfere with other equipment on the satellite and with ground activities. In order to avoid such interference, the International Telecommunication Union allocates frequency ranges for use by space-borne SARs. Space-borne X-band SARs, for example, have been allotted the 9,500–9,800 MHz frequency window. Instrument design must take these frequency regulations into account. The ITU has also established maximum limits on the power density that radars can use to illuminate the ground.

8.2.5. In situ instruments

Antennas or coils that react to the characteristics of the electrical field around the satellite can be used as electric field sensors. Special sensors can detect magnetic fields. Mass spectrometers can be used to detect chemical composition. There are also detectors that identify the nature and energy of ions or high-energy particles such as cosmic rays. Accelerometers detect atmospheric density by measuring the acceleration of masses located on the satellite. Gravity fields can be detected by monitoring the differential movements between reference proof masses. The proof masses can be located on the same satellite, or on different satellites located hundreds of thousands of kilometers away.

Gravity field sensors are complex instruments and they completely shape the host satellite's configuration. Gravity sensors require an extremely quiet satellite that minimizes parasitical disturbance accelerations and relative displacements within the satellite. This requires the use of special attitude and orbital control tools, as well as the provision of sophisticated thermal and structural concepts to minimize dimensional distortions. In situ instruments for detecting the speed and density of the solar wind or space dust; chemical composition of particles; density of matter or different types of ions; and the frequency of energetic particles are normally small and their interfaces with the satellite can be defined in terms of clear accommodation requirements. The manufacturing of these instruments is highly specialized and requires very specific skills. The scientist intending to utilize the instrument's output often directly fabricates them. Instruments capable of detecting electric and magnetic fields are generally small but their satellite interfaces are complex. These instruments are strong drivers of the satellite's configuration and overall design. Electric and magnetic field sensors need to be located far from the perturbing fields generated by the host satellite's electronics. This demands the use of large deployable booms for situating the instruments at their tip. Large deployable are flexible and they will couple with the attitude control system of the satellite making high-accuracy pointing complex. Magnetic field sensors also force strict requirements on the satellite's overall design: materials able to generate magnetic fields should be either avoided altogether or strictly controlled, and the electronics, as well as the satellite's electricity generation, storage, and distribution networks should be designed to minimize their magnetic signatures. Electric and magnetic field sensors also require that the satellite's external surface be kept at a constant potential in order to avoid the generation of magnetic or electric fields on this surface. Instruments for detecting very-low-frequency variations of the electrical field may be very large, up to tens of meters or more.

As said before, many kinds of in situ instruments are extremely specialized and the scientists themselves frequently manufacture them. If that is the case, the scientist must also provide formal interface documentation for the accommodation of the instrument. As a consequence, the function, not just of the architect but of the entire satellite design team, will be reduced to accommodating the payload in line with the formal interface requirements, accounting for all possible design constraints imposed by the other payload and platform requirements. The big exception is gravity sensing, where the boundary between the instrument and the rest of the satellite becomes blurred: the instrument is the satellite and the satellite is the instrument. Every feature of the gravimeter will have an impact on the rest of the satellite, so instrument and satellite must be designed together.

8.2.6. Communication payloads

The designation *communication satellite* is used for satellites that act mainly as relays of information previously sent to them; *communication payloads* are the payloads carried by communication satellites. There are many kinds of telecommunication missions: they can be global or local; they can feature unidirectional or bidirectional information flows; the communication lines they provide can be rigidly established or flexible, according to specific needs; they can fly in different orbits. The design of space communication missions is highly specialized and beyond the scope of this book. Furthermore, communication satellites are often based on existing designs providing little leeway for the work of architects. Nevertheless, communication missions can serve as useful examples of satellite mission design, especially when discussing issues such as orbits, constellation design, satellite configuration, and satellite launcher interfaces. This book will use communication satellites as example of design several times in the following chapters. To the degree that all satellites receive and send information, they are all communication missions. The communication components carried by all types of satellites will be discussed in Chapters 11 and 12, which cover overall satellite data flow.

8.3. Instruments examples

8.3.1. Aeolus

8.3.1.1. Mission aim

ESA's ADM-Aeolus[4] mission will be the first mission to provide global wind profile measurements throughout the atmosphere. These profiles will find application in weather forecasting and lead to a deeper understanding of weather and climate processes. The provision of detailed wind profiles will benefit the accuracy of numerical modeling of weather over the whole Earth, especially in the tropical regions, and will help in prediction of severe storms. Aeolus carries a Doppler wind Lidar, which emits pulses of light and registers the signals scattered back to the instrument by the clouds and dust in the atmosphere. The Doppler shift of the backscattered signal provides the relative velocity between satellite and air in the direction of the Lidar's line of sight.

[4] SDM-Aeolus ESA's Wind Mission, ESA BR 236, ESA February 2005.

Figure 8.6. Aeolus Configuration (ESA's Achievements, BR-250, ESA 2005).

The mission will be capable of providing global wind profiles up to altitudes of 30 km. For winds within the lower 2 km of the atmosphere, the wind recovery accuracy will be better than 1 m/s. For winds between 2 km and 16 km altitude, the wind speed will be better than 2 m/s. The instrument will provide wind averages along a strip of 50 km and a total of 120 wind profiles per hour. The mission will be launched in late 2010. The mission will last 3 years with a possible 1-year extension. It will be launched in a Vega or Rockot launcher. It will fly at 408 km altitude, in a Sun-synchronous orbit with an inclination of 97.06° and with a local time of the ascending node of 6:00 (Section 9.4.2). The satellite's total mass will be 1,350 kg (650 kg of platform, 250 kg of propellant, and 450 kg of instruments). The central body of the satellite is a cube 4.6 m high, 1.9 m long, and 2 m wide. The solar array is able to provide 2.2 kW while the average power demand of the satellite is 1,400 W. The satellite will point toward Earth but will be tilted by 35° from the vertical in the across-track direction (Figure 8.6).

8.3.1.2. Instruments

The Atmospheric Laser Doppler Instrument (ALADIN) consists of a powerful laser system for emitting short laser pulses into the atmosphere; a large telescope for collecting the backscattered light signal; and a very sensitive receiver for analyzing the Doppler shift of the signals received from different layers of the atmosphere. The laser pulses are generated by a diode-pumped Nd:YAG laser at 355 nm, with an energy of 150 mJ. The pulses are not continuous; rather 7 s of pulses are emitted at 100 Hz every 28 s. This provides a wind velocity profile averaged over 50 km every 200 km. The Cassegrain telescope is dominated by its main mirror, which measures 1.5 m in diameter. ALADIN aims 35° in the across-track direction to avoid Doppler shift from its own velocity. Two optical analyzers measure the Doppler shift of the molecular (Rayleigh) scattering, and the scattering from aerosols and water droplets

(Mie), respectively. Highly sensitive photodetectors then transform the light signals into electric signals, which are amplified and stored on board until transmitted to the ground for processing.

8.3.2. JWST

8.3.2.1. Mission aim

The James Webb Space Telescope (JWST) [5] was introduced in Section 7.2.1 and will be discussed further in Section 10.3.3 as an example of satellite configuration. It was conceived as a follow-up to the Hubble Space Telescope, and will allow scientists to see the most distant first generations of stars. These early stars and galaxies are so remote that the extremely faint signals and light they produce is strongly red-shifted to the infrared part of the electromagnetic spectrum. To provide proper observation of the faint targets in infrared wavelengths, a very large diameter telescope must be used, and it must be cooled to very low temperatures. The mission will observe the first light after the Big Bang, the assembly of young galaxies, and the birth of stars and planetary systems. Astronomers will use JWST as an operational facility, which will receive observation requests, prepare observation schedules, perform the observations, and distribute the data. In accordance with operational mission aims, it has been designed for a long life of some 10 years. The satellite, with a mass of 6,800 kg will be launched in 2013 by an Ariane 5 ECA rocket into a heliocentric halo orbit around the Lagragian L2 point (Section 9.4.6). In this orbit the distance to the Earth will vary from 1.2 to 1.8 million km with a periodicity of 6 months. L2 is located at a fixed point relative to the Earth and Sun, and from this location the Earth and Sun are in the same direction. The attitude of JWST is inertially stabilized. In order to satisfy astronomers' observation requests, the satellite's telescope must be able to change pointing direction. However, if it is to avoid overheating by the Earth or Sun, it must adhere to strict pointing limitations (Figure 7.12). The impact of these limitations is severe and the telescope is protected by a large deployable Sun-shield that dominates the deployed configuration of the satellite. The deployable Sun-shield is 19 by 13 m.

8.3.2.2. Instruments

The 6.5 m aperture telescope has a hinged, multifaceted, adjustable primary mirror, and a deployable secondary mirror with its support structure .The primary mirror is made up of 10 hexagonal facets that are stowed at launch in three panels: the central panel, which is fixed, and the two side panels which are hinged and will be deployed in orbit. Another mechanism is used to deploy the telescope's secondary mirror that is also stowed during launch (Figure 8.7). The primary mirror of JWST has six times the area of the Hubble mirror but its density per unit of area is nine times lower than Hubble's. The telescope's primary mirror with its 10 mirror elements can be adjusted in orbit to improve the resulting alignment after launching as well as the instrument's cool-down. The mirror distortions will be reduced to less than 0.15 μm. Due to the sophisticated deployable Sun-shield and to the constant geometry with respect to the

[5] James Webb Space Telescope Project Overview, P. Sabelhaus et al., IEEE-AC paper #1464 2006.

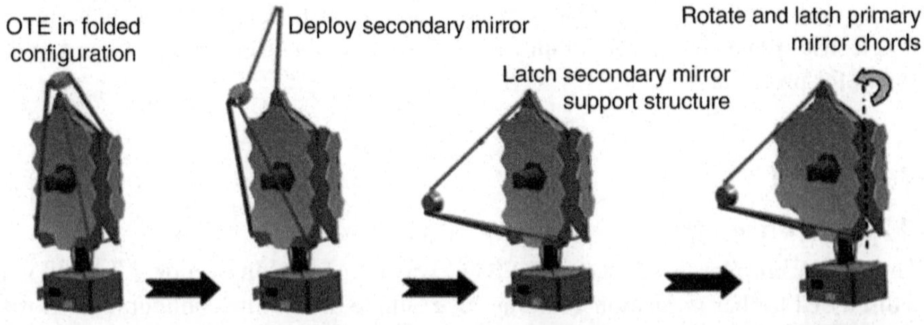

Figure 8.7. JWST Deployment Sequence. Interface with the Launcher Is at the Bottom of the Structure. (James Webb Space Telescope (JWST) Observatory Architecture and Performance, John Nel, AIAA-2004-5986, American Institute of Aeronautics and Astronautics 2004).

Sun, the entire instrument can be passively cooled to temperatures as low as 40 K. This provides sufficient sensitivity for adequately detecting the faint infrared radiation produced by the nascent stars and galaxies. For optimal performance, a mechanical cooler further reduces the temperature of the mid-infrared detector assembly to 7 K. The telescope observes in the 0.6–28 μm range and light from the collecting telescope is subsequently sent to four different instruments: a near-infrared multiobject spectrometer; a near-infrared camera; a near-infrared tunable filters camera; and a mid-infrared camera and spectrometer. The four instruments will be packed in a special module that will also provide structure, thermal control, control electronics, and data handling.

8.3.3. Sentinel-3

8.3.3.1. Mission aim

The Sentinel 3 mission[6] is part of the Global Monitoring for Environment and Security (GMES[7]) initiative whose overall objective is to support Europe's goals regarding sustainable development. The mission is dedicated to the operational (Section 4.5.4) provision of medium resolution global ocean and land observation, with consistent quality and a high level of availability. The implementation phase started in 2007, and launch will take place in 2012. The mission consists of a constellation of two satellites separated by 180° and flying Sun-synchronous orbits with 98.6° inclination. The orbits provide full Earth coverage and uniform conditions of illumination throughout the year. The orbit altitude is 800 km, and the orbital period is such that the satellite performs exactly 385 orbits in 27 days. This allows an optimal coverage pattern for land and ocean applications. The satellite will be qualified for a minimum operational mission life of 7 years but is being designed to function

[6] GMES Sentinel-3. A Long Term Monitoring of Ocean and Land to Support Sustainable Development, Y. Baillion, M. Aguirre, IAC -07-B.1.2.04, IAF 2007.

[7] http://www.gmes.info/.

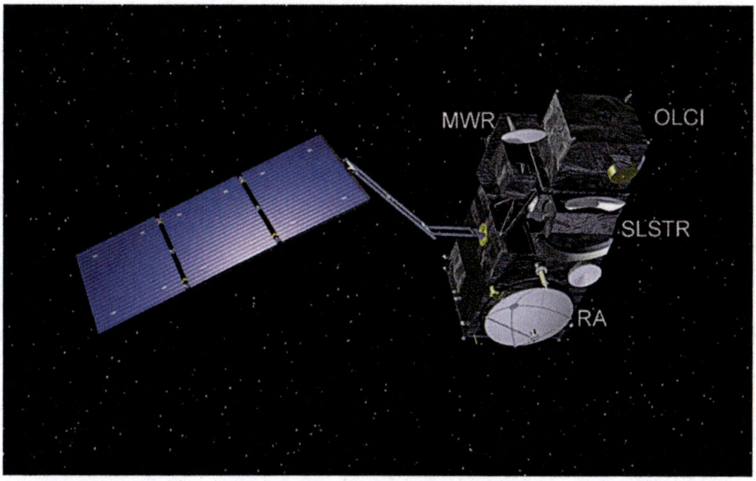

Figure 8.8. Sentinel-3 In-Orbit Configuration (Courtesy Thales Alenia Space).

for up to 12 years. The satellite's total mass is 1,400 kg, 100 kg of which is fuel. The satellite's average power consumption is 1,100 W. It has a lower cubic part carrying the platform subsystems and an upper part where most of the instruments are located (Figure 8.8).

8.3.3.2. Instruments

Besides its large instrument complement, the satellite is still compact and compatible with the volume and mass limitations of small launchers (Figure 8.8). To provide the different types of observables, Sentinel-3 carries a large swath/medium spatial resolution optical visible and near-infrared instrument (OLCI), an optical visible and thermal imager (SLSTR), and a set of four instruments constituting altogether a radar altimeter system. Sentinel-3 will focus on four different types of observables:

- Ocean color products used to derive biophysical data provided by the OLCI instrument
- Ocean topography data provided by the radar altimeter instrument, supported by the microwave radiometer MWR, the laser retro-reflector LRR, and a GPS receiver
- Surface (land and sea) temperature provided by the SLSTR instrument
- Land color products used to derive biophysical information about land and vegetation, produced by the synergetic use of SLSTR and OLCI instruments

OLCI is an optical medium resolution (300 m) wide swath (1,250 km) spectrometer providing 21 bands in a frequency range going from 0.4 to 1.02 μm. The instrument is made up of five pushbroom small cameras, each one providing a part of the overall swath. The SLSTR instrument is an optical mechanical scanning instrument providing nine bands from 0.55 to 12 μm. The resolution provided goes from 500 m to 1 km. It has two scanning mechanisms providing two views: one looking down and with a swath of 1,800 km and the second one pointing ahead and providing a swath of 750 km. This double view allows improved atmospheric correction and accurate recovery of the surface temperature. The radar altimeter (RA) sends pulses to the surface of the water; the time needed for their return provides, after corrections, the distance from the satellite to the water. To correct for delays produced by ions located in the radar signal's path, the RA emits pulses in two bands: Ku and C. The difference in delay between the two signals permits the derivation of the ion density in the path traveled by the radar pulses. To adequately calculate this distance it is necessary to correct errors in the measurements; to that end, the satellite carries a passive microwave radiometer (MWR) to correct for water vapor induced delays, and a navigation sensor to establish an absolute reference position for the satellite, and to derive an absolute actual ocean level. The comparison of this actual level with the reference level provided by the equipotential gravity of the Earth's surface – the Geoid – provides a record of ocean currents. The satellite also carries a cluster of mirrors (LRR) to reflect back the light sent by ground-based lasers, providing a supplementary tool for orbit determination.

8.3.4. Megha-Tropiques

8.3.4.1. Mission aim

Megha-Tropiques[8] is a satellite whose aim is to improve our knowledge of the water cycle and atmospheric convection dynamics in the tropical atmosphere.

The project is a cooperative effort between the French and the Indian space agencies (CNES and ISRO, respectively). The mission provides simultaneous measurements of water vapor, clouds, condensed water, precipitation, and evaporation. Convection in the tropical region is a rapid and very energetic process that involves the transport of water – whether as vapor, liquid, or solid – and that includes strong exchanges of energy between the different layers of the atmosphere. The mission will study the characteristics and life cycle of tropical convection weather systems, and will provide products such as: cloud condensed water content; cloud ice content; precipitation rates; latent heat release into the atmosphere; integrated atmospheric water vapor contents; radiative fluxes in the upper atmosphere; and sea-surface wind measurements. Megha-Tropiques is planned to be launched in late 2009 by an Indian PSLV launcher into an 870 km altitude orbit with a low inclination of 20°, so as to be permanently flying over the tropics. The satellite is small with a total mass of 900 kg.

8.3.4.2. Instruments

The satellite carries three instruments: the larger size MADRAS and the smaller SAPHIR and SCARAB.

The conical scanning microwave imager MADRAS is aimed at studying precipitation and cloud properties. The instrument's swath is 1,700 km. It features a parabolic antenna with a diameter of 0.65 m, rotating at 25 rpm. MADRAS measures the rising

[8] CNES Website http://132.149.11.177/MEGHAT/index.htm.

Figure 8.9. Megha-Tropiques Satellite (http://132.149.11.177/lcMEGHAT/megha_tropiquesp.gif).

radiation produced by water and clouds in five frequencies; from 10.7 to 157 GHz. All but one of the frequencies detects the horizontal and vertically polarized radiation. The spatial resolution of the three lowest frequencies is 40 km; the spatial resolution of the highest is 6 km. The instrument contains a calibration mirror used for registering energy from the cold sky, focusing it onto the receiving horns. With the exception of the cold space calibration mirror, the instrument rotates with respect to the satellite (Figure 8.21), and carries protective structures to improve the thermal stability of the components in the path of the measured radiation. The lateral protective surface features a large opening to allow visibility of deep space. The instrument is relatively large and dominates the configuration of the satellite.

SAPHIR is a multichannelled passive microwave humidity sounder that provides humidity profiles by measuring the brightness temperature in six different channels situated close to the 183.31 GHz water vapor absorption line. The swath of SAPHIR is identical to that of MADRAS, but employs an across-track scanning pattern.

SCARAB is a cross-track passive multispectral scanning radiometer capable of sensing upper atmosphere infrared radiation emitted by the Earth. SCARAB measures the energy emitted in four channels: the first provides the reflected solar irradiance from 0.2 to 4 μm; the second, Earth emitted thermal radiation from 10.5 to 12.5 μm; a third channel registers the total irradiation from 0.2 μm to 200 μm. The last is a visible light channel, which provides a geographical reference. Due to the scanning geometry, points directly under the satellite's orbital track possess better spatial resolution (35 km) than other points. At the edge of the swath the resolution diminishes to 90 km (Figure 8.9).

8.3.5. Ulysses

8.3.5.1. Mission aim

Ulysses was launched in 1990 and it has been orbiting around the Sun since that date. It has helped the scientists to understand the Sun and its heliosphere, and the Sun's influence on the Earth. It is the first satellite doing in situ measurements of the inner heliosphere from the solar equator to the solar poles. The orbit of Ulysses is heliocentric with 1.35 AU of perigee and 5.4 AU of apogee. It is inclined with respect to the ecliptic plane by 79.1°. The orbital period is 6.2 years. To allow the observation of the solar poles, the satellite required a large amount of energy to allow it to depart from the ecliptic plane near to which all the planet's orbital planes are located.

8.3.5.2. Instruments

Ulysses is equipped with nine scientific instruments – some of them can be seen in Figure 8.10 – able to detect and measure solar wind ions (SWOOPS) and electrons (HI-SCALE), magnetic fields (VHM and FGM), energetic particles (EPAC), radio and plasma waves (UARP), dust, gas, and X and gamma rays (GRB). All these instruments perform in situ measurements, and are small, with masses in the range 2–15 kg. Large monopole (7.5 m) and dipole (72.5m) antennas detected lower-frequency electric and plasma waves. The magnetic field is measured by two magnetometers, a quantum detector (VHM), and a set of coils (FGM). Both magnetometers are located on the tip of a long deployable boom in order to reduce the magnetic noise generated by the satellite platform electronics.

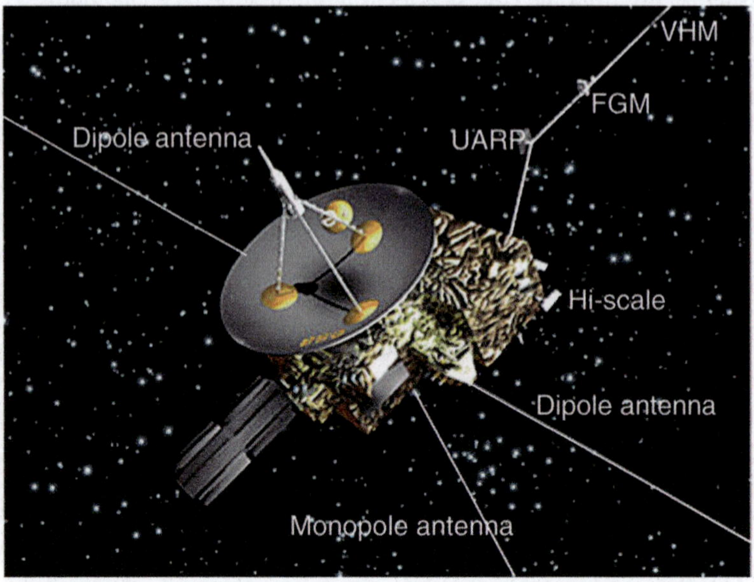

Figure 8.10. Ulysses Satellite (Image from ESA http://www.esa.int/).

8.4. Observational needs as design drivers

This chapter will analyze the whole spectral range of signals; from very low frequency, to microwave, optical to high-energy radiation.

8.4.1. Observation frequency and atmosphere

With very few exceptions – gravity sensing is an example – observables are detected by measuring electromagnetic effects. Electromagnetic effects can be differentiated by their frequency.

Instruments observing different frequencies have different designs, require different features, and belong to different engineering disciplines. Figure 8.11 provides an overview of the observation frequencies of the five missions described in Section 8.3 and of the five astronomical observatory missions introduced in Section 7.2.1. These two sets of missions provide a good selection of observations over the whole electromagnetic spectrum. Ulysses radio and plasma wave experiments observe very low-frequency variations of the surrounding electric field with a 72.5 m long wire antenna. Madras in Megha-Tropiques observes in the microwave range and has a large rotating antenna requiring careful balance and dimensional stability to avoid perturbation in the host satellite platform. All the infrared instruments (ISO, JWST Herschel, and Sentinel-3's SLSTR instrument) observe in the infrared and require cooling to very low temperatures. JWST also requires a very large deployable structure to protect the instrument from the heat of the Sun and the Earth. Hubble, which observes in the visible and infrared ranges, was dominated by the necessity for in-orbit servicing by astronauts, and the consequent obligatory compatibility with the Space Shuttle's orbit. Aeolus, which observes in the visible range, is shaped by the need for a telescope with a large aperture. IUE and XMM-Newton both observe X and gamma rays

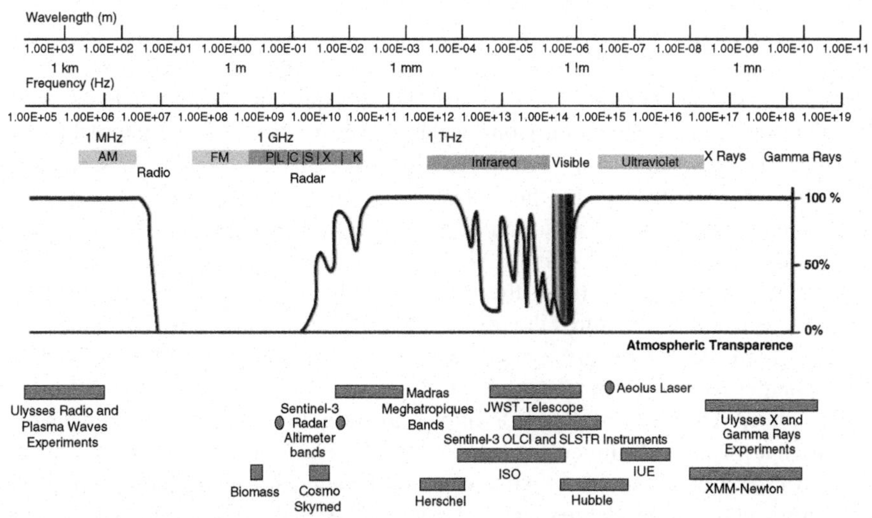

Figure 8.11. Radiation Spectrum and Instruments Observing at Different Frequencies.

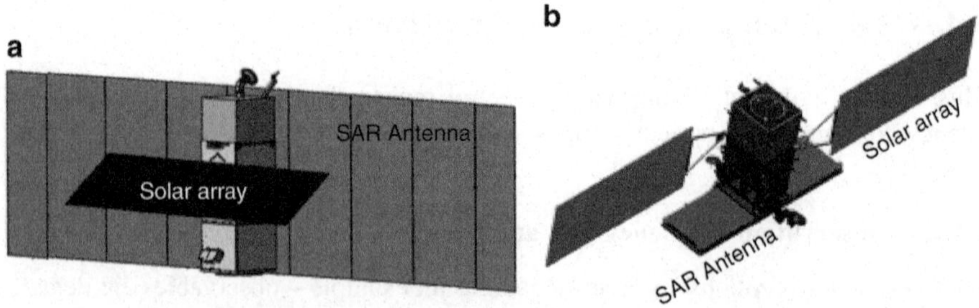

Figure 8.12. P-Band (Biomass) (a) and X-Band (Cosmo-SkyMed) (b) SAR Satellites (Cosmo-Skymed Figure Courtesy ASI. Biomass Figure Courtesy Thales Alenia).

and are driven by the need for very high dimensional accuracy, a characteristic that becomes more important as the observed frequency increases – the grazing incidence mirrors of XMM require dimensional accuracies better than 0.2 nm.

Earth's atmosphere is transparent to radiation in some regions of the electromagnetic spectrum and opaque in others. In order to observe Earth from space, it is necessary to work at frequencies in which the atmosphere is reasonably transparent. Figure 8.11 shows the ranges available for doing so. Frequency ranges in which the transparency alters rapidly can be used to study the vertical structure of the atmosphere: as the transparency decreases the radiation reaching the instrument will originate from higher altitudes, and as the transparency increases the radiation will originate at lower altitudes; this permits the study of the variations along the observation path of parameters like atmospheric humidity and temperature. For astronomical observations space instruments represent the only option in frequency ranges where the atmosphere is opaque. This is the case with the Herschel, ISO, and XMM missions. Satellite-based astronomy also avoids atmospheric distortions at all frequencies. This is why there are space-based astronomy missions that observe over frequencies even where the atmosphere is transparent; Hubble is a prime example of this.

To analyze the impact of observation frequencies on space mission design, it is instructive to compare similar instruments working at different frequencies. Biomass and Cosmo-Skymed (Figure 8.12) are two SARs requiring large antennas. Cosmo-Skymed is an Italian X-band mission for the high-resolution observation of Earth. It can obtain images with a spatial resolution of approximately 1 m. Biomass is a proposed P-band ESA mission for the global monitoring of woodlands and would provide images with a spatial resolution of 50 m. The wavelength of P-band is around 69 cm; the wavelength of X-band SAR is 3.1 cm. The dimensions governing both designs are scaled by the ratio of these wavelengths. A particular concept proposed for Biomass at P-band requires the use of a 17.9 × 4.5 m antenna and its solar array produces 1 kW, while Cosmo-SkyMed needs a much smaller 5.7 × 1.4 m antenna, but needs a larger solar array producing 3.8 kW.

The lower frequency of Biomass allows a lower level of antenna dimensional stability but the antenna dimensions are very large. The higher frequency of Cosmo-Skymed uses a much smaller SAR antenna but requires a much larger amount of emitted power requiring a larger-sized solar array. The resulting satellite configurations and configuration-related problems are totally different for each mission.

8.4.2. Data quality

Before discussing the different parameters measuring data quality it is necessary to differentiate between accuracy, precision, and resolution.

Accuracy can be defined as the difference between the exact parameter under observation and its recovered value. Precision can be defined as the difference between repeated acquisitions of the same observation and resolution can be defined as the smallest variation of the observation that the instrument is able to detect. Some applications, for instance long-term-sea-level change monitoring will require the determination of the ocean level with high "absolute" accuracy. Nevertheless, the recovery of other observables – for example, the detection of anomalous ocean dynamic topography (Section 8.5) associated with ocean circulation – requires detecting relative changes, and they will need good resolution but do not need good accuracy. Missions requiring high accuracies will be difficult and demanding to implement because it will be necessary to implement tools ensuring the constant and absolutely correct behavior of the instruments during the whole duration of the mission. This will require complex instruments and complex calibration of the instruments. Missions asking for high precision will require keeping the instrument's behavior constant during long periods, but it will be less demanding in terms of absolute calibration. Missions requiring only good resolution are the least demanding and will tolerate slow drifts in the response of the instruments to the observables.

It will be always compulsory to differentiate when the consumers demand absolute accuracy and when it will be enough to provide them with good precision or resolution. The architect is in the right position to understand the real needs of the consumers and the limitations of the supplier's design. From this central perspective the architect shall be responsible for correctly specifying the design so that underdesign and overdesign are both avoided.

Data quality is determined by a mission's aims. There are the following classes of quality parameters:

- Radiometric accuracy
- Spectral quality
- Spatial resolution
- MTF
- Temporal resolution: coverage and revisit

The following paragraphs describe these classes.

8.4.2.1. Radiometric accuracy

Radiometric accuracy is an instrument's capacity for providing correct readings of the radiation intensity emitted by the observable. To increase radiometric accuracy the ratio between the signal to be measured and the noise generated during the measurement process must be increased.

Increasing the aperture of the instrument can boost the signal strength; however, as we have seen, large apertures give rise to its own problems, namely, increased size and cost. The use of multiple rows of detectors to observe the same targets increases the integration time as well as the radiometric quality. Nevertheless, the use of this technique requires that the satellite's attitude be precisely monitored during the integration period. This generates demanding attitude control and recovery requirements;

attitude control aspects will be examined in Chapter 9. To reduce noise, the detector's quality must be improved, and the temperature of both the instrument and the detector must be reduced. High-quality detectors are expensive and can be risky to develop. Detectors operating at low temperature require complex thermal design and satellite configurations. This will be discussed in detail in Chapter 10, but can be seen clearly in the configurations of JWST (Figure 7.10).

To further improve radiometric quality the relationship between the radiation intensity reaching the instrument and the read-out of the detectors must be accurately determined. Calibration devices that generate a well-known radiative input to the instrument determine this relationship. High-accuracy calibration devices are complex and may require special lines of sight to the Sun or outer space. This complicates the instrument and satellite configuration.

8.4.2.2. Spectral quality

Spectral quality refers to the accuracy, width, and number of spectral bands that the instrument is able to recover: the better the spectral quality the better the capability of recovering useful information on the observable.

Spectral quality requires the use of spectrometers or filters internal to the instrument. Spectrometers disperse incoming radiation, which is then directed toward an array of detectors with different rows for different frequencies, allowing the sampling of the dispersed radiation. Filters are used to divide the incoming radiation and direct it to different focal planes: one focal plane per filter. Filters are small, but they require one focal plane per band. Spectrometers can provide a large number of small-bandwidth bands over a single matrix of detectors, but they are bulky and complex. In both cases, instruments observing in many bands will be large. Instruments observing a large number of bands will also produce a large amount of data. Chapter 11 will analyze the system-level implications of handling a very large data throughput.

8.4.2.3. Spatial resolution

Spatial resolution is the minimum distance between two adjacent features or the minimum size of a feature that can be detected by a remote sensing system. Each spatial resolution ground pixel is produced by the read-out of a single detector during the selected integration time. In the case of pushbroom focal planes, to achieve a high spatial resolution with a large swath, a long array with a large number of individual detectors must be employed. In the case of whiskbroom focal planes (see Figure 8.4), high scanning speeds are crucial. This entails complex scanning mechanisms capable of moving at the necessary speed throughout the duration of the mission, implying a very high reliability mechanism. In both cases the instruments become more complex. The risk and required development time of the mission will increase in those cases where the detector array would involve dedicated development.

Spatial resolution is also directly linked to the aperture diameter and instrument size. The diffraction limit defines the minimum aperture for an instrument able to provide a given resolution at a given frequency. In microwave instruments the diameter of the aperture limits the resolution. The Madras microwave instrument provides a spatial resolution of 40 km at 10.7 GHz and of 6 km at 157 GHz. Instruments working in the optical bands are not normally diffraction limited: Ikonos provides images with 0.8 m resolution in the visible range from low Earth orbit with an aperture diameter of 0.5 m.

8.4.2.4. MTF

The modulation-transfer function (MTF) expresses the capability of an optical system to provide good contrast between progressively finer maximum contrast stripes. Resolution and contrast are in conflict because higher resolutions provide unavoidably lower contrast. The contrast would be a perfect 1 at zero resolution but it decreases as the width of the stripes – hence the requested spatial resolution – becomes finer. The quality of an optical system is indicated by its MTF value at the finest resolution of the system. MTF goes from 1 (perfect contrast) to 0 (no contrast). Values of MTF of the order of 0.3 at the maximum spatial resolution are considered reasonable.

8.4.2.5. Temporal resolution: coverage and revisit

To be able to understand the events under observation the mission should be designed to track their dynamics. Many events may be considered stable and only need to be observed once during the duration of the mission. In other cases the events are variable and need to be observed often enough to characterize how they change over time. The weather changes quickly and meteorology requires refreshed global views of the atmosphere every couple of hours, while global ocean circulation changes slowly and oceanographers requires refreshed global ocean views every couple of days.

The time elapsed between two consecutive observations of the same area will define the temporal resolution of the observations provided by the mission.

Some missions require the systematic delivery of synoptic views covering large areas of interest. In extreme cases the area of interest can extend to the whole Earth or celestial sphere. In these cases, temporal resolution must provide repetitive whole coverage of the designated areas. To deliver rapid coverage of large areas, the mission must provide a wide swath. If a single satellite is unable to provide the required coverage in the required time, several satellites operating simultaneously will be needed. This is the case with Sentinel-3 (Section 8.3.3), which requires two satellites to fulfill its coverage requirements. Design for optimal coverage is an important topic that will be discussed in detail in Section 9.9.3 of the next chapter when discussing orbit and attitude. The coverage requirement directly shapes the instrument swath and mission data throughput.

In other cases, the consumer might only be interested in observing specific and localized, swiftly evolving events such as floods, forest fires, or energetic astronomic events. In these situations the instrument can feature a small swath within a wider possible pointing range and the instrument must be as long as possible pointed in the appropriate direction. In such cases good temporal resolution will require frequent revisit of the specific area of interest, and the satellite must have an optimal orbit and pointing range, though a large swath is not necessary. Design for optimal revisits is also discussed in Section 9.9.3. Revisit missions require smaller swaths and produce smaller amounts of data, but their operation is very complex and interactive as required by the need to command the satellite to point in the varying required directions. This entails intense dialog between the satellite and mission operations segment on the ground. The implications of this dialog will be discussed in Section 8.4.5 below.

Optical instruments observing the Earth will be able to operate only without clouds and with good Sun illumination. In these cases it will be necessary to differentiate between the theoretical revisit or coverage and the actual one that takes into account the clouds and the illumination conditions.

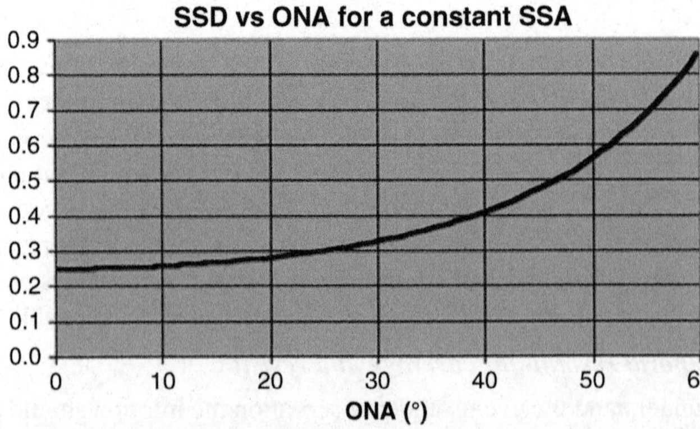

Figure 8.13. Relationship Between Spatial Sampling Distance (SSD) at Nadir and at Different Off-Nadir Angles (ONA) for an Orbit at 800 km Altitude.

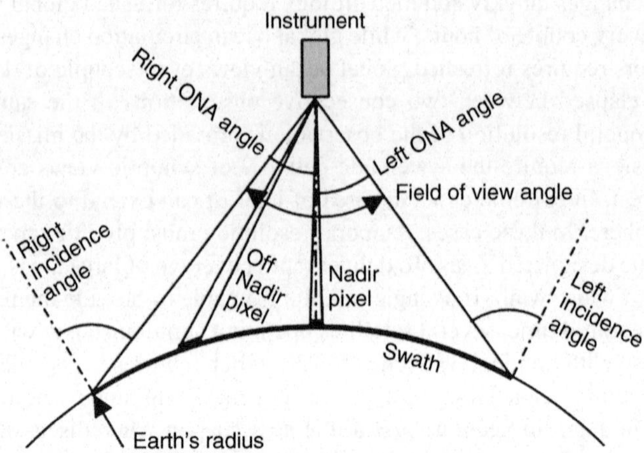

Figure 8.14. Being the Earth a Sphere, a Large Swath Instrument Will Have Progressively More Distorted Images as the Observing Area Moves Away From the Nadir Sub-Satellite Direction.

8.4.3. Image distortion

Instruments observing the Earth and having a large swath or a wide pointing range will inevitably scrutinize areas in which the geometry of the observation is distorted. Satellites covering horizon to horizon can observe the entire Earth in a single day, but this rapid revisiting will be paid for with degraded radiometric quality and diminished resolution at the edge of the swath. Nonetheless, in situations where the time resolution of the mission is of utmost importance, the quality degradation associated with the image distortion can be tolerated. This situation occurs in operational meteorological observations from low Earth orbit flying at an altitude of about 800 km. At this altitude, the ground pixel size – the spatial sampling distance – will be degraded by a factor of two with off-Nadir angles (Figures 8.13 and 8.14) of the order of 47°. At this height, the Earth's horizon will be perceived at an angle of 62.7°; at the horizon the distortion of the pixel will be infinite.

Figure 8.15. A Geostationary Satellite Observing With an Instrument Able to Provide a field of View of 300 km by 300 km Over the Equator Will Produce Progressively More Distorted Images When Observing Nearer the North Pole. The Lattice of Dark Lines Depicts the Progressive Distortion of the Area Seen by the Instrument When Observing at High Latitudes. The Distortion Will Produce Also a Progressive Degradation of Space Resolution and Image Quality.

Observations near the edge of the swath are degraded not only as regards spatial resolution but also as regards radiometric quality. This is because the radiation being monitored reaches the instrument after having traveled through a much longer distance within the atmosphere than those originating near the center of the swath. So signals reach the instrument contaminated by large amounts of solar radiation scattered by the atmosphere. Illumination from the Sun will also vary from one edge to the other, and from these to the center of any large instrument swath. This introduces a supplementary element of quality degradation to any image. A satellite observing the Earth from the geostationary position (Section 9.4.4) will provide poorer images at higher latitudes than nearer the equator. The geostationary orbit is located over the equator and at an altitude of 35,786 km. At this distance the satellite will point to the edge of the Earth with an off-Nadir angle of 8.7°, the Earth latitude coinciding with the horizon is seen by the satellite at 81.3°. As the pixel approaches nearer to the horizon, it becomes more and more distorted (Figure 8.15).

The maximum acceptable distortion will depend on the application. Figure 8.16 demonstrate that incidence angles as high as 60° provide "useful" images where shapes can be still detected. Nevertheless in applications requiring a very high radiometric quality, for example, to detect different types of vegetation, the usable incidence angle will be smaller. The experts will determine the maximum value acceptable for each application.

8.4.4. Data quantity

Instruments carried by a satellite produce an output data stream. The amount of data produced depends on the number of pixels covered per unit of time, as well as on the

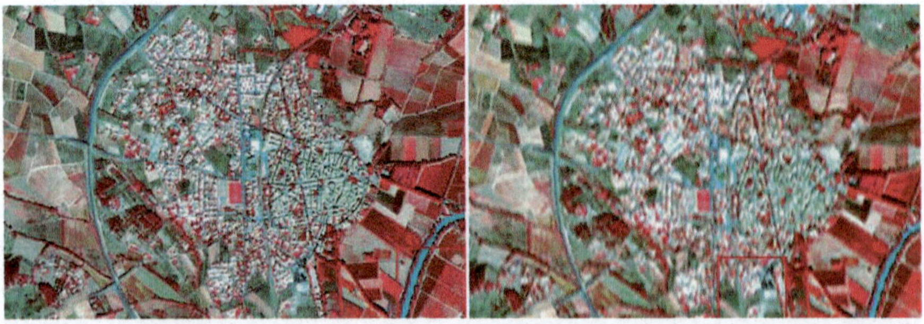

Figure 8.16. High Definition Observation of the Earth With the Same Instrument at the Same Distance and With 0° (*Left*) and 60° (*Right*) Line of Sight Incidence Angles. The image of the Right is Clearly Blurred With Respect to the Image of the Left. This Illustrates the Degradation of Quality Produced by the Large 60° Incidence Angle.

amount of data recovered from each pixel. Hence, instrument data throughput will increase with:

- Is the information acquired over a point or over larger or smaller areas? This defines the area of the observation, and the operational swath of the instrument.
- Are the phenomena under study subject to rapid spatial changes, so that changes occurring over small distances must be resolved in order to understand them? This defines the spatial resolution of the observation.
- Is the phenomena characterized by a highly variable intensity, so that highly accurate measurements of both very small and very large quantities are necessary? This defines the dynamic range of the observation.
- Do the phenomena under study emit radiation at many different frequencies, and is it necessary to characterize the phenomena to study this frequency behavior in detail? This defines the spectral ranges and spectral resolution of the observations.

The instrument's swath depends on coverage requirements, as discussed in the previous section. Observing global, dynamic targets demands rapid coverage and implies the generation of large amounts of data. The necessary spatial resolution will also be determined by the nature of the observable. Some observables can be adequately characterized by observing them with just one frequency, while other events must be observed in more than one band. A single-band "black and white" image is capable of resolving man-made objects such as roads, buildings, or aircraft, but many bands are necessary to derive complex products like the amount of chlorophyll in coastal waters, or to differentiate between young and mature woodlands. Higher radiometric resolutions increase data throughput because enhanced resolution requires more bits per pixel. High spatial and radiometric resolution, high swath, and the recovery of many bands are desirable features in any instrument; but reducing the data throughput directly impacts these desirable features. The consumer's desire for the highest possible data throughput can easily be interpreted as a request for higher than possible data throughput. High-resolution imagers are notoriously heavy data generators, and a modern high-resolution SAR can produce up to 1 Gb/s. The transmission to Earth of such a quantity of data is difficult and the downlink capacity of the satellite can turn into a bottleneck for the whole mission. The only solution would be to limit the instrument duty cycle, that is, the percentage of time that the instrument actually functions. Data downlink problems are compounded by international regulations on bandwidth availability and by the relatively brief ground contacts available from low Earth orbit. Sentinel-3 is not a high-resolution mission; nevertheless, it employs nearly all the data downlink bandwidth capacity allowed by international regulations, and requires one contact with the ground on every orbit (Section 12.3.3). Satellites in higher orbit achieve more contact time, but the longer distance from satellite to ground is an additional factor limiting data throughput. These limitations will be discussed in Chapter 12. One possible solution for overcoming a data throughput bottleneck is to increase the number of satellites. This was discussed in Section 7.3.1 above. One supplementary advantage of providing improved coverage by increasing the number of satellites is that it becomes possible to fulfill the consumer's desires with instruments requiring a smaller swath, smaller incidence angles at the edge of the swath, with less spatial distortion, and more efficient correction of atmospheric effects. All this will increase image quality. Another possibility to reduce data throughput is to implement data compression. This will be discussed in Section 12.4.4.

In situ instruments as in Ulysses (Section 8.3.5) and sounders as in Aeolus (Section 8.3.1) do not have any swath and produce moderate amounts of data. In situ instruments characterize the observable's spatial variation by monitoring different points as they travel along their orbit. If the observables are dynamic in both space and time the consumer will have difficulties distinguishing between space or time variations. The only way to properly discriminate between them is to use multiple satellites. With multiple satellites it becomes possible to measure spatial variations by comparing the simultaneous measurements of the different instruments, such as with the Cluster mission (Section 11.4.1), which consists of four satellites studying swiftly changing events in the magnetosphere of the Earth.

8.4.5. Systematic versus interactive observation

As regards the study of observables, the specific areas of interest to be observed will be predictable in some cases while in others they will be unpredictable. The observing patterns of operational meteorology or oceanography are perfectly predictable and the instrument's task is to repeatedly cover the entire globe or ocean. This same degree of predictability applies to missions whose aim is to survey larger or smaller areas such as an entire planet, the dry land, or known areas of the celestial sphere where some astronomical feature needs to be studied. Examples of events that cannot be predicted beforehand include floods, forest fires, or the sudden appearance of a supernova. In cases like these, the mission must be able to quickly react over the consumers' possibly conflicting requests.

For predictable targets the observational approach will be uniform, repetitive, and simple. For unpredictable ones the approach must be interactive, that is, it becomes necessary to periodically decide where and how to carry out the observation and to command the satellite to do so; hence, operations will be more complex. With the exception of JWST, all the missions described in Section 8.4 are systematic, while the astronomical observatory missions discussed in Section 7.2 are all interactive. In the latter case, astronomers will make requests to observe specific targets; a scientific review team will judge the requests according to their scientific merits; satellite operations engineers will analyze them from the point of view of satellite resources and pointing capabilities. Together they will draw up an observation plan accepting some requests and rejecting others. Finally, the observation plan will be sent to the satellite for execution. These steps involve intense collaboration among the science and operations teams and the users.

Both swiftly changing and slowly changing or static targets can entail complex interaction. What makes a mission interactive is not so much the observable's speed or relative activity, as its unpredictably. Most astronomers making use of the missions described in Section 7.2 are not studying quickly changing events; however, these missions' operating plans undergo constant modification. Their organizational complexity arises from the need to allocate a scare resource – observation time – among a large number of competing users with conflicting observation needs. Quickly changing dynamic observables requiring global monitoring, on the other hand, demand swift, global coverage. As a consequence, these missions could require many satellites with large swaths. Operations, however, will be relatively simple, with continuously functioning instruments. Operational meteorology offers a good example of space missions with demanding coverage requirements but straightforward systematic observational operations. There are some missions that combine both types of observation, such as a celestial survey of the entire sky in the first part of the mission, in which operations are systematic, followed by imaging specific targets of interest at the consumer's request

The main design criteria for systematic observations are:

- In order to study the dynamics of the observable the mission must be capable of surveying the area of interest – for example, the whole Earth or celestial sphere – within a specified period of time. This is the speed of coverage requirement. The speed of coverage will determine the instrument's swath.
- The satellite and instrument configuration, as well as the orbit must provide the necessary coverage of the target area within the required interval. The instrument's observing pattern may be broad or narrow, but will always be repetitive.
- Operations will be simple, observing pattern updates rare and under normal circumstances the uplinking of commands infrequent.

The main design criteria for interactive observations are:

- The system must be capable of responding quickly to specific observation requests. Two parameters must be specified: a speed of response, "reactiveness" or response time, i.e., the time interval from reception of the request to the first observation and a "revisit" requirement for regulating the frequency that information is updated.
- To track the dynamics of the event under observation, the mission must be able to revisit within a specified period of time. This will define the ideal orbit and instrument pointing range.
- The satellite's swath is governed by the size of the dynamic event or by the need for a wider context area around the event.
- Instrument and satellite operations involve the creation and updating of complex acquisition plans reconciling consumer needs with satellite constraints.
- Instrument observing patterns will be highly variable. Instrument configuration and satellite attitude control must be designed to provide pointing flexibility.

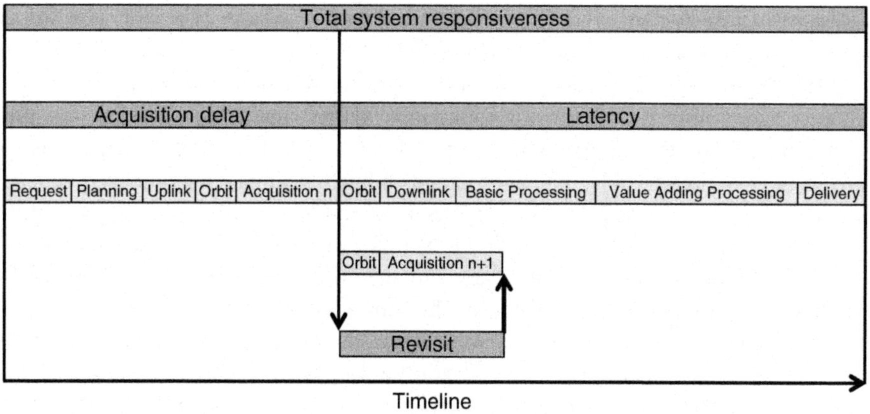

Figure 8.17. Timeline Events and Figures of Merit in the Acquisition of Observables.

8.4.6. Responsiveness, acquisition delay, and latency

Instruments acquire information about events. Consumers would like to avail themselves of this information as quickly as possible, but delivery delays are unavoidable. All information passes through various phases: it must be requested; planned, uplinked; acquired; downlinked to the ground; processed on the ground; and distributed to the consumers. These steps can be faster or slower depending on the design of the mission. Figures of merit, assessing speed of data delivery to consumers must be formulated. Parameters like revisit and coverage were already discussed but three further parameters must now be defined (Figure 8.17):

- Latency
- Acquisition delay
- Total system responsiveness

The time elapsed between data acquisition and data delivery is the data's latency. The existence of data latency requirements and the time allocated to this latency will depend on the nature of the observable and the utilization of the data. Some applications will demand the fast delivery of acquired data while others can tolerate longer data latencies. Operational observation of dynamic events demands short data latency, because it will be used for real-time decision making but scientific utilization, which requires deliberate study of the data, does not normally have demanding latency requirements. Aeolus (Section 8.3.1) data could be used either to improve our understanding of the physics of the atmosphere or to help predict the actual evolution of the weather. In the former case, long data latency would pose no problems for scientists; in the second case, however, the data needs to be made available as quickly as possible if it is to be integrated into weather forecasting models, so short data latency is crucial. Sentinel-3 (Section 8.3.3) provides synoptic views of the Earth every couple of days, but the data is delivered within 3 h of being acquired. These delays are compatible with the study of dynamic processes of moderate speed such as ocean circulation or vegetation growth. Data latency requirements are fundamental to the design of instrument output data downlink architecture, processing, and distribution. All these aspects will be covered in Chapter 12.

The "acquisition delay" time of Figure 8.17 includes the time needed to: accept, process, and uplink the necessary commands to the satellite, as well as the time needed for the satellite to arrive to the orbital position and attitude required to properly point to the observable. The acceptance and processing of the demands depends on the design of the operational ground segment and the uplink to the satellite of the request for data acquisition depends on the choice of operational ground stations and of the orbit. The operational ground segment and ground stations will be discussed in Chapter 11, and the orbit in Chapter 9. The time needed for the satellite to arrive to the right orbital position depends exclusively on the orbit and the time necessary for the satellite to point correctly to the target will depend on the performances of the agility of the attitude control of the satellite. Scientific missions studying fast transient events need also fast reaction; they required a very low acquisition delay. The Italian gamma ray observatory "Agile"[9] has a very short delay time by the nature of its targets, i.e., gamma ray bursts. Its delay time is typically a few minutes from request to acquisition of the requested image.

The total system responsiveness from request to delivery is computed by the addition of the delay; acquisition; and latency intervals. The total responsiveness is determined by the addition of the requirements in each one of its components. For some kinds of missions one or another of these time intervals could be absent: systematic missions will have no acquisition delay, while missions requiring a single observation will have no revisit. Nevertheless, all observations require downlinking and processing, which means that all missions feature some form of latency.

To improve total responsiveness it is necessary to have efficient operations and fast data processing and distribution. Nevertheless the delay will also depend on the opportunities to communicate between the ground stations and the satellite, and the opportunities for the satellite to observe the areas requested. All these opportunities are a rigid function of the orbital geometry, the number of satellites, and the choice of ground stations.

8.4.7. Observations and rotation of the line of sight

Remote sensing instruments acquire images over areas of interest. In most cases the area of interest is too large to be captured by an individual image, so to properly cover the whole area the mission must regularly produce a mosaic of images. To accomplish this, all the images must be properly orientated. This means that not only must the instrument point in the direction of the target but the satellite's orientation with respect to the target line of sight shall allow the proper concatenation of the different images or strips of images.

This additional constraint on the satellite's attitude limits the design possibilities of the satellite's configuration. All instruments producing regular synoptic views possess this constraint of orientation.

[9] AGILE a Satellite for High-Energy Gamma Ray Astrophysics, Bignami and Palumbo, Advances in Space Research Volume 31 Issue 2 2003, Elsevier Science 2003.

8.4.8. Instrument interfaces

A satellite's platform subsystems provide all essential services to the instruments; after all, it is the instruments that perform the actual observations that justify the mission. In order to formalize this concept, a document must be drawn up specifying the services that the satellite will provide to the instruments. During the early design stages it will be necessary to modify and adjust the design of the instrument, the satellite itself and the mission's operational processes. Once a convergence has been reached among these three elements, it becomes possible to produce a definitive payload interface requirements document. A comprehensive set of instrument interface requirements will include:

- Permitted instrument mass and mass properties, for both the launch and in-orbit operational configurations.
- Instrument volume envelope for both the launch and in-orbit operational configurations. The instrument's aperture will be the fundamental driver of the overall instrument size and resulting envelope.
- Structural interface requirements between the instrument and:
  - The spacecraft structure.
  - Any mechanical ground support equipment used for on-ground handling of the instrument.
- Viewing and pointing:
  - The need or not for the satellite to provide some or all of the instrument's scanning and pointing functions.
  - The size and orientation of the required field of view. This will affect the satellite's overall configuration and will determine the areas of the satellite that the instruments will occupy.
  - Viewing and pointing to the Moon, the Sun, or deep space for calibration purposes, or avoidance of these to prevent damage to sensitive detectors.
  - The primary pointing direction with respect to the Sun, the Nadir (Section 9.3), and the stars.
  - Pointing range. If the satellite carries its own scanning mechanisms, this will require that the satellite provides a clear field of view to the instrument. If the satellite is responsible for pointing the instrument, the pointing range will drive requirements for the satellite's attitude control.
  - Pointing accuracy, pointing knowledge, pointing stability, and pointing stability knowledge. These will affect the satellite's attitude control system as well as its thermoelastic stability.
  - The tracking or scanning rate, and the torques and forces affecting the satellite as a result of the pointing and scanning of the instrument.
  - Pointing or tracking duration and duty cycle.
  - Instrument to spacecraft external alignment reference interfaces.
- Electrical power and energy:
  - Voltage. Normally a basic service common to the whole satellite, the platform will indicate the working voltage and voltage range to the instrument, and the instrument will accommodate itself to the request.
  - Average power and peak power and peak power duty cycles. The difference between average and peak power will depend on the nature of the instrument.

- Limits on induced current ripple in the mains power bus. SAR can cause this due to their pulsed power characteristics.

- Telemetry and Commands:
  - Number of command and telemetry channels.
  - Command memory size.
  - Data on the instrument's status.
  - Instrument data output rate and instrument output memory size. This figure is normally much larger than the memory size required by the commands.
  - Time measurement synchronization requirements.

- Thermal control:
  - Temperature limits for the whole instrument.
  - Temperature limits for special components of the instrument such as detectors that could require extreme cooling.
  - Heat exchange limit between the spacecraft and the instrument external environment.
  - Any need for heater power and thermal control services provided by the platform; for example, temperature measurements and thermostats.

Special instruments may introduce additional interface requirements. Instruments sensing the electromagnetic field will require magnetic or electric limitations on the satellite, and gravity sensing instruments will require limitations in the accelerations within the satellite.

During a project's early stages it is unnecessary and even unwise to establish a rigid set of interface requirements that the instrument must request from the satellite. Before a project is approved, mission requirements and mission design interact; likewise, instrument design that of the overall satellite design must interact and find mutually acceptable solutions. Only when this preliminary accommodation process has ended will it be necessary to formalize the interfaces between the instruments and the rest of the satellite.

8.5. End-to-end performance as design driver

Section 3.4 introduced the concepts of value engineering, mission utility metrics, and end-to-end mission performances simulators. The present section will retake the concept of end-to-end performance and illustrate it with examples built around the instruments discussed in this chapter.

Although the present chapter has moved from a focus on observables to a focus on instruments, in order to provide a true estimation of mission value, we must return to the observables. To provide "the set of information that defines a system's value, cost and risk sufficiently for the purposes of the sponsor" it is necessary to develop a tool for measuring the quality of the final products derived from the instrument's data. This tool is the end-to-end mission performances simulator.

The end-to-end performances simulator should include modules that model the behavior of the instruments, the rest of the satellite, and the on-ground processing; but it should also include modules for analyzing how the instrument's data can be transformed into information valuable for the mission's consumer. This requires the collaboration of experts from widely different space engineering fields, as well as from the sciences related to the observables and to the consumer.

The end-to-end mission simulation tool has a dual aim:

- **It allows the breakdown of the overall mission performance into derived requirements allocated to the mission's elements and components**. That is, by employing the simulation tool top-down, it acts as a design tool allowing the determination of the allocation of requirements corresponding to the optimal designs.
- **It allows deriving end-to-end performance from the performance of the mission elements and components**. That is, by employing the tool bottom-up, it acts as a verification tool, providing a measure of overall mission value.

The existence of an end-to-end model permits the swift assessment of "what-if" scenarios, by analyzing the impact on final mission performances of design modifications made with the intention of reducing costs. The end-to-end model also facilitates the analysis of alternative concepts that fulfill the same requirements with widely different architectures; for example, radiometric accuracy provided by means of sophisticated on-board calibration tools in contrast to their provision by extensive on-ground processing.

Radar altimetry for ocean circulation offers a good example for understanding end-to-end modeling because the process is long and complex and involves widely different science and engineering disciplines. Yet, after two decades of radar altimetry, it is a domain in which both the basic science and detailed implementation procedures are well settled. In ocean circulation, the oceans can be divided into a large number of small interacting cells. The laws of physics related to the dynamics of fluids govern the interactions between these cells. These cells will have properties such as temperature and salinity, together with dynamical characteristics governing the exchanges between cells, such as the water's velocity, among many others. If it is possible to define the properties of all the cells at an initial instant, a numerical model may be employed to predict the evolution in time of these properties. Based upon data acquired at frequent intervals, such models are able to predict the probable evolution by simulating the behavior of the cells, under constraints imposed by the laws of physics, the characteristics of the ocean's initial state derived from measurement data, and by boundary conditions like the geography of the coast and the intensity and direction of winds or currents.

By redistributing water around the globe, ocean currents distort the ocean surface from a position of equilibrium that it would adopt were the whole ocean at rest. This theoretical surface is called the "geoid" and it constitutes the absolute reference or equipotential surface of the Earth's gravity field. Any differences between the local water level and the reference geoid represent sea-surface anomalies. These anomalies in sea level may be mapped, and the shapes of contours or slopes of the ocean surface may be used to determine the movement of the ocean at any point, and thus determine the ocean's currents. This implies that the fundamental observations required for predicting the dynamics of ocean currents are precise observations of local sea-surface anomalies. Today satellite radar altimeters can provide measurements of these sea-surface anomalies.

Radar altimeters determine sea-surface height and sea-level anomalies by sending a radar signal to the ocean surface and measuring the time needed for the signal to return. This interval provides the distance from the satellite to the water level.

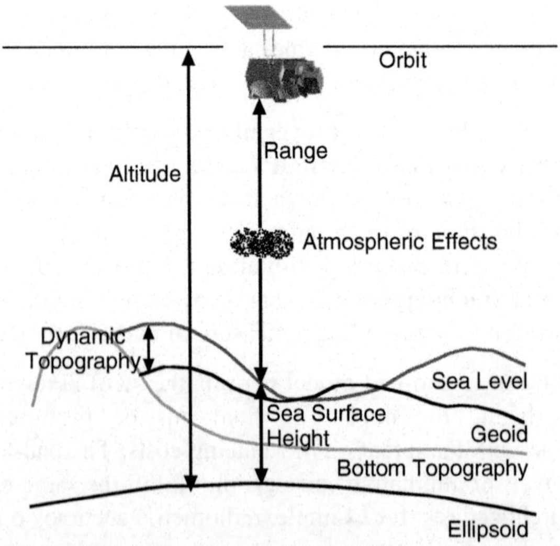

Figure 8.18. Factors Affecting the Level of the Ocean.

The radar altimeter specialist will provide algorithms to simulate the performance of the radar. Knowing the radar signal's speed it is possible to transform the delay into distance (Figure 8.18).

The speed of the signal depends on the ion density in the ionosphere and the atmosphere's density and water vapor content, and determination of these parameters is essential. In Sentinel-3, the microwave radiometer (MWR) instrument recovers the water vapor content of the local atmosphere. The ion density is measured by the comparing the return delay – which is a function of the band and of the ion density – in two bands: Ku and C. The behavior of these supplementary components shall be modeled in the end-to-end simulation, which should also include realistic models of atmospheric pressure and atmospheric humidity. The next step entails correcting for the inverse barometer effect; that is, the impact of low or high atmospheric pressure on local ocean sea levels. This correction requires excellent high-resolution models of the global atmospheric pressure. Meteorologists shall deliver these models. The interaction between the ocean surface and the radar pulses depends on wave activity, and the correct appraisal of ocean levels requires a priori knowledge of waves. Oceanographers and atmospheric modelers are responsible for providing this a priori information. After all these corrections have been made, the true distance from the satellite to the water surface can be calculated. An on-board high-accuracy GPS receiver or another tool for calculating precise orbital coordinates will permit the derivation of the satellite's orbit and derive the ocean sea-surface anomaly with respect to an absolute reference frame. The last step will be to correct water-level alterations generated by the tides using models of the tide. The final output of this long cascade of calculations will be the end-to-end performance as regards sea-surface anomalies, which are the observables for oceanography.

A clear problem for any end-to-end performance simulation model is the handling of interfaces between the different submodels that it includes. Different teams with different technical know-how and different approaches will create each one of these submodels. The prompt identification of all formal interface requirements that would allow all the submodels to run and communicate concurrently inside the same end-to-end model and in the same machine is a crucial early step.

8.6. Allocation of functions

The instrument is the heart of the mission and shapes the design of all other mission components. Given this fact, it seems logical to leave the analysis of the impact generated by the instruments for future chapters that will discuss the different possibilities for answering the instrument-generated design drivers.

This section will proceed to analyze the allocation of functions and design decisions that cannot be discussed properly in any of the future chapters.

8.6.1. Scanning

Instruments need to cover an area of interest. The area of interest might be much larger than the instrument's instantaneous field of view, so it must be capable of scanning to achieve total coverage. Scanning and pointing were introduced in Figure 8.3 of Section 8.2.1, as was the concept of the "image cube" with its three dimensions: along-track swath; across-track swath; and spectral dimension. Figure 8.3 provided an example of how it is possible to point or scan by pointing the whole satellite or by using pointing mirrors. Figure 8.4 indicated that the movement of the satellite along its orbit provides along-track swath for the instrument. These topics will be expanded in this section (Figure 8.19).

The scanning function can be performed by the instrument, the satellite or the movement of the satellite along its orbit. If the instrument provides the scanning, then design decisions on scanning, i.e., selection of bearings, motors, or control electronics, become internal to the instrument and of interest mainly to the specialist. However, two aspects have important system-level implications: size increases and mechanical perturbations.

The pointing mirror must be able to illuminate the aperture of the instrument in all the positions required for the delivery of the swath (Figure 8.20). This obliges the mirror to be larger than the aperture, and space must be allocated for the enlarged mirror and to allow movement between the points corresponding to the two extremes of the swath. Due to the need for additional space, pointing mirrors are only advisable for instruments with small or moderate apertures. Large swath, moderate resolution optical instruments, on the other hand, use mirrors because moderate resolution is compatible with moderate aperture. Sentinel-3's SLSTR instrument is a good example of this type of instrument, which uses pointing mirrors for scanning. Nonetheless, all the telescopes described in Section 7.2 – IUE, Hubble, ISO, XMM, Herschel, and JWST – have a large aperture and are pointed by the whole satellite. The decision in favor of pointing with the instrument or pointing with the satellite has a major impact on satellite configuration.

The scanning of the mirror will introduce mechanical perturbations in the rest of the satellite. These perturbations must be limited to avoid disturbing the pointing

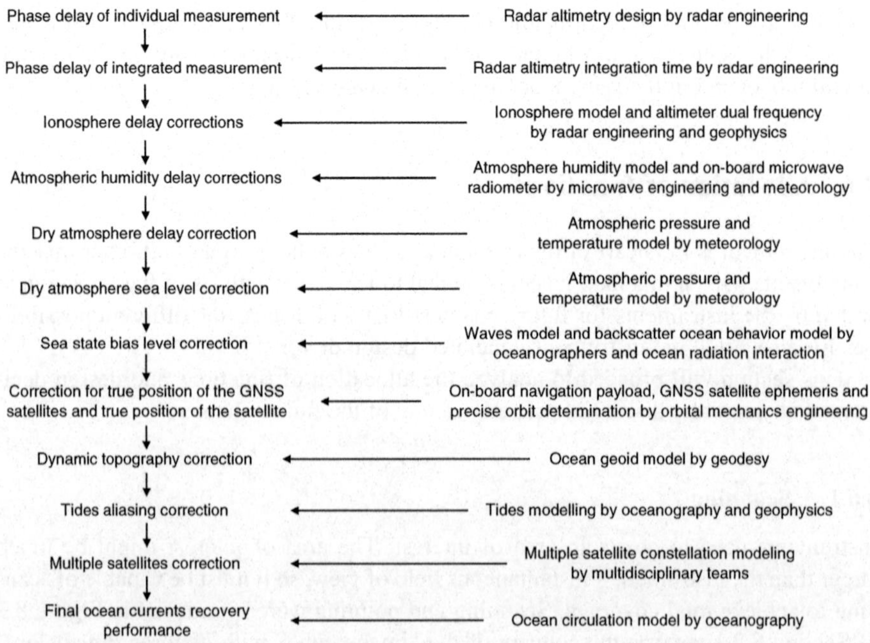

Figure 8.19. Radar Altimetry End-to-End Modeling Steps and Know-How Involved in Each Step.

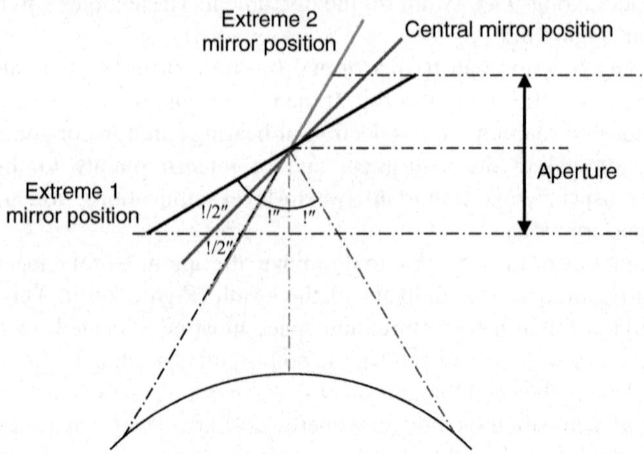

Figure 8.20. Size of the Instrument Aperture with Respect to the Size of the Pointing Mirror.

stability of the whole satellite, of the instrument itself or of any other instrument carried by the satellite. The resulting mechanical perturbations budget is mainly an attitude control issue and will be discussed in detail in Section 9.8. Perturbations can be especially severe in cases where there is an oscillating movement of the scanning mirror.

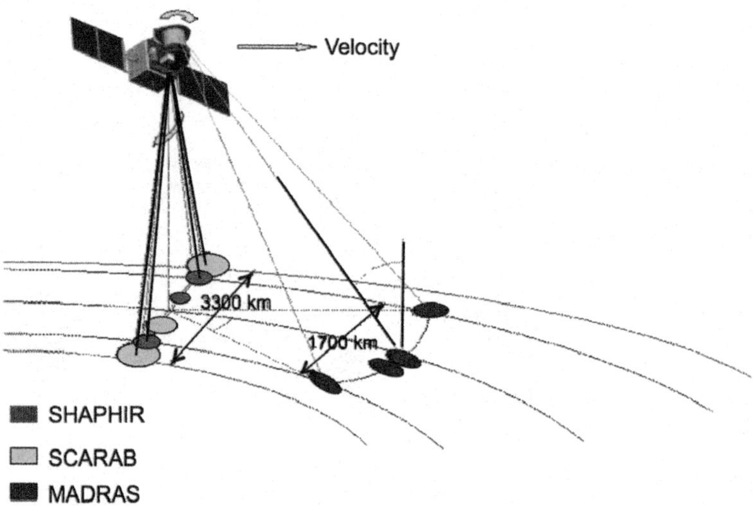

Figure 8.21. Megha-Tropiques Instrument Observing Geometry of the Surface of the Earth (CNES http://132.149.11.177/MEGHAT/GP_satellite.htm). SAPHIR and SCARAB Have Across-Track Scanning, MADRAS Has Conical Scanning. SHAPIR and SCARAB Images Are Distorted at the Edge of the Swath, MADRAS Images Are Not.

The rotation axis of the instrument's scan can be centered on the satellite's velocity vector, or around the Nadir direction. In the former case, the swath line will be straight and perpendicular to the satellite's ground track; the instrument will be an across-scanner. This is the case of SAPHIR and SCARAB, instruments aboard Megha-Tropiques (Section 8.3.4). MADRAS has a conical scan axis around the Nadir direction. In the case of conical scanners the distance and the geometry from the instrument to the ground pixel is the same for all the points of the swath and the shape and size of the all the ground pixels of the swath are always the same, resulting in better data quality (Figure 8.21). Across-track scanners are easier to implement but produce image distortions when pointing away from the Nadir direction (Section 8.4.3).

Geostationary satellites (Section 9.4.4) have a fixed geometry with respect to the Earth's surface, and the movement of the satellite along its orbit does not produce swath. To cover the area of interest, geostationary satellites need to provide physical scanning in two dimensions. Figure 8.22 provides two possible implementations. In the first, there are two sets of pointing mirrors: the first set provides scanning by varying the longitude at constant latitude. This provides an image strip roughly parallel to the equator. The second mechanism changes the latitude of the strip obtained by the first mechanism. This implementation requires a set of linear detector arrays: one per frequency of observation. In the second alternative the focal plane of the instrument is a large matrix of detectors. The different bands are recovered by using a rotating wheel with filters that intercept the optical path to the focal plane. In this case the scanning will be accomplished through the satellite changing its attitude for each image. The result will be the creation of a mosaic from the relatively large individual images captured by the large matrix array. The mosaic can be created by a "step-and-stare"

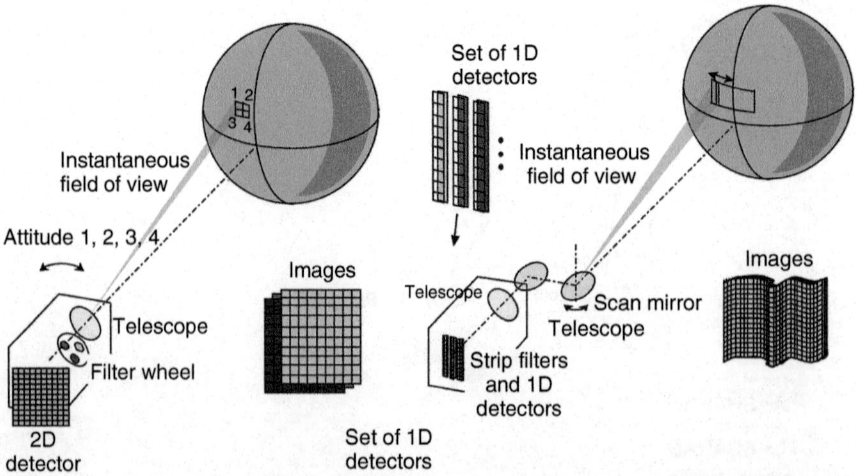

Figure 8.22. Two Possible Scanning and Frequency Bands Implementation Concepts for an Instrument Flying in a Geostationary Satellite (Courtesy EADS Astrium).

operation of mechanisms or by changing the satellite's attitude. Spinning provides a third possibility; where the spin of the satellite can be used to provide scan along the latitude lines. This is the case of MSG (Figure 7.3).

Of the five missions discussed in Section 8.3, Aeolus does not scan, JWST scans by pointing the whole satellite in the required direction, and Ulysses does not need to scan because it carries in situ instruments. Megha-Tropiques and Sentinel-3 have instruments that scan by using mechanisms internal to the instrument.

The satellite as a whole can also do scanning and pointing. This simplifies the instrument but complicates the configuration of the satellite. The resulting issues will be analyzed in Section 10.5.5.

8.6.2. Internal and external calibration

Calibration is directly related to the quality of the instrument output and to the end-to-end mission performance. The calibration approach to the instrument is another design decision that needs to be taken by the architect the consumer and the instrument design specialist. The architect needs to be aware of the influence of the calibration on the overall mission design, and which interfaces it generates.

Instruments requiring superior radiometric quality need to include calibration hardware as part of the instrument. Calibration components include sources that produce a well-known, stable signal that illuminates the detectors. The comparison between the measured output and the known input serves to calibrate the instrument throughout the mission. In some cases the source of well-known radiation will be external: the Sun; the Moon; deep space; or uniform parts of the Earth's surface. These external calibration sources could require special periodic pointing attitudes by the instrument or the satellite. This has implications for the physical design of the satellite; for example, the need to provide a clean field of view to the Sun for realizing the calibration.

Sometimes, different instruments from different missions are observing the same events. Comparison of the distinct results can be of benefit to all of them. This approach is sometimes called "external validation" to distinguish it from "internal calibration" which operates inside the bounds of the space system. The architect shall be aware of any need for external validation because it requires the establishment of agreements with the external sources, or the assurance that the external sources are really stable over the duration of the mission.

8.6.3. Solid aperture versus deployable versus synthetic aperture

As repeated many times in this chapter, the key parameter in any remote sensing instrument is the size of its aperture. The dimensions of a lens, mirror, or antenna define the aperture. If this component is built in one piece, its dimensions must fulfill the aperture-size requirement. If the required aperture is larger than the space available in the launcher, it is possible to use mechanisms that deploy the necessary hardware required to attain the required operational aperture after launch. The choice of fixed versus deployable is one of the most important factors in the configuration of the instrument and the satellite as a whole. The comparison of the Hubble Space Telescope and JWST configurations (Section 7.2.1) demonstrates the radical changes in satellite design as a consequence of this decision. This will be discussed extensively in Section 10.5.2. Synthesis of an aperture is another technique for increasing the dimensions of the instrument aperture. Due to the longer wavelengths they observe, aperture synthesis has developed faster in microwave than in optical observation. Synthetic aperture radars were introduced in Section 8.2.4. Synthetic and real aperture instruments performing similar observations create different types of design and interface problems for the architect. It would be instructive to compare two microwave instruments: one synthetic aperture (SMOS) and the other real aperture (CMIS) (Figure 8.23 below).

CMIS, weighing 348 kg, was flown on the NASA Coriolis satellite. This instrument provided microwave radiation measurements in 77 channels between 6 GHz and 183 GHz. Its configuration was very similar to the MADRAS instrument described in Section 8.3.4, but it had a much larger antenna aperture of 2.2 m. As in MADRAS, conical scanning around the Nadir direction provided the swath. SMOS was launched in 2009 by ESA. It is being used to derive soil moisture and sea-surface salinity by performing microwave radiometry in a single band at 1.4 GHz. The synthetic aperture technique employed by SMOS permits an instrument with a 4 m aperture, which can perform electronic scanning with no mechanisms or physical movements, but it is able to provide observations only at one frequency.

8.6.4. Resolution versus altitude

The spatial resolution of an image is one of its fundamental quality parameters. There are two possible situations:

- The observable is located far away in deep space and its distance is independent of the orbit. In these cases good spatial resolution is only possible by means of large apertures and, sophisticated detector arrays in the instrument's focal plane.

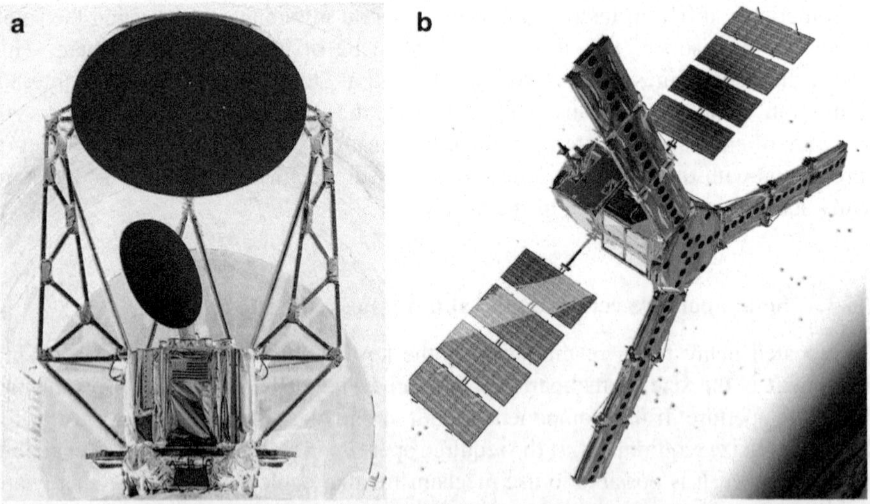

Figure 8.23. Real Aperture Microwave Radiometers CMIS in Coriolis Satellite (a) and Synthetic Aperture Microwave Radiometer MIRAS in SMOS Satellite (b) (CMIS Instrument in Test Chamber Image from USA Navy NRL. SMOS in Orbit Image from ESA).

- The observable is located on the celestial body being orbited. In this case reducing the orbit altitude will increase the spatial resolution of the image.

The same instrument flying at lower altitude will increase resolution, but the instrument's swath will be proportionally reduced. However, in applications focusing on nondynamic observables having a large swath will not be a significant design factor. Nevertheless, there are limits in this approach because the atmospheric drag associated to low altitudes will perturb the satellite orbit. The atmosphere becomes denser at lower altitudes and low orbits are impractical, as they require complex orbit control systems. The GOCE mission (Section 10.3.1) uses electrical propulsion to maintain an orbit altitude of just 250 km. The mission concept is defined by the need to fly this low, and the price to be paid is a high one. Flying below 400 km becomes progressively less practical. Military satellites use low Earth orbits in order to gain the maximum possible spatial resolution. The American military satellite KH-11 had an apogee of some 443 km and a perigee as low as 300 km.[10] Only a small portion of the Earth coincides with the perigee, and the opportunities for acquiring images at such a low altitude were infrequent. This scheme cannot be used to observe targets that require frequent revisits and ample coverage.

[10] Jane's Space Systems and Industry 2007-2008. Janes's 2007.

8.7. Allocation of budgets

In the previous sections of the present chapter the following requirements, which are associated with budgets, have been mentioned:

- Radiometric quality and MTF in Section 8.4.2
- Data quantity in Section 8.4.4
- Data delivery and latency in Section 8.4.6

The problems associated with the implementation of large data throughput will be one of the main topics of Chapter 12. Chapter 11 (Section 11.5.3) and Chapter 12 (Sections 12.4.2 and 12.5.3) will analyze in detail data delivery and data latency issues. The two remaining budgets – MTF and radiometric quality – will be introduced in Sections 8.7.1 and 8.7.2. Section 8.7.3 will provide supplementary information on the budget resulting from the end-to-end data quality requirement. The fulfillment of the instrument data quality budget will be just a part of the larger end-to-end budget. Section 8.7.3 will expand on the analysis provided in Section 8.5 on radar altimetry end-to-end performance calculations.

8.7.1. Radiometric quality

Radiometric quality is associated with the instrument signal-to-noise budget. To increase the signal it is necessary to provide a configuration (Chapter 10) that will allow accommodating a large instrument aperture size. To reduce the noise, it is necessary to have good detector technologies and a sufficient cooling of the detectors. Cooling alternatives will be discussed in Section 10.5.4. To increase the time allocated for the signal to be integrated will also improve the signal-to-noise ratio. Nevertheless, during the time allocated for signal integration it will be necessary to keep a stable line of sight. This will derive requirements on attitude control (Section 9.8). To allow longer integration times will also reduce the amount of data produced by the mission. This means, the fulfillment of the instrument's signal-to-noise ratio budget generates a cascade of design alternatives and of derived sub-budgets that need mutual balancing.

8.7.2. MTF

MTF has been defined in Section 8.4.2 as the capability of the instrument to properly reproduce contrast at increasingly fine spatial resolutions. The specified MTF value shall be allocated between the following sources:

- The MTF that is allocated to the instrument itself. These errors include the detectors, aperture, diffraction limit, and telescope dimensional accuracy errors.
- MTF associated to the perturbations of the instrument's line of sight during the integration time necessary to detect the signal. These perturbations will have a slow, low-frequency component "drift" and a fast, high-frequency "jitter" component. Instrument line of sight and satellite attitude perturbations and accuracies will be studied in detail in Section 9.8.2.

The handling of the MTF budget associated with the instrument will be in the hands of the instrument specialist. The handling of the MTF budget associated with the line of sight perturbations will be the responsibility of the attitude control stability specialist. The architect shall keep a balance between them.

8.7.3. The end-to-end performance

Section 8.5 described end-to-end mission performance simulation and examined radar altimetry in some detail. We saw that appraising overall mission performance from a top-down perspective offered opportunities for optimizing the mission's design. Sentinel-3's mandate was to recover sea-surface height with an error on the order of 3–5 cm. The availability of an end-to-end performance simulation for Sentinel-3 allowed the creation of what-if scenarios in which the error allocated to each individual error source could be increased or decreased. This permitted alternative mission architectures to be examined, with different designs for and different allocations to all the contributors to the mission's overall performance. The goal was to achieve a mission design in which the budget was effectively apportioned, so that the most difficult design features had proportionally larger error allocations, thus reducing the mission's overall difficulty. In Sentinel-3's case, the end-to-end simulation was also used to study greatly simplified radar altimeter options, including a design featuring one single-frequency radar which increased the ionospheric error contribution and another design without a microwave radiometer, which increased the errors caused by the echoes passing through a wet troposphere. The results of the analysis advised keeping both design features because their elimination led to unacceptable error levels. The final budget allocation was as follows (Table 8.1).

The addition rules for errors have been introduced in Section 5.5.4. In the case presented in the table above, the different error sources are uncorrelated; the worst case of ionosphere perturbation will not coincide with the worst case of troposphere water vapor content. Therefore, the total sum of the error sources would be a root mean square addition of the individual contributions. The outcomes show that the final performance would be driven by the larger contributors, and a moderate relaxation of requirements with respect to the middle and lower contributors would not

Table 8.1. Altimetry Error Budget.

Error type[a]	Error
Altimeter random error	1.3 cm
Sea state bias/altimeter bias	2.0 cm
Ionosphere propagation correction error	0.7 cm
Dry troposphere correction error	0.7 cm
Wet troposphere correction error	1.4 cm
Total range error (root mean square)	3 cm
Radial position GPS error	2 cm
Total sea-surface height error	3.6 cm

[a]The Sentinel-3 Mission and its Topography Element. C. Mavrodordatos et al., Geoscience and Remote Sensing Symposium, IGARSS 2007, IEEE International.

significantly impact the overall mission performance, but could appreciably reduce development risks and costs.

In an end-to-end simulation, each of the lines in the table above would represent a budget to be allocated within a given design area. For example, the acceptable altimeter random error would be allocated to all the internal instrument factors that contribute to the correct functioning of the radar altimeter. Decisions related to this second level of apportionment are internal to the instrument, and fall within the domain of the instrument specialists. The architect's responsibility is to ensure that the fulfillment of the value originally apportioned to this particular error source is in fact adequate for the design of the instrument and in line with the cost and risk guidelines provided by the sponsor. If the original error allocation pushes the concept beyond the cost and risk guidelines, it will be necessary to improve and refine the design, as well as to allocate a larger portion of the overall error budget to that particular error source. This requires ascertaining whether any other lines in the error budget can be tightened. This process must be repeated until the design's level of difficulty and resulting performances are balanced. This will then define the optimal error allocation levels. If the requested end-to-end performance is too demanding, this equilibrium can be impossible to reach, and various design components would appear as incompatible with the cost or risk guidelines. In that case, it would be necessary to request a reduction of mission ambitions from the consumer, or for more money and development time from the mission sponsor. The availability of the end-to-end simulation will allow for properly justifying any request for reduction of the mission's ambitions.

Chapter 9

The Orbit and Attitude Domain

Keywords Satellite Orbit Design, Satellite Attitude Design, Space Environment, Satellite Orbit Types, Satellite Attitude Types, Space Mission Phases, Satellite Pointing Perturbations, Satellite Pointing Control, Coverage, Revisit, Geo-Location, Co-Registration, Satellite Agility

This domain was introduced in Section 7.1.2. This chapter starts with a brief introduction to the elements and components included in this domain. It continues by analyzing the space environment and how it drives the orbit and attitude of the spacecraft. The two following sections (Sections 9.3 and 9.4) describe the different types of attitudes and orbits. Section 9.5 describes mission phases and satellite modes. The next section describes four space missions:

- The Earth observation mission Sentinel-3 introduced in Section 8.3.3. It orbits Earth at an altitude of 800 km with an inclination of 98.65°. As it is Earth pointed, its orbit and attitude are determined by the need to provide systematic and efficient Earth coverage for its three instruments.
- The communications constellation Iridium consisting of 66 active satellites – plus in-orbit spares – in six orbital planes, flying 780 km above Earth with an inclination of 86.4°. The satellites are Nadir pointing. The constellation is intended to provide communication links across the globe, anywhere, anytime.
- The solar science Ulysses satellite introduced in Section 8.3.5. It flies in a heliocentric orbit with 1.35 AU of periapsis and 5.4 AU of apoapsis, and is inclined with respect to the ecliptic plane by 79.1°. The satellite is stabilized by rotation and it is kept Earth pointed. Its mandate is to study the Sun's magnetosphere at all latitudes.
- Pleiades, which is a constellation of two agile satellites devised for high-resolution, fast-reaction Earth observation. Pleiades flies at an altitude of 694 km with an inclination of 98.2°. It is Earth pointed but can be maneuvered to point up to 43° away from Nadir. It was designed to provide daily access to high-resolution views of any point over the Earth.

These missions provide a wide selection of distinct orbit and attitude control concepts as well as a range of applications; nevertheless, in order to adequately cover the entire spectrum of design alternatives, other missions will be referred to when

M.A. Aguirre, *Introduction to Space Systems: Design and Synthesis*, Space Technology Library 27,
DOI 10.1007/978-1-4614-3758-1_9, © Springer Science+Business Media New York 2013

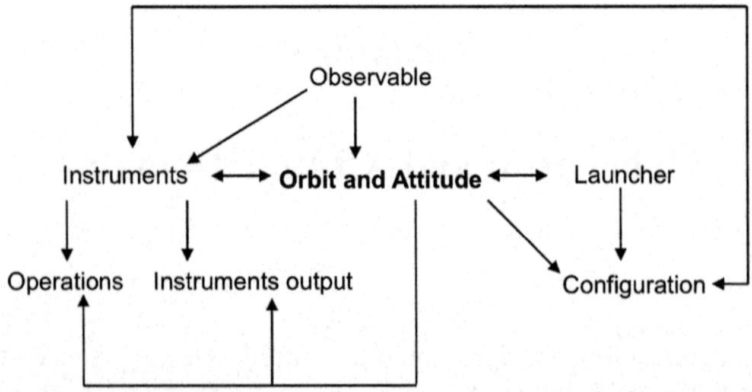

Figure 9.1. External Interrelations of the Orbit and Attitude Domain.

needed. Section 9.7 studies the geometry and visibility of the observed bodies – Sun, Earth, the observable, or deep space – located on the celestial sphere around the satellite and how this geometry is different for different orbits and attitudes. The next section (Section 9.8) introduces pointing and perturbations to pointing. The next two sections – as in all the chapters of this part of the book – analyze design decisions related to the allocation of functions and to the numerical allocation of budgets. The last Section (9.11) deals with the implementation and maintenance of constellations.

As indicated in Section 7.1, the determination of the satellite's attitude and orbit are the first design decisions directly governed by the requirements of the observables and instruments. The interactions involved in the orbit and attitude design indicated in Figure 7.1 are recalled and depicted in detail in Figure 9.1 below. This domain is rich in interrelations. It interfaces with:

- The instruments, because attitude and orbit define the observables' conditions of visibility
- The launcher, because the launcher carries the satellite to the selected orbit
- The satellite configuration – location of the antennae, solar arrays, and heat radiators, location of the instruments – because the attitude and the orbit define the visibility of the satellite
- The satellite operations and instrument data downlink architectures, because the orbital geometry defines the pattern of contact with ground stations

As Figure 9.1 demonstrates, the interrelations of the orbit and attitude are especially complex. In order to define these interrelations correctly a good architectural overview is needed to derive the best system concept. Within this domain, there are two areas of special difficulty and that require detailed and long discussion:

- What is the geometry and visibility around the satellite? This will be analyzed in Section 9.7. Section 10.4 will discuss how the configuration of the satellite deals with the drivers generated by this geometry.
- Pointing as an end-to-end issue. This is analyzed in Section 9.8.

The above-mentioned complexities justify the length of this chapter, which is the longest of the book.

9.1. Elements and components involved in the domain

This section will look at the elements and components included in this domain. It is divided into three subsections according to the functions to be performed:

- Launchers
- Orbit determination and control tools
- Attitude determination and control tools

9.1.1. Launchers

The launcher is not designed for a specific mission but selected from available existing designs and as such forms part of a mission's design constraints. In commercial projects the architect may freely select the optimal launcher. In noncommercial projects the launcher must often be selected from a number of politically preferred alternatives provided by the sponsor to the architect.

The launcher's owner establishes the launcher's rules of use in formal documents. The mission architect can thus disregard the internal details of the launcher apart from its payload interfaces, and depend mainly on the information contained in the formal launcher documentation, typically referred to as the "Launcher User's Manual" or "Launcher User's Guide." For this reason, in the place of a description of the components and technologies involved in the highly specialized domain of launcher design, which is outside the aims of this book, this section will provide an overview of launcher functions and how launchers and launching sites shape satellite design and the choice of orbit.

Launchers lift the satellites from Earth and put them into orbit. While the launcher travels through the atmosphere, the satellite remains enclosed within a protective launcher fairing that isolates it from direct aerodynamic forces and heating. Once outside the atmosphere the fairing is jettisoned. During every phase of the launch the launcher provides mechanical and electrical interfaces with the satellite. The mechanical interface secures the satellite during launch while the electrical interface provides electrical power. Some launchers are able to provide multiple firing of their rocket engines, which allows orbital corrections or the launching of several satellites into moderately different orbits. Some launchers also feature provisions for the simultaneous accommodation of two or three medium-sized satellites or a larger number of smaller satellites.

Once the injection orbit has been reached, the launcher severs its physical connection to the satellite and ejects it. Any asymmetry in the ejection forces imparted by the launcher will produce a satellite rotating speed. This speed should be as slow as possible so that the satellite can rapidly acquire a safe attitude and establish communications with the satellite control center on the ground.

Launchers always introduce some errors into the satellite's intended trajectory during injection. The launcher documentation should indicate the worst-case values of these errors. The injection errors will determine the amount of fuel to be carried by the satellite for orbital-error injection corrections.

The launch site provides the necessary services for the launch. The launcher requires supplementary services for fuelling, assembly, and launch control. The satellite requires final checkout, fueling, and attachment to the launcher.

The first decisions are which launcher and which launch site are to be employed. As said above, the launcher is usually selected from a limited range of possibilities. The launcher must be as cheap as possible, but also reliable, and obviously must provide performance specifications compatible with the mission's needs. It also needs to be decided whether the launch will be dedicated or shared. A dedicated launch provides optimal orbital injection as well as a freer choice of launch date, but must bear the overall cost of the launch. Shared launches reduce costs but entail reaching a compromise on the launch date and launch orbit between the missions sharing the launcher.

Placing constellations in orbit usually requires launching many satellites at once, and it may be necessary to provide special structural interfaces between the various satellites and the launcher. An agreement with the launcher authority on a strategy for ejecting the satellites according to prearranged trajectories and established time delays is crucial for avoiding the risk of mutual collisions.

The choice of launcher imposes requirements on the satellite. If more than one launcher is to be used, the requirements imposed on the satellite will be derived from the worst-case values specified by all the launchers.

The launch site defines what services it will provide for the satellite's final verification and its installation inside the launcher. The launch site authority can also identify duration times for these activities, which will be included as part of the mission's schedule. A further constraint arises because not all launch sites are suitable for all orbits; for orbital injection into the equatorial plane – as is the case with geostationary communication satellites – a launch site near the equator will maximize the added rotational speed of the Earth to the energy provided by the launcher, hence a launch from near the Equator is more efficient.

Launcher stages are discarded as they are used, and there is a risk of debris falling over inhabited areas; this limits the azimuths that can be used from each launch site. All these orbital limitations are included in the launcher's documentation. In some cases there is a limited period for which the launch site may be used each year.

Normally a long time elapses between the design of the mission architecture and official approval of the contract by the launcher authority. To avoid unforeseen issues regarding launcher choice, it is advisable to make the satellite compatible with more than one launcher, in case the first choice increases in price or its availability becomes problematic.

9.1.2. Orbit determination and correction

Satellite orbits frequently require correcting. Orbital corrections are of two types:

- Orbital change, which implies a relatively powerful, long thrust to shift the satellite path from one orbit to another
- Orbital control, which is a short and careful push to control the path of the satellite when it drifts from its nominal one

Several situations require proportionally large orbital changes:

- To compensate for injection errors introduced by the launcher; may require large initial orbit corrections.

- When the satellite is released into an orbit that is not the final one. For example, many communication satellites are injected by their launchers into a highly eccentric orbit, and then move into the required circular geostationary orbit using their own on-board propulsion.
- In cases – such as interplanetary missions – that call for several major orbital changes in order to follow the complex trajectory required by their missions. This can be accomplished using the on-board propulsion subsystem, though careful rendezvous with planets can also provide large orbital changes.
- Collision avoidance maneuvers; these may arise, particularly in low Earth orbit missions, due to the ever-increasing number of objects in orbit around the Earth or. In effect they amount to two small orbit changes for each maneuver cycle, the first to avoid the object that risks colliding with the satellite, and the second to recover the original trajectory.
- Deorbiting is also an increasingly frequent requirement aimed at clearing space from satellites that have already performed their functions and have become space junk, which represents a danger to future space missions. Low Earth orbit satellites must reduce their altitude, so that atmospheric drag will ensure reentry and burning of the satellite in the atmosphere. Satellites in geostationary orbit must also place themselves in "graveyard orbits" at the end of their operational lives.

External perturbations affect the nominal orbit of all satellites, and their trajectories will drift with respect to the nominal ones. These external perturbations include: atmospheric drag; the higher-order values of Earth's gravitational potential; and the gravitational pull of the Sun, Moon, and the other planets in the solar system. To correct this drift the satellite must have its own propulsion subsystem capable of delivering careful and controlled amounts of thrust that shall be provided in the most fuel-efficient manner to avoid drift away from the nominal orbit by more than is allowed by mission requirements. On the other hand, external forces can also be employed to perform useful orbital changes; for example, the Sun-synchronous orbit (Section 9.4.2) uses the nonspherical values of Earth's gravitational potential to provide a natural change of the orbital plane with the required speed.

To be able to perform any orbital change, it is first necessary to determine the orbit it is following. The ground segment flight operations team traditionally performs orbit determination, but lately low Earth orbit satellites can perform the task autonomously by using receivers that detect signals from navigation satellites such as GPS or Galileo.

To be capable of producing an orbital change the satellite must expel fuel at high speed. The energy source for providing the necessary speed depends on the nature of the satellite's on-board propulsion system, which can take the following forms:

- A pressurized inert gas will be ejected at a speed which is a function of the storage pressure.
- A monopropellant such as hydrazine will decompose in the presence of a catalyst, and the chemical reaction will generate energy which increases the speed of ejection.
- Two chemical substances, an oxidizer and a fuel will react with each other and the energy produced will increase the temperature and kinetic energy of the ejected mix.

- In ion-thrusters the propellant is ionized and accelerated in an electrical field. In this case, the acceleration energy is provided by the electrical power components of the satellite.

Pressurized inert gas systems include a tank for storage of the gas, pipes and filters, valves to regulate the flow of the gas, and nozzles to allow the pressurized gas to be expelled, providing thrust. Pressurized inert gas is easy and cheap to implement, but the energy generated solely by the expanding gas will provide only a small amount of impulse per unit of fuel mass. Pressurized inert gas is adequate for missions requiring orbit control, but is insufficient for those demanding large orbital corrections. The propellant used by these systems is inert and not dangerous. This makes them easy to handle on the ground during satellite manufacturing, assembly, and testing. Due to its small size and ease of use, pressurized inert gas is a common choice for small missions.

Monopropellant systems use the same components as pressurized inert gas does, but monopropellants require the addition a catalyst, which produces the chemical breakdown of the fuel. Monopropellants like hydrogen peroxide or hydrazine decompose and generate high amounts of heat, and the resulting ejection speed of the propellant is high. Monopropellant technology is well established and is used extensively, but the fuels are highly volatile, toxic, and hazardous, requiring special on-ground storage and handling procedures and facilities.

In bipropellant propulsion subsystems the oxidizer and the fuel are stored in different containers and are mixed in a combustion chamber. The chemical reaction increases the temperature of the combustion products, which are expelled at high speed through a nozzle. In the case of liquid bipropellant systems, the components are similar to the components necessary for monopropellants, but different tanks, valves, filters, and pipes are required for the handling of the oxidizer and the fuel. The chemical reaction produced by liquid bipropellants is more energetic than that generated by the exothermic decomposition of monopropellants. Bipropellant systems can be used to advantage for high-impulse thrust, as demanded by major orbital changes, or for moderate thrusts, as required by orbit control or small orbital changes, but they are more complex and expensive than monopropellants. Again, there are special on-ground fuel-handling procedures and facilities required for the use of bipropellants.

As in the previous cases, electric propulsion requires containers for the storage of the gas to be ejected, pipes, and valves, but it also requires the ion thruster itself. In thrusters of this kind, the gas is ionized, accelerated, ejected, and electrically neutralized. Electric propulsion also requires all the electrical components necessary for the acceleration of the ions, including high-power high-voltage electronics, control electronics, and the supplementary solar array and electric power production hardware for providing the energy needed for accelerating the ions. The mass and volume associated with electric power production, conditioning, and control are large and could reduce a large part of the mass advantages associated with the higher ejection speed of the propellant. These thrusters are suitable for providing small, continuous thrusts over long periods of time.

Swing-by is a maneuver in which a satellite performs a rendezvous with a planet en-route to its final destination. The satellite approaches the planet, flies near to it, and then soars away, its trajectory now changed with respect to its initial one. This is a specialized method of orbital change using external perturbations that are used extensively in interplanetary missions. Swing-by changes a satellite's orbit by transferring momentum from the planet to the satellite, and allows the performance of large orbital changes with zero fuel. The use of this method is by no means cost-free, because it increases the length of the mission and involves complex operations, but it may frequently be the only practical way to accomplish the proposed trajectory.

Orbit determination has traditionally been performed using range-ranging in the ground-to-satellite communications link. This link includes a well-defined frequency tone sent to the satellite, which is then returned by the satellite to the ground station. The returned signal frequency will change as a function of the speed between the satellite and the ground station from which the actual orbit of the satellite can be derived. Recently, in-orbit GPS navigation receivers carried by satellites in low Earth orbit are able to autonomously determine the position and derive the orbit of the satellite. It is also perfectly possible to use both tools at the same time.

The establishment of the orbit and the overall propulsion architecture is simultaneous because the establishment of one entails the establishment of the other. This double decision is taken very early in the project and has a heavy impact on nearly all key mission parameters: the instruments, mission life and mission costs and risks.

The selected orbit and the mission needs will establish the orbital change and orbital control required, including parameters such as: the use or not of orbit correction; the use of swing-by or powerful on-board propulsion; the use of a single propulsion system, or one system for large orbital correction maneuvers and another for finer orbital maintenance. The choice of the orbit correction architecture also shapes the establishment of an overall mission time-line; if, for example, the orbit requires swing-by the mission will last longer than if it includes a powerful upper stage.

Once the fundamental architecture has been established, the next step will be to decide the nature of the orbit correction subsystem: monopropellant versus bipropellant; and all the other detailed design decisions, i.e., the number and location of tanks, valves, filters, thrusters, and pipes. It is crucial to establish an orbit correction concept, and to determine how often orbit correction will be needed. Another important point to consider is how orbit correction maneuvers will affect the mission operations and the satellite's observational availability; satellites are frequently unable to fulfill their observing functions while performing maneuvers. The orbit control subsystem can also used to control the satellite attitude in special circumstances like the safe mode (Section 9.5) or when it is necessary to change the attitude for major orbital changes. The requirements derived from this need shall also be taken into account.

The design of rocket engines, either for large launcher engines or for small satellite thrusters, is highly specialized and rather disconnected from the other space engineering disciplines. The propulsion components of a satellite are almost never designed specifically for the mission but selected from within a range of already available components that appear suitable to fulfill the mission's needs. From this point of view the propulsion engineer plays a relatively small role on the overall process of architectural definition. On the other hand, the simultaneous decision regarding the orbit and the architecture of orbital control is a fundamental one, to be taken early in the project and having major impact throughout the mission design; this means it is a decision in which the mission architect plays the leading role. The architect shall also be active in identifying the implications for the mission of design decisions, such as the reduction of instrument availability during orbit corrections, or the choice between satellite autonomy and ground intervention for orbit determination and control, to be discussed in Chapter 11 as part of satellite operations. It is also necessary to take into account the impact of the orbit correction subsystem concept on:

- Center of masses position and center of masses position changes as the fuel is used up along the mission. This will drive the design of the attitude control subsystem of the satellite.
- The thruster axis location and possible plume impingement on satellite components sensitive to contamination.

9.1.3. Attitude determination and control

Attitude control includes determining the control approach to be used and the associated attitude control modes. Selection of the control approach includes identifying the appropriate attitude sensors, actuators, perturbation torques, and the control logic to be used.

There are five possible attitude control approaches: uncontrolled; gravity-gradient; spin-stabilized; dual spin/momentum bias; and inertial control. These approaches will be described in detail in Section 9.3.

The satellite's attitude can be determined by locating it with respect to familiar objects such as the Sun, the Earth's limb, the Earth's magnetic field, or stars. Sun sensors determine the satellite's attitude by detecting the direction of solar radiation; they are crude, but simple and reliable. Earth limb sensors detect the temperature change between the Earth's limb, and the chill of deep space. They are of intermediate accuracy and complexity but have the advantage that the satellite's instruments are frequently targeted in the direction of Earth as well.

Magnetometers measure the Earth's magnetic field vector, and derive attitude by comparing their measurements with a map of the field. They possess low accuracy but are robust and reliable. Star trackers are small telescopes that detect the direction of well-characterized stars with great accuracy, providing an absolute reference of the satellite's attitude in inertial space. Two star trackers, viewing in two different directions, are needed to provide a three-axes attitude reference for the satellite. Modern star trackers can also recognize the satellite's attitude in a "lost in space" situation, and are able to derive angular speed by measuring the changes in the position of the stars over time. Gyroscopes of various kinds are also able to measure the satellite's angular velocity. They complement the attitude sensors, which determine the angular position, providing more robust and accurate control logic.

The satellite's attitude control logic depends on various factors, including the attitude control concept, the sensors, the actuators and attitude accuracy and agility requirements. In the past, the control loop was implemented using "hard-wired" analog control logic, but these days it is digital and is implemented in either the satellite's central computer or in a dedicated one. The digital control always includes feedback loops, but may also feature feed-forward ones to assist in attitude control by permitting the satellite to initiate control from a well-placed initial point.

The actuators are able to provide torques for changing the satellite's attitude. Gravity-gradient booms are deployable structures with a mass at the tip which generate a gravitationally induced torque for the satellite to align itself with the local vertical or gravity vector; as soon as the boom strays from the local vertical, the differential pull of the gravity acting in different parts of the satellite will attempt to align the boom with the gravitational field direction, thus providing a reference attitude for the satellite. If the satellite oscillates, passive dampers dissipate the energy produced. They contain fluids that are forced to interact with balls or holes when the satellite oscillates, dissipating the energy associated with the movement, and eventually causing it to stop. The inertia properties of the satellite are driven by its shape and mass distribution and the resulting inertia ratios can provide inherent gravitational torques that act as nondeploying gravity-gradient booms. A magneto-torque is a current loop that interacts with the Earth's magnetic field, introducing a torque into the satellite. Regulating the current in the loop can regulate the strength of the torque. The torque is produced by the interaction between the current loop and the Earth's magnetic field and is always orthogonal to the plane formed by the magnetic dipole vector of the control loop and the Earth magnetic field vector at the satellite position. Nevertheless, using several magneto-torques it is possible to provide torques around different axes. Although they are relatively simple and can be accurate, and are able to operate for long periods, the instantaneous torque that they provide is of only moderate strength. Momentum wheels consist of an inertia wheel, an electric motor, bearings, and control electronics. The control electronics can direct the motor to accelerate or decelerate the inertia wheel's rotation speed. Any change in the wheel's speed will produce a torque on the rest of the satellite. Momentum wheels are flexible tools providing small or large torques with notable accuracy; they are limited, however, by a maximum allowable speed, and must be unloaded – that is their speed must be reduced – by external torques to prevent them from reaching their maximum allowable speeds. To provide torques in all directions, at least three wheels are needed. A fourth one is usually provided for redundancy. Control momentum gyros are momentum wheels whose rotation axis can be changed. Changing the rotation axis of a momentum wheel rotating at high speed will introduce strong torques on the rest of the satellite, allowing fast variations in its attitude. Control momentum gyros are expensive, high performance items with complex control electronics, but they provide highly accurate, highest-strength attitude actuation. The firing of thrusters whose line of action does not pass through the satellite's center of mass will also produce torque. Thrusters can be used for large or small attitude changes, as well as for unloading momentum wheels.

A number of satellites requiring the utmost accuracy use the instruments themselves as supplementary attitude sensors. GOCE (Section 10.3.1) uses its gravity-sensing accelerometers to derive the satellite's angular accelerations. Astronomical telescopes use the images acquired by the telescope itself as a supplementary, extremely high-accuracy star tracker.

The first fundamental design decision is the choice between a fully active control concept and the use of some additional passive control component, such as the gravity-gradient boom or the spinning of the satellite. Section 7.1.3 compared the configuration of the three-axis stabilized GOESS with the spin-stabilized MSG. The choice of one or the other type of attitude will have a major impact on the satellite configuration and on the entire mission design. The choice is influenced by many complex factors, among them previous experience with the alternative approaches; instrument requirements; attitude control requirements; cost and risk. Once the decision has been made, the selection of sensors, control logic, and actuators for the different control modes must be undertaken. It is especially important to design a strategy for the satellite's initial attitude acquisition after its ejection by the launcher.

The attitude control bandwidth, which determines the frequency at which the control authority starts to decrease, is very important. The pointing error is inversely proportional to the bandwidth, and an accurate pointing system requires a large bandwidth. Ideally the attitude control loop will require that the satellite behaves as a rigid solid within its control bandwidth and that any large flexible deployable structures should have their natural frequencies outside the control bandwidth. Nevertheless, this will not always be possible; some satellites require high pointing accuracy and carry large low-stiffness deployables; JWST (Section 8.3.2) is an example of this. Mission design will call for specifying not only the satellite's pointing accuracy but also its pointing stability and accuracy in the recovery of both parameters.

For agile satellites, the speed of repointing is crucial; this requires the use of very large momentum wheels, control momentum gyros, or powerful thrusters to provide the necessary large torques. In other cases, the mission will call for slow and continuous attitude changes; momentum wheels are ideal for providing well-controlled continuous torques around any axis by ordering changes in their rotation speeds. Momentum wheels are the most flexible and widely used attitude actuators; they are able to provide either moderately small or large torques as needed during either short or moderately long periods of time.

Attitude control is highly specialized discipline with well-developed analytical tools to derive the satellite behavior when confronted with its varying environment. These tools are simulations of the dynamic environments of the satellite and of the reaction to them of the attitude control components. The models shall be able to represent the changing environment with the necessary level of accuracy. Attitude control is central to the overall satellite design and will drive many of the design decisions to be taken by other engineering disciplines. This duality of high specialization but strong influence upon the overall design process produces a situation where the attitude control is often designed by the same entity that is responsible for the delivery of the whole satellite. The attitude control sensors and actuators are selected from already available models, provided by specialized suppliers.

The architect takes a leading role in balancing the impacts of the attitude control design on the mission as a whole. The relationship of the attitude control system with the satellite's instruments and modes of operation are especially important.

In the case of high-accuracy pointing it is important to control the micro-perturbations generated by the satellite, including those generated by the attitude control actuators. It is necessary to establish and administer a micro-perturbation budget to be allocated between all possible sources. Pointing budgets are another factor in which the attitude control interacts with other subsystems such as the structure or thermal control and the attitude control could be relaxed as long as the dimensional stability was tightened. The need to allocate conflicting requirements between different engineering domains is another area where the architect's lead is crucial.

9.2. The space environment as orbit and attitude design driver

The present section will describe the space environment in general and will analyze how the external environment drives the satellite's orbit and attitude design. Section 10.2, which complements the present discussion, will analyze the space environment as a configuration driver. The space environment is described in specialized literature such as ECSS-E-10-04A.[1] The following description makes extensive use of this reference; for a more comprehensive description of the space environment please refer to the text cited. There is a book specifically dedicated to the Space Environment: *The Space Environment and Its Effects on Space Systems* from Vincet Pisacane, AIAA Education Series 2008. This section analyzes mainly the space environment around the Earth, which is still today the environment for most space missions. Satellites flying beyond Earth orbit to other planets and near-Earth objects are specifically designed to withstand these interplanetary environments. This topic is introduced in Section 9.2.6.

9.2.1. Gravity field

Satellites move around the gravity field of a central body. As a provisional approximation, the central body can be considered as perfectly spherical and uniform. In this case, the satellite will adhere to Newtonian laws, and its orbit will trace a hyperbola, parabola, or ellipse, with the central body at one of its foci. In reality the central body's mass distribution is never uniform and the orbit around it must be correspondingly modified with respect to the ideal. Whatever the shape of the central body the gravity field will be a potential field, fulfilling Laplace's equations, and can be described in terms of a central force augmented by a series of spherical harmonics expressed by Legendre functions of two parameters: degree n and order m. These spherical harmonics, which are a function of the mass distribution of the body being orbited, can be determined by analyzing the trajectory of satellites. High-accuracy models such as JGM-2[2] provide harmonics up to degree and order 70. The GOCE (Section 10.3.1) mission's aim is the high-accuracy determination of Earth's gravity field that will provide harmonics up to degree and order 200 and more.

Missions requiring highly accurate determinations of the satellite's position require the use of highly accurate models to determine their orbit, but most missions only call for consideration of the few most important harmonics. The most important harmonic is J2, which represents the 22 km of Earth's polar flattening. J2's order of magnitude is 0.0005 of the main Newtonian gravity term. The perturbations generated by the J2 term can be used positively to provide controlled and useful changes to the orbit. The Sun-synchronous (Section 9.4.2) and Molniya (Section 9.4.3) orbits owe their existence to J2. The J3 term represents the fact that the Earth is not only flattened but is slightly pear-shaped, with a polar bulge of 16.5 m. J3 is of the order of 10^{-6} of Newtonian gravity and is used to control "frozen orbits" (Section 9.4.2).

[1] Space Environment, ECSS-E-10-04 ECSS ESA.

[2] Gravity Model Development for TOPEX/POSEIDEN: Joint Gravity Models 1 and 2, Nerem R.S. et al., J. Geophys. Res. Vol. 99, December 1994.

The harmonic J22 represents the fact that the equatorial radius of the Earth is not uniform. J22 produces long period (800 days) east-west oscillations on satellites located in the geostationary position (Section 9.4.4). To control the position of the satellite it will be necessary to perform orbit corrections. The gravity field is stronger when closer to the center of the planet and the resulting difference in gravitational pull, the gravity-gradient, acting upon the different parts of a satellite, will introduce torques, which will align the satellite in the direction of minimal energy, with its long side pointed along the vertical direction. In some satellite attitudes, this gravity-gradient torque represents a perturbation to be compensated for; in other cases it will be of help because it provides a fixed attitude with respect to the central body.

The Earth's gravity field dominates orbits around the Earth; nevertheless the gravitational force of the Sun and Moon will disrupt the nominal path of the satellite, which must be compensated for in cases where careful control of the orbit is necessary for the fulfillment of a mission's aims. Geostationary (Section 9.4.4) orbits are located in the plane defined by the Earth's equator; the Sun and Moon will tend to pull the satellite away from this plane. The design of geostationary orbits requires the provision of on-board propulsion to compensate for these perturbations. As the satellite moves away from the Earth, its gravity field becomes weaker. The Lagrangian point (Section 9.4.6) is located in the place where the Earth's and Sun's gravity fields have the same strength.

9.2.2. Earth's magnetic field

The Earth has a magnetic field that, like the gravity field, is potential, and which can also be described as a main dipole term plus an expansion of progressively smaller terms. The Earth's dipole axis is tilted with respect to its rotation axis, and changes direction over time, with its strength decaying. To cope with these variations there are several regularly updated reference fields; the International Geomagnetic Reference Field deserves special mention.[3] The Earth's magnetic field does not affect a satellite's orbit, though it does affect its attitude by generating torques that interact with the electrical currents inside the satellite. The Earth's magnetic field can also be used as an attitude control tool by generating currents on board, and as an attitude determination tool by measuring the magnetic field and deriving the satellite's attitude from it. If the Earth's magnetic field is to be used to determine and control a satellite's attitude, the satellite must carry on-board maps of the field. The magnetic field's higher spherical harmonic terms will normally be of interest only to missions focusing on the measurement of the field. These terms depend on magnetic features of the Earth's crust, mantle, and core, and their measurement can be used to study the Earth's geophysical properties. A precise description of the magnetic field as a function of latitude, longitude, and radius is also crucial for analyzing the Earth's ionosphere, essential for deriving the satellite's radiation environment (Section 9.2.5).

The above description refers to the magnetic field generated by the solid Earth; nevertheless; there are external factors that also influence the magnetic field around

[3]International Geomagnetic Reference Field, the Third Generation, Peddie N. W., J. Geomag Geoelect 34, 1982.

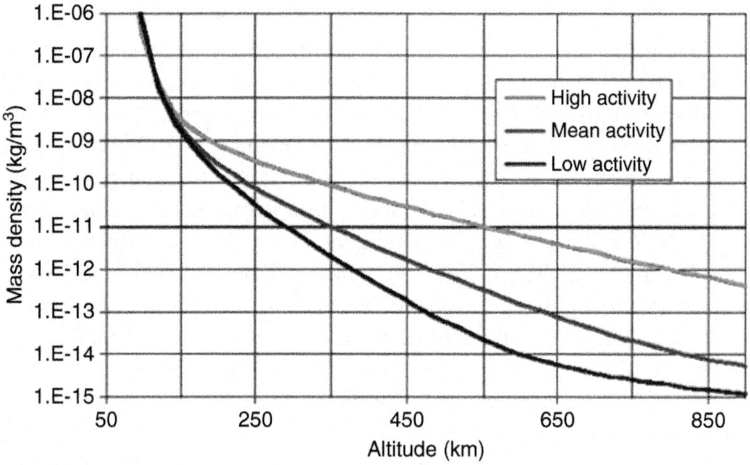

Figure 9.2. Atmospheric Density as a Function of Altitude and of the ≈ 11 Years Cycle of Solar Activity (Space Engineering Space Environment, ECSS-E-10-04, ESA 2000).

the satellite: the solar wind will cause distortions of the field at high altitude. Electrical currents inside the ionosphere will also alter the magnetic field. At short distances from the Earth, the impacts of these magnetic field sources is small and do not normally need to be taken into account. At greater distances from the Earth, the strength of the main dipole term decreases and the external fields become important. In the geostationary position, for example, the Earth's magnetic field has a strength of 100 nT, but is subject to an approximately 50% modulation from the diurnal cycle, and can even become reversed in severe geomagnetic solar storms.

9.2.3. Neutral atmosphere

Aerodynamic drag generated by the remaining atmosphere still present at the satellite's orbit altitude will disturb its path. This effect will affect the design of any satellite flying below 1,000 km. Atmospheric drag is a significant design factor for altitudes below 500 km and becomes the key design driver for satellites flying below 300 km. GOCE, which flies at an altitude of 260 km (Section 10.3.1), is an example of this. The aerodynamic forces caused by the orbital velocity and high-altitude atmospheric winds will, unless corrected, produce changes in the satellite's orbital parameters. These forces call for compensation to avoid orbital decay. Atmospheric drag, on the other hand, can be used at the end of a mission to reduce the satellite's altitude in preparation for full atmospheric reentry and satellite self-destruction. This will eliminate in-orbit debris that would otherwise be a danger for other spacecraft. The amount of atmospheric drag strongly depends on both orbit altitude and solar activity. Atmospheric density decays exponentially with altitude. Atmospheric density can change by up to two orders of magnitude between solar maximum and minimum. Figure 9.2 provides an overview of atmospheric density as a function of altitude for different levels of solar activity, which increases and decreases along a solar activity cycle with a period of around 11 years.

Reference models exist that permit the determination of atmospheric density and wind currents. ESA ECSS[4] recommends the use of MSISE-90,[5] which is administered by NASA Goddard. MSISE provides altitude profiles of temperature, molecular and atomic concentrations, total density, and molecular weight as a function of the "F10.7" solar activity index and of the Earth's magnetic "Ap" activity index. The 10.7 index reflects the solar radio flux at 10.7 cm, which correlates strongly with the intensity of ultraviolet sunlight. The Ap index describes the fluctuations of the Earth's geomagnetic field and is obtained by averaging the results of ground station measurements. For low flying satellites with large transversal areas it is also necessary to take into account the effect of lateral winds. ESA ECSS recommends the use of the HWM-93[6] model. The satellite's orbit correction and attitude control hardware must be dimensioned to compensate for atmospheric drag. The compensation of external forces aimed at orbit maintenance calls for the performance of possibly frequent orbit control maneuvers in the case of large satellites flying at low altitudes during solar maximum periods, and requiring a fine control of orbital trajectory. Frequent maneuvers of this kind will be a burden to the delivery of mission products and to satellite operations in general. The aerodynamic forces are mainly produced by the satellite orbital velocity relative to the nearly steady atmosphere. The atmospheric winds at satellite altitude play a minor role, apart from the rotation of the whole atmosphere dragged by the Earth rotation. It is worth mentioning that the aerodynamic drag is also a source of important disturbance torques for low flying satellites.

9.2.4. Solar radiation

Satellites receive electromagnetic radiation from many sources, the Sun being the dominant one among them. Radiation from the Sun is approximately equal to radiation from a black body emitting at a temperature of 5,762 K[7] (Figure 9.3). This energy will illuminate and heat the satellite and will provide electrical energy to those satellites carrying solar cells. The intensity of the illumination is fairly constant, varying less than 0.1% throughout the ~11-year solar cycle. The Earth is ~1.5 10^8 km from the Sun and at that distance, the integrated energy of the Sun is 1.371 kW/m^2. The Earth's orbit is not completely circular, and due to this, the Sun's energy reaching a satellite orbiting the Earth varies by ±3.5 % throughout the year.

Besides the uniformity of the Sun's overall radiation, its ultraviolet component varies with the ~11-year solar cycle as well as with the 27-day equatorial solar rotation. This variation ranges from about 50% for the slower near ultraviolet frequencies to a factor of 2 for ultraviolet's faster ones, and to many orders of magnitude for X-rays frequencies. This high-energy radiation is the most important factor in the

[4]Space Engineering Space Environment, ECSS-E-10-04 A, ESA January 2000.

[5]Extension of the MSIS Thermosphere Model into the Middle and Lower Atmosphere, Hedin A.E., J. Geophys. Res., Vol. 96, No. A2, pp. 1159--1172, Feb.1, 1991 and ftp://nssdc.gsfc.nasa.gov/pub/models/msise90

[6]Revised Global Model of Thermosphere Winds Using Satellite and Ground-Based Observations, Hedin A.E. et al., J. Geophys. Res., Vol. 96, No. A5, pp 7657–7688, May 1, 1991.

[7]Space Engineering Space Environment, ECSS-E-10-04, ESA 2000.

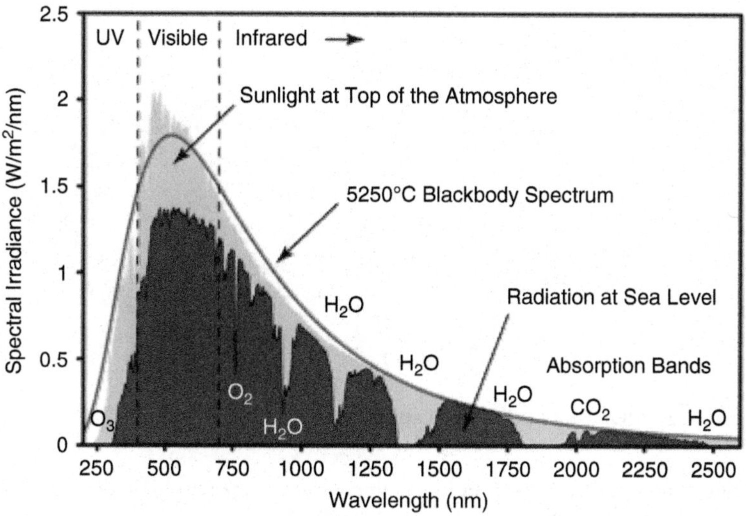

Figure 9.3. Solar Radiation Spectrum at the Top of the Atmosphere and at Sea Level (Wikipedia Commons Image Solar Spectrum.png).

heating of the atmosphere's highest layers, which explains the large variations in atmospheric density between solar maximum and solar minimum at altitudes normally used by satellites.

The Earth reflects a fraction of the sunlight reaching it. The ratio between total and reflected sunlight is called "albedo." The Earth's average albedo is around 0.3, but depends on surface properties, and can vary between 0.05 and 0.9. The highest values correspond to the presence of clouds and snow, and the lowest to vegetation and ocean water. The Earth also produces its own infrared radiation, emitting the equivalent of a black body with a temperature of 288 K.

The geometry of the Sun with respect to the satellite is a major design driver. This geometry will be described in Section 9.7 of the present chapter. The impact of the Sun's geometry on the satellite configuration and thermal design are also quite important and will be examined in Sections 10.2.3 and 10.4 of the next chapter.

The pressure of the solar radiation acts on the satellite and perturbs its orbit and attitude. The effect is very small but it is cumulative and acts continually and shall take into account in the design of geostationary satellites. The deployable solar sail, symmetric with respect to the solar array, depicted in Figure 7.2 was introduced to render the satellite symmetric and compensate the torque produced by the solar radiation pressure on the solar array.

9.2.5. Ionosphere radiation

The Sun, Earth's atmosphere, and Earth's magnetic field interact in a complex way. As a result of this interaction energetic ionized particles are generated and trapped in specific regions of the near-Earth environment. The lower part of Earth's atmosphere contains ions from an altitude of about 80 km to several thousand kilometers high. At higher altitudes the magnetic field shape traps a high density of ions in regions

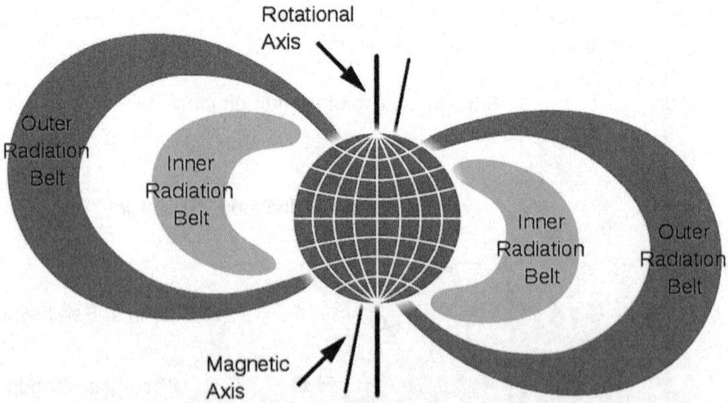

Figure 9.4. Van Allen Belts (Wikipedia commons).

called the Van Allen radiation belts, which are two regions of electrically charged particles created by the Earth's magnetic field. The outer belt is at a height of between 25,000 and 32,000 km. The inner belt reaches from 1,000 to 10,000 km. Electrically charged particles will bombard and damage satellites flying through these belts, causing electronics to fail; latch-ups and single-event upsets will gradually harm electronic components. Radiation in the Van Allen belts is intense and satellites must avoid them as far as possible. The intensity of ionized particles in the near-Earth environment depends on both the Earth's magnetic field and the Sun's activity (Figure 9.4).

Computer models can provide an integrated equivalent of potentially absorbed radiation dosages for a specific orbit, mission time, and mission life. These and other models predict the effects of ions and neutral particles and are used to determine the amount of shielding necessary for the protection of a satellite's electronic components. High radiation dosages require the use of heavy radiation shielding and sophisticated radiation-resistant components, with their corresponding higher costs. As the satellite travels through the plasma it can become charged with an electrostatic potential. If different parts of the satellite become charged at different voltages, there is a danger of a possible discharge between them, which could damage sensitive on-board equipment. Uniform, conductive coatings on the satellite's exterior, and grounding between its different parts will help avoid this.

9.2.6. The space environment outside the Earth

The previous sections have described the environment around the Earth, which is still the environment of most space missions. Nevertheless, missions traveling through interplanetary space or orbiting other planets are designed for the special environment that they will encounter during their mission. The space environment specialist will provide the available information on the environment that the space mission is going to withstand. Tracking satellites orbiting the Moon and Mars has allowed deriving the gravity field of these celestial bodies with a reasonably good level of knowledge. Satellites visiting other planets have allowed the determination of magnetic or radiation fields around them. The mission design of landers and rovers on other celestial bodies depends on the atmosphere of the target body. Venus has and a surface pressure 92 times denser than the Earth but Mars atmospheric pressure less than 1% of the pressure of the Earth. Jupiter has a very strong radiation belt and missions traveling near Jupiter must withstand this harsh environment. Conversely, the radiation environment of interplanetary space is rather benign and satellites on interplanetary missions like Ulysses (Section 8.3.5 and 9.6.2) have demonstrated a very long life. The intensity of the solar radiation is inversely proportional to the square of the distance to the Sun. When flying far away from the Sun, the reduced solar strength results in the mission requiring large solar arrays or other sources of power that must be carried on the spacecraft; Rosetta (described in Section 11.4.2) is an example of this. However, satellites flying near the Sun must withstand very intense high solar fluxes. The solar radiation level for satellites orbiting Mercury it is between 6,249 W/m^2 and 14,389 W/m^2 depending on Mercury's position along its orbit. This will have major impacts in the design of any mission trying to visit Mercury.

9.3. Attitude and attitude types

Satellites are physical objects moving freely in space, and a terminology must be agreed upon to define its attitude, or orientation, with respect to a reference frame. The most natural reference frame is the satellite's own orbit. The orbit defines three directions (Figure 9.5):

- The velocity or ram direction
- The Nadir toward the center of the orbit
- The perpendicular to the orbital plane

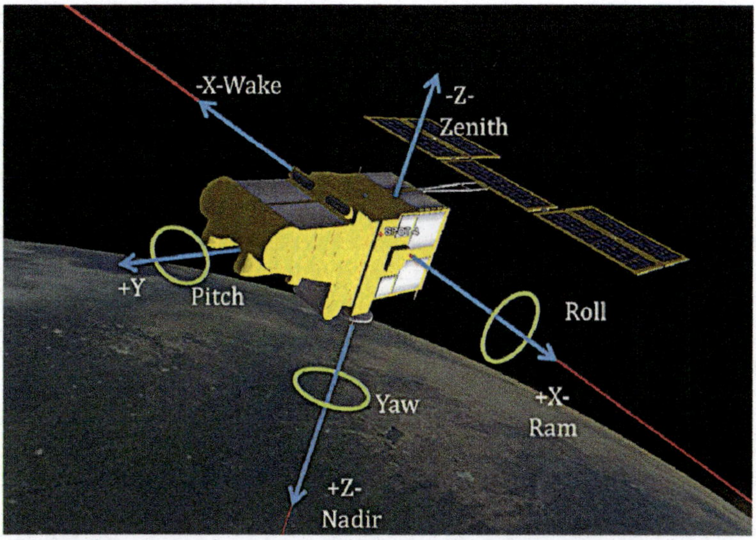

Figure 9.5. Satellite Local Orbit Reference Frame and Satellite Attitude.

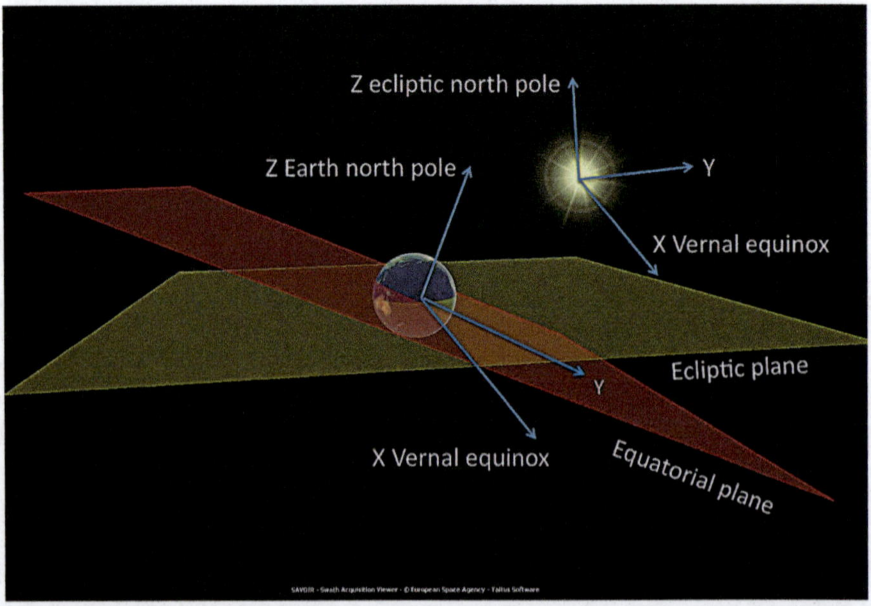

Figure 9.6. Earth and Sun-Centered Inertial Reference Frames.

There are also three directions opposite to these: the antivelocity or wake; the Zenith; and the antinormal to the orbital plane. The satellite attitude and its rotations can be defined by reference to this frame. The Nadir direction defines a natural "down" orientation for the mission, as well as a direction for pointing the instruments in cases where they are surveying the body being orbited, which is normally the Earth. It will be also necessary to refer the satellite orbit with respect to a permanent "inertial" reference plane. In the case of satellites orbiting the Earth, the best reference plane is the Equatorial plane (Figure 9.6). The center of the Earth will be the origin of the reference frame, the Earth's North Pole will be the Z-axis and the Y-axis will be the Aries Vernal Equinox direction that defines the intersection between Ecliptic and Equatorial planes. In the case of satellites not orbiting the Earth it is better to use the Sun's center of mass as a center of the reference frame and the Ecliptic plane as reference plane. The Z-axis will be perpendicular to the Ecliptic, and – again – the direction of the Vernal Equinox will provide the X-axis. That means, the Earth and Sun-centered inertial reference frames share the X-axis direction. There are three specialized terms to define rotations:

- Roll, for rotations around the X-axis
- Pitch, for rotations around the Y-axis
- Yaw, for rotations around the Z-axis

The following sections will provide examples of different types of satellite attitudes.

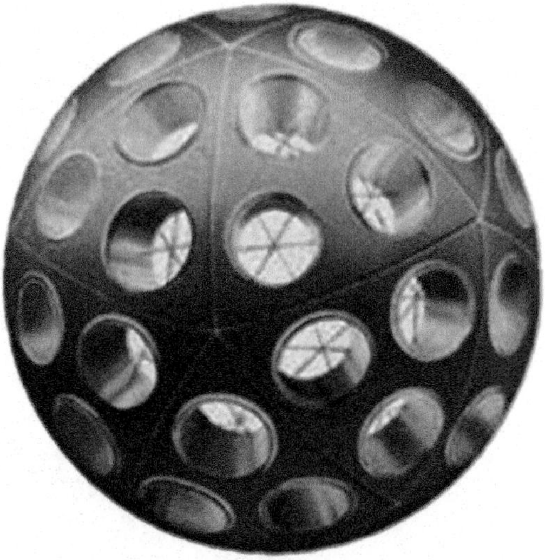

Figure 9.7. **Uncontrolled Attitude Gravity-Sensing Satellite Starlette (http://directory.eoportal.org/get_announce.php?an_id=9867).**

9.3.1. Uncontrolled satellite attitude

Some missions do not require any specific attitude for the satellite. The two French Starlette satellites, launched in 1975 and 1993, are good examples. Starlette was a small but heavy spherical craft fully covered with mirrors that reflected laser pulses sent from the ground. The time needed for the signal's return determined the satellite's orbit with high precision, which in turn allowed measurements of the Earth's field of gravity. These satellites where totally passive, and being fully covered by mirrors, did not need to maintain any particular attitude with respect to the ground in order for the laser light to reflect correctly back to Earth (Figure 9.7).

9.3.2. Gravity-gradient attitude control

If the satellite has a long axis, the long axis will align itself naturally along the Nadir direction, passively providing a Nadir-oriented face. This is a convenient alignment because it provides a reference to the "up" and "down" directions for instrument observations and communications. To be stable, the axis aligned with the local vertical shall be a minimum inertia axis, and to avoid continuous oscillations, the satellite shall include oscillation dampers.

Gravity-gradient stabilization provides one reference direction and constrains two attitude axes, but it provides no control around yaw. In order to control rotations around the yaw-axis supplementary attitude detection and actuation tools must be provided. A momentum wheel that rotates around pitch will provide gyroscopic pointing stability around yaw and roll. In other cases magnetometers and magnetotorques are used to provide attitude sensing and actuation, supplementing the passive gravity-gradient control.

This approach provides an accuracy of the order of ±5°, which though not high is sufficient for many applications. A recent example is provided by the Orbcomm

Figure 9.8. Orbcomm Gravity-Gradient Stabilized Satellite (http://www.satnews.com/cgi-bin/display_image.cgi?1908629056).

constellation of 28 satellites (Figure 9.8) providing worldwide messaging services. Each satellite weights 42 kg and carries a 6 m-long boom that simultaneously provides gravity-gradient stabilization and a good Nadir-pointed structure for locating Earth-oriented communication antennas. Orbcomm uses magnetic torquers for attitude damping.

9.3.3. Stabilized by rotation attitude control

A rotating satellite possesses a gyroscopic inertia that will maintain its orientation in inertial space unless perturbed by external torques. If the satellite's spin speed is high, the steadiness will be high as well, thus ensuring high pointing stability. The pointing accuracy provided by satellite spin is on the order of 0.1° or better. This stability, however, can be a liability in cases that require reorientating of the satellite's pointing axis because large torques acting during long periods will be needed to overcome the gyroscopic inertia.

To achieve stable rotation the satellite must rotate around the axis with the largest inertia momentum. This will govern the satellite's configuration and produce drum-shaped satellites; ESA's ISEE-2 (Figure 9.9) is an example of this. Spin-stabilized satellite attitudes are measured by Sun or Earth sensors. Thrusters are normally used for changing the spin rate and for corrections to the direction of the spin axis. If the satellite's angular momentum vector is not perfectly aligned with the principal inertia axis, nutation movements will result. Nutation can be avoided by using passive nutation dampers. Ulysses is stabilized by spin as described in detail in Section 9.6.2.

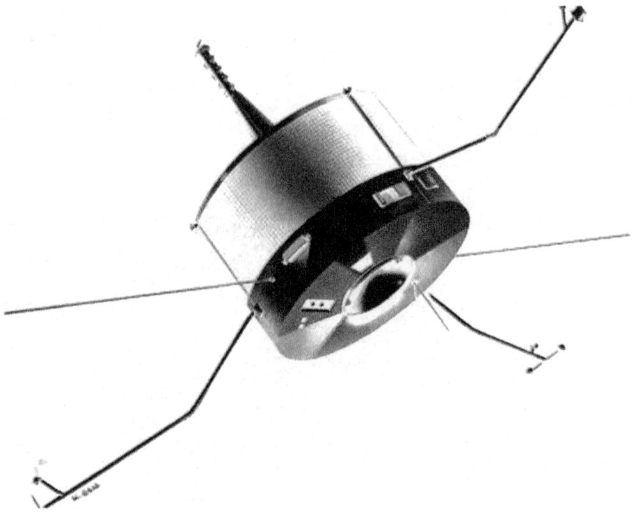

Figure 9.9. Spin-Stabilized Satellite ISEE-2 (ISEE-B Executive Summary of He Phase C/D Proposal by Dornier System and the STAR Consortium 1974).

As in the case of gravity-gradient stabilization, stabilization by rotation provides a reference direction and constrains two attitude axes. Spin stabilization provides an easy method for highly accurate, low cost, uncomplicated, and reliable attitude control systems, and has been used extensively in the past. Nevertheless its rigid observation geometry makes them inadequate for some application and it introduces constraints on the satellite's configuration. This will be studied in detail in Section 10.4.2.

9.3.4. Dual spin and momentum bias attitude control

In dual spin satellites one part of it rotates with respect to the other. This double rotation can provide the gyroscopic inertia advantage of the spin-stabilized satellites while providing flexibility of pointing for communication antennae or instruments that would be located in the nonrotating part of the satellite. In the limit, the rotating part of the satellite can be reduced to just a single momentum wheel. In this case the satellite, as seen from outside, does not rotate but the permanently rotating momentum wheel provides the necessary momentum bias and its associated good pointing stability and gyroscopic stiffness. Iridium, described in detail in Section 9.6.3, provides an example of this design. In some cases the combined satellite momentum can be reduced to zero; the satellite will not benefit from the intrinsic pointing stability provided by gyroscopic inertia, but it will enjoy maximum pointing flexibility. The Defense Support Satellite (DSP) (Figure 9.10) is continuously pointed toward Nadir, observing Earth from the geostationary orbit and is used to monitor ballistic missiles with a powerful infrared telescope. The satellite body rotates at 5.7 rpm and the telescope boresight is tilted 7° with respect to the Earth-pointed rotating axis; the satellite's rotation allows the telescope to scan the surface of the Earth. To compensate for the large gyroscopic torque produced by the satellite's large rotating body, DSP satellites feature an exception-

Figure 9.10. DSP Configuration (http://upload.wikimedia.org/wikipedia/commons/5/5d/DSP_Phase3.jpg).

ally large momentum wheel weighing 80 kg, counter-rotating in the opposite direction and offsetting the satellite's overall momentum.

9.3.5. Inertially stabilized attitude control

Inertially stabilized satellites can point in any direction and keep or change the pointing direction, as the mission requires. This provides the highest possible flexibility for the satellite attitude. In order to achieve this, the satellite's attitude control features a servo-system possessing sensors, actuators, feedback control loops, and control logic. This is the broadest attitude control strategy and it uses a large range of attitude sensors, actuators, and control logic as required by attitude agility, accuracy, and stability requirements. In the past this approach was considered to be complex and expensive, but advances in digital technology have allowed the use of inertial attitude control on more inexpensive satellites. The low cost DMC family of satellites (Section 11.4.4) is an example of this. It is possible to divide inertially stabilized satellites into three groups:

- *Nadir-pointed*: Nadir-pointed satellites face the central body they are orbiting, usually Earth. This is the required attitude for Earth-observing satellites and is used frequently. Sentinel-3 (Section 9.6.1) offers a good example of this kind of mission.
- *Agile, Nadir-pointed*: In some cases, Earth-observing satellites require rapid repointing within a large area around the Nadir direction. This provides the necessary pointing flexibility for observing small areas of special interest in exceptional detail. Pleiades (Section 9.6.4) provides an example of this.
- *Inertially pointed:* These satellites can be pointed in any direction within inertial space. All the astronomical observatory missions mentioned in Section 7.2 are of this type.

9.4. Orbits and orbit types

The selection of a satellite's orbit necessarily entails the selection of all the parameters defining it. Orbit definition consists of six parameters:

- The longitude of the ascending node and the orbit inclination, which define the plane of the orbit within inertial space
- The semi-major axis, the eccentricity, and the argument of the periapsis, which define the orbit within its orbital plane
- The satellite anomaly, which defines the satellite's position within its orbit

The longitude of the ascending node (Ω) is the angle from the origin of latitude to the direction of the ascending node of the orbit. The ascending node is the point where the orbit crosses the reference plane in the ascending direction. For orbits around the Earth the reference plane is the equatorial plane, and the Aries direction (the direction of the crossing of the equatorial and ecliptic planes) is the origin. For heliocentric orbits the reference plane is the ecliptic and the Aries direction is also the reference.

Inclination (i) is the angle between the plane of the orbit and the reference plane.

Eccentricity (e) defines the shape of the orbit, which, in the case of Newtonian orbits around a point mass, is a conic section. Circles have an eccentricity of 0; ellipses of increasing elongation have eccentricities from 0 to 1; parabolas have an eccentricity of 1; and the hyperbola an eccentricity larger than 1.

The *semi-major axis (a)* defines the size of the orbit. In elliptical orbits this is one-half of the longest diameter or the major axis of the ellipse.

The *argument of periapsis (ω)* is the angle from the periapsis (the point of closest approach to the central body) to the ascending node (Figure 9.11).

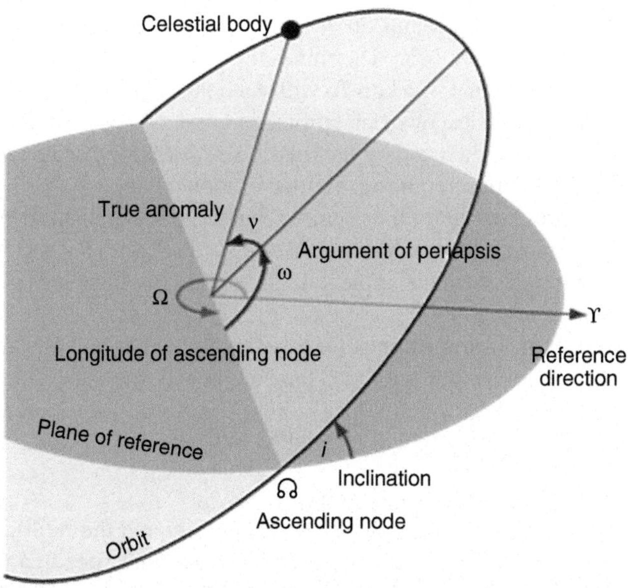

Figure 9.11. Orbital Parameters Definition (http://en.wikipedia.org/wiki/Image:Orbit1.svg).

The anomaly (υ) is the angular distance from the satellite to the periapsis.

In addition to the six parameters defined above, there are three useful supplementary parameters that are functions of these six:

Local time of the ascending node defines the angle of the orbit with respect to the Sun by providing the local solar time on the ground just below the satellite when it crosses over the equator. This parameter is used to define the orbit of Earth-observing satellites.

Orbital period is the time needed for the satellite to complete an orbit and is a function of the semi-major axis.

Altitude over the Earth is variable due to orbit eccentricities and the Earth being nonspherical. Normally the altitude over the Earth is defined as the semi-major axis minus the equatorial radius of the Earth. This parameter is useful for understanding the observing conditions of satellites in Earth orbit.

9.4.1. Low Earth Orbit (LEO)

9.4.1.1. General

Low Earth Orbits (LEO), as the name implies, is any orbit around the Earth at relatively low altitude. Launchers can reach LEO orbits with moderate amounts of energy. LEO orbits are also suitable for missions studying the Earth in detail because the distance from satellite to surface allows the observation of fine details without large instruments. Satellites studying the LEO local environment – for example, those studying the Earth's magnetic field or its atmospheric density – also use LEO. The typical altitude range for these orbits is from 500 to 900 km. At lower altitudes the atmospheric density is high, and missions flying below 500 km need progressively more complex orbit control components, including large amounts of fuel. Aeolus (8.3.1) orbits at 400 km to simplify its active optical Lidar, but it needs to carry 266 kg of fuel for a satellite dry mass (total mass excluding fuel) of 1,100 kg. ESA's GOCE (10.3.1) is able to fly at 250 km but requires continuous electric propulsion for drag compensation. The harsh radiation environment of the Van Allen belts determines the LEO's upper limit. CNES's Jason mission flies at an altitude of 1,336 km. To withstand the resulting harsh radiation environment Jason requires the use of special protection.

Orbit altitude is also influenced by the mission's observing needs; lower altitudes are preferable for missions requiring high-resolution image of small areas, while higher altitudes are preferable for those employing large swath instruments of moderate resolution. A typical high-resolution mission like Pleiades (Section 9.6.4) flies at an altitude of 694 km, while a typical large swath mission such as Sentinel-3 (Section 9.6.1) orbits at 800 km.

Orbit altitude has clear implications for mission operations and contact opportunities with Earth; the higher the altitude, the longer the contacts with the ground. A typical 90-min low Earth orbit provides around 10 min of visibility for each ground station. Ground station visibility will be studied in Chapters 11 and 12.

9.4.1.2. High inclination

In these orbits, the angle between the satellite velocity and the rotational speed of the Earth is close to 90°; therefore, the Earth's rotational speed cannot help the

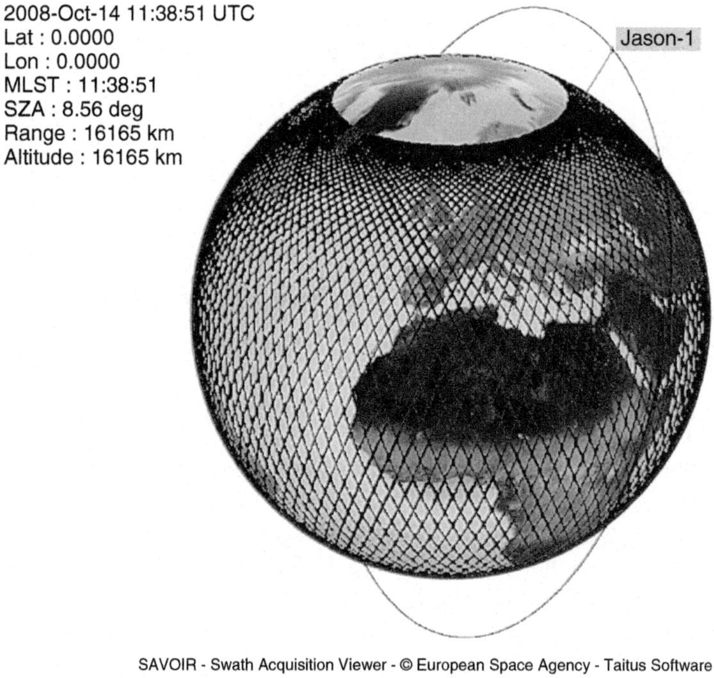

Figure 9.12. Orbit, Orbital Plane, and Orbit Track of Jason.

launcher and reaching high inclination orbits requires more energy than reaching low inclination ones. The small European launcher Vega, launched from Kourou, Guyana, near the Equator, can put ~2,000 kg into a circular orbit at 800 km with an inclination of 0°, but only ~1,500 km at a 90° inclination. High inclination LEO orbits, on the other hand, provide global or near-global coverage of the Earth, which makes them ideal for missions studying the entire Earth. High inclination orbits also ensure frequent contact with ground stations located at high latitudes near the Poles, facilitating good communications between satellite and ground.

Figure 9.12 provides the track pattern of the radar altimetry mission Jason. The orbit, as mentioned above, has an altitude of 1,336 km and an inclination of 66°. Its orbital period is adjusted so that the orbit repeats its ground track pattern every 127 orbits, in slightly less than 10 days. Jason acts as a reference mission for determining sea levels and ocean circulation with extreme accuracy. The mission's altitude is driven by the need to minimize gravity and orbit drag perturbations in order to provide the highest orbital determination accuracy. The inclination is driven by the conflicting needs of covering a majority of the oceans while ensuring a good geometry in the crossing points that act as reference tie-downs and shall be located as well as possible. Jason's orbit also acts to minimize the effects of the tides on the measurement of average sea levels. The communications constellation Iridium (Section 9.6.3) also utilizes LEO high inclination orbits, in the case of Iridium in order to provide true global communications coverage for Earth.

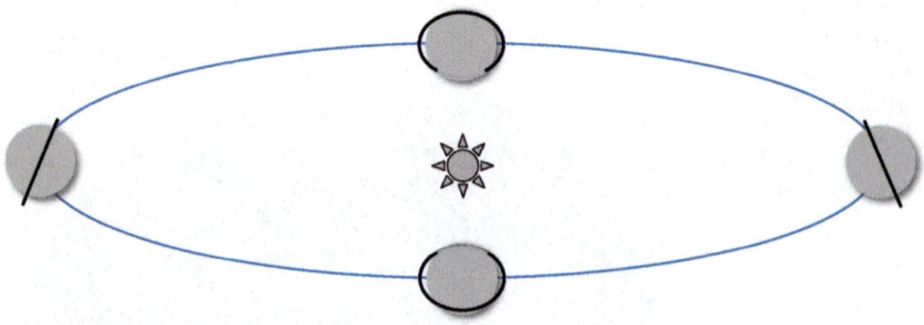

Figure 9.13. Sun-Synchronous Orbits and Sun Geometry Along the Year.

9.4.1.3. LEO low inclination

Low inclination LEO orbits maximize the launcher's mass injection capability, making the orbit ideal for applications not requiring any specific observation geometry, for instance, microgravity or technology demonstration missions, or those with crews. Low inclination satellites do not provide coverage of the entire Earth but they do offer rapid coverage of the limited areas they fly over. This has clear advantages for missions whose area of interest is limited to equatorial or tropical regions, such as those studying hurricanes or other intense, dynamic tropical events. NASA's Tropical Rainfall Monitoring Mission (TRMM) and CNES-ISRO MeghaTropiques (Section 8.3.4) study tropical weather patterns. TRMM's orbit has an inclination of 35°, and that of MeghaTropiques, 20°. Summer forest fires are common in areas with a Mediterranean climate located around 40° of latitude; so an inclination of about 40° would be optimal for a mission aimed at detecting and monitoring forest fires in these areas.

9.4.2. LEO Sun-synchronous orbit

Sun-synchronous orbits have an inclination such that the J2 term of the Earth's gravity field (Section 9.2.1) generates a precession of one turn per year of the orbital plane; that is, the orbital plane rotates at the same speed as the Sun. This ensures a fairly constant angle between the orbital plane and the direction of the Sun. The angle is only "fairly" constant due to the seasonal variations produced by the angle between the equatorial plane of the Earth and the ecliptic plane of the Sun (Figure 9.13).

Sun-synchronous orbits have five strong advantages:

- The inclination corresponding to LEO Sun-synchronous orbits is close to the polar one, allowing almost full Earth coverage. For the typical range of Earth-orbiting altitudes between 500 and 1,000 km, this inclination varies from 97.4° to 99.5°.
- This fairly constant Sun geometry produces uniform lighting conditions of the observation area. The uniform illumination of the observation area allows a correspondingly consistent quality in the images produced, and all images of the same area taken during the mission are easily compared.
- The geometry of the Sun with respect to the satellite is constant, allowing one side of the satellite to be consistently lit by the Sun, and another that will never be, thus simplifying the spacecraft's configuration and thermal design.

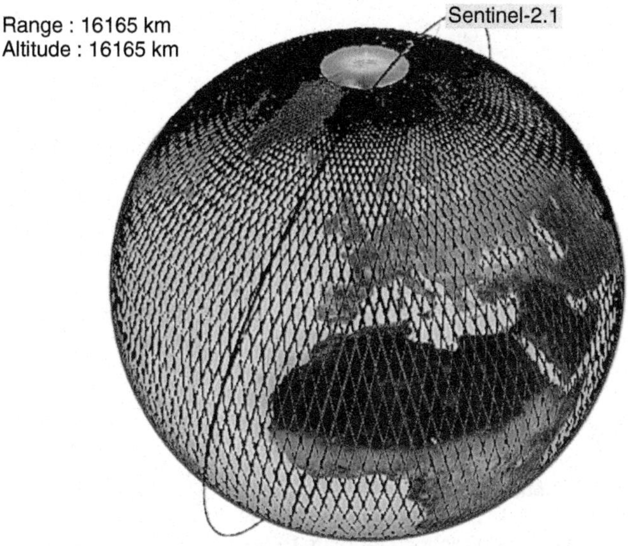

Figure 9.14. Sentinel-2-1 Orbit. The Orbital Period Is such that the Orbital Tracks over the Earth Repeat Every 10 Solar Days and 143 Orbits.

- The high inclination orbits allows for frequent contacts with ground stations located at high latitudes near the poles.
- The Nadir-oriented face is pointed toward the planet being orbited and is the logical place for locating the instruments and communication antennas.

The first three advantages are relevant to the mode of observation, while the last two are directly related to the configuration. All these advantages explain the popularity of this kind of orbit for providing global observations of the Earth. Earth observation normally requires regular orbital patterns for optimal Earth coverage, revisits, and repetitive contact patterns with ground stations. This is why most Earth observation missions exactly repeat their orbital track patterns, where m orbital periods coincide with n solar days. This is the case of Sentinel-2 (Figure 9.14) where 143 orbits last 10 solar days.

The local time of the ascending node was introduced at the beginning of Section 9.4. The selection of this parameter, which defines the geometry of the Sun with respect to the satellite and the area being observed, must take into account the mission's objectives. Local times near 6:00 or 18:00 define an almost constant geometry with the solar direction close to the perpendicular of the orbital plane. This simplifies the design of the satellite configuration, allowing for the easy production of solar energy and the easy dissipation of heat. Lighting conditions, however, are poor because the Sun is just above or below the horizon of the areas being observed and cannot properly illuminate them. Hence, these orbits are only suitable for missions in which solar illumination is unnecessary for the effective observation of their targets. This is the case of Synthetic Aperture Radars (SAR's). Local times around 12:00 provide adequate illumination of the solar array with a single axis rotation perpendicular to the orbital plane, but images taken from the satellite right at Zenith are flat and of poor quality, so this orbit is rarely used. Intermediate local times from

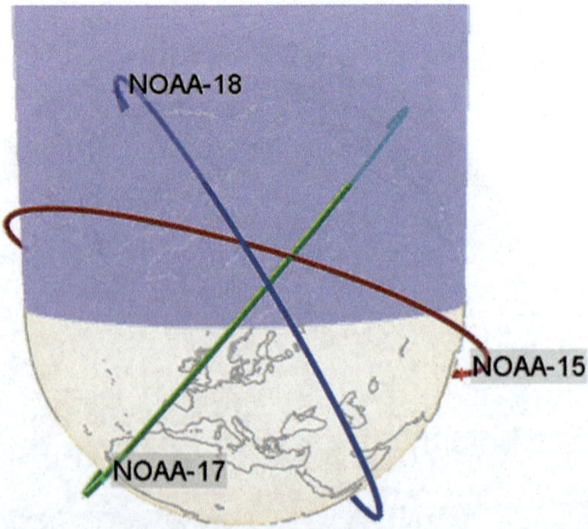

Figure 9.15. Constellation with Three Well-Distributed Solar Times: NOAA NPOESS 16 at 17:30, NOAA 17 at 21:30, and NOAA 18 at 13:30. The Gray Umbra Area Indicates the Direction Opposite to the Sun.

8:00 to 10:30 are considered ideal for optical Earth observations, and most missions of this sort fly around 10:00; these include Ikonos, Spot, Landsat, Sentinel-3, and Sentinel-2. An additional advantage of this local time is that any early morning mist has dissipated but late afternoon tropical clouds have not yet formed, permitting clear viewing with minimum cloud cover. Observations from each Sun-synchronous orbit are made just twice a day: one during the ascending and another during the descending orbital track, and only one of the two will be made with solar illumination. If the mission requires frequent observations, a constellation of satellites flying at different local times offers a solution. The operational meteorological constellation NPOESS (Figure 9.15) is an example, and employs three satellites, which renew the information covering the Earth every 4 h.

To ensure the uniformity of observing conditions, Earth-orbiting satellites should maintain as constant a distance to the surfaces being observed as possible. As the Earth is not perfectly spherical, however, actual orbits cannot be perfectly circular and the distance between the satellite and ground changes all along the orbital path (Figure 9.16). The situation becomes even more complex when taking into account that the higher-order gravity terms J2, J3, and J12 (Section 9.2.1) will affect the eccentricity "e" and argument of perigee "ω" of the orbit. This will further increase the variations in the distance from satellite to ground. Nevertheless, certain combinations of initial eccentricity and argument of perigee help stabilize the long-term evolution of both parameters, minimizing these variations in the orbital altitude of the satellite. These latter orbits are known as frozen orbits[8] and Sun-synchronous missions employ them extensively. For the commonest altitude range of around 800 km, the initial parameters providing a frozen orbit correspond to an eccentricity around 0.001 and an "ω" around 90°. Sentinel-3 (Section 9.6.1) and Pleiades (Section 9.6.4) fly in Sun-synchronous frozen orbits.

[8]Chobotov Orbital Mechanics, AIAA Education Series.

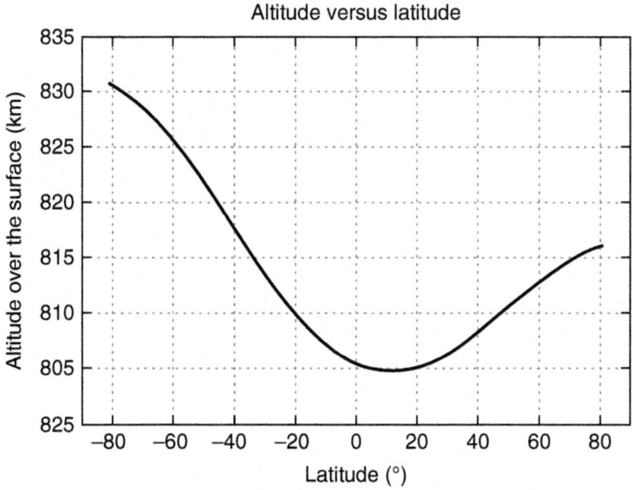

Figure 9.16. Satellite Altitude Variation Along the Frozen Sun-Synchronous Sentinel-3 Orbit.

In the case of Sentinel-3, which has an operating altitude (defined as the semi-major axis minus the Earth's equatorial radius) of 800 km, the actual distances from the satellite to the ground can be seen in Figure 9.16.

9.4.3. MEO

Mid-altitude Earth's orbits (MEO) can be defined as comprising from the 1,200 km altitude of the higher LEO orbits to the 35,786 km altitude of the geostationary orbits. In terms of orbital period, this includes from 100 min to 24 h. This is a broad range, but missions utilizing these altitudes are infrequent because the radiation of the Van Allen belts creates an unfriendly environment for satellites, limiting orbital possibilities to the area around 20,000 km above the inner belt, but still below the outer one. The long periods of MEO orbits provide for relatively slow transits of the satellite in relation to the Earth's surface, offering prolonged uninterrupted visibility of large parts of the Earth. This is especially advantageous for navigational satellite constellations like GPS or Galileo, and both these constellations fly circular MEO orbits. Their high altitude also offers low levels of gravity and atmospheric density perturbations, allowing precise determination of each satellite's orbit, so crucial for navigational satellites. The GPS satellites fly in six uniformly distributed orbital planes at an altitude of 20,200 km, with an orbit inclination of 55° (Figure 9.17). Molniya orbits are another type of MEO. These orbits have an inclination of 63.4° or 116.6°. At this inclination the Earth's J2 component produces no drift of the right ascension of the ascending node. Molniya orbits are highly eccentric and feature periods in resonance with the Earth's rotation, of either 12 or 24 h. These orbits cross the Van Allen belts, and satellites employing them must be designed to withstand the harsh radiation environment. These orbits afford long periods of visibility between

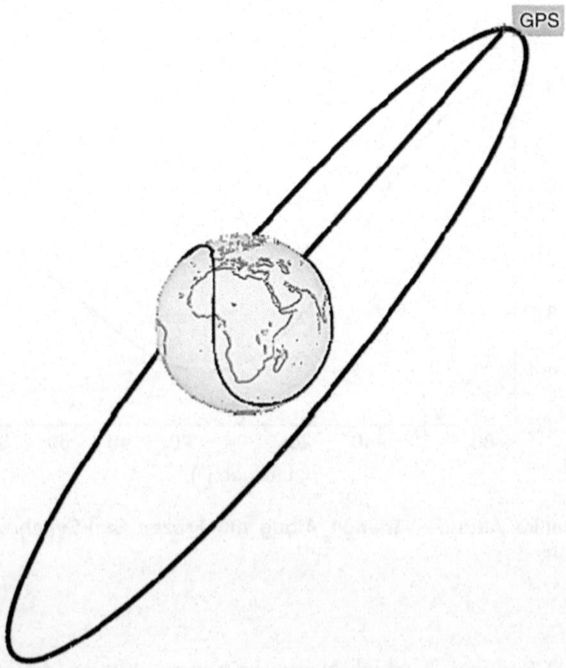

Figure 9.17. Orbit and Track over the Earth of a GPS Satellite.

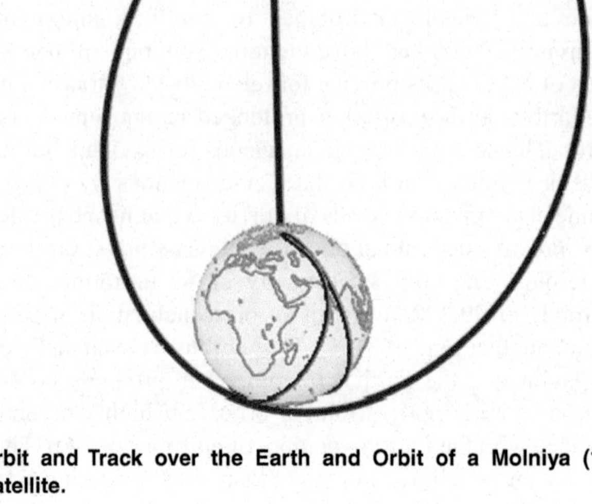

Figure 9.18. Orbit and Track over the Earth and Orbit of a Molniya (12 h of Period) Satellite.

the satellite and regions of the Earth at high latitudes coinciding with the apogee perigee of the orbit: one zone in the case of 24 h orbits; two zones in the case of 12 h ones (Figure 9.18). This is useful for missions calling for the effective monitoring of these regions. Molniya orbits have been used extensively by Russia to provide communications over its territory.

9.4.4. Geosynchronous and geostationary orbits

Geosynchronous orbits are orbits with a period of 24 h. Geostationary (GEO) orbits are geosynchronous orbits with an eccentricity and orbit inclination of zero. GEO orbits provide a constant relative geometry between Earth's surface and the satellite, affording permanent monitoring of wide areas of the Earth and permanent communication links from them to the satellite. The geostationary orbit is the orbit of choice for communication satellites. GEO satellites are also used for operational meteorology. A single GEO satellite only provides visibility over one area; to provide global visibility over most of the Earth, three or four satellites are needed. Nevertheless, GEO satellites, whose orbit is in the equatorial plane, cannot observe or be observed from the poles. The present GEO meteorological operational constellation is made up of four GEO satellites, two provided by the U.S., one by Europe and one by Japan.

The distance from a GEO satellite to the ground is 35,786 km. The communications link between the satellite and Earth must be strong enough to compensate for this great distance. In the case of Earth observation satellites the spatial resolution obtained from GEO orbits will be correspondingly inferior to that normally obtained from LEO orbits. The best communications and observation geometry is over the equator where the satellite is directly overhead; the geometry degrades when moving toward the Poles. At 76.3° latitude geostationary satellites are just 5° above the local horizon. This low elevation is considered a practical limit for good quality communication links, because the signal becomes increasingly affected by the thickness of atmosphere. The observation geometry of high latitude regions of the Earth will also be severely distorted by the wide angle between the line of sight of the satellite and the ground horizon. The geometrical distortion of images was studied in Section 8.4.3. GEO observations above 60° of latitude are of questionable quality, due to the distorted geometry, the thickness of the atmosphere being traversed by the signals and the low level of solar illumination at these high latitudes. GEO orbits are located in the equatorial plane and the Sun is in the ecliptic plane (Figure 9.19), so for half the year the Sun will be "above" a GEO satellite's orbital plane, and half the year "below" it.

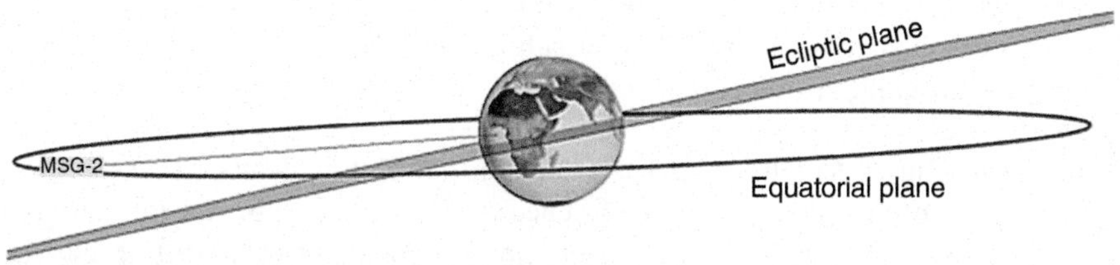

Figure 9.19. Meteosat Second Generation Orbit and Its Geometry with Respect to the Ecliptic Plane.

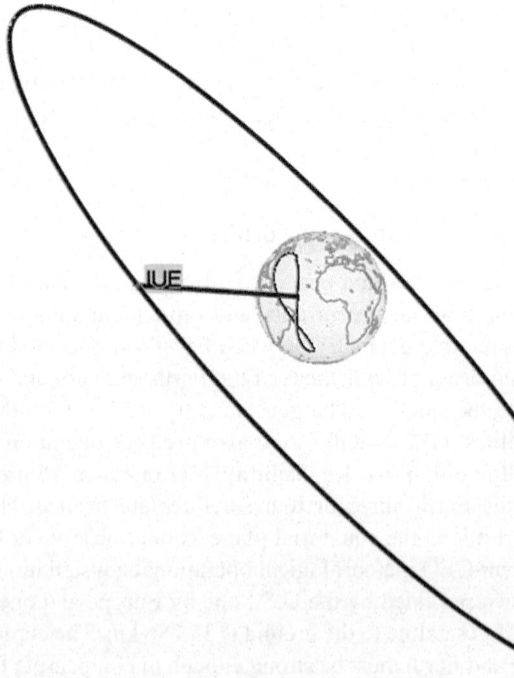

Figure 9.20. Ground Track of the Geosynchronous IUE.

Geosynchronous nongeostationary orbits are of interest for some applications. A high inclination, circular geosynchronous orbit, for instance, provides good visibility at higher latitudes, such as those over the poles. Even so, while observing one pole, the other will not be visible to the satellite, so a single satellite will not afford continuous observing capabilities, which is precisely the strongest asset of geosynchronous observations. Three or four satellites are needed for continuous observation of both poles; as this is generally unpractical, Molniya orbits (Section 9.4.3) are preferred for high latitude specialized satellites. IUE (Section 7.2.1) used a geosynchronous nongeostationary orbit to provide continuous communication links with either of the two ground stations used for both missions: one in the USA and one in Europe (Figure 9.20). In order to precisely maintain their orbit, satellites in GEO must struggle against two kinds of disturbances:

- Those that alter the orbit inclination, shifting the orbital track over the Earth. They are caused by the gravitational pull of the Sun and Moon nudging the satellite away from the equatorial plane. The result is a 1° increase of the orbit inclination per year, reaching a maximum of 15° in 17 years, when the effect will start to reverse itself.
- Those causing the satellite's track over the Earth to shift with an east-west orientation. The Earth's equatorial circumference is not perfectly circular (a fact represented by the J22 gravity term). J22 will produce a slow motion drift of the GEO satellite track around stable longitudes, coinciding with the minor radius of the Earth and located at longitudes 75.3° E and 255.3° E.

In most missions, it is necessary to control these orbital perturbations. The MSG satellite depicted in Figure 7.3 uses around 11%[9] of its propellant load of 976 kg for north-south correction, and 4% for east-west correction. The remaining 83% is needed to inject the satellite into the final GEO orbit from the intermediate highly elliptical geosynchronous transfer orbit provided by the Ariane launcher. A geosynchronous transfer orbit (GTO) is an intermediate orbit used to reach GEO. It is highly elliptic with an apogee at the GEO altitude and a perigee at a few hundred kilometers of altitude. Many launchers, like Ariane, inject the satellites they carry into GTO and the satellites themselves should carry enough fuel to allow their own transfer from the GTO to GEO.

9.4.5. Longer period Earth orbit

Very long period orbits (more than 24 h) are useful for situating satellites far from Earth and its perturbations, affording long uninterrupted observation periods of the celestial sphere. This is especially beneficial for astronomical research; XMM-Newton (Figure 7.8) is a prime example; its orbit features 7,365 km at perigee, 113,774 at apogee, an orbit inclination of 49° and an orbital period of 2 days (Figure 9.22). While near the apogee, the orbit permits up to 30 h of uninterrupted observations, always pointing toward the same direction in space, and remaining well above the Earth's radiation belts and their potential impact on the instruments. Cluster (Section 11.4.1 and Figure 9.21) is a constellation of four satellites carrying in situ measuring instruments in slightly different orbits with 25,500 km of perigee, 125,000 km of apogee, and at inclination of 90°. The orbits were chosen for optimal analysis of the Earth's magnetosphere and the Sun-Earth interaction environment. The highly inclined, very elliptical orbit permits measurements through broad areas of the Earth's magnetosphere, the instruments sampling various distances to the Earth and different latitudes of the magnetosphere, from those nearest the Earth's equatorial plane to those closest to the poles.

9.4.6. Lagrangian points

Lagrangian points are locations in space where a satellite's orbital motion is balanced by the gravity of two large objects, and hence remains stationary with respect to them. They are stationary solutions of the circular "three-body problem." The Lagrangian points of the Earth-Sun system are of particular importance because satellites orbiting at these points will maintain a constant geometry with respect to the Earth and the Sun. There are five Lagrangian points in the Earth-Sun system, and two of them, L1 and L2, are useful for space missions. L1 is the point between the Earth and Sun where their gravity is equal and is situated along the Earth-Sun line and around 1.5 million km from the Earth. This orbit is ideal for continuous monitoring of the Sun free of perturbations caused by the Earth. ESA/NASA's Soho mission, who carries Sun observing instruments, is situated at the L1 point, as was NASA's ISEE-3, which

---

[9]Meteosat Second Generation. The Satellite Development, ESA BR-153, ESA November 1999.

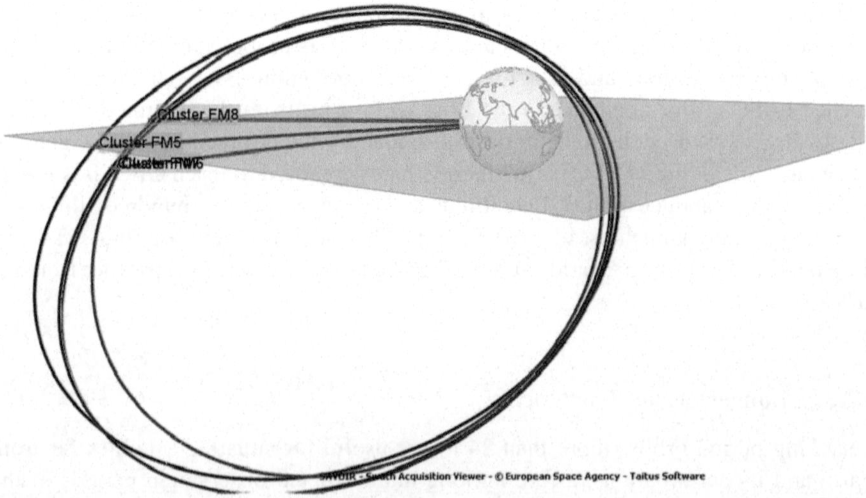

Figure 9.21. Orbits of the Four Cluster Satellites. The Ecliptic Plane Is also Provided.

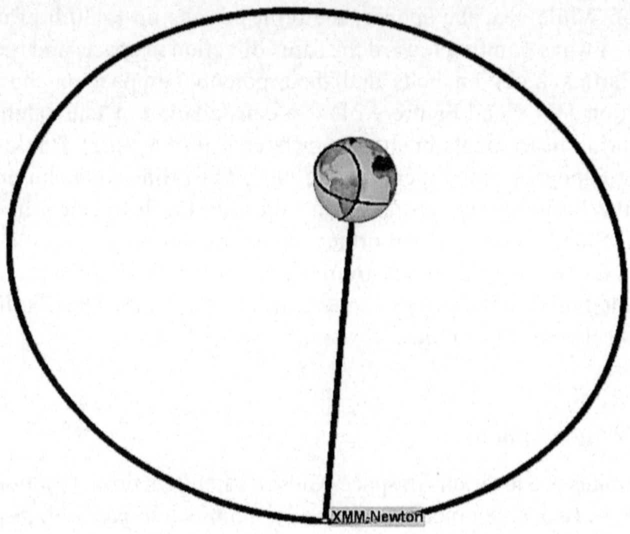

Figure 9.22. XMM Orbit and Ground Track of the Satellite over the Earth.

studied electromagnetic disturbances reaching Earth from the Sun. The L2 point is also situated about 1.5 million km from Earth, but behind it, on the side opposite the Sun. JWST (Sections 8.3.2 and 10.3.2) uses this orbit. Satellites in L1 and L2 not only maintain a constant geometry with respect to the Earth and the Sun, they also enjoy extreme protection against the optical and thermal perturbations caused by the Earth and Sun for instruments needing high sensitivity and accuracy, such as those detecting remote or weak astronomical signals. Notwithstanding the above description, orbits located precisely at the L1 or L2 positions do suffer from instability, and satellites located at either of the two points will gradually drift away. There are trajectories just to the side of L1 and L2 called "halo" orbits, which only demand a moderate level

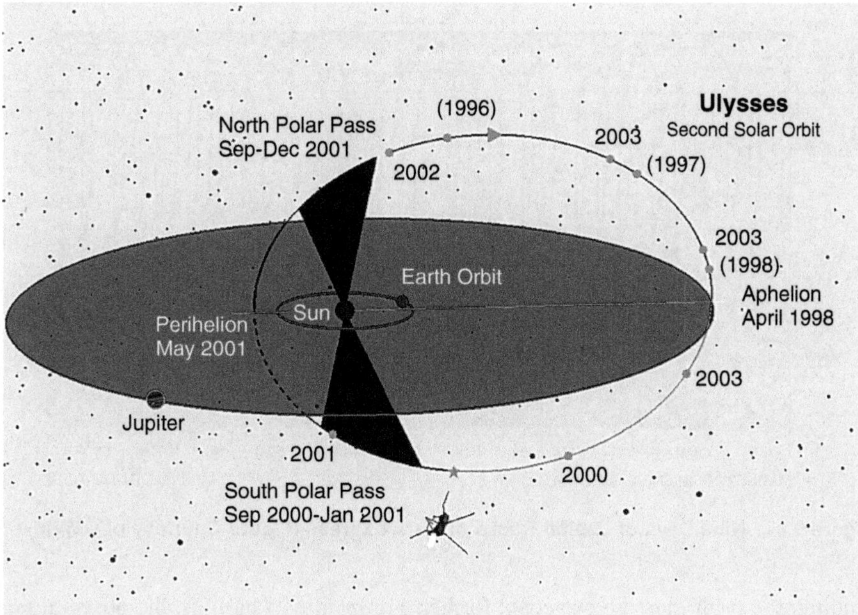

Figure 9.23. Ulysses Second Solar Orbit (http://ulysses.esa.int/science-e-media/img/ba/Ulysses).

of orbital correction to be maintained, and these are the orbits actually used for space missions employing the Lagrangian points.

9.4.7. Interplanetary orbits

Interplanetary orbits are necessary for missions requiring the observation of "in situ" events located in interplanetary space, or when the study of remote celestial bodies requires close-up examination.

Ulysses (Sections 8.3.5 and 9.6.2) is an example of a mission carrying out in situ observations of regions remote from the ecliptic plane and that requires an orbit able to propel the satellite through the high solar latitude zones of the heliosphere (Figure 9.23). Ulysses used Jupiter to perform a massive change of orbit inclination in order to reach its targeted orbit. The satellite employed the two dedicated large rockets stages IUS and PAM (Figure 9.27) to attain its goal. Rosetta (Section 11.4.2) is another mission following a complex interplanetary path in order to observe two asteroids on its way to a rendezvous with the comet that is its final target. Rosetta's orbit is especially complex. It was launched from Kourou in March 2004 and includes three Earth swing-by maneuvers, one swing-by maneuver with Mars, two along-the-way swing-bys with asteroids Lutetia and Steins, and the final rendezvous with the comet in November of 2014. The mission will last 10 years.

9.4.8. Orbits around other planets

Satellites reaching other planets arrive with an excess of velocity that must be reduced to allow their capture. In order to undertake observations of the surface or to make a

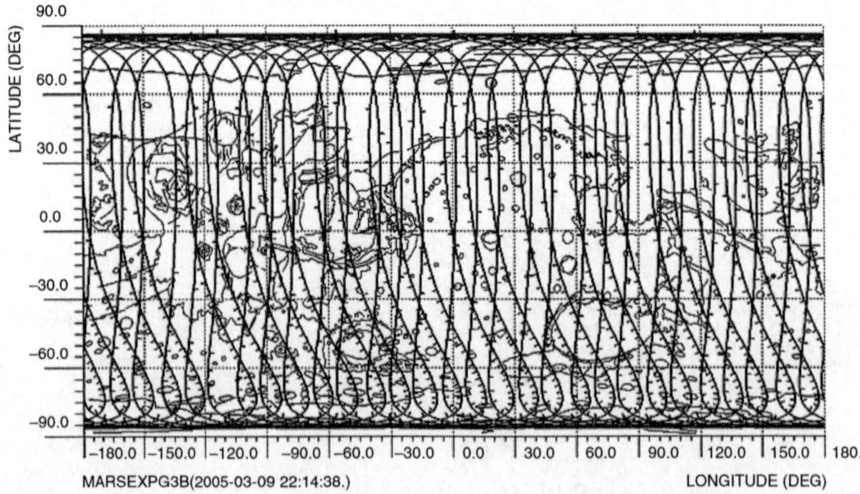

Figure 9.24. Nine Days of Orbital Tracks of Mars Express (Figure Courtesy of GMV).

landing, the orbit must be corrected further. For orbiting satellites, the observational patterns of the craft over the planet will depend on the planet's rotation speed, and, as with orbits around the Earth, on the planet's degree of deviation from the spherical. Mars has a very thin atmosphere, so orbit altitudes can be much lower than those around the Earth, allowing high-resolution observations with proportionally smaller optics. The final orbit of ESA's Mars Express around Mars (Figure 9.24) had an inclination of 86°, 10,107 km of apogee over the surface of the planet and 298 km of perigee, with an orbital period of 6.7 h.

9.5. Mission phases and modes and satellite attitude

Throughout the life of a satellite, it passes through phases requiring different attitudes or attitude control approaches. Each of these attitude control approaches is called a "mission mode." The following phases and their corresponding attitude modes are typical:

- Initialization. This phase begins when the launcher releases the satellite and ends when the satellite has acquired a safe attitude that ensures adequate power production, thermal dissipation, and communications. During this phase, the satellite will have also switched-on all the components necessary for ensuring its own unlimited survival. That is, the satellite finishes in a "safe mode" where the attitude and attitude control ensure the satellite's secure operation during an unlimited period of time.
- Commissioning. This begins when the satellite reconfigures itself by deploying all the mechanisms not yet deployed during the initialization phase and progressively switching-on, calibrating, and characterizing all the satellite's platform and instrument components. This phase ends with the satellite in a "nominal mode" and attitude, ready to perform all its functions.

- The nominal operations phase lasts, with interruptions, for the duration of the mission. The satellite begins to acquire observation data as specified by the mission's aims. During this phase the satellite maintains its nominal mode with its nominal attitude.
- Orbit control. This mode will be activated regularly during short periods. The satellite will suspend data acquisition in order to perform orbit correction, firing its thrusters as needed. The satellite may need to perform large attitude change maneuvers to orientate the thrusters in the direction required by the orbital control needs.
- In case of major satellite failure the satellite enters into safe mode. In this mode the satellite has a safe attitude that can be maintained permanently without danger and with most of the satellite inoperative.
- Dormant mode. Missions that require being minimally active during long periods of time will require this mode. For example Rosetta (Section 11.4.2) is in a dormant mode during most of its long fly to arrive to its intended comet target.
- Deorbiting. This phase is activated at the end of the mission. The satellite could be required to perform dedicated orbit correction maneuvers to ensure reentry into Earth's atmosphere. This step is taken to prevent the no-longer operational satellite from posing a danger to future space missions in the form of space debris. During the "deorbiting" mode, much of the spacecraft's electronics will be switched off and fuel purged from the tanks.

These phases and modes can vary according to the needs of each mission; for example, missions orbiting Mars will not require deorbiting; some missions may not need an orbit control mode. Nevertheless, in order to ensure the satellite's safety, all missions require the safe mode in which the most important parameter is design robustness, and the nominal mode, guaranteeing the degree of accuracy in the delivery of products demanded by the mission. These two modes require different sets of sensors and actuators, as well as different attitude control loops. The robust attitude control of the safe mode requires the use of reliable performance sensors, actuators, and control logic. This permits the attitude control to continue functioning even when a large part of the satellite's components are disabled or intentionally switched off. Sensors suitable for the safe mode include magnetometers or Sun sensors; suitable actuators include magneto-torquers and wheels utilized solely in open loops to provide gyroscopic stiffness. Control logic can be digital or analog, as long as it can function adequately when the satellite's main digital computers are inactive. The more accurate attitude control of the nominal mode can be achieved by using Earth or star sensors and momentum wheels managed by digital control loops. The control loop is likely to be digital and reside in the satellite's central computer or in one dedicated solely to attitude control.

Figure 9.25 provides the on-ground and in-orbit satellite modes of Sentinel-3. The off mode, test mode, and stand-by mode are ground modes to allow satellite testing. The launch mode is the mode of the satellite while attached to the launcher. All the other modes have been already explained. Modes transitions can be performed under telecommand (TC) or autonomously (Auto).

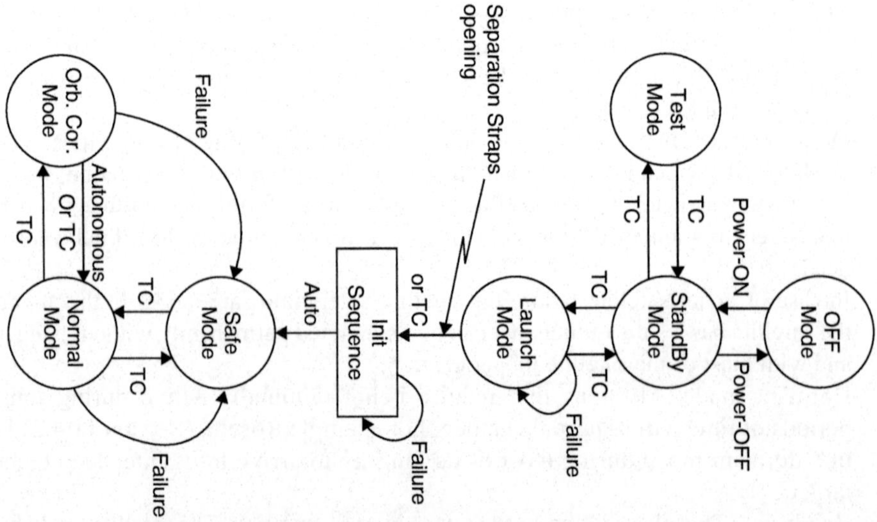

Figure 9.25. Example of On-Ground and In-Orbit Satellite Modes and Mode Transitions (Sentine-3 Definition Phase, AOCS Technical Justification, S3-RP-AAF-SC-375, Alcatel Alenia Space 2006).

9.6. Orbit and attitude examples

9.6.1. Sentinel-3

The aims and instruments of this operational Earth observation mission were introduced in Section 8.3.3. This section will examine Sentinel-3's orbit and attitude.

9.6.1.1. Orbit

Sentinel-3[10] flies in a Sun-synchronous orbit with an inclination of 98.65°and a semi-major axis of 7,178 km, which is equivalent to an orbital altitude of 800 km. The orbital period is 6,059 s. To afford optimal coverage for land and sea images and for ocean altimetry, the satellite has a repetitive orbital pattern of 385 orbits every 27 days, which is equivalent to 14+7/27 orbits per day. As 7/27 is near to ¼, the pattern repeats itself almost exactly every 4 days. This permits optimal coverage for optical observation in 4 days – or in 2 days using two satellites in the same orbit but separated by 180° – and in 27 days for ocean altimetry. The local time of the ascending node of the orbital plane is 10:00. The orbit is frozen with an eccentricity of 0.001148 and with an argument of the perigee of 90° (Figure 9.26).

The satellite needs to control its orbit so that the actual ground track does not deviate from the nominal one by more than ±1 km. For all its orbit control needs, Sentinel-3 carries 120 kg of hydrazine and two sets of four 1 N thrusters. The thrusters are used for the correction of initial orbital injection errors, for orbit control and for deorbiting the satellite at the end of its operational life, when the thrusters will

[10] Sentinel-3 a Long-Term Monitoring of Ocean and Land, Y. Baillion, GMES IAC-07-B.1.2.04.

Figure 9.26. Sentinel-3 A and B Orbit and 1 Day of Ground Track of both Satellites.

reduce the satellite's orbit altitude to 500 km, where atmospheric drag will ensure reentry and destruction in the atmosphere within a reasonable period of time. Sentinel-3 carries a high quality GPS navigation sensor whose primary function is to support the radar altimeter by providing high precision orbit determination accuracy – after accurate on-ground processing – of the order of a few cm. The GPS is also used to provide real-time satellite position with an accuracy in the order of 10 m. Orbit correction maneuvers are only undertaken by ground command.

9.6.1.2. Attitude

Sentinel-3's attitude and orbit control system is Nadir-pointing, three-axis stabilized with "yaw steering." The latter feature is needed because in the nominal attitude mode the satellite points downward, keeping the instruments perpendicular to the Earth's surface. In yaw steering the satellite steers itself around the yaw direction to compensate for the image distortion caused by the Earth's rotation. This requires a yaw rotation of 4° at the equator and progressively decreasing until 0° at the poles, these changes occurring around the orbit with twice the orbital frequency.

In the high-accuracy nominal mode, the attitude control is based on four reaction wheels. By varying the speed of the wheels it is possible to provide carefully controlled torques around any desired direction. An optical head with three star trackers provides the satellite attitude and attitude rates. If one of the three star sensors were to fail, attitude and attitude rates could still be provided, albeit with a correspondingly degraded performance. The speed of the wheels is kept at acceptable levels by the use of magneto-torquers. The correct magneto-torque settings depend on the Earth's magnetic field, whose strength is derived from an on-board model of the field along with the satellite's position, provided by the GNSS sensor. In the nominal mode, the pointing error is better than 0.1° and the attitude measurement error is better than 0.01°. The satellite possesses a nonnominal safe mode that points the satellite toward the Sun, providing power and thermal conditions compatible with its long-term survival. In safe mode, Sentinel-3 employs a set of less accurate but highly reliable attitude sensors and actuators. Magnetometers, coarse Sun sensors, and a coarse gyroscope provide attitude sensing. The magnetotorquer provides attitude control and the wheels provide gyroscopic stability, but they stay in open loop and are not used for attitude correction. In this mode the strength and direction of the magnetic field is provided by magnetometers. The satellite also has an orbit control mode, which is used to perform large out-of-plane orbital correction maneuvers. During this mode the thrusters are used for attitude control and the wheels do not operate. The attitude is provided by the star trackers supplemented by the coarse rate sensors.

9.6.2. Ulysses

The scientific mission Ulysses, which was launched in 1990 and studied the Sun and the heliosphere, was introduced in Section 8.3.5. Ulysses was decommissioned in 2008 after 18 years of life.

9.6.2.1. Orbit

As mentioned in 8.3.5 Ulysses' instruments made in situ measurements and therefore had to travel through the heliosphere in order to study it. To provide a comprehensive profile of the heliosphere at all solar latitudes the satellite's final orbit was heliocentric, with 1.35 AU of perihelion, 5.4 AU of aphelion and 80° of inclination with respect to the solar equator (Figure 9.23). The satellite flew from one solar pole to the other along all the latitudes around the Sun, with an orbital period of 6.2 years.

In order to accomplish the principal objective of orbiting the solar poles at up to 80° of heliocentric latitude, the mission called for a very special orbit injection strategy, which required large amounts of energy. Ulysses, which weighed just 370 kg, needed an entire space shuttle to accommodate the satellite and one each of its IUS and PAM supplementary stages (Figure 9.27), which were fired to reach a final velocity of 15.4 km/s, making Ulysses at the time the fastest man-made object in history. The satellite also required a Jupiter swing-by to shift the original injection orbit from an inclination of 23° to the final 80° demanded by the mission. Ulysses took 16 months of interplanetary cruising to reach Jupiter.

The satellite featured an on-board propulsion subsystem used for small orbital correction maneuvers. The satellite's position was determined by the mission control center, which also commanded all necessary thrust operations. The propulsion subsystem included two sets of four thrusters, giving each of them a thrust of 2 N. The thrusters were fed with 33.5 kg of hydrazine contained in a single spherical pressurized tank.

9.6.2.2. Attitude

The satellite (Figure 8.10) is spin-stabilized and rotates at 5 rpm around the axis defined by the boresight of the antenna, which is pointed continuously toward Earth.

Figure 9.27. Ulysses on Top of the IUS and PAM Stages (ESA achievements, BR-250, ESA 2005).

Due to its 6.2-year orbital period, this direction changes so slowly that it can be considered as fixed in inertial space. The primary functions of the attitude and orbit control subsystem (AOCS) are to maintain the spacecraft's spin axis pointing toward Earth and to maintain its spin rate at 5 rpm. Additional functions are dictated by trajectory control needs, nutation damping, and by the measurement of the attitude for scientific reasons. The spacecraft's Earth-pointing attitude is measured and controlled by a system that derives spin-rate, spin-phase, and solar-aspect-angle information from redundant Sun sensors. A related AOCS operation is the periodic precession maneuvering for correction of the gradual Earth direction drift with respect to the spin axis. These operations are performed either on board in closed loop or in open loop via ground or time-tagged commands. The hydrazine thrusters also provide the necessary spin speed and direction control corrections. The attitude control is supported by a passive nutation damper, which is a tube containing a ball in the middle of a dense fluid that moves and dissipates energy when the rotation axis nutates with respect to the satellite nominal rotation axis. The satellite has one pair of 72.5 m-long wire booms kept in tension by the centrifugal force created by the spin of the satellite. Each wire boom has a passive tubular damper at its root, which reduces oscillations between the boom and the spacecraft. The AOCS includes failure-mode-detection and protection functions, which result in fail-safe operation and a reacquisition capability in both automatic and ground-initiated recovery sequences. The AOCS also provides autonomous system capabilities for safe spacecraft reconfiguration. This is required during unexpected and/or predicted periods of nontracking and because of the long signal travel time between ground and spacecraft. The satellite also has a search-mode to reacquire the Earth if no command is received after a specified time of up to 30 days. As explained above, to reach its final injection orbit, Ulysses required the firing of its two IUS and PAM stages. To provide stability, the firing was done at very high spin. This high spin was reduced to the nominal 5 rpm by deploying the mentioned wire booms.

9.6.3. Iridium

9.6.3.1. Aim

Iridium[11] is a 66-satellite constellation serving global mobile communications that provides true worldwide connectivity for cell-phones. The emitting Iridium telephone uplinks its signal to any visible satellite in the constellation, each of which has on-board switching technology that directs the signal from satellite to satellite toward the most conveniently situated neighboring satellite in the constellation. The signal will eventually be downlinked to either a regional gateway, which provides connectivity to the overall communication network, or directly to the receiving Iridium telephone. Iridium's satellite-to-satellite capability affords broad flexibility in the location and number of regional gateways, which connect the network directly to the public telephone system. Each satellite can direct up to 3,840 voice circuits. Signals from the Iridium emitting cell phone to the receiving satellite, and from the final emitting satellite to the receiving Iridium cell phone are all transmitted in the L band. Steerable Ka-band terminals located on the satellites transmit the signals between the satellites themselves and to and from the regional gateways. Iridium sells its services through communication companies that also provide value-added products and services that employ Iridium functionalities. Each satellite's mass is 680 kg. The original design life was 5–8 years. The satellites were launched in 1997 and 1998 but they were still functioning properly in 2009, exceeding with ample margin the design life requirement. The constellation will provide service until 2014 and an update for Iridium is presently under development.

9.6.3.2. Orbit

The Iridium constellation is distributed in six orbital planes. The baseline constellation has 66 satellites plus 12 in-orbit spares. The Iridium satellites fly in a LEO circular orbit at an altitude of 780 km and with an inclination of 86.4°. The altitude was selected to optimize the communications link between the satellite and the ground-located cell-phones. The inclination was selected to provide maximum orbital stability while maintaining an adequate separation between the satellites over the poles, avoiding danger of collision. The orbital period is 100 min and 28 s. The main requisite for the orbit design was the need to provide continuous total coverage of the Earth. Each Iridium satellite has a large ground-pointing antenna set providing 48 L band overlapping spot beams (Figure 9.28). The footprint's total size is approximately 4,400 km in diameter. The orbit is controlled with an accuracy of 6 km across-track and 5 km along-track. The satellites' orbital parameters are propagated on board and updated once a week. The position of neighboring satellites is communicated to each satellite as a guide for the steerable cross-link antennas. Seven hydrazine thrusters are used for initial attitude control after launcher separation, for Earth capture, and for spin-up, ascent, and descent. An electrothermal hydrazine thruster is used for large orbital changes.

9.6.3.3. Attitude

Iridium's attitude and orbit control system provides attitude control and orbit and constellation maintenance. In orbit, the attitude is Nadir-pointed with the long satellite

[11] An Overview of the Iridium Low Earth Orbit Satellite System, Carl Fossa et al., 0-7803-4449-IEEE 1998.

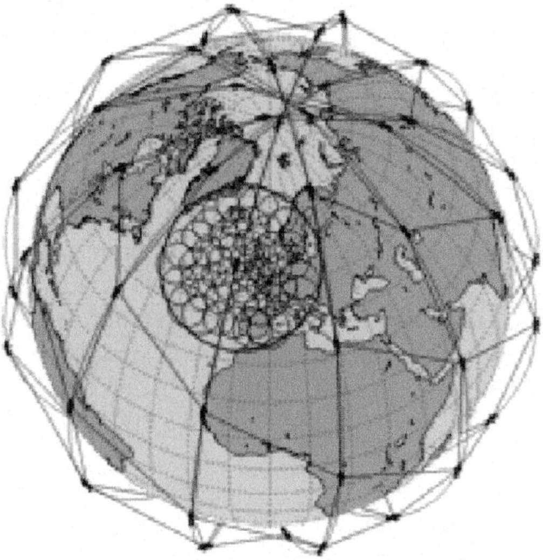

Figure 9.28. Iridium 66 Satellite Constellation. The Figure also Provides the Footprint of the L-Band Antennae Array over the Ground (Image Courtesy from Iridium).

axis kept oriented toward Earth. The satellite is momentum bias stabilized. The attitude[12] is driven by the need to provide the correct visibility among all the antennas:

- The three L-band antennas to connect the satellite with the ground-based Iridium cell-phones
- The Ka-band feeder link to connect with the ground-based Gateways
- The Ka-band cross-links to connect each satellite with the four nearest satellites of the constellation: those ahead and behind in the same orbital plane, and those to the right and left in adjacent orbital planes

In the nominal attitude mode, the satellites' primary attitude control is achieved by the gyroscopic stability of a single pitch-axis momentum wheel. A fine Earth horizon sensor, which is the primary attitude reference, detects the infrared signature of the Earth's horizon. Pitch errors are removed by commanding changes in wheel speed; roll and yaw errors are controlled by commanding torques from two magneto-torquers bars. The direction of the magnetic field is determined by a three-axis-magnetometer. In the nominal attitude the pointing accuracy is 0.2° in roll, 0.3° in pitch and 0.4° in yaw. Just after separation from the launchers, a three-axis gyro is used for attitude reference, a wide field-of-view low accuracy Earth sensor is used for initial Earth acquisition, and the hydrazine thrusters cited above are used to control the attitude. This same arrangement of sensors is used for attitude control during orbital maneuvering, including satellite in-orbit storage, orbit ascent, and descent. The satellite uses a blow-down monopropellant hydrazine subsystem for all propulsive functions.

[12] Systems Engineering Trades for the Iridium Constellation, T. P. Garrison, AIAA Journal of Spacecrafts and Rockets 1997.

9.6.4. Pleiades

9.6.4.1. Aims

Pleiades[13] is a constellation of two Earth observation satellites that provide images of very high spatial resolution in five bands: wide-band panchromatic; blue; green; red and near-infrared. The images have a spatial resolution at Nadir of 0.7 m. Each image provides a field of view of 20 km. The satellite is agile, enabling rapid pointing and flexible image acquisition. With the two satellites in orbit the constellation provides daily access across the globe. The mission is ideal for general cartography and for critical military and civilian security needs. The program is led by CNES in cooperation with Sweden, Belgium, Spain, and Austria. The program began in 2001 and the first satellite will be launched in 2010, with the second following a year later; the design lifetime is 5 years. Each satellite weighs 1,000 kg and the solar array can generate up to 1,500 W of power.

9.6.4.2. Orbit

Pleiades (Figure 9.29) flies in a Sun-synchronous frozen orbit with an altitude of 694 km (Figure 9.30). The constellation's two satellites are phased at 180° in an orbit of 14+15/26 revolutions per day and a Sun-synchronous inclination of 98.2. The local time of the ascending node is 10:30. This orbit track affords daily access to any point on Earth. Each of the satellites possesses autonomous orbit determination using a Doris receiver, which can determine its position with an accuracy of around 1 m. The satellites[14] control their orbit using a propulsion subsystem grouped in a module that contains all the relevant equipment. The hydrazine system is based on four 1 N thrusters and a tank with 104 l of hydrazine. These thrusters are used for orbit control, but not for attitude control.

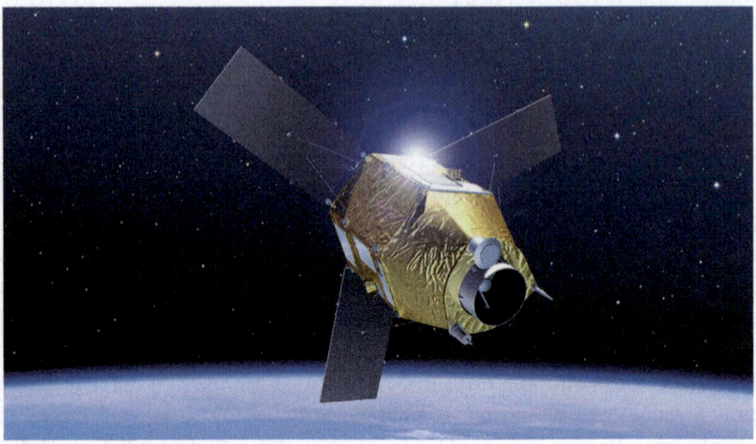

Figure 9.29. Pleiades Satellite (CNES http://132.149.11.177/PLEIADES/Fr/index.htm).

[13]CNES http://132.149.11.177/PLEIADES/GP_satellite.htm

[14]Pleiades a Multi Mission and Multi Cooperative Program, A. Baudoin et al., Acta Astronautice Vol %! No 1-9 2002, 52 IAF Congress, Pergamon Elsevier.

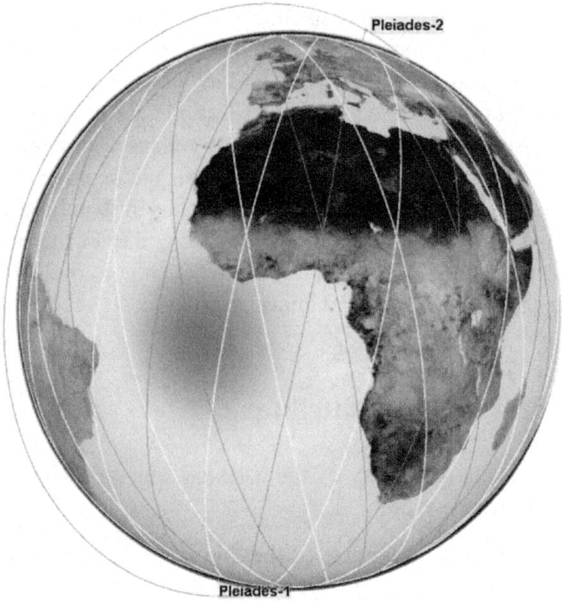

Figure 9.30. One Day of the Track of Both Pleiades Satellites.

9.6.4.3. Attitude

In the nominal mode, attitude determination is performed by a gyro-stellar system, including gyroscopes providing angular velocity and star trackers providing angular position. Solid-state fiber optic gyroscopes are used to ensure highly accurate attitude determination while maneuvering. All attitude determination units feature optical heads located on the instrument structure, to minimize thermal distortion of the instruments' line of sight. In line with the satellites' high agility, the attitude control system can determine the satellite's attitude from a "lost in space" condition, and is able to recognize its initial attitude by distinguishing stellar patterns. Pleiades is a very high-resolution mission and its pointing requirements are correspondingly very demanding. The satellite is able to properly locate images on-ground with an accuracy of 10 m. At the 694 km altitude of the satellite this is equivalent to a pointing knowledge of 14 μrad. In order to allow the fast repointing needed for optimal maneuvering, Pleiades uses a cluster of four Control Moment Gyros (CMGs) as attitude actuators. Each actuator provides a momentum of 15 Nms. Special guiding techniques are used to avoid the usual drawbacks of CMGs; instead of following a predefined attitude profile that can lock the cluster in singularities, a cluster reorientation is realized, taking into account the satellite's history and trajectory, in order to globally optimize the system. This approach allows the use of the complete angular momentum capacity envelope, which is about 3.2 times the elementary CMG momentum, and results in simplification of GMG cluster management flight software. Pleiades also has a safe mode, called the "Acquisition and Safe Hold" mode, which is based on the use of variations in the Earth's magnetic field. In this mode the satellite rotates at two revolutions per orbit around the axis perpendicular to the plane of the orbit.

As noted above, Pleiades is an agile satellite able to maneuver along and across-track. This allows the creation of "mosaics" of individual 20 × 20 km images that together provide large images up to 120 × 120 km. The satellite is able to point off-Nadir up to ±43°, which permits revisits in less than 1 day (Section 9.9.4) over the entire Earth with a constellation of just two satellites. The satellite is compact, with the telescope located at the center. It features a deployable solar array, but the deployed panels are solidly attached to the spacecraft's body and have supplementary stiffening stringers. This minimizes inertia, maximizes stiffness, and optimizes the agility of the satellite. The satellite's powerful CMG cluster allows it to turn 5° in 8 s and 60° in 25 s. Pleiades' agility allows its satellites to acquire up to 450 images per day.

9.7. Geometry around the satellite

This section analyzes the geometry of the celestial sphere around the satellite. The following elements will be examined:

- The Sun. At 1 AU the angular size of the Sun is about half a degree. The Sun is visible only when not eclipsed by the Earth.
- The Earth. The angular size of the Earth depends on the orbit altitude. In a LEO orbit with an altitude of 800 km the angular size of the Earth is 125°; in a GEO orbit it is 17.4°.
- Deep space, which occupies the remaining celestial sphere centered on the satellite.
- For remote sensing satellites, the direction of the observable will define the pointing direction of the instruments. The observable will be located over either the Earth or deep space.
- Ground stations for communications located above the Earth.

The geometry around the satellite depends on the satellite's orbit and attitude. The attitude is the more significant of the two because it defines the orientation of the bodies around the satellite, while the orbit defines their angular size. The celestial sphere geometry around the satellite strongly shapes its configuration; see Section 10.4 for a detailed discussion of how the satellite configuration answers to the design constraints imposed by the geometry around the satellite.

Section 9.3 listed five possible attitude strategies; Section 9.3.1 looked at uncontrolled satellites lacking a geometry that could be systematically analyzed; Section 9.3.2 examined gravity-gradient satellites and pointed out that from the point of view of geometry, they are Nadir-pointing. Their special geometry will be studied in this section. Section 9.3.3 analyzed stabilization by rotation; satellites stabilized by rotation have their own specific observing geometry and will also be analyzed in this section. Section 9.3.5 divided inertially stabilized satellites into three groups: Nadir-pointed, Agile, and inertially pointed. Nadir-pointed, inertially stabilized satellites have the same geometry as Nadir-pointed gravity-gradient spacecraft, and they will be discussed together. Agile satellites possess a widely variable geometry that shall be analyzed separately, as will the special geometry of inertially stabilized satellites. Section 9.3.4 examined dual spin and momentum bias attitudes: the satellite's spinning segment will have the geometry of a spinning satellite, while the nonspinning portion's geometry will be inertially pointed or Nadir-pointed. Thus, an exhaustive analysis of the geometry surrounding a satellite will consider four distinct situations:

- Satellites pointed to Nadir
- Spinning satellites
- Satellites pointed in an inertial space
- Agile satellites

These four categories will be covered in the following subsections.

9.7.1. Nadir pointing

9.7.1.1. General

In Nadir-pointing satellites the direction of the Earth is fixed with respect to the satellite, and the Earth is always in the Nadir direction. Almost always both the observable, which defines the orientation of the instruments, and the ground stations, which determine antenna orientation, will be located on the Earth.

The solid angle occupied by the Earth is a function of the satellite's altitude (Figure 9.31). At very low altitude, the Earth dominates almost half the celestial sphere around the satellite, and higher orbits correspond to smaller solid Earth angles. At the most frequently employed altitude, around 800 km, the Earth occupies about 1/3 of the celestial sphere. In GEO the Earth occupies a solid angle with a diameter of 17.4°. The solid angle occupied by the Sun, on the other hand, is equal to half a degree and can be considered constant for all seasons and for all orbit altitudes around the Earth.

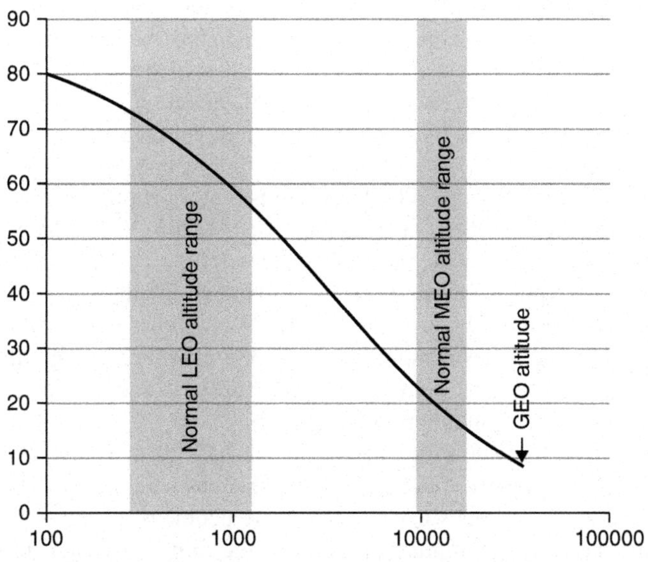

Figure 9.31. Earth Solid Radius Angle in Degrees as a Function of Satellite Altitude in Kilometer.

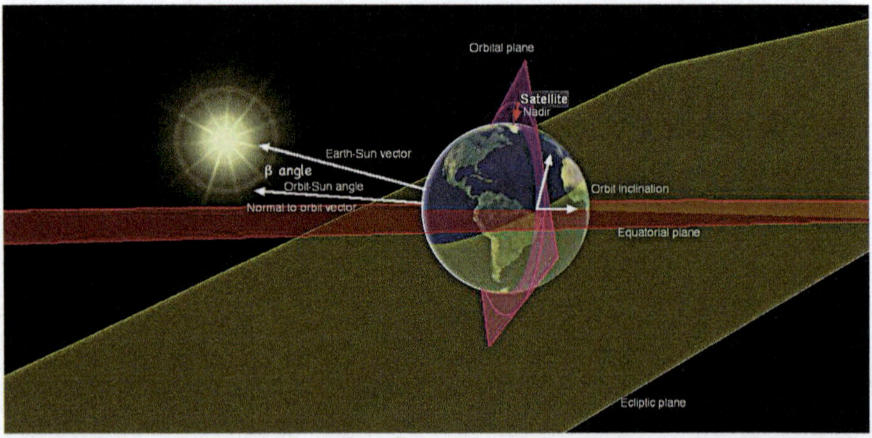

Figure 9.32. Orbit-Earth-Sun Geometry in an Earth-Centered Inertial Reference Frame.

The Sun/orbit geometry depends on the orbit inclination and on the direction of the Earth-Sun vector (Figure 9.32). The Earth-Sun vector moves with the seasons but is always in the ecliptic plane, but the orbit inclination is fixed with respect to the equatorial plane and the ecliptic and equatorial planes make an angle of 23.4°. That means the Sun/orbit geometry will change with the seasons as the Sun/orbit direction moves up and down 23.4° with respect to the orbital plane. The Sun/orbit geometry is defined by the angle between the Sun direction and the normal to the orbit direction (Figure 9.32), termed the $\beta-$angle. In Nadir-pointing attitude the satellite rotates once per orbit with respect to inertial space, and the Sun likewise rotates around the satellite once per orbit in a satellite-fixed reference frame. That means the Sun describes a cone around the satellite during each orbit (Figure 9.33). The axis of the cone is the vector normal to the orbital plane, and the cone angle is the β-angle, which changes with the seasons; the cone is wide open when the Sun is near the orbital plane and more acute when the Sun is near the perpendicular to the orbital plane. When the Sun is nearly perpendicular to the orbital plane, its direction will be almost constant along the orbit, and when the Sun lies in the orbital plane, its direction will revolve around the satellite.

When the solid angle occupied by the Earth is larger than the angular distance of the Sun to the Nadir axis, the Earth will eclipse the Sun. When the Sun is nearly perpendicular to the orbital plane the cone angle will be very acute, and the Sun will always be far away from the Nadir direction, and the orbit will experience no eclipse. When the Sun lies in the orbital plane, eclipse time will be maximized and proportional to the part of the celestial sphere occupied by the Earth. At normal LEO altitudes the Earth's solid radius is around 60° (Figure 9.31) and the maximum eclipse will be about 1/3 of the orbital period.

This means that the Sun's geometry with respect to a Nadir-pointed satellite changes along the orbit, describing a cone, and changes with the seasons that change the β cone angle. This defines a widely variable geometry that will have a heavy impact on the satellite's configuration because the Sun's direction determines the location of the solar arrays, thermal radiators, and instruments. The speed with which the β-angle changes will be the rotational speed of the Satellite-Sun direction with

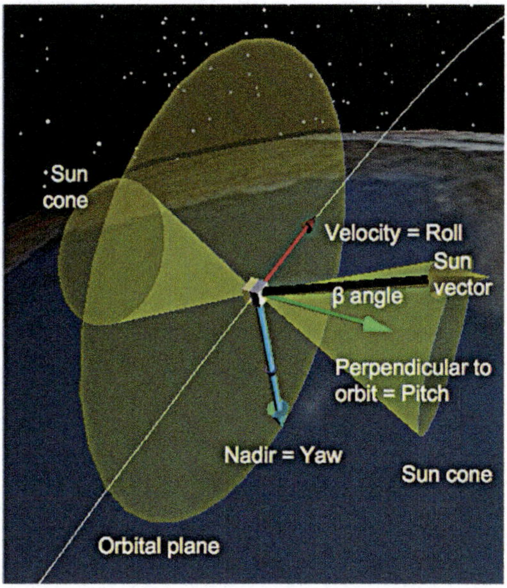

Figure 9.33. Orbit-Earth-Sun Geometry in a Satellite-Centered Reference Frame.

respect to the orbital plane. In inertial space, the Earth-Sun direction rotates once per year, and the orbital plane rotation speed is a function of the gravity anomalous J2 parameter (Section 9.2.1), the orbital inclination and the orbit altitude and eccentricity. For near circular orbits at a typical altitude around 800 km, this differential speed (Figure 9.34) varies from a maximum of one turn every 47.5 days for equatorial orbits, to one turn per year at 90° inclinations (where the orbital plane is fixed in the inertial space), to a maximum of infinite duration for Sun-synchronous inclinations (Section 9.4.2) around 98°.

In order to fully understand the Earth and Sun geometry around a Nadir-pointing satellite, distinct Sun-Earth-satellite geometries, which are a function of the orbit inclination and altitude, must be thoroughly examined. Various scenarios will be analyzed in the next paragraphs.

9.7.1.2. Sun-synchronous orbits

Sun-synchronous orbits (Figure 9.35) were introduced in 9.4.2. That section described their main characteristics and their advantages for Earth observation. One of the advantages was that they have fairly constant Sun-orbit geometries. The geometry of Sun-synchronous orbits is not exactly constant:

- Throughout the seasons the Sun moves up and down the orbital plane due to the angle of 23.4° between the Ecliptic and Equatorial planes.
- Sun-synchronous orbit inclination is between 97° and 99° – depending of the altitude – and the excess over 90° will combine with the previous 23.4°.

Throughout the seasons the Sun cone angle β oscillates. The oscillation will depend of the vector addition of the 23.4° and the 7–9° indicated above. This vector addition will be function of the local time of the descending node (Section 9.4), which

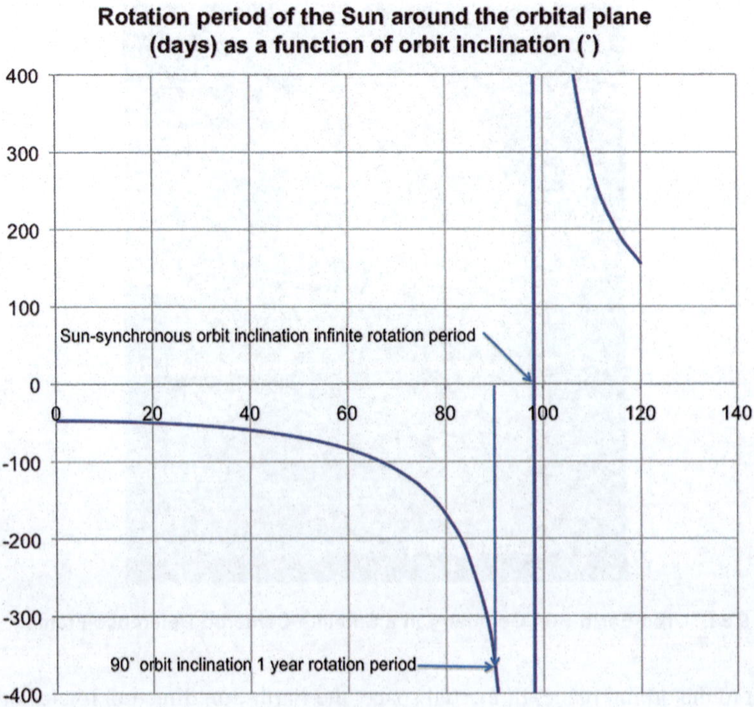

Figure 9.34. Relative Rotation of the Sun Direction and of the Orbital Plane for an Orbit at 800 km Altitude.

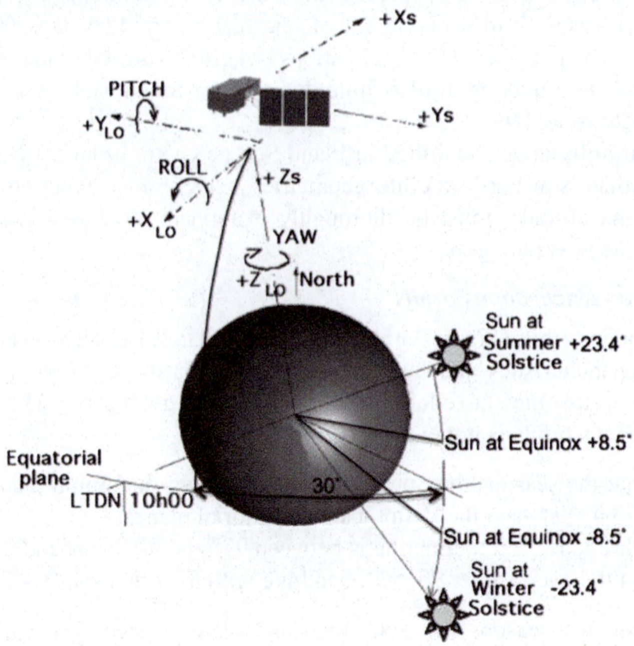

Figure 9.35. Orbit-Earth-Sun Geometry in an Earth-Centered Inertial Reference Frame for a Satellite with a Local Time of 10:00 and Orbiting at 800 km Altitude.

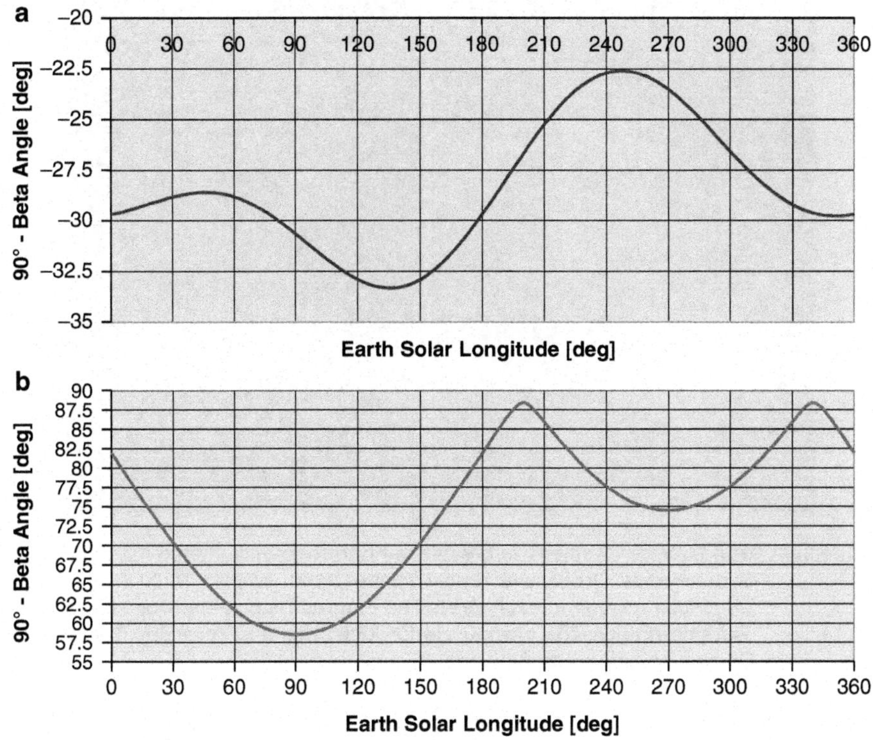

Figure 9.36. Sun to Orbital Plane Angle over 1 Year for Two Local Times: 10:00 (a) and 18:00 (b).

defines the angle between the orbital plane and the Sun direction when the Sun crosses the equatorial plane in a descending direction. Therefore, it is necessary to analyze the different situations for different orbital local times. Figures 9.36 and 9.37 provide the geometry for two local times 10:00 and 18:00.

In Sun-synchronous orbits with a local time not far away from 12:00 – like the 10:00 orbit of Sentinel-3 – the Sun is never too far away from the orbital plane. If the local time is 18:00 the Sun is never far away from the normal to the orbital plane. Figure 9.37 depicts the geometry in both cases. The seasonal variations of eclipse duration for these two orbits can be seen in Figure 9.38. The eclipse duration experienced by the satellite for the 10:00 orbit is fairly constant along the year and lasts around 33 min; equivalent to one third of the orbit. Sun-synchronous satellites needing to minimize eclipse duration must fly with a 6:00 h angle with the Sun always close to the perpendicular to the orbit. At 800 km of altitude, the 06:00 orbit is eclipse-free except for a period of around 80 days near the summer solstice when eclipses lasts up to 15 min.

9.7.1.3. High inclination non-Sun-synchronous orbits

In non-Sun-synchronous orbits, the Sun rotates around the orbital plane. If the orbit inclination is high, the Sun direction will be sometimes in the orbital plane and sometimes close to the perpendicular to the orbital plane; that means the β angle will take

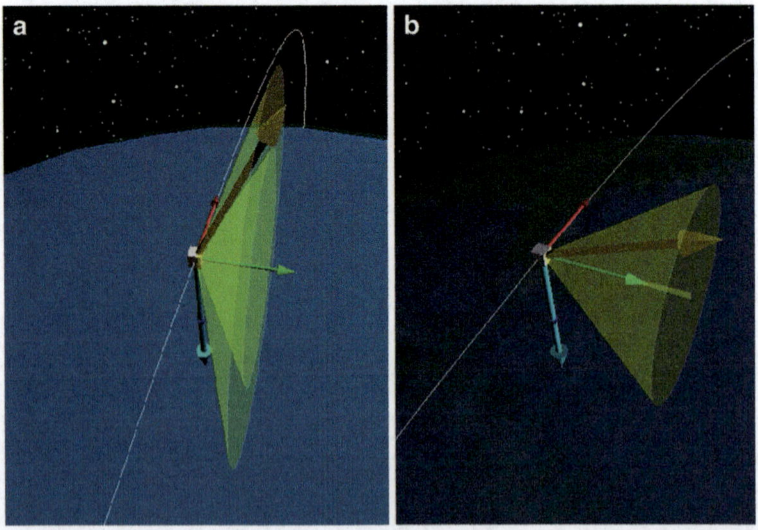

Figure 9.37. Sun Satellite Geometry in a Satellite-Fixed Reference Frame Along Two Sun-Synchronous Orbits with Local Times 10:00 (a) and 18:00 (b). The Sun Describes a Cone in each Orbit. The Figures Display the Cones of the Two Extreme Angles. The Inner Cone for 18:00 has $\beta \approx 0°$ near the Solstices, as Depicted by Figure 9.36b.

any value between 0° and 90° and Nadir-pointing, non-Sun-synchronous orbits with high orbit inclination will have a highly variable Sun-orbit-satellite geometry with sunlight reaching the satellite from almost any possible direction. This makes the thermal design more complex. The eclipse duration will vary from 1/3 of the orbital period when the Sun is near the orbital plane, to no eclipse when the Sun is near the perpendicular to the orbital plane (Figure 9.39).

In most situations the satellite will have faces that should always be illuminated by the Sun and faces that should totally avoid solar illumination; nevertheless, as the Sun rotates around the orbital plane, it will pass from illuminating one side of the satellite to illuminating the other. If the satellite was rotated around its yaw-axis by 180° when the Sun passes through the orbital plane, the leading face becomes the trailing face and vice versa, and the face that was not Sun illuminated continues to avoid direct solar illumination. The observing geometry of Nadir-looking instruments will also not be altered. This maneuver (Figure 9.40), which is called yaw-flip, shall be repeated every time the Sun transverses the orbital plane and it is common in Nadir-pointing satellites that do not fly Sun-Synchronous orbits and that need a face of the satellite always avoiding the Sun.

9.7.1.4. Low inclination orbits

As the orbit inclination becomes smaller the area of possible Sun illumination is correspondingly reduced. In case the inclination was zero the maximum Sun to orbital plane angle will be 23.4° in the equinoxes and 0° in the solstices (Figure 9.41).

In this case, the Sun will never be too far from the orbital plane and each orbit it will describe a shallow cone around the satellite, progressively illuminating the wake,

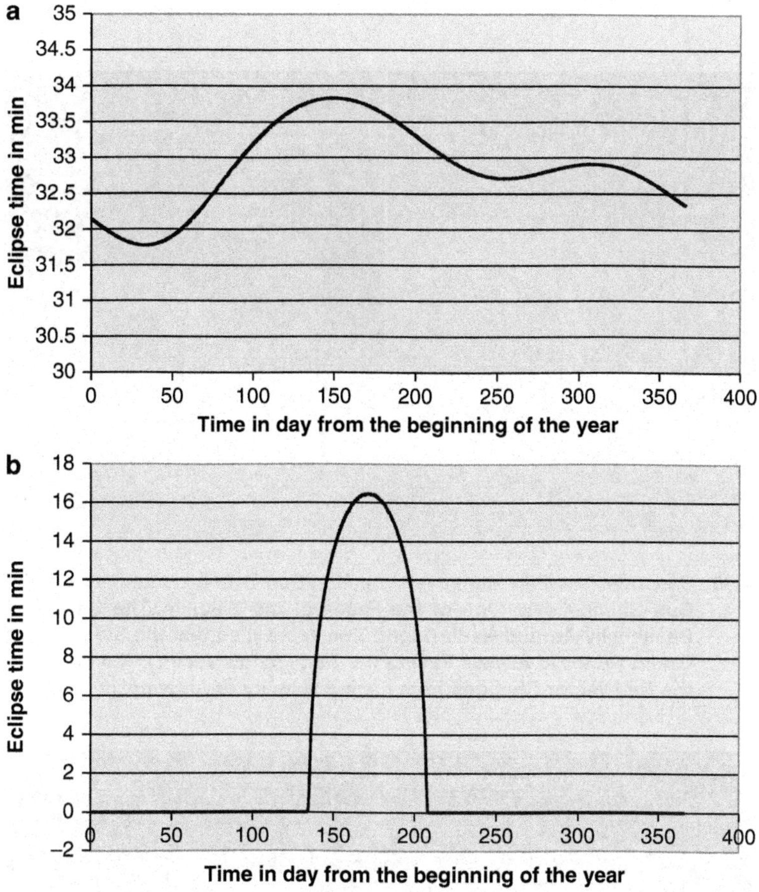

Figure 9.38. **Eclipse Duration for a Sun-Synchronous Orbit at 800 km Altitude and with Two Different Local Times (a) at 10:00 and (b) at 18:00. Vertical Axis is in Minutes of Eclipse per Orbit.**

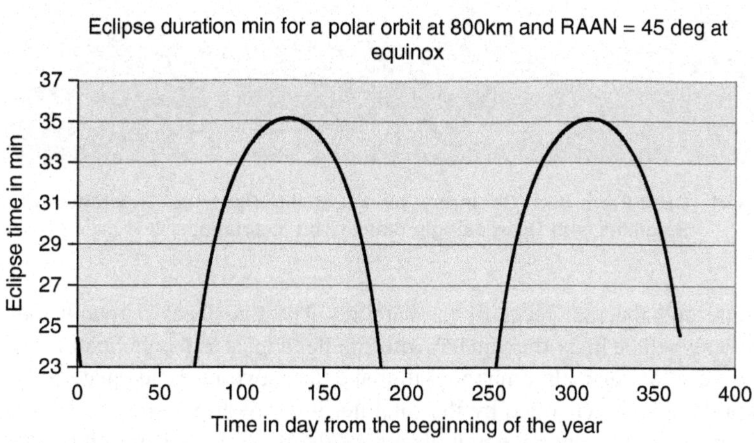

Figure 9.39. **Eclipse Duration During the Year for an Orbit with an Inclination of 90°. Vertical Axis Is in Minutes of Eclipse per Orbit.**

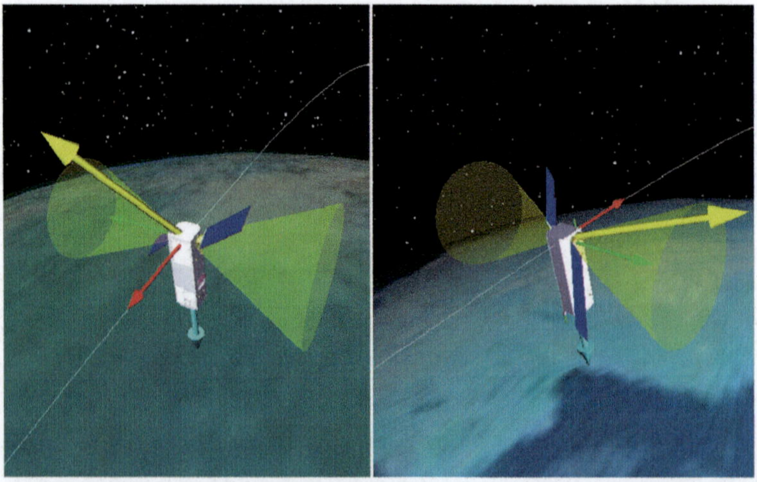

Figure 9.40. Seasonal Yaw-Flip Maneuver in a Non-Sun-Synchronous Orbit to Avoid the Sun Shining over One of the Faces of the Satellite. The Satellite Rotates Seasonally Around Nadir (Light Blue Vector); so that the Side Marked by the Green Vector Is Always Facing the Sun (Yellow Vector). The Face Marked by the Red Vector Changes from Facing Velocity to Antivelocity.

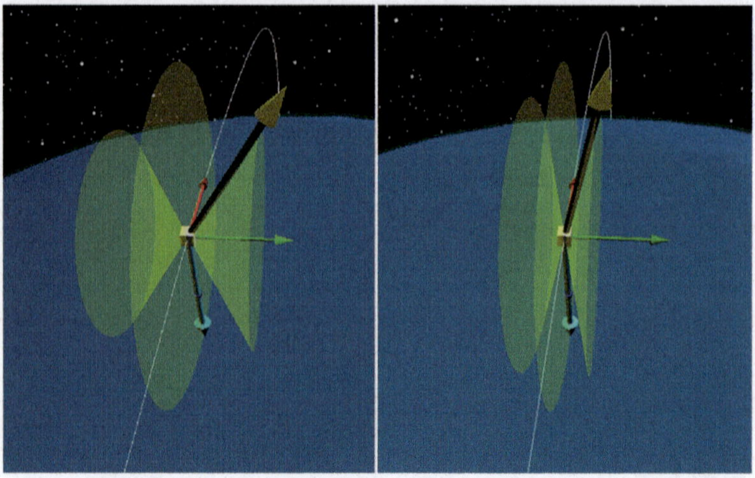

Figure 9.41. Orbit-Earth-Sun Geometry in a Satellite-Centered Reference Frame for Satellites with Decreasingly Small Orbit Inclinations.

the zenith, and the ram faces of the satellite. The two faces perpendicular to the orbital plane will be lit by the Sun but with smaller angles as the inclination is smaller. This low level of solar illumination will make these two faces adequate for radiators dissipating the heat generated by the satellite. If a yaw-flip maneuver is compatible with the mission then one face will always be totally protected from direct sunlight.

The eclipse duration will be fairly constant during the whole year. Figure 9.42 shows the eclipse duration for a 10° inclination orbit. The duration is modulated by

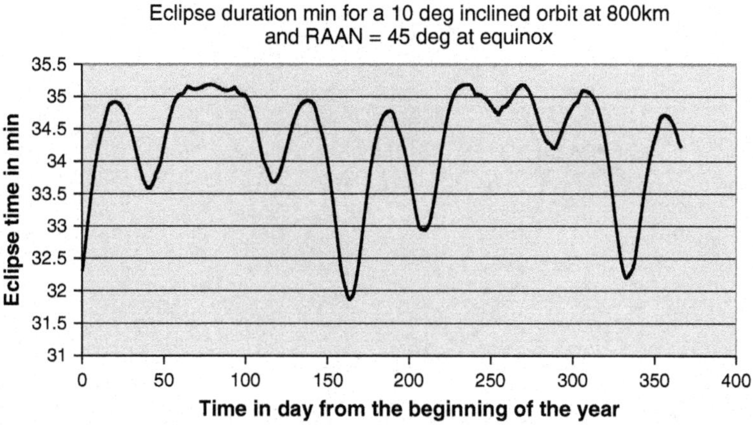

Figure 9.42. Eclipse Duration for 10° Inclination Orbit.

the yearly cycle of the Sun and by the roughly 50 days period that the orbital plane needs to rotate one turn in inertial space.

9.7.1.5. Geostationary orbits

A special example of a low inclination orbit is the geostationary one (Section 9.4.4). In geostationary orbits, the Sun will be located in the northern hemisphere during half the year and in the southern hemisphere during the other half. If the satellite makes a yaw-flip maneuver at both equinoxes, the North Pole and the South Pole faces flip. This ensures that one face will always be protected from direct sunlight, which is crucial for improving thermal control of the satellite and its instruments.

The solid angle radius of the Earth from the geostationary orbit is 8.7°. The satellite is in the equatorial plane and the Sun, which is in the ecliptic plane, will shift up and down by 23.4° with the seasons with respect to the orbital plane. This means that GEO orbits are eclipses-free for most of the year. Nevertheless, during the periods around both equinoxes the Sun will be close to the equatorial plane, and when the angle of the Sun above the orbital plane is below 8.7°, eclipses will occur. These eclipses will be longest during the equinoxes; lasting up to 69.4 min. GEO satellites must be able to cope with these long eclipses. When the Sun is near Earth but not yet hidden by it, the satellite's Nadir facing side will receive direct sunlight (Figure 9.43). This can be detrimental to the satellite's instruments or payload, which will require protective shutters, and the instruments will be inoperative when the shutters are closed.

9.7.2. Rotating satellites

Compared with the complex geometry of Nadir-pointed satellites, the geometry of rotating satellites is simple. The positions of the Earth, the Sun, the observable, and the deep space in the satellite-centered celestial sphere are defined by the satellite's rotation axis, and the rotation axis is fixed in inertial space. If the rotation axis is pointed perpendicular to either the ecliptic or the equatorial plane, the satellite will

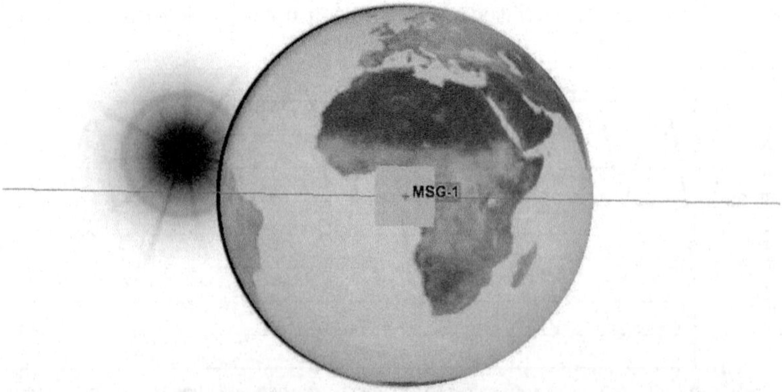

Figure 9.43. The Sun Entering into Eclipse in a Geostationary Orbit on 16 September near the Autumnal Equinox. The Sun Blinds the Earth-Observing Instrument When Pointing near the Edge of the Earth.

have top and bottom sides that do not receive sunlight, and a lateral surface that is sequentially lit by the Sun as the satellite rotates. The satellite's rotation speed – and hence the Sun's rotation speed relative to it—is normally rapid enough to render sunlight and thermal conditions uniform on all the lateral surfaces. Thus, the lateral surfaces are the logical place for locating solar cells, and the top and bottom faces for radiating the satellite's thermal waste energy. The Cluster (Figure 10.3) constellation of four satellites (Figure 9.21) is an example of this kind of configuration. The satellite features a cylindrical symmetry, with the top and bottom faces used for waste heat radiation and the cylindrical sides for solar cells. This type of geometry and the resulting configuration is common in satellites used to study the Earth's magnetosphere. Section 7.1.3 already commented the special configuration of rotating satellite and Figure 7.3 provided the configuration of MSG. Ulysses (Section 9.6.2) has its own radio-isotopic power source and has no need to point the solar cells toward the Sun. In Ulysses the spin direction is pointed toward Earth to allow communications via the large-gain antenna located on the Earth-pointed face.

9.7.3. Inertial satellites

Inertial satellites point toward the celestial sphere and are almost without exception astronomical observatories. Their instruments have no need to point toward Earth, permitting more freedom in the selection of attitude. In the case of inertially stabilized satellites, the celestial sphere does not rotate around the satellite once per orbit as it does for Nadir-pointed satellites. For inertial satellites, the changes in geometry are seasonal and slow. Because of the Sun's intense radiation, pointing either directly at it or within a large cone around the solar vector must be avoided. The orbital plane will rotate seasonally with respect to the Sun, which defines a situation where some areas of the sky are accessible in some seasons and inaccessible in others. One way to protect the telescope from solar radiation is to have it pointing perpendicular to the Sun, and target observables by rotating the satellite around the Sun axis, allowing the telescope to scan a belt perpendicular to this rotation axis.

This was discussed in section 7.2.2 and in Figure 7.12. If the satellite is orbiting the Earth however, the Earth will rotate around the satellite once per orbit blinding changing areas of the spacecraft as the Earth rotates, and the pointing and operations of the satellite shall take this into account. Hubble (Section 7.2) is one example of an inertially stabilized astronomical observatories flying in LEO. In the case of Hubble, which has an orbital inclination of 28°, it takes 8 weeks for the celestial sphere to rotate around the orbital plane. The CNES mission Corot is another example. Corot has an orbit inclination of 90° takes 1 year. In both cases users would have preferred higher orbits and more flexible pointing possibilities, but Hubble was forced into its orbit by the need for astronaut servicing, and Corot due to budget constraints. To reduce the Earth's occultation period requires orbiting at high altitudes, where the angular size of the Earth will be smaller. Most of the missions discussed in Section 7.2 – IUE, XMM, and ISO – orbit far above Earth.

9.8. Pointing control, pointing perturbations, and pointing corrections

The two subsections of this section analyze the interrelations between: satellite attitude control, instruments pointing, instruments data acquisition quality, and pointing perturbations. That means, this section will analyze pointing as an end-to-end issue. This book concentrates on the analysis of interrelations between space systems components. Instrument pointing is perhaps the richest area in interactions and Section 7.2.3 and Figure 7.13 already selected this area to illustrate cascades of complex interacting requirements and design decisions.

The needs for data quality will drive the requirements on pointing. As a result of consumer's needs, it will be necessary to create budgets on: absolute pointing, pointing knowledge, pointing stability, pointing stability knowledge, geo-location, and co-registration (Section 9.10). In case where there was a very strict requirement on pointing and several sources of mechanical disturbances, it will be also necessary to create an independent budget on mechanical disturbances. The apportioning of these budgets will generate requirements all over the space system: satellite, instrument, and ground processing. Furthermore the interrelations related to pointing are especially complex:

- There is the need to differentiate between probabilistic and nonprobabilistic errors.
- There is the need to differentiate the behavior of the different errors in the frequency domain.
- There is the need to differentiate between errors that can be estimated, errors that can be calibrated and corrected and errors that cannot be estimated or corrected.

The handling of pointing is in the hands of the attitude control specialist. Nevertheless, the system level implications of pointing single it out as an area of special interest for the system architect. In line with this complexity, and in line with the need for the architect to understand this complexity, this section and their subsections are correspondingly thorough. The approach used by this chapter follows the approaches developed in ESA and the European space industry during many years.

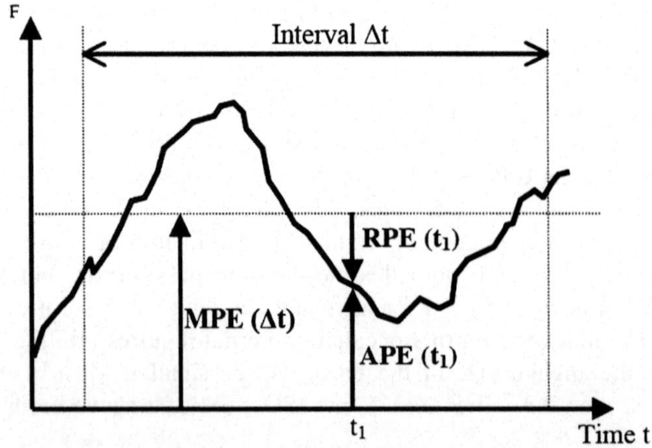

Figure 9.44. During the Instrument Integration Interval Δt, the Mean Pointing Error (MPE) Combines with the Relative Pointing Error (RPE) to Provide the Instantaneous Absolute Pointing Error (APE).

This approach was started in the 1990s by the "ESA pointing handbook"[15] and is now consolidated in the "Control Performance" ECSS-E-ST-60-10 document.[16]

9.8.1. Satellite and instruments pointing and pointing perturbations

The payload justifies the mission. The correct understanding of satellite attitude shall be in terms of the quality of the pointing of its payload:

- The satellite shall point the payload toward the intended target. This requires the attitude of the satellite and the pointing direction of the payload to be correct. This will define an Absolute Pointing Error requirement (APE in Figure 9.44).
- Imaging instruments produce data by integrating the radiation received by the focal array detectors during a given interval of time (Δt in Figure 9.44). During this integration interval the satellite must remain stable to avoid smearing the signal reaching the satellite from a single direction across different detector pixels. This defines a requirement on Relative Pointing Error (RPE in Figure 9.44). RPE is defined within the instrument's integration time Δt. RPE will specify limits to the satellite and instrument attitude change rates.
- The satellite shall be able to provide information allowing the correct identification of the direction of the signal detected by the payload. That means, to correctly locate the source of the signal. This requires the attitude of the satellite to be correctly known. This will generate requirements on the knowledge of the attitude "Absolute Knowledge Error" (AKE) and "Relative Knowledge Error" (RKE). RKE is also relative within the instrument integration time Δt and it specifies the knowledge of attitude change rates.

[15]ESA Pointing Handbook, TIDC-CR-5522 February 1993.

[16]Space Engineering Control Performance, ECSS-E-ST-60-10, 2008.

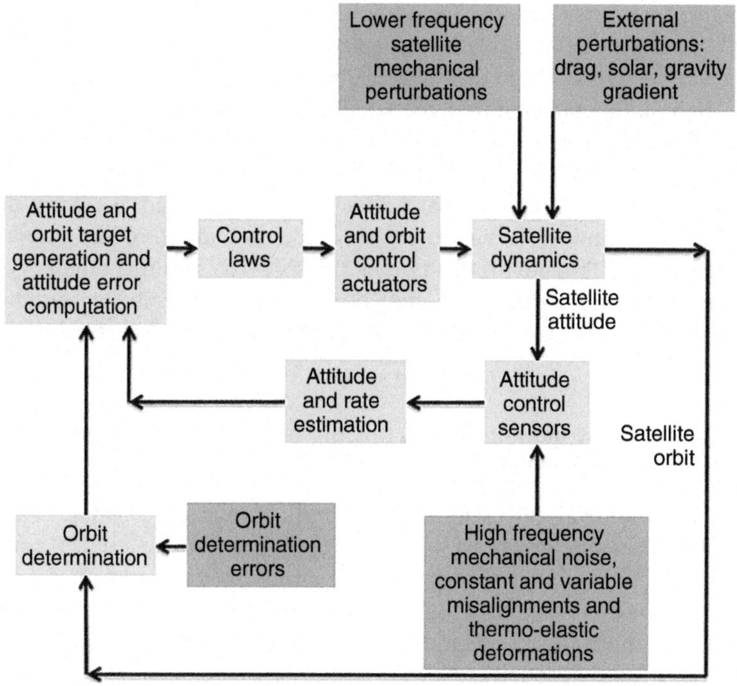

Figure 9.45. Satellite Attitude and Orbit Control Loops.

Therefore the satellite and their instruments must be pointing in the right direction, this direction shall be known, and the pointing direction rate of change shall be small and known.

The determination of the alignment between satellite attitude and instrument line of sight can be estimated by the help of specialized analysis – for example thermoelastic deformations analysis – and by comparing the attitude information provided by the satellite's attitude determination devices with the true line of sight of the instruments as determined by the location in the images of points, like rivers or stars, whose location is very well known. This will allow characterizing and correcting misalignments between the satellite attitude – as defined by the attitude sensing devices – and the instrument's line of sight. Figure 9.45 presents an overview of the processes involved in orbit and attitude determination and control. In the case of attitude control, external and internal perturbations will affect the satellite attitude and orbit and the pointing of its instruments. The attitude control system will be able to compensate external and internal perturbations that will be slow enough for its sensors to measure them and for its actuators to counteract them. That means the attitude control system will only compensate the "lower frequency" slower perturbations. Fast, high frequency perturbations and other lower frequency perturbations that the attitude sensors cannot detect – for example the above-mentioned thermoelastic deformations – will also impact the line of sight of the instruments. The consideration of the frequency signature of pointing and perturbations is of the maximum importance and will be treated in detail in the next two sections. The maximum allowable value of these perturbations that the attitude control system cannot control shall be specified to keep their impact within acceptable levels. The mission's final products need to be correctly located to their source above the Earth or in the celestial sphere, and that observations gathered from the same locations must be co-registered together as such. This will generate geo-localization and co-registration requirements and budgets. This means that attitude control system generates numerous and interrelated budgets that need to be controlled and allocated. To know and to control the exact pointing of the instruments it is also necessary to know and control not only the satellite attitude but also the satellite position in the orbit. The resulting requirements will specify the on-ground and in-orbit satellite position and velocity sensors; the accuracy of the orbit determination algorithms and the orbit control strategy.

9.8.2. Pointing control, pointing perturbing torques, image acquisition, and frequency ranges

The perturbing internal forces and torques will change the satellite attitude and attitude rates. These perturbations will be fought by the satellite attitude control subsystem. The attitude control subsystem shall ensure that the satellite points in the right direction and that the line of sight of the instruments is stable during the interval Δt when the instruments are integrating information over the observable. The measurements and control capabilities of the attitude control, the integration interval of the instrument and the internal sources of perturbations can be characterized and grouped with respect to their frequency:

- The attitude control determination devices will be able to detect only actual pointing variations that are slower that a given limit. That means there is an upper limit to the attitude sensing frequency.
- The attitude control actuators and control loop will need time to react to the detected pointing variations. This will establish an upper limit in the control of the attitude.
- The perturbations acting over the pointing could be slow acting, fast acting, or even instantaneous jumps. Internal vibrations of the satellite could enhance or reduce perturbations depending of their frequency.
- To provide adequate radiometric quality (Section 8.4.2) the signal arriving to the instruments shall be integrated during a specified period of time Δt (Figure 9.44). This interval will also define an associated instrument related frequency.

The different time and frequency features of the attitude control system, the perturbations, and the period of time Δt of the instruments will make them to interact in complex ways that shall be studied in a systematic fashion. An overall perspective of the frequency domain interaction of pointing perturbations and pointing control in a satellite can be seen in Figure 9.46. A perturbation of constant strength and increasing frequency produces decreasing attitude disturbances because a force needs time to produce attitude change and increasing frequencies provide shorter and shorter times for the forces to produce changes in attitude. This explains the downward slope of the curve that provides the "transfer function" relationship between a disturbing force/torque and the resulting disturbance in attitude.

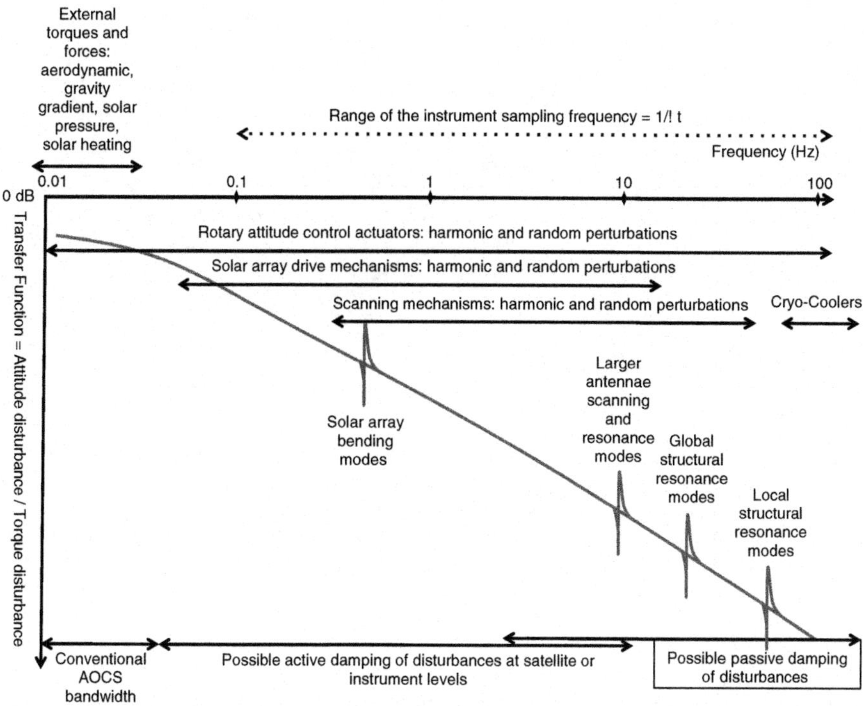

Figure 9.46. Pointing Control Pointing Perturbations and Instrument Integration in the Frequency Domain (Courtesy Thales Alenia).

Some perturbations will be at a very specific frequency. This will be the case of the resonances of the solar array or other satellite structures, or from harmonic perturbations associated with rotating elements like pointing mirrors, antennae, or rotating attitude control actuators. An especially dangerous situation will develop if a harmonic perturbation was in resonance with the resonance mode of some satellite structure. Other perturbations will be wide frequency "mechanical noises." This will be the case of bearings or mechanical cryo-coolers. The activation of mechanical actuators will also produce wide-bandwidth high frequency noises produce by mechanical jolts. The design of the AOCS could limit itself to a conventional low frequency bandwidth or it could try to control some of the higher frequency perturbations as produced by the solar array drive bending or by the rotation of any large size antenna. The instrument integration time can also be adjusted to make it higher or lower frequency. Lower frequencies will be associated to longer integration times and better signal to noise ratio (Sections 8.4.2 and 8.7.1) but lower frequencies will make more difficult the design of the rest of the satellite because it will be required to maintain the instrument line of sight stability during longer times.

Comparing the frequencies associated with the perturbations, attitude control system, and instrument integration interval three situations are possible:

(a) Low frequency line of sight "drift" perturbations whose frequency is lower than the frequency of the attitude control system and lower than the frequency associated with the instrument integration time Δt; lower frequency curve of Figure 9.47a.

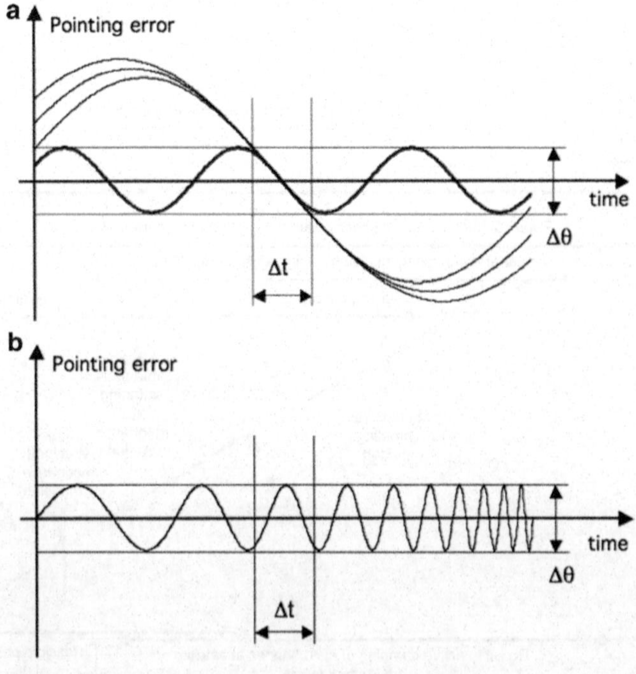

Figure 9.47. (a) Drift Perturbations Have a Frequency *f*, Which Is Slow with Respect to the Δt Instrument Integration Interval. It Is Necessary to Limit the Attitude Rate of Change of the Line of Sight. (b) Jitter Perturbations Have a Frequency *f*, Which Is Fast with Respect to the Δt Integration Interval, and Then It Will Be Necessary to Limit the Attitude Change Amplitude $\Delta\theta$ of the Line of Sight. (Courtesy Thales Alenia).

To counteract these perturbations it will be necessary to specify the line of sight pointing stability RPE/Δt. To reduce the complication of the attitude control system, it will also be necessary to limit the magnitude of these slow frequency perturbations. Slow movements generated by vibrations of a very large – and unavoidably flexible – solar array would be a perturbation of this type.

(b) Perturbations of frequency higher that the upper limit of the bandwidth of the attitude control system but lower than the frequency associated with the instrument integration time Δt. (higher frequency curve of Figure 9.47a). To control these perturbations, it will be necessary to specify the maximum line of sight stability value RPE/Δt that these perturbations will produce. Medium velocity scanning devices, for example large scanning mirrors or antennae, will generate this type of perturbation.

(c) Perturbations faster than the integration time of the instrument. To limit their effect, the amplitude of these "jitter" perturbations shall be specified (Figure 9.47b). This will require specifying the maximum jitter amplitude value $\Delta\theta$. Medium and high velocity mechanical devices, like momentum wheels or fast scanning devices, will produce this type of perturbation. High frequency noise sources of any type will also produce perturbations of this type. As frequency goes up, increasing forces will be needed to produce the same attitude change amplitude $\Delta\theta$. That means very high frequency perturbations shall be very strong to have a noticeable effect on pointing. The decreasing slope of the transfer function line of the Figure 9.46 has already depicted this.

To reduce the instrument integration time Δt makes easier the task of the attitude control. If a long Δt was absolute necessary, it could be possible to acquire many independent fast images of the observable that could be superimpose together until the necessary long integrated time is reached. To superimpose these images it is necessary to have data on the actual pointing of the satellite when each image was acquired. This will change the situation from (b) to (c) relaxing the requirement on the RPE but the requirements on attitude knowledge could still be very demanding. This approach will also reduce the amount of images produced by the mission and complicate the processing of the data produced by the instruments.

Operating mechanisms produce perturbing forces and torques. Attitude control mechanisms like momentum wheels or control momentum gyros will also inevitably produce perturbing forces and torques when operating, and it is necessary to specify the allowable perturbing forces and torques produced by the different attitude control actuators. In general, space mechanisms are meticulously designed and smooth running and produce low levels of disturbance; nevertheless, these perturbations become design drivers when the instrument has very demanding pointing or pointing stability requirements or when the satellite has very large rotating devices. Any Earth-observing instrument providing meter-level accuracy from low Earth orbit will require strict control of perturbing torques. The same will apply to high angular resolution space telescopes for astronomy. Limiting perturbations requires that all rotating masses are well balanced and that rotation bearings and actuators run smoothly. In the case of high frequency perturbations, it should be taken into account that they are locally produced and they reduce in intensity – are damped – as they travel along the satellite structure; mounting the perturbation sources on active or passive dampers can reduce these perturbations.

Perturbations produced by very large rotating structures, such as MeghaTropiques' microwave instrument (depicted in Figures 8.9 and 8.21) represent a special case. This instrument produces a perturbation peak at the rotation frequency, which might be within the control bandwidth of the attitude control system. In that case the attitude control will try to compensate for the disturbances. At any rate, all large rotating components need to be well balanced, and in some cases can incorporate in-orbit actuators for in-flight balancing. This would complicate the design of the instrument, but the design of the attitude control system would be correspondingly easier.

9.8.3. Pointing error types

In order to specify requirements based on an analysis of the errors acting on an instruments' line of sight, it is necessary to understand their types. Errors can be divided according to different classification criteria:

- By being deterministic or probabilistic:
  - The behavior of deterministic errors can be predicted or estimated to some extent. These estimations can be incorporated into the error descriptions.
  - The specific time histories of probabilistic errors cannot be predicted but their behavior can be defined in a statistical manner, ideally in terms of power spectral density, mean, and variance (Section 5.8.4).

- By frequency behavior:
  - They can be constant.
  - They can be very slow varying drifts.
  - They can be harmonic perturbations at specific frequencies. Of special importance are the orbital frequency and the frequencies associated with scanning mechanisms.
  - They can be "broad band," jitter like noises.
- By the possibilities for corrections:
  - Some might be calibrated, characterized, and – partially – corrected by ground processing.
  - Some might not be correctable.

The ESA and ECSS guidelines mentioned at the beginning of Section 9.8 divide errors into the following four types:

- Bias errors. These errors are related to permanent once-off (zero frequency) events. Permanent misalignments produced by assembly errors or by launch loads are examples. These errors are probabilistic. These errors can be measured and partially compensated for by in-orbit characterization of the satellite after launch.
- Drift errors. These are very slowly changing errors. The gradual change in alignments produced by the evaporation of water vapor from a carbon fiber structure is an example. These errors are deterministic and can be analyzed; the analysis can be further validated by periodic in-orbit characterization of the evolution of the effects. To the degree that they are either predictable or measurable, these errors can be also corrected.
- Harmonic errors. These errors act at very specific frequencies, For example, the changing thermal environment around a satellite while orbiting will produce variable thermoelastic deformations at the orbital frequency. This error is deterministic, and can also be measured and corrected by analysis, testing, and by characterization of the satellite at the beginning of its life. The already mentioned slow vibrations of a large solar array or the rotation of a large unbalanced mechanical device will be also harmonic.
- Noise errors. There are many error sources that behave like high frequency broadband jitters. From a probabilistic point of view they are Gaussian noises. The read-out accuracy of sensors or the high frequency torque noise produced by ball bearings are of this type.

This classification is crucial because it will define how individual error sources add together. Probabilistic errors are not directly added, but instead add the root mean square of their corresponding standard deviations; that is, two errors with a standard deviation of 1 will combine to produce a standard deviation of 1.41. Correlated deterministic errors, on the other hand, are added directly when they are originated by the same cause; for example thermoelastic misalignments produced by temperature variations. Uncorrelated deterministic errors should also root mean square added.

As said above deterministic errors can be calculated and – as far as the calculation method is validated – they can be compensated. Bias error can also be determined by in-orbit characterization of the satellite. This will be further analyzed in Section 9.10.4.

9.9. Allocation of functions

9.9.1. Orbit selection

Chapter 7 introduced space mission design; and Section 7.2 analyzed the fundamental design decisions involved in astronomical observatory missions. That analysis demonstrated that the selection of the mission orbit is the highest-level decision made in the earliest stages of the design process, and shall be based on a thorough consideration of all aspects of the mission. The varied orbits utilized by the astronomical observatories described in Section 7.2 – LEO for Hubble, Geosynchronous for IUE and ISO, highly elliptic long period for XMM and Lagrangian for Herschel and JWST – were driven by the need for long, unencumbered periods of observation and efficient operations and were constrained by the limitations provided by launchers and by the level of autonomy allowed by technology. Section 9.4 offered an introduction to the most common orbits and their typical uses, and provided a detailed overview of the criteria driving orbit selection.

Some missions call for the lowest possible orbit altitudes; Aeolus (Section 8.3.1) is a good example of this. Simplification in the instrument design favored a low Aeolus orbit, which complicated the satellite's orbit maintenance components, and the final orbit altitude choice was the result of tensions between the laser power, the telescope aperture, the mission life, and the amount of fuel to be carried on board. GOCE (Section 10.3.1) orbits at an altitude of 250 km. This altitude resulted from a compromise between the need of measuring weak gravity signals as near to Earth's surface as possible and the need to provide increasing amounts of energy for the ion propulsion as the orbit altitude decreased.

In the case of remote sensing missions, the orbit altitude, inclination, and orbit repeat patterns are also a compromise between conflicting needs. Sentinel-3 (Section 9.6.1) could have flown with the same instruments at a lower altitude, thus affording higher spatial resolutions, but to provide the same coverage it would have been necessary to use either a large number of satellites or wider field-of-view instruments which would have provided the same swath at a lower altitude, but the wider field of view would have produced degraded image quality and distorted spatial resolution near the edges of the image (Section 8.4.3). The final result of this trade-off was that an 800 km orbit was selected. Pleiades (Section 9.6.4) has an instantaneous observation area of 20 × 20 km, and its main mission driver was its very high 0.7 m spatial resolution, favored by lower flying altitudes. But Pleiades must also have access to any part of the Earth within 1 day, which demanded off-Nadir pointing with large angles that would have to become even larger if flying at lower altitudes.

Sentinel-3 orbits with a local time of 10:00. Coastal zone observations benefit from local times closer to midday when there are better lighting conditions over the sea – a rather dark target – while missions targeting the land try to avoid midday for its lower image quality over land. Data from the radar altimeters would be more accurate with each satellite flying at different local times, yielding better ocean time sampling. Climate studies, however, demand a long and consistent series of data and continuity with respect to previous missions is crucial. Various ESA satellite missions have flown near 10:00, including ERS-1, ERS-2, and Envisat. The final orbit selection was again a balance between conflicting needs.

The Iridium (Section 9.6.3) constellation's orbit and orbit altitudes were governed by the need to provide contact anytime anywhere with mobile telephones. The chosen low altitude of 780 km offers short distances to the ground-based low-power, portable telephones, without the need for excessively large or powerful satellite antennas. Nevertheless, for complete Earth coverage at all times a constellation of 66 satellites, distributed in 6 orbital planes in near-polar orbits, was necessary. Thuraya[17] is another constellation that like Iridium provides mobile communications to hand-held telephones, but it is not a global system and flies in geostationary orbit. To compensate for the great distance, the satellites carry a powerful communication system, which depends on a large 12.5 m diameter deployable antenna. This antenna offers satisfactory communications to the telephones from the 35,786 km of the geostationary orbit, but only provides service to areas covered by the satellites, one located over 44° East and the other over 98.5° East. This is allowed by the fact that Thuraya does not specify whole Earth coverage because of its specialized on communication for Asia. As with the widely different orbits of the astronomical observatory missions reviewed in Section 7.2, these examples underscore the fact that similar, albeit not identical services can be provided through the use of very different orbit architectures.

The orbits of missions carrying in situ instruments are directly determined by having to travel through the areas being studied. Ulysses is one such mission (Section 9.6.2), studying the solar environment at high latitudes by flying through them with in situ instruments. The Cluster constellation's orbit (Figure 9.21) and separation between satellites are driven by the need to sample the different regions of the Earth's magnetosphere; the 90° orbit inclination provides a full selection of latitudes, a fixed orbital plane affords shifting Sun-Earth geometries throughout the year and the high eccentricity offers a varied selection of distances to Earth. Rosetta's orbit (Section 11.4.2), like that of Ulysses, is interplanetary and makes use of planetary rendezvous. The choice of orbit was governed by the mission's aim of meeting the targeted comet, which required high amounts of impulse for the satellite. The final trajectory resulted from a compromise between the complications of providing powerful on-board propulsion and the difficulties involved in performing multiple planetary rendezvous, which simplifies the hardware but complicates the satellite's operations and prolongs the mission. It should not be forgotten that orbit selection strongly depends on the availability of launchers, frequently a system constraint in which the architect has little freedom of choice. Ulysses was made possible by its very special launching conditions, which included use of the Space Shuttle, plus two supplementary stages. This implementation would have been impossible if the mission were not collaboration between NASA and ESA.

[17]http://www.thuraya.com

9.9.2. Attitude selection

Attitudes were introduced in Section 9.3, which analyzed the use of different types of attitude control and the reasoning behind specific choices of various missions. That section recalled the different tools used for different types of attitude control, including uncontrolled; gravity-gradient; rotation-stabilized; momentum-biased; inertially stabilized Nadir-pointed; and inertially stabilized inertially pointed. Section 9.5 underscored the fact that most missions need more than one attitude control loop: one complex high-accuracy loop and one simple robust loop. It was also recalled that these two loops are performed by different sets of attitude sensors, actuators, and control components. The choice of attitude control has a major impact on the entire mission, and the architect plays a leading role in selection process because it involves balancing the decision's conflicting impacts on many aspects of the mission. The attitude control architecture depends on the mission's specific observational needs. Most satellites observing the planet they orbit will be Nadir-pointing, while those making in situ observations will be spin-stabilized, and astronomical satellites will be inertially pointed. In some cases a trade-off between classes of attitude is possible. Earth-observing geostationary satellites can be spin-stabilized or inertially stabilized; MSG (Figure 7.3) is spin-stabilized while GOESS (Figure 7.2) is inertially stabilized. Nadir-pointing can be implemented with an inertially stabilized architecture, as with Sentinel-3 (Section 9.6.1), or with a gravity-gradient plus momentum bias approach, such as Iridium (Section 9.6.3). In the past, inertially stabilized systems were considered expensive but reasonably accurate inertially stabilized systems are now affordable. Surrey DMC constellation (Section 11.4.4) is an example of cheap, inertially stabilized, Nadir-pointed satellite. Observational needs also determine the level of accuracy required. Higher observational accuracies will require higher accuracy sensors and actuators as well as faster acting wider bandwidth control logic. Especially rapid attitude changes – as required by agile satellites – require powerful devices such as control momentum gyros or exceptionally large momentum wheels. Some high-accuracy systems require the use of the instrument itself as an attitude sensor device. Telescopes that are used to study the sky can act also as star sensors and determine the attitude of the satellite. GOCE's gravity-sensing instrument (Section 10.3.1) is used to determine the satellite's angular accelerations. For very high-accuracy systems, the choice of attitude control tools will also be influenced by the need to minimize internal disturbances, which could exclude the use of attitude actuators that cause mechanical perturbations.

9.9.3. Coverage and revisit

9.9.3.1. General

Coverage was defined in Section 8.4.2, and Section 8.4.5 analyzed the relation between coverage and the systematic acquisition of data about a mission's area of interest, which may be anything from a specific area of the Earth to the whole Earth or entire celestial sphere. Systematic acquisition provides regular synoptic views of the area of interest and must meet requirements to fully cover the area within a specified period of time. For observations of an orbited body, the coverage performance is a

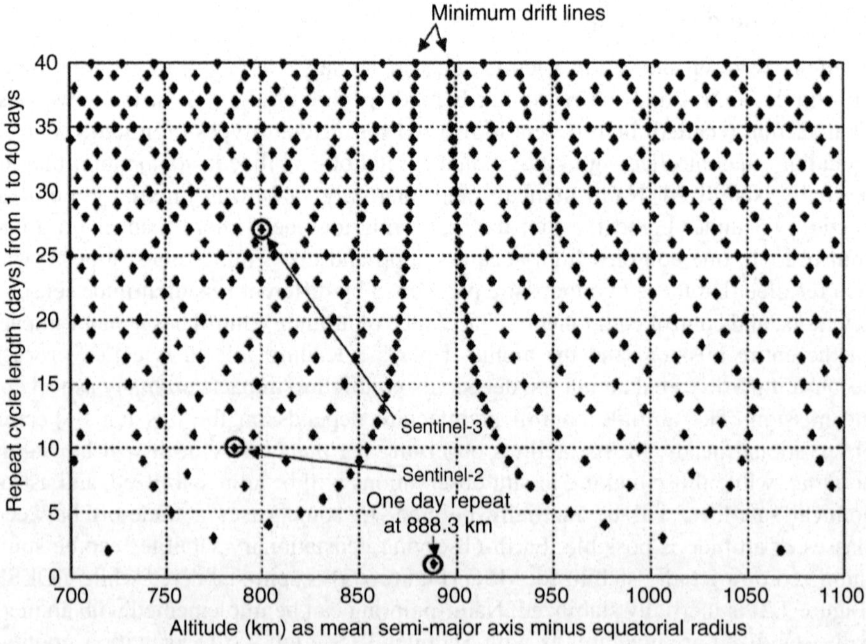

Figure 9.48. Repeating Orbits with a Cycle Lower than 40 Days in the Altitude Range Between 700 and 1,100 km.

function of the instrument swath and the orbital track pattern of the satellite or satellites involved. In order to provide full coverage, the orbit track must traverse the surface with a repetitive and regular pattern separated by a distance smaller than the instrument swath, and the repetition cycle must be equal or smaller than the specified revisit period. Revisit was discussed also in Section 8.4.2, and Section 8.4.5 examined the relationship between revisit and interactive data acquisition. Revisit criteria are similar to those for coverage: the optimal orbital track must be identified, but with the instrument's pointing range replacing that of instrument swath. For celestial sphere observations the orbit is irrelevant to coverage. To provide a synoptic view of the whole celestial sphere the instrument swath must be larger than the angular separation between the tracks produced by the changing satellite attitude with respect to the celestial sphere. This will afford full coverage of the celestial sphere.

9.9.3.2. Repeat orbits

As explained above, the best way to provide optimal coverage and revisit over the Earth is to fly a uniformly distributed orbital track. The criterion for a repeat track orbit is that after a specific integer number of orbits the satellite-Earth geometry returns to its original point. In the case of Sun-synchronous orbits this means that n orbits coincide with m solar days. In the case of non-Sun-synchronous orbits it means that the differential angular speed between the Earth and the orbital plane multiplied by the time necessary to complete exactly n orbits shall be equal to m times 360°. Figure 9.48 provides the possible repeat patterns for Sun-synchronous orbits. In the case of $m = 1$ the orbital path would cover the Earth with regular repetitive tracks every day. In the range of altitudes covered in Figure 9.48, at 888.3 km of altitude, the orbital period is 6,171.4 s and 14

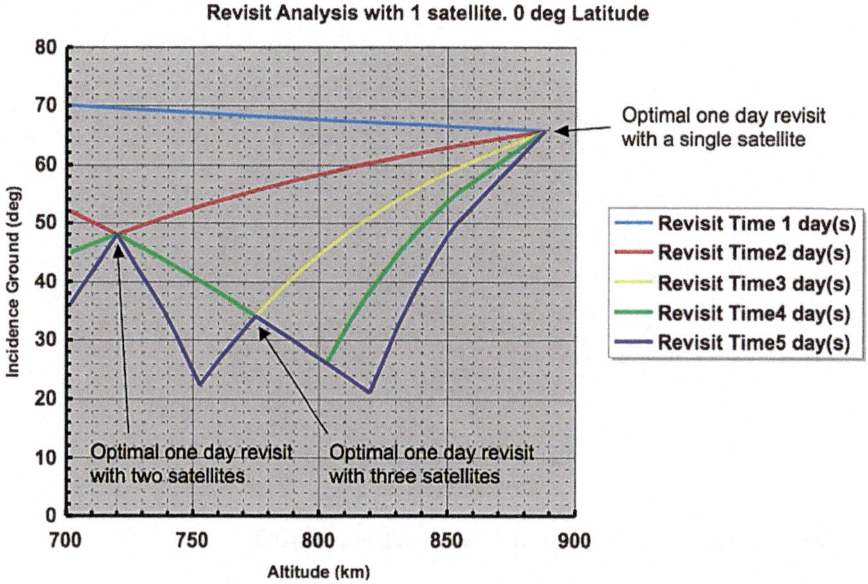

Figure 9.49. Ground Incidence Angles Allowing Full Earth Coverage Between 1 and 5 Days and for Altitudes Between 700 and 900 km. The Optimal 2 and 3 Days Orbit Positions Can Be Used to Provide Optimal 1 Day Coverage with Two or Three Satellites.

orbits require exactly 1 day. The figure also provides the location of the Sentinel-2 and Sentinel-3 track patterns. As explained in 9.6.1, Sentinel-3's track pattern corresponds to an orbital period in which 385 orbits are performed in 27 days. This orbit provides exact repeats every 27 days and near exact repeats every 4 days. Sentinel-2's track pattern corresponds to 143 orbits in 10 days, and was chosen to provide optimal coverage of the Earth in 5 days for two satellites or 10 days with one satellite. Complete observation of areas between tracks requires that either the swath or the pointing range of the instruments is larger than the separation between tracks. Figure 9.49 provides the ground incidences and orbit altitudes corresponding to such situations. According to Figure 9.49, in order to cover the Earth in 1 day with a satellite flying at 888.3 km of altitude, the incidence angles on the ground at the edge of the swath must be ±65.7°, the required pointing range ±53.2° and the swath itself 2,789 km. With this large incidence angle, images will be distorted and the quality of the images obtained at the edge of the swath will be inferior (Section 8.4.3). Additional time for full Earth coverage will improve image quality. Again according to Figure 9.49, if 2 days were allowed, one satellite flying at 720 km of altitude would need a maximum incidence angle of just 48°. This same orbit would allow two satellites to provide 1-day coverage. The two satellites would fly in the same orbit, with each of them daily covering the surface missed by the other. If one satellite fails, the remaining satellite would take up the slack, offering full coverage every second day. This arrangement provides robustness, as the system continues to function, albeit with reduced temporal revisit. Sentinel-2 uses this approach, with two identical satellites flying in the same orbit at an altitude of 786.1 km and anomalies separated by 180°. The orbit repeats exactly every 10 days. The two satellites (Figures 9.50 and 9.51) fly interleaved tracks covering the Earth every 5 days. Sentinel-2 gathers highly accurate data on vegetation and biophysical parameters and requires the

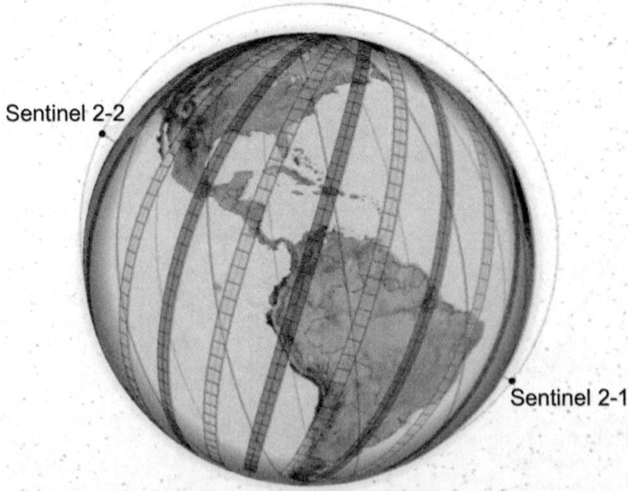

Figure 9.50. One Day of Coverage with Two Sentinel-2 Satellites.

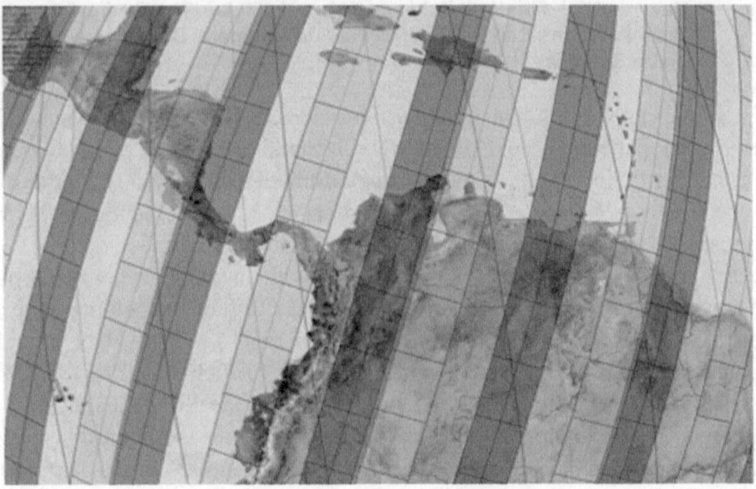

Figure 9.51. Four Days of Coverage with Two Sentinel-2 Satellites, 1 Day More Will Fill the Gaps and Ensure Full Earth Surface Coverage.

highest radiometric quality. The instrument's swath is 274 km and the maximum incidence angle at the edge of the swath is 11.1°.

The concept of exact orbital repeat pattern and optimal coverage also applies to instruments without swath. Radar altimeters provide sea-level anomaly point measurements (Section 8.5) used to provide data for global dynamic models of the ocean. In this case providing coverage means gathering a network of measurements every *m* days dense enough to allow the correct functioning of the global dynamic ocean models. The ocean possesses very strong inertia; dynamic ocean currents and eddies span around

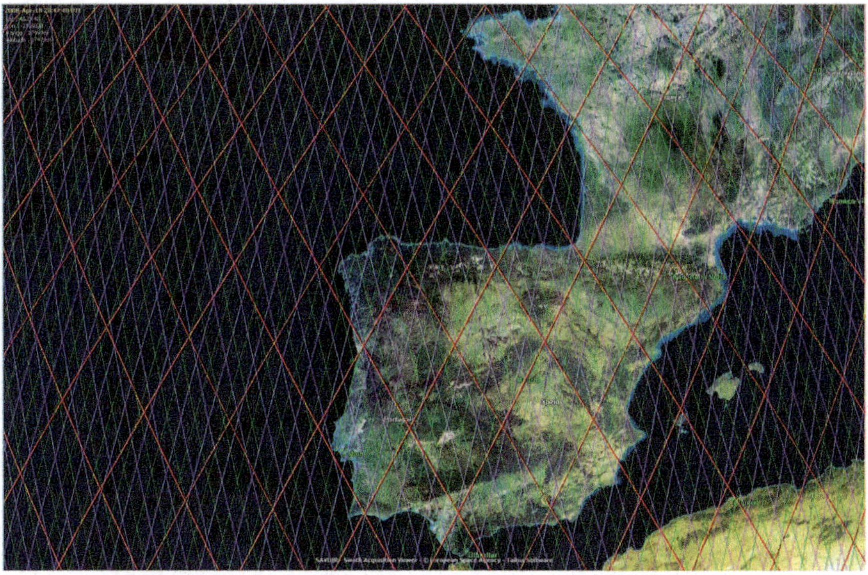

Figure 9.52. Combined Tracks 27 Days of Track of Two Sentinel-3 (More Dense Thinner Net of Measurement Lines) and One Jason (Less Dense Thicker Net of Lines) Satellites.

10–100 km, and persist for weeks or even months. Consistent with these facts, a repeat period of around 1 month and measurement density on the order of 50 km would be adequate. Figure 9.52 provides the 27-day track network of two Sentinel-3 satellites plus 10 days of the radar altimeter mission Jason. The tracks of the two Sentinel satellites combine to produce the required measurement density, while Jason's less dense track is dense enough to act as a reference for the whole set of measurements.

9.10. Allocation of budgets

Many key requirements of orbit and attitude are values that must be apportioned between all the factors contributing to those requirements; thus, each of the key requirements generate a budget that needs to be allocated among all possible error sources. The budgets related to instrument pointing and satellite attitude are especially complex. Section 9.8 has already introduced them. Specifically, Section 9.8.1 and 9.8.2 already established the need to establish requirements for parameters like the absolute and relative pointing errors, and the absolute and relative knowledge errors. Those sections also established the need to create requirements on pointing rate stability. The present section is going to analyze the design alternatives associated with all these budgets.

9.10.1. Satellite location

All missions demand that the satellite's location be measured and controlled with a given level of accuracy. This will be required by the need to control the regular pattern of the orbit's ground track (Figures 9.50) or by the obvious need to observe the actual areas of interest for the mission, making it necessary not only to have the correct attitude but also the correct orbital position. Errors in determining satellite location are most often caused by:

- Errors in the ephemeris of real-time navigation satellites (GPS or Galileo) in case they are used for orbit determination
- Errors in the relative velocity between satellite and ground station in cases where the orbit is determined from the ground by measuring the Doppler shift of the satellite signals
- Errors in the determination of atmospheric drag and other orbit disturbance factors, such as lunar-solar gravity pull, or errors in gravity field data
- Errors in the orbit determination algorithms
- Orbit ephemeris projection errors
- Timing errors

As explained in Section 9.2.1, in low Earth orbit, autonomous orbit determination is now possible using GPS receivers on board the satellite, but almost all missions also perform on-ground orbit determinations and uplink the orbital predictions to the satellite. The use of navigation satellites for deriving real-time orbit determinations on board can provide satellite positions to an accuracy of 10 m or better. In most cases the requirements for satellite position determination are moderate, but very high-resolution Earth observation calls for a correspondingly accurate orbit determination.

Some applications, such as radar altimetry (Section 8.5) or geodetic satellites, demand detailed knowledge of the satellite's position for the derivation of the observables. Precise "a posteriori" orbit determination requires the use of high-accuracy navigation receivers and complex processing. Orbit determination to within a few centimeters is now possible.

9.10.2. Instruments line of sight pointing and recovery

The instrument line of sight error is identical to the "APE" of Section 9.8.1. The line of sight recovery error will be identical to the "AKE" of the same section. Recalling the discussion of Section 9.8.1, 9.8.2, and 9.8.3, the factors contributing to the APE are:

- Attitude pointing errors of the attitude control system:
  - Low frequency external perturbations such as atmospheric drag or gravity-gradient torque
  - Low frequency internal perturbations such as large unbalanced slowly rotating antennas or the uncorrected part of the thermoelastic deformations affecting the attitude control system
  - Measurement errors of the attitude sensors, including star trackers, magnetometers, or gyros
  - Attitude control actuator errors: momentum wheel, control momentum gyros, thrusters, and magneto-torques
  - Control loop errors and bandwidth, control law and control, and commanding algorithm errors
  - Timing errors.

- Thermoelastic deformation, from the attitude control reference frame to the instrument line of sight reference frame
- Scan realization errors in scanning instruments
- Residual alignment bias and drifts after in-orbit commissioning, between the attitude control reference frame and the instrument's line of sight

Attitude pointing performances depend on the attitude control subsystem, as well as on internal and external perturbations produced by the environment, including atmospheric drag, gravity-gradient torque, and solar wind pressure torque, among others. The satellite's attitude pointing error will depend on the specified quality of the attitude sensors, actuators, and the design of the control loop. A star tracker, for instance, will experience random Gaussian errors produced by sources such as instrument quantification or electronic noise. It will also experience harmonic errors and drifts that might be caused by the sensor variations in temperature, and might also have biases produced by assembly or integration errors.

To observe the area of interest the line of sight must be pointed in the right direction with an accuracy commensurate with the satellite's swath. In instruments producing large-scale mosaics with many individual images the APE will be driven by the need for a small overlapping between images to avoid noncovered gaps. But to properly locate the observations the line of sight must be known with an accuracy commensurate with the instrument's resolution. The spatial resolution is much smaller than the instrument field of view. This means that the requirements for line of sight knowledge "AKE" will always be distinct and more demanding than the requirements for line of sight realization "APE."

For example the high-resolution Pleiades mission (Section 9.6.4) produces individual images of 20 km that are combined to produce large-scale or wide area mosaics of images. To be able to produce an efficient 10 % overlapping of images it will be necessary to ask the satellite to point the instrument with an APE = 2.8 mrad. However, Pleiades is required to properly locate autonomously the images on-ground with an accuracy of 10 m. This requires the pointing knowledge to be better than 14 μrad. That means the APE requirement is 200 times more relaxed than that for the AKE. Any AOCS subsystem able to provide the pointing knowledge required by Pleiades will provide the required absolute pointing without any difficulty. This will be the case for most missions.

In some special cases, some missions could have demanding line of sight realization requirements. To ensure the superior quality of its ocean altimetry mission products, Sentinel-3 (Section 9.6.1) imposed rigorous specifications on the altimeter's actual line of sight. The required actual pointing must differ by less than ±0.2° from the perpendicular to the Earth's ellipsoid. After readjustment trade-offs this highest-level error was allocated in the following way:

- Attitude control system pointing error: 0.10°
- Radar altimeter boresight pointing error: 0.04°
- Satellite structure pointing harmonics: 0.1°
- Satellite structure pointing bias: 0.05°

This defined a set of balanced requirements for the attitude control system and for the stability of the satellite. The requirements differentiate between bias and harmonics errors as discussed in Section 9.8.2.

As said before most missions need the correct referencing of the observations to their place of origin. This will impose strict requirements on the knowledge of the line of sight. Factors contributing to line of sight knowledge are:

- Attitude measurement errors of the attitude control system:
  - Measurement errors of attitude sensor: star trackers, magnetometers, or gyros
  - Attitude estimation errors
  - Timing errors
- Residual errors in the knowledge of the thermoelastic deformations (bias, drift, and harmonic) of the instrument's line of sight
- Scan measurement errors in scanning instruments
- Residual errors in the knowledge of the bias and drifts of the instrument line of sight reference frame

In the case of imaging instruments, during the satellite's commissioning phase but before the operational use of the instruments begins, it is possible to characterize the static and variable bias, and the harmonics deformations between the attitude and the instrument reference frames. Comparing the actual and the nominal position of well-located ground or space reference control points make this possible, and the instrument's line of sight pointing knowledge errors will be reduced to the residual errors after the introduction of the measured corrections. The residual bias error after in-orbit correction will be a function of the original nonbias error because the existence of a large nonbias error will complicate the accurate determination of the bias by using reference control points. A good rule of thumb is that with good control points it should be possible to reduce the bias to around one-half of the original nonbias pointing recovery errors. Comparing thermal analysis results, measured in-orbit temperatures and measured in-orbit misalignments can reduce harmonics error associated to the varying Sun illumination during the orbit. This approach would reduce the harmonic errors associated with solar illumination and thermal conditions. It is generally accepted that it is possible to reduce the harmonic pointing errors to one third of its original value.

9.10.3. Pointing stability realization and recovery

Sections 9.8.1 and 9.8.2 already discussed pointing stability and how it was driven by the instrument integration time. The time needed for integration will depend on the instrument concept. The few detectors featuring a so-called whiskbroom sight line (Figure 8.4) must perform a quick transversal scan of the orbit track in order to provide the required swath. In this case the integration time is very brief. So-called pushbroom instruments (Figure 8.4), on the other hand, produce the swath by means of a long fixed array with a large number of detectors, and the time allocated for the integration of each detector's pixels can be much longer. This longer period permits the integration of many more photons, resulting in superior radiometric quality, but it also requires more demanding stability to ensure steady instrument pointing during these longer intervals. The pointing stability is directly related to the avoidance of the signal arriving from a single point being smeared between different receiving pixels. That means, the pointing stability requirement shall be commensurate with the angular resolution corresponding to a single pixel ($\Delta\theta$ in Figure 9.47) divided by the integration interval (Δt in Figure 9.47). Pointing stability can be very exacting in cases where very high-resolution instruments need long integration times. The following elements will alter the pointing stability:

- External perturbations such as upper atmosphere winds
- Low or medium frequency harmonics associated with thermoelastic deformations, or with the movement of large items like solar arrays or large rotating mechanisms
- High frequency perturbations produced by rotary or linear mechanical actuators
- High frequency attitude control residuals associated with flexible satellite structures that the attitude control system is unable to compensate for
- Possible high frequency structural perturbations produced by step events such as the thermal transients from eclipses to solar illumination, the operation of solar array drive mechanisms, or the firing of thrusters

As discussed in Section 9.8.2 the attitude control system will counteract the lower frequency "drift" perturbations that are inside its frequency range. This will generate a maximum drift requirement that will drive the design of the AOCS. At the same time, it will also be necessary to specify the maximum strength of the lower frequency drift perturbations that the AOCS shall be able to counteract. This maximum perturbation shall be apportioned between all the lower frequency perturbation sources.

For higher frequency harmonic perturbations and for broad band "jitter" perturbations outside the capabilities of compensation of the attitude control, the only way to control the pointing stability is to limit the external and internal perturbations that generate pointing instability. This will require establishing supplementary perturbation budgets that shall be apportioned between all the identified sources.

9.10.4. Geo-locating

The observations received by the instruments must be mapped to their place of origin; this process is called geo-locating and requires mapping each observation with respect to an absolute reference frame. This requires:

- Referring observations to the satellite's attitude reference frame
- Transferring the satellite's reference frame to an absolute reference frame in the celestial sphere or over the planet being observed
- Locating the source of the observation within this absolute reference frame

Geo-location requires that both the instrument's line of sight (Section 9.10.2) and satellite position (Section 9.10.1) along its orbit be correctly recovered. The discussion, budgets, and apportionments discussed in Sections 9.10.1 and 9.10.2 are directly applicable to the discussion of geo-location. In addition, geo-location requires that each pixel is allocated to its true geographical origin requiring supplementary reference frame transfers. So timing errors and time coordination error across the system elements are contributors.

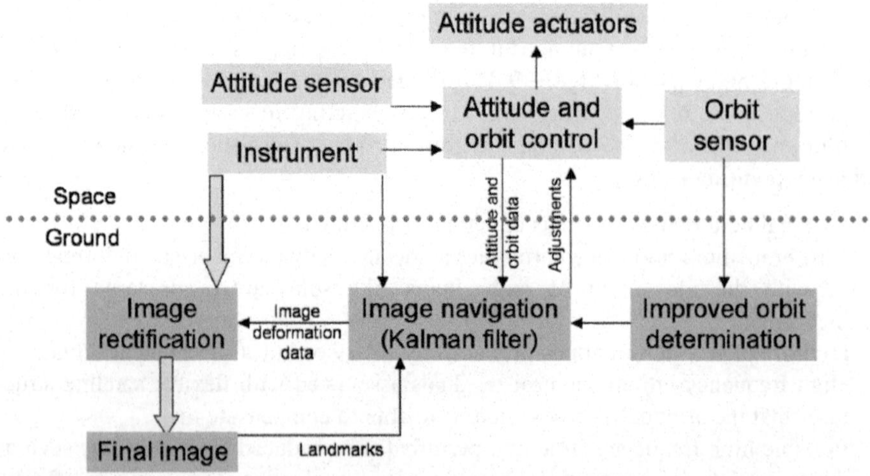

Figure 9.53. Image Geo-location as an Interactive Space and Ground Process.

As in the case of the line of sight pointing recovery "AKE" of Section 9.8.1, geo-location requirements should also be commensurate with the instrument's spatial resolution. In principle, the geo-location requirement should be smaller than spatial resolution requirements. For resolutions of the order of a few hundred meters, geo-locations between 0.2 and 0.4 of the spatial resolution would be a reasonable choice. However, if the resolution is tens of meters or less, providing a geo-locating performance smaller than 1 pixel would be impossible. Pleiades has a resolution of 0.7 m and a geo-location performance of 10 m. Geo-locating data forms part of a process in which the space elements and the ground elements support each other. Section 9.10.2 already discussed the use of ground control points to characterize pointing errors during the satellite commissioning phase. One supplementary possibility will be to use ground control points continuously during the ground processing of the instrument images. Each obtained image will contain clearly defined landmarks of known position that can be used as reference tie-down points to provide feedback on the actual satellite attitude and to correctly geo-locate the whole image. Ground processing algorithms use landmarks such as lakes, rivers, or coastlines; in astronomy, bright stars with accurately known positions can also act as landmarks. The algorithm will integrate and filter the satellite's attitude and orbit information, as well as the landmarks to correct the instrument image and to provide an optimal estimate of the satellite's attitude, and an optimal geo-location for all pixels in the image. The improved attitude information can be relayed back to the satellite for use as adjustment factors by the attitude control system (Figure 9.53).

This feedback between ground and space allows the requirements on satellite attitude recovery to be relaxed and the end-to-end performances to be optimized. If this approach is used the end-to-end performance will now depend on the capacity of the in-orbit geo-location (before ground processing) to be precise enough to act as a good starting point for the ground-based image-processing algorithm. On the other hand, using control points for each image will make the payload data processing slower and more expensive and entails creating large databases of reference tie-down points.

Furthermore, this approach is only feasible when previous information on features of the area in question are available; the first high-resolution images of a planet, for instance, will lack any previous knowledge of the surface; and images of the open ocean, with no land visible, would also lack any recognizable reference points. In situations such as these, the geo-location performance will derive directly from the satellite's performance.

9.10.5. Co-registration

Co-registering is defined as correctly identifying measurements from different bands of the same or different instruments as coming from the same physical source. This calls for identifying "relative" misalignment between and within instruments. There are fewer potential error sources for co-registration than there are for geo-location, because error sources acting in a uniform way over the whole satellite, attitude errors, for example, will not add to the co-registration budget. However, the proper study of physical processes requires the use of information obtained from several bands, perhaps even from several instruments, and this information must be combined with an accuracy far smaller than a pixel, so co-registration requirements of the order of 0.1 or 0.2 of a pixel are frequently called for.

In instruments featuring a single line of sight, co-registration will be internal to the instrument, and will produce requirements on maximum allowable distortions between the different focal planes and between the different optical paths for the same instrument. The total allowable co-registration requirements will be apportioned between all the error sources, and the allocation will generate derived requirements on thermal control, the use of high stability materials, assembly procedures, and on-ground testing. The entire apportioning process will be internal to the instrument, and will normally fall under the responsibility of the instrument specialist. In cases of co-registration requirements between different instruments, on the other hand, the architect will become involved, and the total co-registration budget will be divided between the instruments and the structure connecting them.

Co-registration measurements performed by the same satellite using different instruments with different lines of sight and scan patterns are especially complex. Sentinel-3's SLSTR instrument (Figure 9.54) features two conical scanning lines of sight, and the OLCI instrument has an asymmetric pushbroom across-track field of view. To provide improved information on the status of global plant life, co-registery of data from both instruments to its real physical origin is required. This means that the mechanical stability inside and between each instrument must be specified, as well as the accuracy of the SLSTR scanning mechanisms.

9.10.6. Repointing agility requirements

Instruments with interactive image acquisition (Section 8.4.5) obtain images sequentially over different areas according to regularly updated image acquisition plans. A value must be specified representing the time needed for the satellite to acquire two consecutive images over two different areas. The image acquisition requests will sometimes be known long enough beforehand to allow optimization of the acquisi-

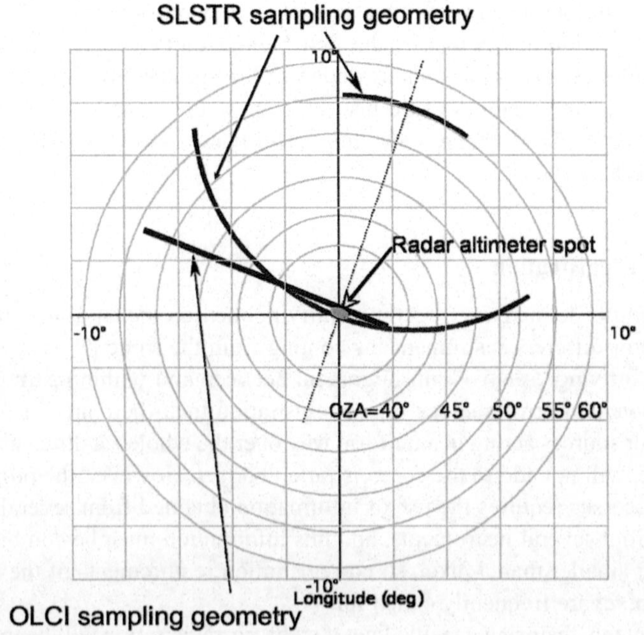

Figure 9.54. Sentinel-3 Co-registration Geometry.

tion strategy, favoring an operation plan that minimizes attitude changes and optimize the number of images able to be acquired. At other times significant attitude changes will be needed to cope with situations requiring urgent observations. In any case an agile satellite that permits rapid attitude changes will be able to provide more images, thus increasing the value of the mission. Superior agility calls for:

- Reducing the time needed for repointing the instrument from one target to the next. This entails implementing powerful attitude control actuators and reducing the satellite's inertia matrix by designing a compact craft.
- Reducing tranquilization time, that is, the time needed by the instrument to recover sufficient stability after repointing to acquire an acceptable image. This means increasing the AOCS bandwidth to allow fast compensation for any remaining oscillations as well as to stiffen the satellite.

The last paragraph of Section 9.6.4 describes the agility performance of Pleiades. As explained there, the satellite has a powerful control momentum gyro, low inertia, and high stiffness. The whole satellite external configuration (Figure 9.29) is driven by the need for compactness and stiffness to allow fast repointing.

Agile missions like Pleiades can provide flexible pointing strategies while producing a large number of images (Figure 9.55).

9.10.7. Delta V and fuel

Satellites have recourse to various sources of energy to meet their propulsion needs. A portion of this energy can be generated by rendezvousing with planets, as Ulysses

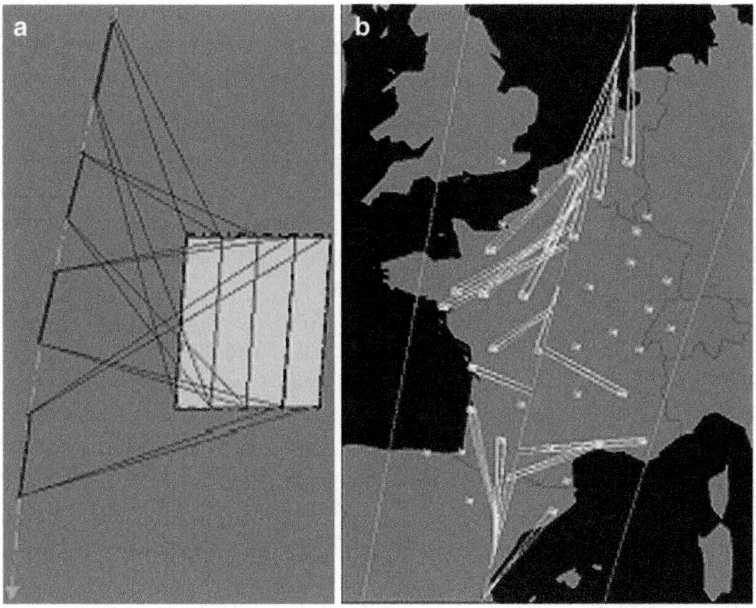

Figure 9.55. Strategy of the Agile Satellite Pleiades to Produce a Mosaic of Images in a Single Path (a) and To Cover a Large Number of Images over an Area of 1,000 × 1,000 km (b) also in a Single Path (CNES http://132.149.11.177/PLEIADES/Fr/GP_systeme.htm).

did (Section 9.6.2 and Figure 9.23). Another option is delivered by specialized sources of thrust, which form part of the mission but are not part of the satellite: the IUS and PAM rockets used to accelerate Ulysses (Figure 9.27) are an example of these. Nevertheless, the satellite will still need to carry enough fuel on board to perform all the necessary remaining orbit correction and maintenance maneuvers. This on-board fuel must be apportioned among all the events requiring it:

- Launchers never actually place satellites into their exact required orbits, and all orbit injection errors must be corrected by means of on-board fuel. The launcher specialist will provide figures for the average and worst-case injection errors, and the satellite must supply the corresponding on-board fuel to compensate for them.
- Some missions are designed to allow large orbit correction maneuvers, and carry sufficient fuel to perform them. MSG for instance, was placed by the launcher in a highly elliptic GTO geotransfer orbit, and the satellite carried enough fuel to reach its final circular geostationary orbit. It needed 800 kg of fuel to perform this large orbit correction.
- Satellites requiring precise orbit control will need to regularly correct any orbital drift produced by external perturbations. In low Earth orbit the main cause of drift will be atmospheric drag. In geostationary orbits, the Sun and Moon's gravitational pull will alter the orbit inclinations and this effect needs to be compensated for.
- The space environment is now densely populated and to prevent close encounters with other orbiting objects some fuel needs to be allocated for possible collision avoidance maneuvers.

- To prevent a satellite from becoming a danger to other missions once its own mission is completed, it must have enough on-board fuel for shifting to an orbit where it will present no danger.

Some missions, on the other hand, are designed to operate without any on-board fuel. This diminishes costs, but prevents them from making any orbit correction maneuvers. This approach is acceptable for most technology demonstration missions and for some science missions not demanding high-accuracy orbits. Given that drift from the nominal orbit increases with time, operational missions, which normally call for long lives, do need some form of orbit control.

The sum of all the fuel allocated for each requisite maneuver will determine the overall fuel needs. The calculation of each maneuver's fuel needs usually includes its own safety margin. The launcher specialist provides worst-case orbit injection errors that are generally larger than the actual one, and the overall budget also includes a system-wide safety margin on the order of 10% or more. This combination of safety margins means that the "most likely" fuel expenditure will be less than the reference design case, and as a result, most satellites' fuel lasts much longer than originally planned, offering the possibility to extend the mission's life.

9.10.8. Mechanical perturbations

Mechanical disturbances alter the pointing and pointing stability of the instruments and to establish pointing and pointing stability requirements will automatically establish requirements for the mechanical perturbations on board. Nevertheless, on missions where there are very strict pointing requirements and many sources of possible internal mechanical perturbations – for example, many mechanisms – it will be good to have a specific budget of mechanical perturbations. The resulting mechanical perturbations budget will provide a reference for the design of all the sources of mechanical perturbations.

In line with the discussion of Section 9.8, the budget shall specify in the frequency domain. It is possible to identify three frequency ranges (Figure 9.56):

- A low frequency range, where as frequency goes up the perturbation requirements will become tighter
- A medium frequency range where the requirements will be constant
- A high frequency range where as the frequency becomes higher, the perturbation requirements will be relaxed

At low frequency the perturbations will be compensated by the attitude control of the satellite. As the frequency becomes slower and slower the attitude control sensors and actuators will have more and more time to measure and to compensate these perturbations. This will allow tolerating larger perturbations as the frequency of these perturbations decreases. At very high frequency the attitude control system will be unable to control the perturbations. Nevertheless, high frequency perturbations will be unable to have a large impact in pointing because the fast changing perturbations will not have time enough to alter by much the pointing. This is the reason why the specified allowed perturbations curve can increase as frequency increases. Between

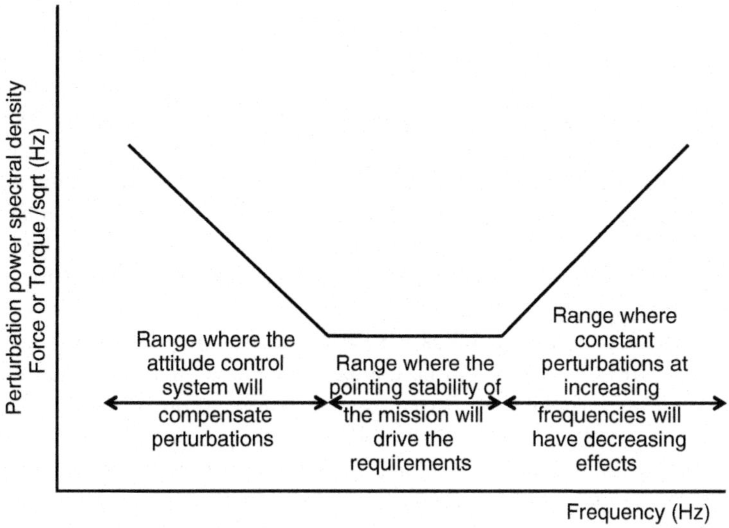

Figure 9.56. Template for Allowed Mechanical Perturbations as a Function of the Frequency.

these two extremes, there will always be a central frequency range where the allowed mechanical perturbation requirements will be minimal.

The resulting – frequency dependent – maximum perturbation envelopes can be used to specify the values allocated to the different sources of mechanical perturbations, which will be also frequency dependent. The addition in the frequency domain of all the specified values shall be smaller than the maximum allowed disturbances at all frequencies.

9.11. Implementation and maintenance of constellations

Some missions require that numerous satellites maintain a specific geometry in relation to each other; these satellites constitute a constellation. The use of multi-satellite constellations was touched on briefly in Section 7.3. Section 8.4.2 examined issues associated with using constellations to avoid deficient image quality at the edges of large swath instruments, and Section 9.9.3 discussed the process of choosing the right orbit altitude and number of satellites to provide the desired coverage and revisit. In addition to fulfilling mission requirements, constellations also provide a robust architecture able to cope with the failure of individual satellites without threatening the viability of the mission as a whole. This robustness is especially advantageous for operational missions, and it is not accident that the three examples provided are all operational missions.

The present section will present supplementary discussions about constellation design and maintenance. Section 9.6 included descriptions of three constellations: Sentinel-3, Iridium, and Pleiades. Sentinel-3 and Pleiades both employ two satellites while Iridium employs 66.

Establishing a clear policy for the launch strategy of a constellation is crucial to mission design: to both initiate and replenish it. For missions consisting of just a few satellites, such as Pleiades or Sentinel-3, it is a question of how and when to launch them. Ideally they should be launched as close in time to one another as possible in order to begin fulfilling the mission's aims. However, in order to maintain the production and industrial organization at a reasonable size, the sequential manufacturing and delivery of the satellites have to be optimized, even though it could mean delays between consecutive satellites. In some cases funding for the mission is a step-by-step process and may mean employing fewer satellites than required for extended periods of time.

A single Pleiades or Sentinel-3 satellite can still provide good quality images, but its coverage or revisit figures will be double the desired values. Though this is not ideal, it could be acceptable in certain circumstances. Iridium, on the other hand, needs all of its satellites if it is to provide continuous communications coverage, which is the mission's fundamental aim, and a constellation with only half its satellites providing service just half the time, would be totally unacceptable. Hence, this kind of mission demands the speedy launch of all the constellation's constituent satellites.

Launchers can accommodate multiple satellites. The change of orbital inclination, however, requires a very large amount of energy and multiple satellites in a single launch are only feasible if all the satellites are to be injected into the same or very nearby orbital planes. Conversely, placing satellites at different points along the same orbital path only requires small amounts of energy. Iridium uses several types of large launchers able to inject five (in the case of Delta) or seven (in the case of Proton) satellites simultaneously into the same orbital plane. One or more replacement satellites must also sometimes be launched to compensate for eventual failures; small launchers, such as Rockot and Long-March, used by Iridium, can fulfill this replacement function. This need for flexibility is a major configuration driver and will be discussed further Chapter 10 (Section 10.2.1 and Figure 10.5).

Operational meteorology also requires the use of constellations. The atmosphere is dynamic and a system monitoring the weather needs to provide very rapid coverage. The World Meteorological Organization[18] requires that synoptic coverage of the Earth to be refreshed every 4 h. This requirement can be fulfilled by a constellation of LEO satellites in three equally distributed orbital planes separated by 4 h of local time. The NPOESS meteorological LEO constellation (Figure 9.15) fulfills these requirements. Each of the three NPOESS satellites provides horizon-to-horizon images, and each can provide full Earth coverage in a day; the three together fulfill the 4 h requirement. This requires a maximum incidence angle at the edge of the swath of around 68°. The resulting image distortion is not ideal but the users consider it acceptable.

Not all missions call for whole Earth coverage; when the area of interest is limited, orbits can be chosen which optimize revisits over the area using a moderate number of satellites. Forest fires can occur at any time; they evolve swiftly with dangerous implications for life and property. The Fuego constellation was designed to provide

[18]Final Report of the CGMS-WMO Optimization Workshop, Optimization of Satellite Observation Missions, WMO 2006.

30 min of revisit over two bands around the Earth between 35° and 45° of latitude. This limitation was acceptable because the monitoring of fires in Mediterranean forests is the fundamental aim of Fuego[19] and all Mediterranean forests are located at these latitudes. To fulfill this very demanding revisit schedule the mission features 12 satellites in a Walker constellation distributed equally in three orbital planes flying at an altitude of 700 km with an orbit inclination of 47.5°. The actual revisit figure over the area of interest was 23 min. Walker constellations[20] contain *n* satellites distributed uniformly in *p* planes. All the orbital planes have identical altitudes and the same inclination with respect to the Earth's equatorial plane. The ascending nodes of the *p* planes are uniformly distributed around the Equator. To completely define the constellation, it is necessary to also define the difference in anomaly between satellites in adjacent orbital planes. This separation is indicated by a third parameter *f*; if a satellite is at its ascending node the next most easterly satellite will be past the node by *f* times 360° divided by the number of satellites in the constellation.

[19] D. Escorial Fuego a Dedicated Constellation of Small Satellites to Detect and Monitor Forest Fires, Acta Astronautica Volume 52 Issues 9–12.

[20] Some Circular Patterns Providing Continuous Whole Earth Coverage, Walker, Journal of the British Interplanetary Society Number 24, 1971.

Chapter 10

The Satellite Configuration Domain

Keywords Satellite Configuration Design, Satellite Configuration Types, Space Environment, Launch Environment, Satellite Configuration Types, Nadir Pointing Satellites, Spinning Satellites, Inertially Pointed Satellites, Satellite Budgets

This domain was introduced in Section 7.1.3. This chapter starts with a brief introduction to the elements and components included in this domain that are related to the "mechanical" space engineering disciplines. It continues analyzing how the external environment drives the configuration of a satellite and then examines the associated design decisions for four missions:

- The gravity-sensing mission GOCE, which is a Nadir-pointing satellite, in a Sun-synchronous orbit 250 km above the Earth, and whose design is driven by the need to minimize atmospheric drag.
- The Ulysses mission studied before in Chapters 8 and 9, whose configuration is driven by large communication antennas, its spinning attitude, and its long booms.
- JWST, an inertially stabilized astronomical observatory flying in a halo orbit around the Lagrangian L2 point, introduced in Chapters 7 and 8, whose configuration is driven by the need to protect its large deployable telescope from the heat of the Sun.
- Iridium, which is a Nadir-pointing communications constellation, already introduced in Chapter 9. Its configuration is driven: by the need to accommodate small or large numbers of satellites in different launchers, by the need to provide good visibility to the different communication antennae it carries, and by the need to cope with very variable geometry with respect to the Sun.

The missions selected are both large and small in scale, perform both "in situ" measurements and remote sensing, and fly in different orbits with distinct satellite attitudes. Hence, they provide a wide range of applications and situations illustrating the different factors affecting the configuration of a space mission. However, the number of configuration drivers is high, and the possible design alternatives are numerous. The examples outlined above are insufficient by themselves to cover the whole spectrum of relevant design decisions, and when necessary reference to other missions will be made.

The chapter continues with a section on the impact of the geometry around the satellite – location of the Sun, Earth, observable, and deep space – and how it acts as a configuration driver. The last two sections analyze the allocation of functions and allocation of budgets included in this domain.

M.A. Aguirre, *Introduction to Space Systems: Design and Synthesis*, Space Technology Library 27,
DOI 10.1007/978-1-4614-3758-1_10,

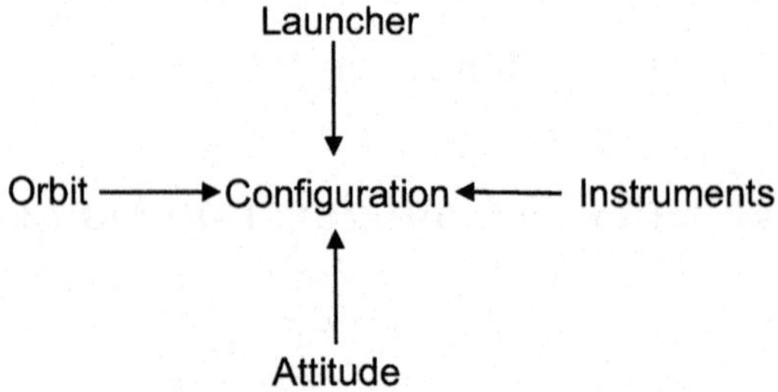

Figure 10.1. External Interrelations of the Configuration Domain.

Figure 10.1 recalls the interrelations of this domain according to the overall scheme presented in Figure 7.1. When compared with the orbit/attitude domain (Figure 9.1) the configuration domain entails a smaller number of interactions. The configuration design decisions are characterized also by the fact that they respond to previous design decisions made in other domains. This explains why all the arrows in Figure 10.1 point toward the configuration domain. This dependency explains why the configuration is undertaken only after the basic instrument concept, the orbit, the attitude, and the launcher(s), are established:

- The physical envelope provided by the launcher will provide the available volume and determine the need for deployable structures.
- The launch loads driven by the launcher and the in-orbit loads driven by the attitude and orbit control will govern the strength and stiffness of the stowed and deployed configurations.
- The constraints imposed by the annual variation in solar illumination, the cold of deep space, and the location(s) of the intended sources to be imaged or measured, all taken together, constrain the choices of orbit and sensor attitude(s) available, and these in turn contribute to the determination of the satellite configuration.
- The attitude control concept will drive the satellite configuration. This was already mentioned in Section 7.1.3.
- The orbital change and orbital control needs will drive the configuration of the propulsion and of the whole satellite.
- External environmental forces including drag, solar pressure, and gravity gradients, which are the functions of the orbit and the attitude, will also drive the configuration.

In line with the heuristic that the mission is designed around the instruments, it is the instrument's complement that must be configured first and as a whole if there are multiple sensors to be accommodated. It is possible that they have quite different, complementary and/or conflicting viewing requirements. These instruments must first be sized so as to optimize their own configuration and minimize their mass and size while being performance compliant. The outcome of applying all technical constraints and requirements of Figure 10.1 is a payload configuration that may be considered the optimum from a technical point of view, but there is one further consideration that may alter the final choice. This consideration is the possible provision of a configuration that simplifies the assembly integration and verification of the spacecraft by providing modularity between a payload assembly and the service module, to aid parallel design, manufacture, and assembly (Figure 10.37).

10.1. Components involved in the domain

The satellite configuration's overall definition will depend on the definition of the following subsystems:

- Primary and secondary structure
- Thermal control
- Deployment and pointing mechanisms
- Solar array as mechanical and thermal element

10.1.1. Structure

Structural engineering was introduced in Section 2.4.1. Structure was defined there as every physical part of the satellite that carries loads and/or supports or protects other parts of the satellite. The structure is usually split into primary and secondary assemblies. The primary structure is the part of the structure subsystem that defines the overall shape of the satellite, carries the main loads, and defines the satellite's fundamental vibration frequencies. The primary structure also provides the interfaces to all ground support equipment and to specific test facilities, used for on-ground assembly, integration, and testing. The secondary structure includes all the parts holding the individual satellite components attached to the primary structure, with negligible participation on the main loads transfer and with no major effect on the fundamental vibration frequencies. Each section of the secondary structure ensures that a satellite component is compatible with the local stiffness and strength requirements applying to that component.

The structure, both primary and secondary, consists of structural components: beams, plates, flanges, shear carrying skins, and longerons. Joining components connects these structural components: screws, nuts, welds, adhesive, and rivets. Structural components for space applications are designed for stiffness, strength, and lightness. To ensure dimensional stability, low thermal expansion is another parameter of high value for space applications. Aluminum and titanium are extensively used for structural space applications. Carbon-fiber reinforced plastics provide strong, stiff, stable, and lightweight structures and are also extensively used. Honeycomb sandwiches with aluminum and Carbon-fiber core and skins are also commonly used.

The most important design decision is the definition of the primary structure. This aspect is mainly driven by the overall external configuration and by the launch loads, which are the most demanding load environment. During launch, the primary structure is clamped to the launcher and is free at the top; so the primary structure at launch is a cantilever beam. The primary structure can be constructed around a central tube, panels, or a lattice of beams. Figure 10.2 shows the internal structure of Sentinel-3; in the lower half the structure is built around a central tube, and the upper half around panels. Figure 10.9 shows the Proteus structural concept, which is an example of a

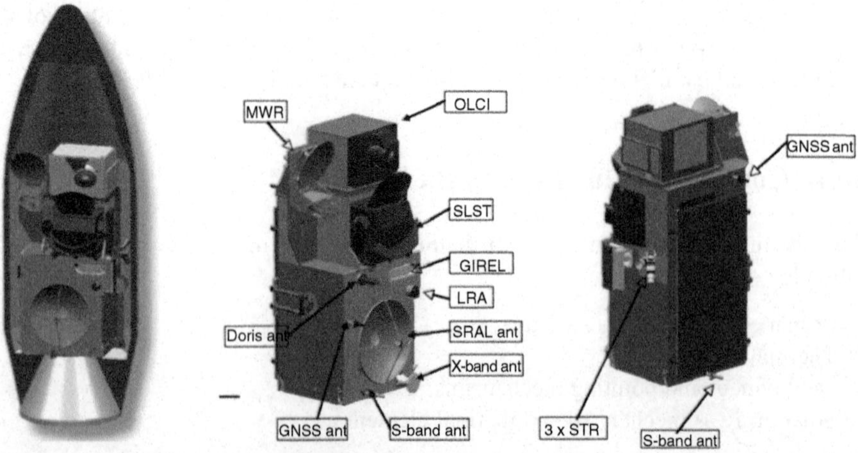

Figure 10.2. Sentinel-3 Stowed Configuration with the Location of the Main Configuration Drivers and Launcher Envelope (Courtesy of Thales Alenia Space).

structure based on a lattice of beams. The primary structural concept includes the choice of material for the structure: metallic or nonmetallic. Metallic structures are heavier and provide lower thermal dimensional stability but they are easier to repeatedly join and separate. They are also usually cheaper than composite structures too. The secondary structure is not normally considered during a project's early design phases. As the mechanical design is strongly dependent on the launcher, the architect must provide specialists with at least two possible launchers, a nominal one and a back-up, so that the structure can be designed against the most demanding load envelope permitted by the launchers under consideration. Another important design decision is whether the satellite will be a single structural block or divided into sections that can be handled separately on-ground. Most medium and large-sized satellite structures are divided into two parts: a lower one where most of the satellite subsystems are located and an upper one were the instruments are located. This structural division allows the parallel assembly and testing of the two halves of the satellite before they are joined. This simplifies and shortens the assembly and test process.

The architect shall follow the advice of the specialist regarding the structural design and the materials to be used; nevertheless, the architect shall take the lead in the selection of the "external" shape of the satellite, because many conflicting factors influence the external shape, including the visibility of the instruments, the environment, the attitude, the orbit, and the launcher. Of particular interest is the stiffness of the spacecraft in both its launch configuration and in-orbit one, in order to ensure compatibility with the launch environment and to assess potential dynamic coupling between the in-orbit configuration and the AOCS. In both cases it is the domain of the structural specialist to aid the architect to achieve such compatibilities. Most satellites have a structural subsystem that is under the responsibility of a specialized supplier, which delivers the primary structure of the satellite. Each piece of secondary structure is normally the responsibility of the supplier that is delivering the components supported by the secondary structure.

10.1.2. Thermal

Thermal engineering was introduced in Section 2.4.2.

The components necessary to fulfill the thermal control functions include: those that produce heat; those used for storing, conducting, and radiating heat; those capable of using external fluid loops energy to transfer heat; components that locally enhance or isolate heat transfer for temperature-sensitive hardware; and those used to control the other thermal components: temperature sensors, heaters, harnesses, and manage the thermal control logic.

The thermal energy received by the satellite is mostly produced by the Sun's heat and is absorbed in the visible part of the spectrum (Figures 10.11 and 9.3), but the energy emitted by the satellite is at a much lower temperature of the satellite itself and is emitted in the infrared range. Additionally, satellites orbiting a planet receive directly emitted infrared radiation and reflected solar radiation (albedo flux) from the planet's surface onto those facets of the satellite that have a view factor to its surface, and these must also be taken into account in the satellite's thermal design, as must any multiple reflections and absorptions of these sources of environmental heat flux between reflective facets of the satellite. Therefore, surfaces that are efficient emitters in the infrared and absorb poorly in the visible range—like white paint or quartz over silver optical reflectors – will cool the satellite, while surfaces that are poor emitters in the infrared but are good absorbers in the visible – like polished aluminum or gold – will heat the satellite. Black paint, which has both a high infrared emissivity ε and high visible absorptivity α properties, is used for good coupling in both directions: cooling and heating. The equilibrium temperature of a surface free from heat from any other source (e.g., such as internal conduction) and using thermo-optical materials for its thermal control depends entirely upon the solar absorptance, to infrared emissivity ratio α/ε, but it must be noted that highly polished surfaces having low values of α early in the mission can soon degrade to much higher values, due to damage from charged particle impacts and exposure to ultraviolet light throughout the mission. This degradation can significantly alter the α/ε ratio over time, so it is necessary to design the thermal configuration to account for such changes by analyzing the thermal balance of the satellite for when the material is new and at the end of the expected life. A warming of the surface will occur as it ages given a constant exposure to the external environment. The emissivity properties are not nearly so sensitive and are usually regarded as constant throughout the mission. To provide surfaces that isolate the satellite from the environment, multilayer insulation is used. Multilayer insulation blankets can consist of alternative layers of poor radiators separated by nets, hence avoiding contacts and conductive leaks between the layers, or multiple layers of crinkled highly reflective sheets. The latter type avoids the mass of the netted layers but risks increased interlayer conductive leaks through the thickness of the blanket. In both cases these blankets must be vented to prevent entrapped air between the layers conducting heat. Consequently, a blanket's insulating performance is a function of the number of layers used. In practice the situation is more complex. As the number of layers increases, the chance of achieving a good interlayer vacuum diminishes and so gas conduction between the layers adds to the heat transmission. This is a risk for short-term missions such as on the space shuttle. Also, such blankets only work with full efficiency on flat or single curvature surfaces, as any compression of the blankets leads to significant increase in the through-blanket conduction rate. Inside the satellite it is also necessary to provide thermal interchanges. This can be done by radiation or by conduction. Black paint is used for radiation coupling. The conduction is provided by specific components that become progressively more complicated as the amounts of heat to be transferred became larger: thermal doublers to transfer moderate amounts of heat; heat pipes for higher amounts; and active fluid loops for the highest heat flows. Some instruments, like thermal infrared detectors, must operate at very low temperatures. Complex, multistage purely passive cooling can meet this requirement, but the resulting arrangement strongly influences the configuration of the whole satellite. An alternative is to provide "active" heat pumps, which require a fluid loop and the provision of mechanical energy. This allows pumping the heat, permitting very low temperatures to be achieved in a specific location. This option will be discussed in detail in Section 10.5.4. Coolant fluids can provide extremely low temperatures. ISO, described in Chapter 7, used 2,100 l of liquid He at 1.8 K to cool the telescope by changing phase from liquid to gas as it was vented into the space. This limited the life of the mission, which ended when the coolant was exhausted. Compared with the numerous components used to cool satellites, there is just a single dedicated component to heat them up: electric heaters. The heaters are mounted in the areas requiring heat, and they are switched on or off according to the current temperature and the overall thermal control strategy. Thermal components are controlled by digital control loops, but they can also use simple analog control approaches, such as on-off switches controlled by thermostats. This robust approach ensures the survival of the satellite in failure situations, even in cases involving inoperative computers. Another simple control approach involves the use of mobile, screen-like blinds with either high thermal radiation or effective insulation properties that can be alternatively exposed to the external environment.

The first decision is to decide what surfaces are going to be used for radiative interchange with the environment and what surfaces are going to be isolated from the environment. This decision is directly driven by the orbital geometry: to allow heat dissipation the radiating surfaces must have a good view of cold space and be protected from strong solar irradiance. Closely linked with the decision above is the need to establish the tools to be used for thermal control; whether they will be simple or complicated. Radiative surfaces and heaters are considered simple. Heat pipes and blinds are considered moderately complicated; heat pumps, complex passive radiators able to cool to very low temperatures, and coolers like liquid He, are considered complicated. If simple tools are used, the satellite configuration and attitude will be strongly shaped by thermal constraints. If complicated tools are used, the satellite configuration and attitude will be far freer but the thermal control becomes more complicated.

The use, or not, of moderately complex or very complex thermal control devices is determined by the requirements of the instruments. Some instruments – such as infrared detectors – must operate at very low temperatures, while others – such as lasers – dissipate very high levels of heat. As mentioned above, the choice of complex thermal control solutions complicates thermal control matters but simplifies the overall configuration. So the final decision regarding these questions rests with the architect.

Design decisions on a satellite with simple thermal control requirements and with fairly constant thermal dissipation geometry are the responsibility of the specialist, who shall also provide the design for the special thermal requirements of the instruments or of any other satellite component requiring special thermal care. That said, the location of the radiative surfaces must be decided upon by the architect, because this decision directly affects the overall configuration, as well as both the satellite's attitude and chosen orbit, all of which are crucial, high-level design decisions.

The satellite has a set of thermal control components that ensure the proper thermal control of the overall satellite. These components are normally joined together into a thermal control subsystem, which is under the responsibility and control of one specialized supplier. On the other hand, the satellite may have dedicated thermal control assemblies to fulfill the thermal control needs of specific parts, such as the scientific instruments. The delivery of these components is normally put under the responsibility of the supplier in charge of the corresponding part.

10.1.3. Mechanisms

Space mechanisms engineering was introduced in Section 2.4.3.

Mechanisms are complex and diverse and are located throughout the varied parts of the satellite that require reconfigurations. Hold-downs restrain satellite components in their stowed configuration. During launch these components behave as structural fasteners: nuts, straps, or locks. Once in orbit, the nuts can be split, the straps cut, or the locks opened using specialized actuators. Mechanical locks are also used to restrain and support the structure in the deployed configuration. Actuators furnish the torque or force required to provide movements and allow the reconfiguration. The most flexible and common device is the electric motor, but stored elastic energy provided by torsion or linear springs are also widely used. If electrical drivers are used, it is necessary to provide movement control electronics and position and speed measuring devices. Elastic energy drivers do not need control electronics, but to avoid uncontrolled deployment, devices are employed that dissipate energy when the velocity increases. Cables and pulleys or linkages transfer the torques or forces provided by the actuators to the places where they need to be applied. Bearings and coatings are used to reduce the friction generated by the movements. Ball and roller bearings are used as well as liquid and solid lubricant layers. Mechanisms can be divided into single and multiple actuation devices. Satellite reconfiguration mechanisms only need to operate once. Pointing devices on the other hand require multiple or continuous actuation. Operational momentum wheels and momentum control gyros, used for satellite attitude control, also operate continuously. A special type of pointing mechanism is used for rotating the solar array toward the Sun in order to maximize energy production.

As in the case of thermal components there is a trade-off between complicating the mechanism and complicating the configuration. Complex deployable mechanisms can be used to reduce the mass and the stowed volume of the satellite, allowing the accommodation of the satellite in smaller and less expensive launchers. Pointing the whole satellite can eliminate pointing mechanisms, and solar arrays can be oversized to make unnecessary the use of solar array drive mechanisms. The trade-off between complicating the mechanisms and complicating the rest of the satellite is a complex issue whose solution depends on the specific circumstances of each mission. It must be decided whether the drive is going to be provided by electrical or mechanical motors. Single-shot mechanisms can normally be implemented using springs that do not require electronic control; but to avoid excessive deployment speeds and end of deployment shock loads, speed control devices are often needed for spring-driven deployments. It is also highly desirable that any mechanism can be reset to its original state without replacing any parts, so that the mechanism tested on-ground is the same as that used in orbit. Such resetting is a frequent requirement on-ground during AIV procedures. Devices using one-shot actuators risk replacing the active unit with one that will not work when required.

Mechanisms are distributed throughout the satellite: large deployable appendages like antennae or solar arrays can have their mechanisms; instruments requiring pointing or scanning have their own internal mechanisms; and rotary actuators are used by the attitude control subsystem of the satellite. All these mechanisms share the same problems and they require from their designers similar types of skills, but each mechanism interacts strongly with the components around it. That means mechanisms are not a natural subsystem.

Mechanism design is a highly specialized activity and the specialist will be responsible for all the associated design decisions. The architect will be directly involved on issues related to overall configuration; for example, the trade-off between a light-weight satellite with large complex deployables compatible with a small launcher, versus a larger, heavier satellite with fewer or no deployables but requiring the use of a larger launcher. Complex mechanisms may make the satellite lighter but they can be unreliable and difficult to develop and test. The satellite's final configuration and the resulting level of mechanism complication is one area where the architect needs to devote an especially high level of attention. The ground testability of large deployable mechanisms is another area that requires the architect's early attention because of its impact on costs and the credibility of the validation approach.

10.1.4. Solar array

Solar cells transform solar energy into electrical power. To fulfill this function the solar cells need structural supporting panels as well as deployment and tracking mechanisms. The structural panels will hold the cells in the stowed and deployed configurations, while the deployment mechanisms will reconfigure the solar panels from the stowed to the deployed configuration, and the tracking mechanisms shall rotate the whole solar array in order to align it as closely as possible with the Sun.

Solar cells transform solar energy into electrical energy. The Sun radiates 1,367 W/m^2 at 1 AU. Solar cells convert this radiation into electricity. The efficiency of the conversion is a function of the cell material, with conversion efficiency rates ranging from 20 % for Silicon to 35 % for the best multijunction Ga-As cells. The actual efficiency of the solar array will be reduced by physical factors such as the percentage of the solar array size that is actually occupied by cells – normally from 80 % to 90 %, and the angle between the solar array and the actual direction of the Sun. Aging cells and cell working temperatures will also reduce efficiency. Taken all together, solar arrays at 1 AU and aligned to the Sun will generate from 150 to 300 W per m^2 of surface. The solar array's structural components are the same as those described in

Sections 10.1.1 (structures), 10.1.2 (thermal dissipation), and 10.1.3 (deployment and pointing mechanisms).

The external surfaces of the satellite, which would be illuminated by the Sun, can be covered with solar cells. This will be the simplest cheapest and most reliable type of solar array. Those solar arrays are called "body mounted." The power provided by the body mounted solar array is limited by:

- The size of the available external surfaces.
- The variable geometry of the Sun with respect to the surfaces covered by cells.
- The operating temperature of the body mounted cells, which being mounted over the satellite body cannot radiate heat easily and could become too hot.

All these factors will limit the total power that these arrays could provide. Nevertheless, these solar arrays are used extensively in small- and moderate-size satellites that do not require large amounts of operating power and where simplicity and low cost are and important design drivers.

The satellite's solar power needs governs the selection of cell material, array size, deployment approach, and Sun-tracking geometry. The satellite's power requirements will be discussed in Section 11.1.1. The solar array size, deployment approach, and Sun-tracking geometry will be shaped by the accommodation constraints generated by the launcher fairing volume and the overall satellite configuration. As the solar array is commonly a large pointable, deployable structure, the fundamental trade-off will be similar to that of the previous section: complications for the mechanisms versus complications for the rest of the satellite. The choice will be between a complex but small, efficient, and light solar array, and a simpler but heavier and larger one. Deployable solar arrays suffer from all the problems typically associated with large deployable structures and mechanisms. However, their design drivers are usually very similar for each mission, more than for other large deployables, and the use of existing qualified designs is a feasible option. In that case the selection and accommodation of the solar array can be left completely in the hands of the specialist. In cases where available space makes that difficult, alternatives include: the designing of a dedicated solar array compatible with the constraints, the allocation of a larger proportion of the available volume to the solar array, or the use of a launcher with a larger fairing size. All three of these alternatives have clear shortcomings. In this situation, the architect possesses the best overall perspective for choosing the optimum way ahead.

With the exception of solar array size and its impact on satellite configuration most design decisions for this subsystem can be safely put in the hands of the specialists. There are no dramatic differences between the distinct design alternatives, and implementation of the satellite could be reasonably undertaken using different approaches; for example, with either regulated or unregulated buses or with different types of cells. The availability of already qualified components and the supplier's experience will be fundamental in orienting the design in one direction or another.

In many cases, and taking into account that the solar array is a large, physically continuous assembly with complex a numerous mechanical interfaces with the satellite, it becomes a small subsystem that is delivered directly to the satellite's supplier, who must take account of any structural dynamic impacts it may have of platform and payload performances.

10.2. The external environment as configuration driver

This section complements Section 9.2 of the previous chapter by describing the environments and analyzing the impact of each external environment on the spacecraft's configuration.

10.2.1. Launcher

The launcher drives the configuration because it determines:

- The volume available for accommodating the stowed configuration and the shape and size of the interface between the satellite and the launcher
- The launch loads and the resulting stiffness and strength requirements that will define almost all the structural elements
- The allowable satellite launch mass and the location of the satellite center of masses

The selection of available launchers is limited, and programmatic constraints – for example, that the launcher shall be from a specific country – limit it even more. This makes launchers a major design driver. Satellites cannot be designed without reference to a launcher, and in fact the design must begin with reference to one or – preferably – several launchers, and design the satellite to fit all of them.

10.2.1.1. Volume limitation

The launcher fairing, whose shape is cylindrical-conical with rotational symmetry about the launch axis, will limit the volume available. It is universally acknowledged that spacecraft design is driven by the importance of reducing weight, but it is equally true that the need for compactness drives the stowed configuration of a satellite, and in the design of LEO satellites volume limitations often come before weight limitations. To answer volume allocation issues, there are three possible strategies: the use of a launcher providing larger fairing, thus providing more volume; the use of deployment mechanisms; or the judicious use of the available volume by optimizing the satellite's shape. All three strategies have limitations: larger launchers are more expensive; even the cleverest solutions to volume limitations may not ultimately work; and the use of deployable mechanisms is risky and expensive. The importance of volume limitation in the design of satellites can be clearly seen in the stowed configuration of Sentinel-3 (Section 9.6.1) as shown in Figure 10.2. The instruments, antennas, and stowed solar array are clearly located for compactness and maximum use of the volume provided by the launcher. In some extreme cases the shape of the launcher determines totally the shape of the satellite.

Snapdragon (Figure 10.3) is an EADS-Astrium patented configuration that optimizes the use of the launcher volume when employing a large planar array antenna on the satellite. The satellite is divided into two half-hexagonal sections that fit perfectly into the cylindrical space provided by the launcher. Once launched the two halves deploy around a common hinge and produce a deployed configuration allowing a large flat surface where the largest planar array antenna compatible with the launcher volume can be located. The configuration also aims for minimum deployment mechanism complexity for maximum antenna size. The interaction between satellite

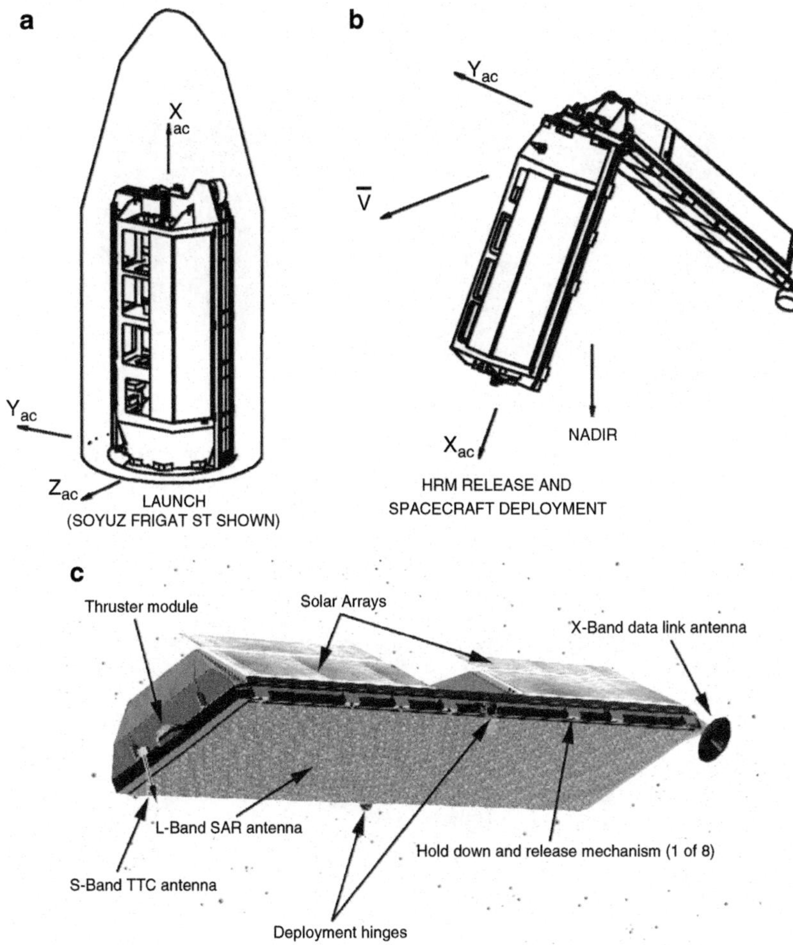

Figure 10.3. Snapdragon (The TerraSAR-L System and Mission Objectives Manfred Zink Proceedings of the Fringe 2003 Workshop, ESA SP-550.) Planar Array Satellite. Inside the Launcher (a), Deploying (b) and in the Deployed Configuration (c).

shape and launcher envelope becomes even more complicated when several satellites are launched simultaneously. In some cases the satellites must be designed with special shapes and supplementary structural adapters must be provided to connect all the individual satellites to the top of the launcher. An example of the accommodation of a constellation of mini-satellites carrying radar altimeters can be seen in Figure 10.4. Eight satellites in two rows of four can be accommodated inside the volume provided by the launcher. The shape of the mini-satellites is entirely governed by the accommodation constraints, and a large central beam for attaching all the mini-satellites must also be accommodated.

The Iridium constellation, introduced in Section 9.6.3 and discussed further in Section 10.3.4, is an extreme case of configuration accommodation to a launcher envelope. To provide flexibility in constellation establishment and replenishment, the

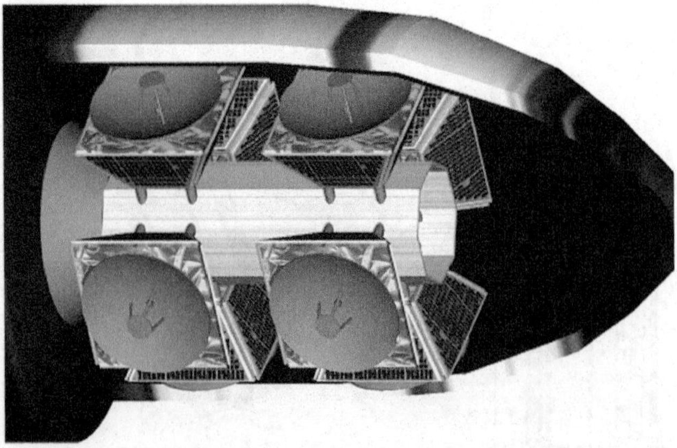

Figure 10.4. Eight Radar Altimeter Satellites Constellation Accommodated in a Single Launch Using a Central Support Structure (Figure Courtesy SSTL).

Figure 10.5. Accommodation of Iridium (Figure Courtesy Iridium) in Delta II (Five Satellites) and Proton (Seven Satellites).

satellite configuration is compatible with the simultaneous launch of a varied number of satellites in different launchers (Figure 10.5). Some launchers provide specialized services for the simultaneous launch of two or three different satellites in a single launch and with minimal shape constraints and with minimal interface problems between them. To provide this service, the space under the launch fairing is split into two or three clearly separated envelopes, each one with an independent interface with the launcher. This double or triple configuration is used extensively for communication satellites into GEO orbits (Figure 10.6).

10.2.1.2. Stiffness and strength limitations

During launch, the launcher will transmit vibrations and accelerations to the satellite structure:

- The launcher thrust will generate an intense slowly changing acceleration along the axis of thrust, which corresponds with the longitudinal axis of the stowed satellite. Figure 10.7 provides an illustration of the chronology of these accelerations.

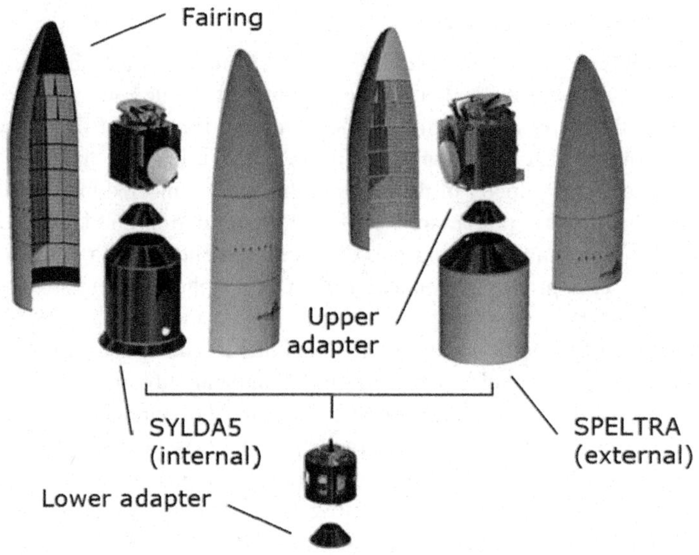

Figure 10.6. Ariane-5 Accommodation for Two Satellites (Ariane-5 Users Manual, Edouard Perez Arianespace 2004). Each One on an Independent Envelope: Upper and Lower.

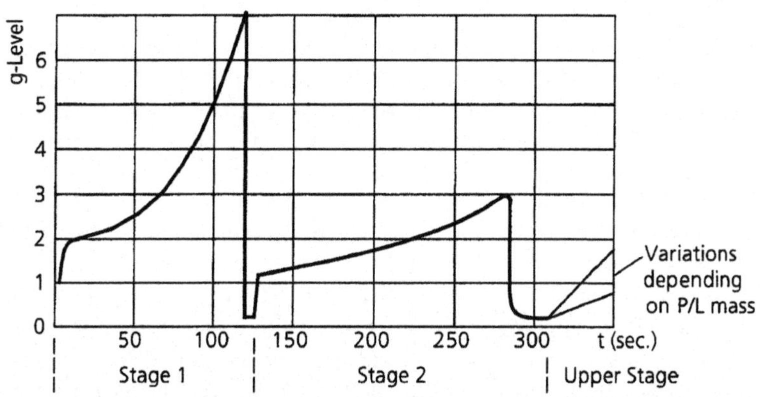

Figure 10.7. Variation with Time of the Longitudinal Static Acceleration During Rockot Flight (Eurockot Users Manual Eurockot 1999).

- The launcher will also produce less intense lateral accelerations generated by lateral winds or by lateral components of the launcher thrust.
- Sudden events, such as the separation of launcher stages, or the switching off or on of thrusters will generate sudden lateral or longitudinal shocks.
- Launcher noise and high-frequency thrust variations will also cause vibrations with varying amplitudes and frequencies.

The vibrations generated by the launcher will interact with the satellite in complex ways. Depending on satellite location and frequency range, some of the vibrations caused by acceleration will be damped and others will be amplified. The resulting load environment will be studied in detail in Section 10.2.2.

10.2.1.3. Launcher mass limit

Injection into different orbits requires different levels of energy and the mass injection capability of a launcher is dependent on the required orbit. The energy needed to place satellites into orbit is also a function of the terrestrial location of the launch center. Launch installations near the Equator are more efficient because they add – free of charge – the rotational velocity of the Earth to the energy injected into the satellite.

Figure 10.8 provides comparisons of the accommodation and launch mass constraints for the same mission using two different launchers: Vega and Rockot. The satellite, which is a proposed concept for an atmospheric chemistry-monitoring mission, needs to fly at an altitude of 820 km with an inclination of 96°. Vega can place about 1,450 kg into that orbit, while Rockot can place around 1,150 kg. Vega will be launched from Guyana, near the Equator, and can directly launch in the right direction northwards toward the Atlantic. The Rockot launcher possesses very similar characteristics, but is launched from Plesetsk, Russia, at 62.7° northern latitude. Unlike launches from Guyana, in order to avoid problems with falling launcher parts, Plesetsk can only launch to very specific orbit inclinations, as indicated in the figure. These launch site limitations must be taken into account when designing a space system.

10.2.2. Load environment

Satellites confront mechanical loads throughout their lives. The satellite configuration, and especially its structure, must be able to withstand these loads without suffering degradation that could compromise the fulfillment of the mission's objectives. The loads are produced: on the ground, during launch, and in orbit.

10.2.2.1. On the ground

On the ground, before launch, the satellite will be handled and transported. This inevitably produces stresses and strains on the satellite. On the ground, transportation and assembly jigs called "mechanical ground support equipment" must be designed and built to support and protect the satellite and its components during handling and transportation, thus ensuring that the on-ground loads environment does not become a design factor for the satellite's structure. The terrestrial transportation and handling environment, however, differs greatly from that in-orbit; for example, during assembly the top part of the satellite might hang from slings attached to a transportation crane, while during launch the satellite will be clamped to its base. This could cause the on-ground environment to become a design driver for some special parts of the satellite. The structure must also guarantee easy assembly and disassembly in order to access and eventually refurbish any component located inside the satellite. The satellite assembly process is long and complex, requiring many assembly and disassembly operations, and the easier and faster the assembly, the cheaper will be mission costs. This need for accessibility forbids structural solutions that could have provided superior strength with low mass but that might be incompatible with easy accessibility. Figure 10.9 depicts the assembly configuration of the Thales-CNES Proteus platform. Clearly, the concept provides easy access, thus minimizing assembly costs and delays.

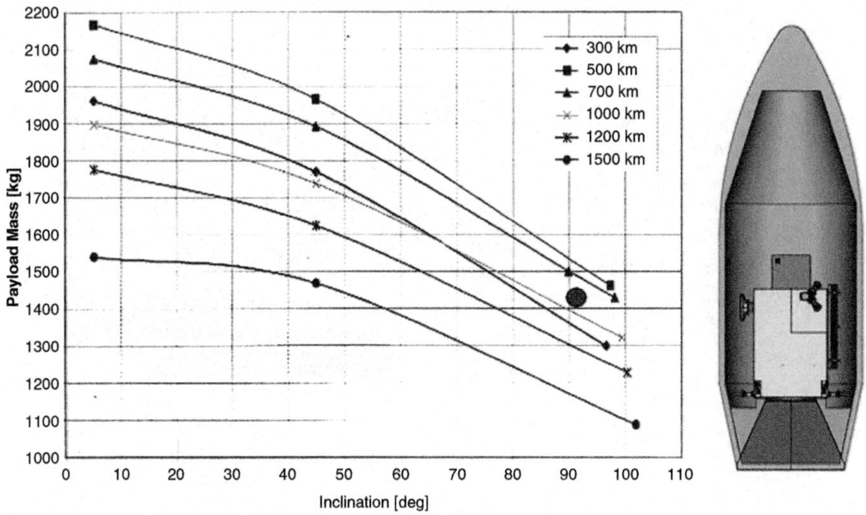

Vega compatibility

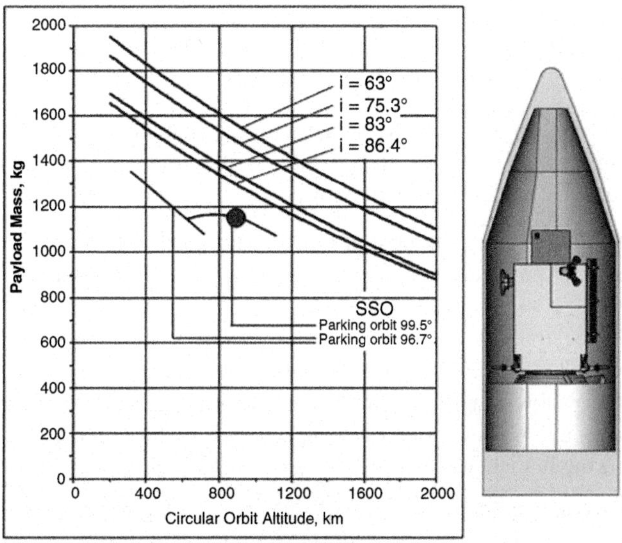

Rockot compatibility

Figure 10.8. Atmospheric Monitoring Satellite Accommodation on Two Possible Launchers. The Dot Indicates the Desired Orbit (Rockot Users Manual, and Vega Users Manual Edouard Perez Arianespace 2002, Configuration Courtesy of Thales Alenia).

10.2.2.2. During launch

Launch loads generated by the launcher were introduced in Section 10.2.1, which analyzed the load environment generated by the launcher. The present section will expand on that discussion and will describe how the load environment becomes a set of requirements that will shape the primary and secondary structure and determine the

Figure 10.9. Primary Structure of Proteus Platform (http://smsc.cnes.fr/PROTEUS/Fr/, Figure Copyright of CNES and Thales) Provides Easy Access to the Satellite Components.

satellite's structural test loads. The owner of the launcher transforms the launch accelerations into a set of strength and stiffness requirements imposed on the satellite that wants to use his launcher. These requirements include:

- Quasi-Static accelerations along and transverse to the direction of launch. Quasi-static means that the acceleration change slowly enough to be treated as static by the designer.
- Sinusoidal accelerations, along and transverse to the direction of launch.
- Acoustic noise level as functions of the frequency.
- Satellite's allowable longitudinal and transversal stiffness. Specifying stiffness it is possible to avoid dangerous couplings between vibration frequencies of the satellite and the launcher.

Two similar small launchers, Vega and Rockot, will serve as examples of how structural components on satellites are specified. In the case of Vega[1] and Rockot[2] their user manuals specify the maximum accelerations along and transversal to the flight direction for the three stages of the launcher, both static and dynamic (Table 10.1):

Vega specified the low frequency longitudinal and transversal sinusoidal accelerations as 1 g at the base of the spacecraft in the 2–100 Hz frequency range. Rockot specifies values in the 5–100 Hz range from 0.8 to 1.5 g in the longitudinal direction and 0.6 g in the transversal one. For noise levels Vega specifies 145 dB in the 20–2,000 Hz range. Rockot specifies 142 dB in the 20–4,000 Hz range.

[1] Vega User's Manual Table 3.1, Issue 0 Rev 0 March 2002.

[2] Rockot User's Manual Table 5.1.2-1, Issue 2 Rev 1 June 1999.

Table 10.1. Maximum Longitudinal and Transversal Accelerations as Specified By Two Launchers.

Acceleration (g)	Longitudinal				Transversal	
	Static		Dynamic		Static + Dynamic	
Flight event	Vega	Rockot	Vega	Rockot	Vega	Rockot
Lift-off	−2	−1.8	±1.5	±1.8	±1	0.3
First Stage	−5.3	7.2	±1	±0.9	±1	±0.8
Second Stage	−6	−3	±1	0	±1	±0.4
Third stage	−5	−1.6	±1	0	±1	±0.5

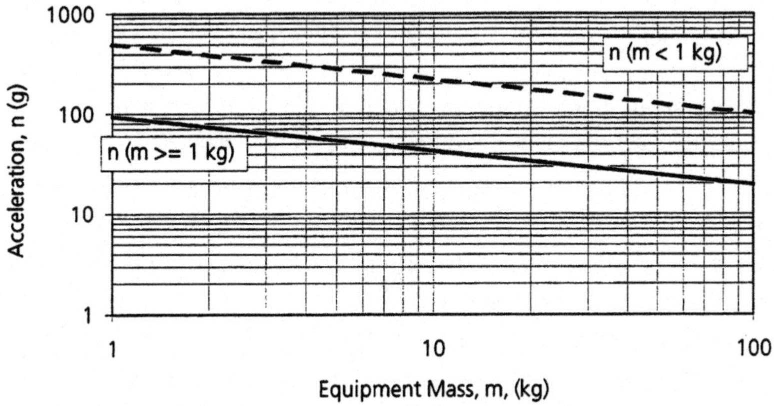

Figure 10.10. Rockot Guidelines for Dimensioning of Secondary Structure (Rockot Users Manual Figure 6.3).

The stowed satellite must be stiff enough to avoid dangerous coupling between the dynamics of the launcher and the dynamics of the satellite. For this reason, the launchers also specify minimum stiffness. Rockot requests that the first lateral natural frequency of the satellite be larger than 15 Hz and forbids longitudinal resonances in the range 16–33[3] Hz. Vega[4] requires lateral frequencies higher than 15 Hz and longitudinal frequencies higher than 35 Hz. The specifications above will be transformed into load cases for the stress analysis and in test cases for the primary structure of the satellite.

The specification for the secondary structure include:

- Sinusoidal acceleration profiles
- Stiffness requirements for individual components
- Random vibration acceleration profiles

Based on the design and stiffness of the secondary structure, and the fact that the satellite will amplify the launcher accelerations, amplification factors need to be taken into account. To cover for all the possible amplification factors, the launcher documentation or the specification of the satellite provides (Figure 10.10) worst-case

[3]Rockot Users Manual, Issue 2 Rev 0 May 1998 Chapter 6.3.4.

[4]Vega Users Manual, Issue 0 Chapter 4.4.2.

envelopes that are a function of the mass being supported by the secondary structure: the lighter the mass the higher the dimensioning accelerations. This specification approach is easy to use because the only data needed is the mass of the item being supported by the secondary structure. It is also conservative enough to be used in the preliminary design.

10.2.2.3. Test loads

Test loads are intended to demonstrate that the satellite is able to withstand launch stresses. Test loads are designed to envelope the worst-case real launch loads multiplied by specified safety margins; so in a well-implemented mission, tests will expose the hardware to the worst environmental stresses that it will ever experience, and the structure will actually be designed to withstand test standards. Test loads are defined in the specifications applicable to the satellite mechanical environment.

In order to validate the primary structure against quasi-constant accelerations, the test consists of:

- Either, the application of static forces onto the structure with hydraulic actuators
- Or, sinusoidal accelerations generated by vibration shakers

Tests for the validation of the primary structure against dynamic accelerations include:

- Simulation in acoustic chambers of the noise environment at launch
- Sinusoidal testing in vibration shakers

The use of an acoustic noise is particularly applicable to a spacecraft that carries lightweight appendages having relatively low mass per unit area, such as large stowed solar array panels. Tests for the validation of the secondary structure are carried out on vibration shakers. The tests verify stiffness; resistance against sinusoidal accelerations; and resistance against random vibration.

10.2.2.4. Loads in orbit

The in-orbit environment generates four types of mechanical loads:

- Thermoelastic loads generated by the varying in-orbit temperatures expanding or contracting the structural components of the satellite
- Deployment loads generated during the reconfiguration of the satellite, including end-of-deployment shock loads
- Loads generated in the deployed configuration by large or small attitude or orbit correction maneuvers
- Loads generated by the centrifugal fields of spinning satellites

In-orbit, the load environment is low and does not affect the primary structure's dimensioning. Nevertheless, in-orbit loads can be a factor in dimensioning large deployable structures and mechanisms like antennas or solar arrays. Large deployable structures dominate any satellite configuration in which they are present. This makes their early analysis and design a crucial step. The whole JWST configuration (Sections 7.2.1, 8.3.2, and 10.3.3) turns around the deployable mechanisms of its telescope and its sun-shield.

10.2.3. Thermal radiation environment: Sun, Earth and deep space

In space satellites will encounter four sources of thermal radiation:

- The Sun, which can be considered a black body at 5,800[5] K. At 1 AU the energy of the Sun is 1,367 W/m^2. The effects of the Sun were introduced in Section 9.2.4.
- Sunlight reflected back into space from the Earth's surface. The fraction of sunlight reflected by the Earth is called albedo. Earth's albedo is highly variable, oscillating from 5 % to 100 %, but its average is 30 %.[6]
- Earth-emitted thermal radiation. The Earth can be considered a black body at 288 K. The average energy emitted is 230 W/m^2, but depending on the temperature of the Earth's surface it can fluctuate from 150 to 350 W/m^2.
- Cosmic microwave radiation produced by deep space can be considered a black body at 3 K.

Planck's Law (Figure 10.11) controls the relationship between the temperature of a radiation source and its wavelength. The Sun's radiation peaks at 0.4 μm; the Earth's peaks at 10 μm; and radiation from deep space peaks at 1.9 mm. Satellites produce heat as a by-product of electrical energy utilization and they dissipate this energy by interchanging thermal flows with the Sun, the Earth, and deep space. Satellites will radiate heat according to the temperature and the properties of its different external surfaces. The thermal control components of the satellite must keep the satellite within temperature ranges required for its correct functioning, and they must do this within the changing geometry caused by the orbit of the satellite throughout the year. To ensure adequate functioning, most satellite components must be maintained at ambient Earth temperatures, around 290 K. The satellite's radiative heat interchange with the Sun, Earth, and deep space will depend on their angular size in the celestial sphere around the satellite. The geometry of the Sun, Earth, and deep space around the satellite was discussed in Section 9.7.

The position of the Sun and Earth changes according to the satellite's orbit and attitude and can be affected by the season of the year. The angular size of the Sun and Earth will also change if the distances to them change with the orbit. The satellite's configuration specifications and thermal design must take into account the changing thermal environment as defined by the changing geometry of the satellite with respect to the Sun, the Earth, and deep space. Section 10.4 will systematically examine the most adequate configurations for different types of orbits and attitudes. The satellite's surface will contain a number of dedicated thermal radiator areas, while the rest of the surface will be thermally isolated from the environment. The radiator areas shall be positioned to provide the required heating and cooling levels for the satellite and all its parts and in all the specified satellite attitudes and orbits. The space mission plan must avoid excessive variations in the radiators' orientation.

---

[5]Spacecrafts Techniques and Technology, Volume 3 Chapter X.3.3.2.2, CNES Cepadues Editions.

[6]ECSS-E-10-04 A Chapter 6.3.

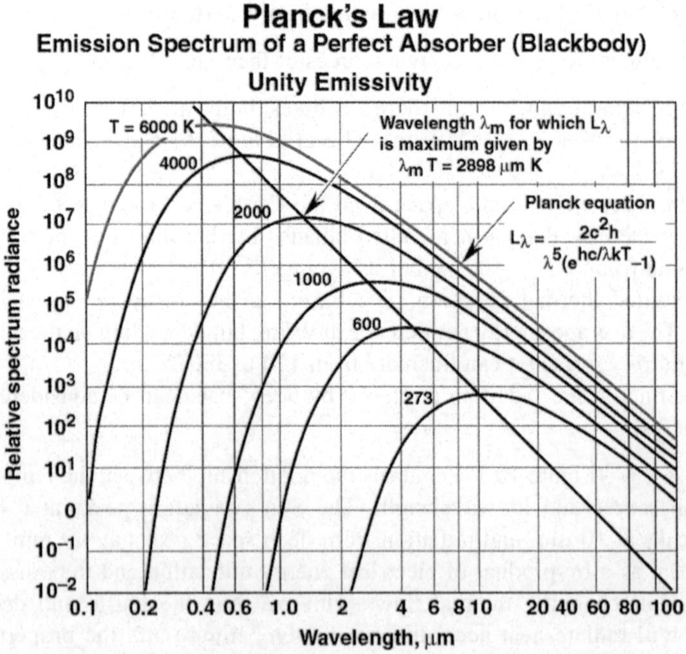

Figure 10.11. Emission and Absorption at Different Temperatures (ESA ECSS).

10.2.4. Space environment generated external forces and torques

The in-orbit configuration of the satellite will be driven by the forces and torques generated by the external in-orbit environment. There are four sources of forces and torques:

- The neutral atmosphere (Section 9.2.3): orbital drag and lateral winds
- Radiation pressure (Section 9.2.4)
- Gravity-gradient torques (Section 9.2.1)
- Magnetic field induced torques (Section 9.2.2).

According to Figure 9.2, the satellite's orbital altitude and solar activity are the most important factors affecting the upper atmospheric density and the consequent resulting forces and torques on the satellite. Variations in altitude or solar activity will produce correspondingly high variations in the aerodynamic forces and torques on the satellite.

The satellite configuration and flight attitude directly affects its cross-sectional area normal to the velocity vector that together with the atmospheric density and orbital velocity determines the forces and torques on the satellite.

The lower the altitude the more significant these configuration drivers become. All satellites orbiting below 1,000 km and requiring precise orbit control need orbit maintenance to counter aerodynamic drag. In the 500–1,000 km altitude range an average-sized satellite only needs to carry moderate amounts of fuel, so the impact on the configuration will be modest. However, if the orbit altitude is lower and if the satellite's transversal area is large, the amount of fuel needed for orbit maintenance becomes significant with respect to the satellite's overall mass, and reduction of the satellite's transversal area becomes an important configuration driver. An extreme case is GOCE (Figure 10.14), which flies at an altitude of 250 km, where the minimization of transversal area is the most important configuration driver. Snapdragon (Figure 10.3) and Starlette (Figure 9.7) are also examples of satellites with a configuration that minimizes the effects of external forces.

As shown in Figure 9.2, during the 11-year solar activity cycle, atmospheric density changes by two or three orders of magnitude. This makes aerodynamic forces much stronger during solar maximum than during solar minimum. Satellites in LEO orbit often require repetitive coverage and revisit orbital patterns (Section 9.9.3). These kinds of missions – and any other LEO mission that require strict orbital control – will specify the maximum acceptable difference between the nominal and the actual orbit. To fulfill this requirement, the atmospheric drag must be compensated for by more or less frequent orbital correction maneuvers. These maneuvers will be much more frequent during the solar maximum phases than near solar minimum, complicating the operations of missions with large transversal areas during solar maxima. Hence, configurations that are perfectly acceptable during solar minimum phases become unacceptable during solar maximums. The cross-sectional area of the satellite might also change during the life of the mission, introducing a supplementary factor of variability in the aerodynamic forces.

The pressure produced by the radiation of the Sun and by the ever-changing solar wind is small but continuous, and this small torque can produce a cumulative effect that needs compensation. This effect becomes a configuration driver for satellites in the geostationary position. GOESS (Figure 7.2) carries a "solar sail" that must be deployed far away from the satellite body to compensate the radiation pressure generated by the single-wing solar array of the satellite.

The force of gravity acting on different parts of the satellite varies. This produces a "gravity-gradient" torque that makes the satellite align along a minimum potential energy direction – generally with the satellite's longer dimension pointing to Nadir. This torque is used for simple and inexpensive attitude control, but in some missions the preferred attitude and configuration does not coincide with any minimum energy direction and the gravity-gradient exerts a constant torque on the satellite. Most of the time this torque is trivial and can be easily compensated for by the satellite's attitude control actuators. The exceptions are satellites with large deployable structures and attitudes that generate high gravity-gradient torques. This means a geometry that is neither along the local vertical or the local horizontal. An example is the P-band synthetic aperture radar (SAR) depicted in Figure 10.12 that carries a 12 $\times$14.7 m large deployable parabolic antenna (LRA).[7] Satellites with large deployable appendages orbiting in LEO must always consider external forces and torques in deriving their configurations. Satellites with large deployable structures in LEO will be affected by atmospheric drag and by torques exerted by the gravity-gradient.

10.2.5. Electromagnetic radiation environment

Satellite configurations must be compatible with the external and self-induced radiation environments. Section 9.2.6 introduced the natural radiation environment. This environment contains energetic particles that could damage the satellite components. Protective metal layers are used to shield these components.

[7]Courtesy from Thales Alenia.

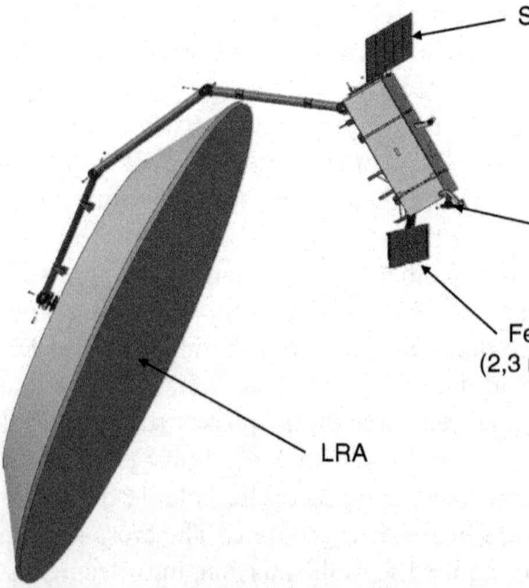

Figure 10.12. Possible Configuration for a Mission Carrying a P-Band Synthetic Aperture Radar for the Monitoring of Forest.

Satellites include electromagnetic radiation emitters and receivers. The emitters are often powerful and the receivers can be very sensitive. This means there is a risk of the satellite emitters affecting the receivers and generating error signals. To avoid this, the design process should include a frequency interference analysis to identify the power of the emitters and the sensitivity of the receivers at different frequencies. The gain patterns of the emitting and receiving antennas, the distance between them, and the possible existence of corners or edges between the emitting and receiving antennas, will all play a role in the quantification of this problem. Simulation tools exist that can predict this effect, but it is a phenomenon difficult to calculate and the simulation tools employ large safety margins, often leading to overdesign. To reduce the electromagnetic interference problem the emitting and receiving frequencies need to be separated, and frequencies that could have close emitting and receiving harmonics, even when well separated, need to be avoided. Antennas with sharp radiation patterns will also mitigate the problem. If feasible, the receiver can be designed to be turned off when a potentially contaminating emitter is active.

In some cases it is necessary to provide deployable mechanisms to separate the radiation source from the detector. Ulysses' 5.5 m radial boom (Figures 10.17 and 8.10) separates the sensitive instruments located on it from the radiation produced by the radioisotope thermoelectric generator used by Ulysses and from the magnetic fields generated by the satellite electronics. A related issue is the reception by the receiving antenna of signals coming from two sources: one from the direction of the event under observation and another reflected off nearby surfaces of the satellite. This effect is known as multipath and will produce error signals from the receiving instruments. The NASA-CNES radar altimetry mission Topex-Poseidon determined the satellite's position by measuring the distance to GPS satellites; but GPS signals could bounce off the satellite's own body before reaching the antenna, thus producing a satellite position error. To avoid this problem, Topex-Poseidon, as did Ulysses, located the GPS antennas on top of a 5.5 m long boom away from the satellite's body. In-orbit experience demonstrated that

Figure 10.13. GPS Antenna Accommodation on Topex-Poseidon (a) and Jason (b) (Figures from CNES Space Site: http://132.149.11.177/).

the problem was smaller than predicted and the following NASA-CNES radar altimetry mission – Jason – eliminated the boom, instead of locating the GPS receiving antenna on a small platform over the satellite. Jason's subsequent good performance confirmed the wisdom of the newer simpler design (Figure 10.13).

10.2.6. Other effects of the external environment

Further environmental factors that impact the design of the satellite configuration include:

- The effect of radiation on the degradation of solar cells
- The effect of radiation on the degradation of the thermo-optical properties of thermal control surfaces
- The erosion produced on the external surfaces of the satellite by ions, mainly by atomic oxygen
- The effect of meteoroids and space debris on the satellite

None of these effects is dramatic in terms of spacecraft configuration, but all of them need to be taken into account in the design of the satellite. The degradation of solar cells and thermal control surfaces is determined through the use of the most appropriate materials and the use of degradation factors and safety margins that are functions of the physical environment and the duration of the mission. The application of these factors will produce a robust, slightly overdesigned satellite concept, ensuring the fulfillment of the mission's aims in the anticipated environment. The impact on satellite configuration is typically the need to accommodate a solar array sized to cope with the degraded cell performance experienced at the end of the mission's life. The physical effect of ions, meteorites, or space debris is normally minor and is evaluated through mathematical models. In some cases, this particular environmental hazard can become an important design factor. GOCE's velocity-facing external surface, which receives the direct impact of atomic oxygen ions contained in the atmosphere at 250 km, is coated with layers resistant to erosion by them. Satellites flying through an especially dusty environment – near comets, for instance – could require special protective structural layers, ESA's mission Giotto was an example of this. Giotto carried a special shield to protect the satellite from the particles around Haley comet nucleus.[8] Of course, satellites are not designed to withstand impacts from substantial pieces of space debris as the energies involved are so high. Instead, a preventative approach is employed, in that predictions of the likelihood of such collisions occurring are made whenever possible, and sufficient fuel is carried to produce the necessary orbit maneuvers required to avoid the collision. The consequences for spacecraft configuration are the need to accommodate a sufficiently large fuel tank.

10.3. Configuration examples

10.3.1. GOCE

10.3.1.1. Mission aim

GOCE's[9] mission is to map Earth's gravity field with unprecedented accuracy and resolution. This will improve our understanding of the Earth's internal structure and provide a more reliable reference for ocean and climate studies, including sea-level changes. The gravitational field is measured by monitoring the differential fall rate of six different masses located within the satellite. The lower the satellite's orbit, the higher the resolution and accuracy of the gravity field determination. This requires as low an orbit as possible. A Rockot launcher launched the mission in 2009 and it is flying above Earth at an altitude of 260 km in a Sun-Synchronous orbit, at 6:00 local time and with an inclination of 96.5°. The mission life includes two gravity-sensing 6 month periods that are eclipse-free, each separated by a period of 4 months with eclipses when the satellite is dormant because the eclipse environment is not compatible with the smooth acquisition of gravity data. Its mass is 1,100 kg, 200 kg allocated to the gravity-sensing instrument.

10.3.1.2. Configuration

The intense drag forces the satellite withstands due to its low orbit influences its configuration, which is designed to provide the smallest transversal section possible, thus minimizing atmospheric drag. The satellite is shaped like an aircraft, with

[8] ESA achievements, BR-250, ESA 2005.

[9] ESA Achievements, ESA BR 250, ESA 2005.

Figure 10.14. GOCE Configuration Sun Looking Side (http://www.esa.int/).

wing-like solar arrays and a rudder-like structure located at the wake end of the satellite. It features a pair of electric ion-thrusters, also located at the trailing end of the satellite, to compensate for atmospheric drag and avoid orbit decay. The ion-thrusters' power demands require the largest possible solar array, which in addition must be rigid in order to minimize gravitational perturbations on the sensitive accelerometers. To attain these two conflicting attributes, the solar array is designed to occupy the entire launcher space, and is able to provide up to 1.3 kW. The rudder-like surfaces are provided to act like wind vanes adding passive attitude stability to the satellite. This configuration is made possible by the Sun-Synchronous dawn dusk orbit and by the craft's attitude. This attitude and orbit means the Sun is always nearly perpendicular to the satellite's orbital plane (Figure 9.37b). One face of the satellite is always facing the Sun and is covered by solar cells (Figures 10.14 and 10.15).

The opposite face, which has a permanent view of cold space, is used for thermal dissipation (Figure 10.16). The satellite's minimal transversal cross-section points permanently in the direction of travel, allowing minimal drag. The internal configuration (Figure 9.15) is dominated by the gravity-gradient sensing instrument, located near the center of gravity of the satellite in order to minimize gravity perturbations. GOCE's instrument axes must be oriented with respect to the Earth's. The instrument makes any internal load path for GOCE impossible, and the primary structures are the craft's external panels. Thick transversal flanges, which are also used to accommodate the satellite's components and electronic boxes, reinforce this external primary structure. The thermal control is made up of a multilayer insulator, optical surface radiators, electrical heaters, and thermistors for temperature measurement. GOCE's electronics boxes are distributed within the different transversal flanges of the satellite. GOCE being slender and with one of its two lateral faces always protected from the Sun, all electrical paths to the thermal radiators on that face are short. The battery is thermally isolated and located on the leading edge side of the satellite.

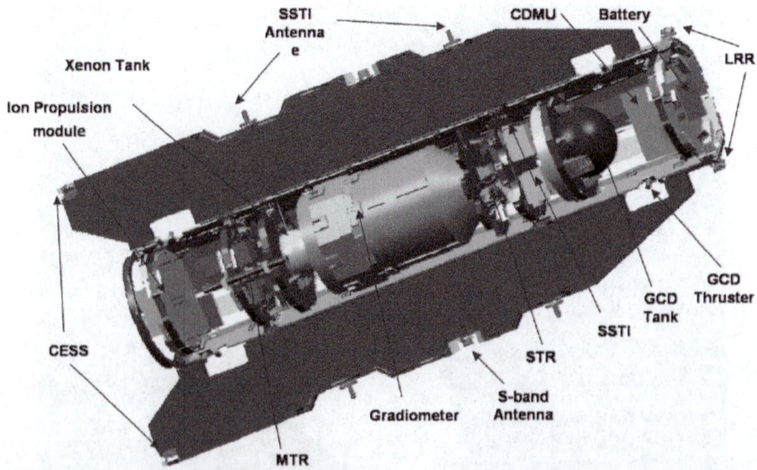

Figure 10.15. GOCE Internal Configuration (Figure Courtesy Thales Alenia).

Figure 10.16. GOCE Anti-Sun Face Configuration (http://www.esa.int).

It has its own radiator, which uses high efficiency radiating optical surface reflectors. To be able to detect the extremely weak gravity signals at the operation altitude, any thermoelastic deformations that could generate relative accelerations between the gravity-sensing accelerometers must be minimized, as they would be interpreted as coming from Earth. To do so, the instrument's primary structure is built of ultra-stable carbon-carbon composite and is isolated from external thermal variations. To achieve this isolation, the gravity-sensing accelerometers dissipate their thermal energy into a chamber, which in turn dissipates its energy into a second, thermally controlled chamber that finally transfers the heat to radiators located on the anti-Sun face of the satellite. The solar array edge that faces Earth holds the communications antenna while the edge facing the Zenith holds the GPS satellites' navigation antenna.

10.3.2. Ulysses

Section 8.3.5 explained the aims of Ulysses while Section 9.6.2 discussed its orbit and attitude control systems. This section will examine its configuration.

Ulysses is spin-stabilized and rotates at 5 rpm around the axis defined by the antenna's boresight, which always points toward Earth. Ulysses is a flat, square satellite dominated by a 1.65 m diameter large-gain communications antenna and by a radioisotope thermoelectric generator that is used to produce electrical energy. Ulysses also possesses four deployable booms supporting and performing experiments. The 5.5 m "radial boom" and the two 75 m long wire booms deploy in directions perpendicular to the rotation axis, their stability aided by the satellite's centrifugal field, while the 7 m "axial boom" is deployed along the rotation axis. This makes demanding requirements on the stiffness and alignment of the latter boom.

Ulysses was launched atop a set of stages – IUS and PAM (Figure 9.27). PAM provides a classical structural interface through the base of the satellite; IUS offers direct supplementary support for the heavy radioisotope thermoelectric generator (RTG). Ulysses is a small satellite with a total mass of just 370 kg, with 55 kg allocated to the payload. The RTG initially provided 284 W (Figure 10.17).

The axial boom is formed by a prestressed, coilable elastic Cu-Be tube anchored in the axial-boom drive mechanism located on the anti-Sun face of the spacecraft. The boom was deployed to a length of 7.5 m by traction force via a set of rollers driven by a stepper motor. The satellite's structure (Figure 10.18) is box-shaped and made from aluminum. The primary structure consists of four longerons; a main platform for the satellite's electronics boxes; and connection struts for the antenna, the radioisotope thermal-electric generator and the external faces of the satellite.

Ulysses' high-gain antenna face always points toward Earth. The spacecraft's orbital parameters makes the direction of both the Earth and the Sun roughly coincident, so the side opposite the high-gain one, facing away from the Sun, is suitable for thermal dissipation. In keeping with this geometry, thermal blankets protect the sunlit antenna side, while the opposite side holds the thermal radiator. Ulysses' thermal control is realized in conjunction with an internal/external power dump that either heats the internal parts of the satellite or dumps excess heat to the outside to compensate for the varying solar inputs.

10.3.3. JWST

The James Webb Space Telescope (JWST) was introduced in Section 7.2.1 and discussed further in Section 8.4.2. The present section will describe the satellite's configuration. Both sections already discussed briefly its structure and configuration.

The two main considerations shaping JWST's configuration are: the necessity to provide a telescope with an aperture larger than the launcher's payload fairing diameter; and the need to keep the telescope cooled to extremely low temperature in order to achieve the required sensitivity. The first requirement calls for a deployable primary mirror for the telescope; the second requires that the telescope be continuously protected from the direct radiation produced by the Sun. From the chosen orbit the Sun and the Earth are in roughly the same direction, and one side of the satellite always points toward them while the opposite side faces outer space. The satellite platform, the solar

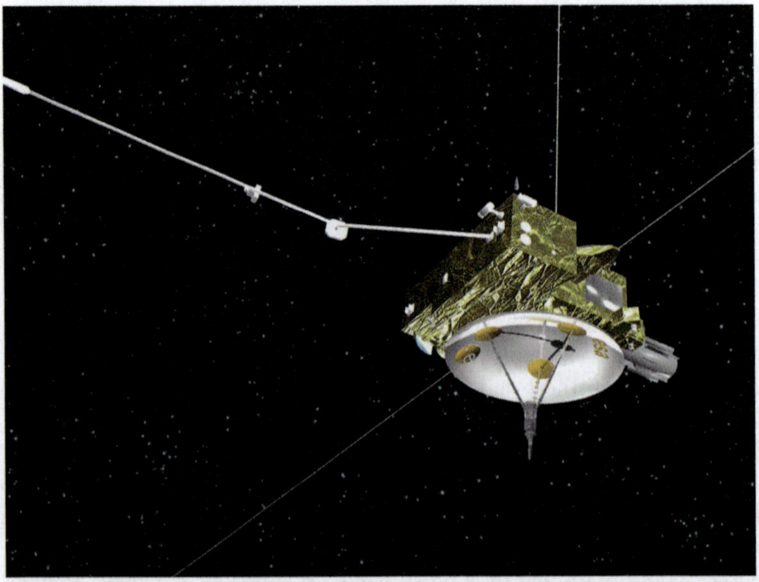

Figure 10.17. Ulysses Sun-Earth Side Configuration (http://www.esa.int).

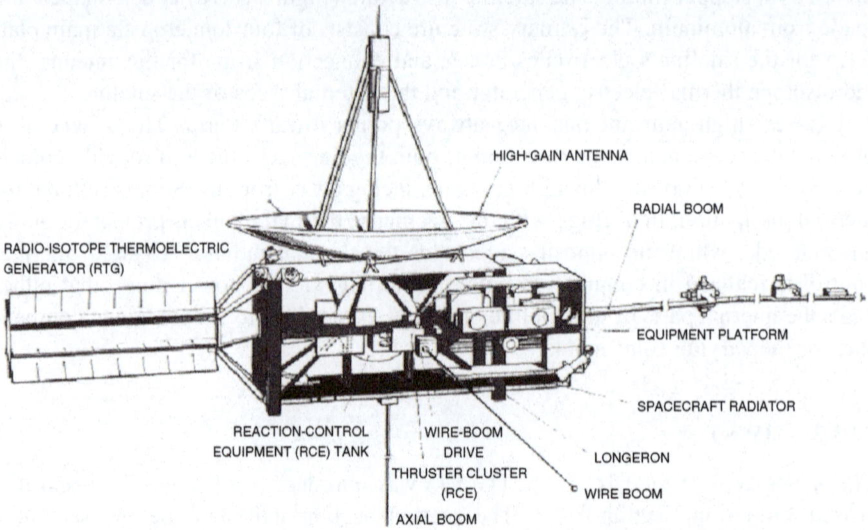

Figure 10.18. Ulysses, Internal Arrangement, Primary Structure (*Lighter Gray*) and Radiator (*Darker Gray*) (ESA Achievements, BR-250 2005).

arrays, and the communication antennas are located on the Earth/Sun-facing side, with the telescope on the opposite side having an unobstructed view of deep space.

A large deployable sun-shield separates both satellite halves. Figure 7.10 depicts the telescope side of JWST and Figure 10.19 above depicts the Sun/Earth-looking side. The configuration is dominated by its two large deployable mechanisms:

Figure 10.19. JWST Configuration (James Webb Space Telescope, P. Jakobsen, ESA Bulletin, ESA 2008. Drawing Courtesy of Northrop Grumman).

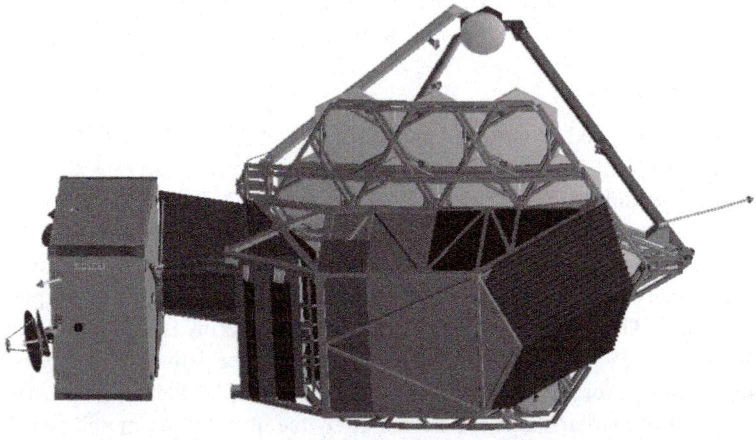

Figure 10.20. Structure of the JWST Satellite (James Webb Space Telescope (JWST) Observatory Architecture and Performance, John Nel, AIAA-2004-5986, American Institute of Aeronautics and Astronautics).

- The 6.5 m aperture telescope with its hinged, multifaceted, adjustable primary mirror and its deployable secondary mirror support structure (Figure 8.7)
- The 19 × 13 m deployable sun-shield

In orbit, a supplementary mechanism deploys a tower separating the warm satellite platform from the telescope. This minimizes thermal leakages from the platform to the instrument. The complex mechanisms of JWST are justified by its two key requirements: prolonged mission life while maintaining instruments as cold as possible (Figure 10.20).

Figure 10.21. View of the Stowed Geometry of JWST Inside Its Ariane 5 Launcher (James Webb Space Telescope, P. Jakobsen, ESA Bulletin, ESA 2008. Drawing Courtesy of Northrop Grumman).

The satellite platform's configuration is standard. Its deployable solar array and high-gain communication antennas are prominent. As for the spacecraft's attitude, the solar array is fixed and always facing the Sun, whereas the high-gain communication antennas must point at Earth with high accuracy and need pointing mechanisms. Because of its large and complex deployable structures and mechanisms, the stowed and deployed configurations of JWST are quite different. Figure 10.21 depicts the stowed configuration. The deployment steps are depicted by Figure 8.7. The principal design driver of the stowed configuration is the Ariane-5 fairing envelope. JWST has been clearly designed to take full advantage of all the space available inside the fairing, to allow stowing an instrument with the largest aperture compatible with the launcher.

10.3.4. Iridium

Section 9.6.3 introduced Iridium and provided a description of the satellite's orbit and attitude. The present section will focus on the constraints driving the design of Iridium's configuration.

Iridium's central portion is long and slender, of triangular transversal shape; its elongated section is orientated along the Nadir direction, providing supplementary stabilizing gravity-gradient torque[10] with a pair of solar arrays deployed perpendicular to the Nadir line. The vehicle is 4.06 m long and the solar array is 8.38 m from tip to tip. The total mass of each satellite is 725 kg, and their power delivery capacity is 620 W each, though the power subsystem can handle power peaks as high as 4,000 W.

The slender triangular shape is determined by the need to accommodate numerous satellites in a single launcher by stacking them in parallel along the launcher's longitudinal axis while still providing each satellite with individual interfaces with the launcher. Each satellite is divided into three modules:

- The upper part contains the satellite's general platform services, including power and solar arrays.;
- The middle section includes the payload electronics and provides interfaces with the three L-band square-flat array antennas linked with the mobile ground telephone.
- The lower one includes the Ka-band antennas and associated electronics.

[10]T. P. Garrison Systems Engineering Trades for the Iridium Constellation, AIAA Journal of Spacecrafts and Rockets 1997.

Figure 10.22. Iridium Stowed and Deployed Configuration (Figure Courtesy of Iridium and Celestrak. Image Obtained Using AGI-STK).

The satellites' configuration is strongly influenced by their communication antennas' field-of-view needs:

- There are four gimbaled Ka-band feeder links for contacting ground-located communication antenna gateways, which require uninterrupted views in the Nadir direction.
- Four Ka-band cross-links for contact with other satellites – two fixed and two gimbaled. The two fixed links are for contact with satellites ahead and behind in the same orbital plane, while the two pointable antennas are for contact with satellites in lateral planes to the right and left.
- Three large L-band antennas (MMA of Figure 10.22) for contacting ground-located hand-held telephones. They require clear views in a large cone around the Nadir direction.
- The satellites' configuration is also strongly influenced by their orbit, featuring an inclination of 86°, with the Sun drifting seasonally around the satellites' orbital plane. As a result of this drift, exposure of the satellites' and their solar arrays to sunlight vary constantly with the orbit and the changing seasons (Section 9.7.1 and Figure 9.33). To compensate for this, each solar array is equipped with two pointing mechanisms: one with a rotation capability of 360° that continuously tracks the optimal solar position along the orbit, and another with a rotation capability of 90° that regularly adjusts itself by small increments so that the first mechanism can maintain the solar array perpendicular to the Sun at all times. This varying geometry is discussed in detail in Section 10.4.1 and Figure 10.29.

As already indicated in Section 10.2.1, the satellites' configuration allows maximum launch flexibility. Their elongated triangular shape allowed for efficient and simultaneous accommodation of five or seven units in Delta and Proton launchers. The design is also compatible with smaller launchers in cases where a malfunction in orbit necessitates replenishment of the constellation with a small number of satellites. Both Long March and Rockot have actually been used for this purpose, and in each case the launchers were able to successfully put two satellites in orbit.

10.4. The geometry around the satellite and the configuration

The geometry of the celestial sphere surrounding the satellite, including the Sun, the Earth, the Moon, observation targets, and deep space, is highly variable and depends on the satellite's orbit and attitude and the nature and direction of the observable. This complex and variable geometry was discussed thoroughly in Section 9.7. The present discussion will begin with the geometries analyzed there and will describe and examine the configurations that have been adopted in response to these geometries. The number of possible geometries is numerous, and the potential configurations responding to them are even more abundant. This section will consider a large selection of alternatives that have been employed in the past. The outline of this section will follow that of Section 9.7 with regard to the types of satellites studied: Nadir-pointing, rotating, inertial, and agile.

10.4.1. Nadir-pointed satellites

10.4.1.1. General

A satellite pointing to Nadir always has one of its faces permanently pointed to the planet being orbited. This face is the most suitable location for the instruments observing the planet due to the unobstructed view it offers of the planet's surface. In the case of Earth-orbiting satellites this face is also the most logical location for communication antennas between Earth and the satellite. The satellite rotational speed with respect to inertial space of one turn per orbital period is low, the resulting in-orbit centrifugal accelerations are weak and Nadir-pointing satellites can use large and slender deployable structures providing a high level of deployed configuration freedom. Nadir-pointing satellite configurations can be as simple as the square box of SSTL's DMC (Figure 11.10) or as complicated as the P-band SAR with a 12 m diameter aperture parabolic antenna depicted in Figure 10.12.

10.4.1.2. Sun-synchronous satellites

Sun-synchronous orbits were introduced in Section 9.4.1, and Section 9.7.1 and Figures 9.35, 9.36, and 9.37 studied their resulting geometry. This type of satellite is widely used because it enjoys the important advantages already discussed: fairly constant solar illumination of both the satellite and observable, and full Earth coverage. The LEO Nadir-pointing Sun-synchronous satellite is a typical Earth observation satellite. The Sun's position with respect to the satellite is defined by the spacecraft's local time (Figure 9.35), and the configuration of the solar array and the satellite as a whole is strongly dependent on the local time. The satellite's movement along its orbit allows it to access the orbited body's surface "along-track," and Sun-synchronous satellites are ideal for surveying Earth from pole-to-pole with Nadir-pointing instruments. As discussed in Section 8.4.5, systematic missions provide regular synoptic

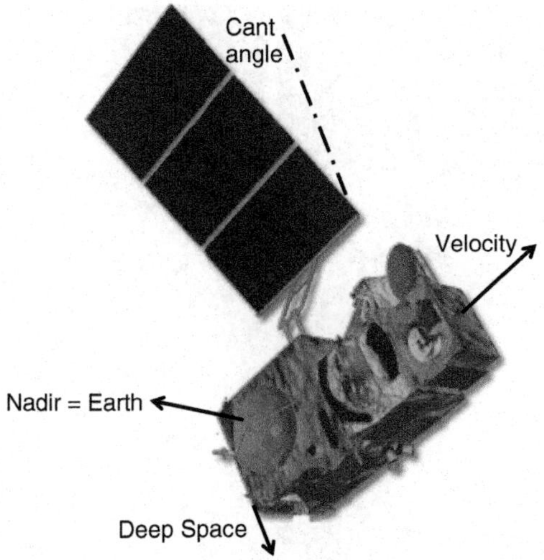

Figure 10.23. Sentinel-3 Deployed Configuration with an Instrument Face (Nadir), Solar Array Face (Sun), and Radiators Face (Deep Space) (Sentinel-3 Phase B-C/D Quotation Executive Summary Thales Alenia Space 2007). Sentinel Cant Angle is 30°.

views of the Earth. Sentinel-3 was introduced in Section 9.6.1. It carries out systematic, synoptic coverage of the Earth's surface. Sentinel-3's in-orbit configuration has a deployable solar array on the Sun side, radiators on the opposite, deep space oriented side and its instruments and communication antennas directed at Earth on the Nadir-oriented face. This is the typical configuration for Earth-observing Sun-synchronous satellites (Figure 10.23). The optimal geometry of the solar array depends on the geometry of the Sun and is defined by the cant angle between the solar array plane and the rotation axis of the solar array. The optimal cant angle of the rotating solar array depends in turn on the angular position of the Sun with respect to the orbital plane; that is, on the local time of the orbit (Section 9.4.2). Sentinel-3 flies at 10:00 local time and the β angle (Figure 9.33) axis of the cone described by the Sun along one orbit oscillates between 22.5 and 33.5° (Figure 9.36a). The resulting optimal cant angle providing maximum solar energy will be around 28–30°.

NPOESS (Section 12.3.4 and Figure 9.15) is a constellation of meteorological satellites providing systematic observation of the atmosphere and their configuration is exactly the same as Sentinel-3's, with a prominent Nadir-looking face for instruments and antennas, a Sun side for the solar array, and an anti-Sun side for radiators. The configuration of the satellites' body is identical in each of the three satellites, with the exception of the solar array cant angle, which is different in order to optimize the geometry of the solar array with respect to the Sun's direction: 22.5° for the 13:30 satellite, 72.5° for the 17:30 satellite, and 37.5° for the 21:30 orbit (21:30 local time is equivalent to 9:30 local time from the point of view of Sun-satellite geometry). That means the satellite orbiting at 17:30 will have a large cant angle with a solar array geometries similar to Figure 10.24a and the satellites orbiting at 21:30 and the 13:30 will have a small cant angle similar to Figure 10.24b.

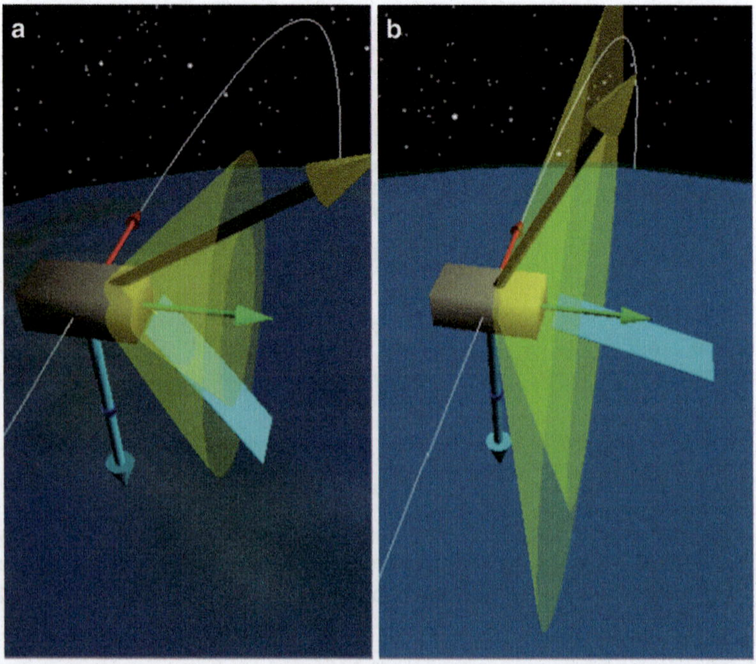

Figure 10.24. Optical Solar Array Cant Angle for Different Local Times. (a) For Local Times at 8:00 (Sun Never Far Away from Perpendicular to Orbital Plane) and (b) for Local Times at 10:00 (Sun Never Far Away from the Orbital Plane).

Spot (Figure 10.25) is an example of a satellite performing interactive observation. Spot obtains large resolution images over areas of the Earth requested by the users. The satellite carries two instruments able to obtain images away from the exact Nadir direction by across-track pointing of mirrors operated by mechanisms located inside the instruments. Each one of the two identical instruments obtains images of 60 × 60 km with spatial resolutions of 2.5 m to 10 m. The two instruments can be used to provide wider coverage, or two stereoscopic, nearly simultaneous views of the same target. The mirror-pointing mechanisms allow viewing of targets within ±27° across-track to the Nadir direction. Spot flies with a local time of the descending node of 10:30. Its solar array cant angle is 22.5°; that means the solar array and the solar array rotation axis are both close to perpendicular to the orbit plane (as in Figure 10.24b).

Besides the obvious similarity of its configuration to that of Sentinel-3, there is a fundamental difference: Spot flies with the longitudinal axis of the satellite perpendicular to the velocity vector, while Sentinel-3 fly with the velocity vector parallel to the longitudinal axis of the satellite. This makes the "bottom" of Spot 5 the Sun-oriented face, and the solar array, once deployed, is located in that area. The anti-Sun side of Spot is totally occupied by the two instruments. This position favors thermal control and protection from the Sun. The platform radiators are located on the three lateral faces, which consecutively receive sunlight along the orbit; the resulting thermal conditions, however are sufficient for heat dissipation of the satellite's platform components. Transversal fliers like Spot have a more symmetrical inertial matrix,

Figure 10.25. Spot 5 Configuration and Flying Geometry (CNES http://132.149.11.177/IcSPOT/spot5.jpg).

which provides some advantages for attitude control, but longitudinal fliers like Sentinel-3 have ample space to accommodate instruments requiring open views of cold space. This additional space is especially valuable for missions flying several instruments that need very low temperature cooling and that will be competing for the limited available space. Most modern Earth observation missions carry several of these instruments, making longitudinal fliers a better option as regards easy instruments accommodation. NASA's Aura (Figure 10.41) is another example of longitudinal flier.

Most LEO Sun-synchronous missions carrying passive optical instruments orbit at altitudes between 500 km and 1,000 km and in order to provide effective solar image illumination they avoid local times around 06:00. In these conditions satellites will experience one eclipse per orbit with a duration slightly over one third of the orbital period (Figure 9.38a). This requires a corresponding increase in size of the solar array in order to absorb enough sunlight for the batteries to remain charged and provide power to the satellite during the periods of eclipse.

GOCE (Section 10.3.1 and Figure 10.14) is another Nadir-pointing satellite in Sun-synchronous orbit, but as it requires no sunlight for its observations it has a local time of 6:00. In this case, rotation of the solar array to provide a reasonable geometry with respect to the Sun is unnecessary (Figure 9.37b). GOCE uses this to its advantage, and features a fixed solar array with wings solidly attached to the satellite. The 6:00 orbit offers brief or no eclipses (Figure 9.38b), affording efficient energy production with minimum-size fixed solar arrays and it is especially suitable for SAR satellites. ESA's Sentinel-1, DLR's Terrasar-X and the Canadian Radarsat-1 and −2 are good examples of this configuration. Sentinel-1 (Figure 10.26) is an operational C-band SAR satellite. It features two deployable solar arrays that do not need to rotate

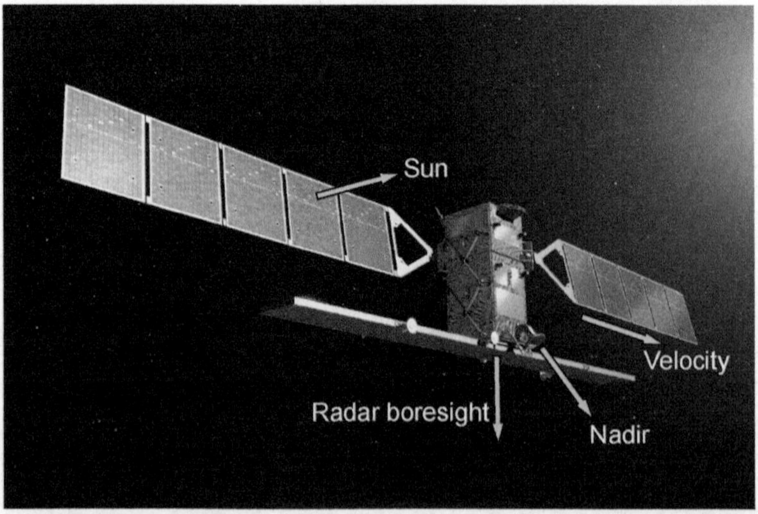

Figure 10.26. Sentinel-1 Configuration and Flying Geometry (Thales Alenia Image).

because the Sun is always nearly perpendicular to them. SAR's are side-viewing instruments and Sentinel-1's C-band SAR, along with the satellite's body, are pointed 30° away from the Nadir, though the communication antenna is aimed directly toward it.

Deployable, rotating, Sun-tracking solar array – as depicted in the previous figures – provide an optimal orientation for the solar cells and ensure optimal power with minimal solar cell area. Nevertheless, low cost, small satellites require normally modest amounts of power and they can be compatible with simple configurations where the solar cells are attached to the external sides of the satellite body. In this case the satellite configuration becomes just like a central body, for example, a cube, without any deployable solar arrays. The different body mounted cells will be progressively illuminated by the Sun as the satellite moves around its orbit. The cubic, body mounted low cost SSTL DMC satellite depicted in Figure 11.10 is an example of this.

10.4.1.3. High-inclination non-Sun-synchronous satellites

In the case of non-Sun-synchronous orbits the Sun will shift widely with respect to the orbital plane. This affects the lighting conditions of the scene under observation and complicates the orientation geometry of the solar array, which needs to face the Sun, and of the thermal radiators, which need to avoid sunlight. To provide the necessary electrical energy there are four possible alternatives:

1. A fixed solar array whose solar panels are pointed in several directions, so that they are able to capture enough solar energy whatever the direction of the Sun. The panels shall be correspondingly overdimensioned.
2. A single-axis rotating solar array that tracks the Sun as accurately as possible, large enough to provide the required energy. This is only effective for moderately inclined orbits.
3. Solar array tracking mechanisms that provide two distinct rotation axes allowing it to track the Sun wherever its relative position to the satellite.

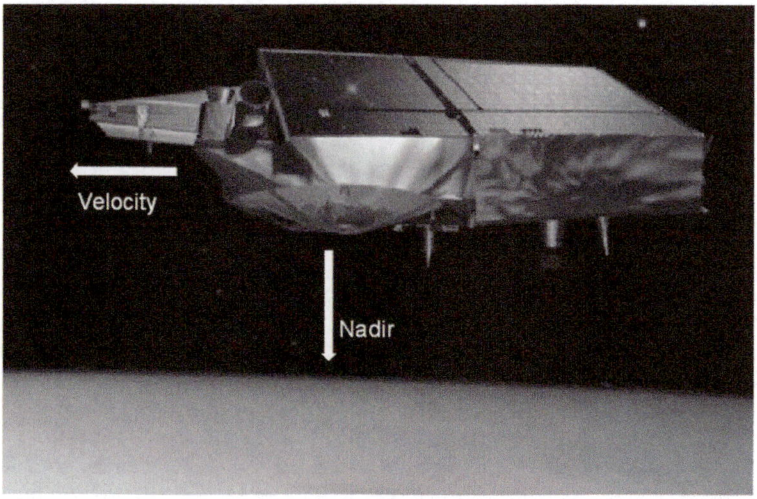

Figure 10.27. Cryosat Configuration (ESA www.esa.int).

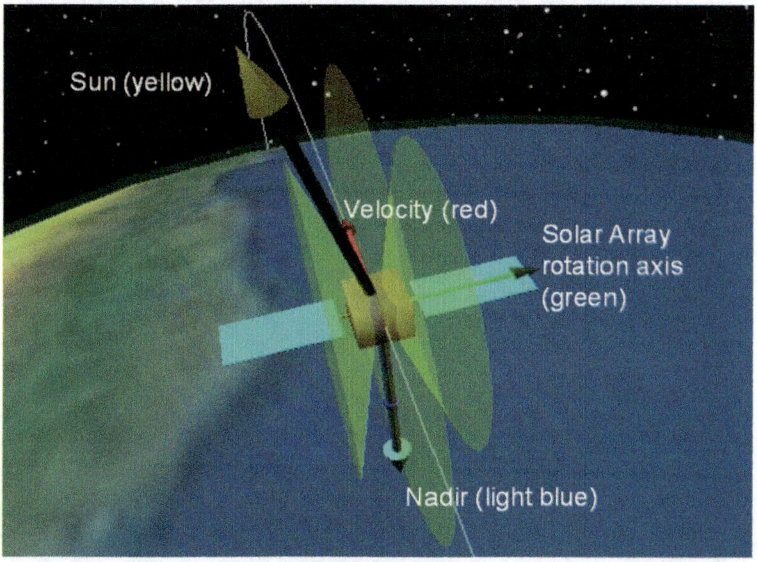

Figure 10.28. MeghaTropiques Sun-Earth Geometry. The Large *Yellow Arrow* Indicates the Sun Direction and the *Yellow Cones* the Sun Direction Evolution Along the Orbit and Along the Seasons.

4. The ability of the satellite as a whole to rotate with respect to the Yaw/Nadir direction, combined with a solar array featuring a second rotation axis, allowing the solar array to remain always perpendicular to the Sun.

Solution 1 was adopted by ESA's Cryosat (Figure 10.27); solution 2 by CNES-ISRO's MeghaTropiques (Section 8.3.4 and Figure 10.28); Iridium (Section 10.3.4 and Figure 10.29) adopted solution 3 and ESA's Galileo (Figure 10.30) adopted solution 4. The following discussion will analyze each one of these cases.

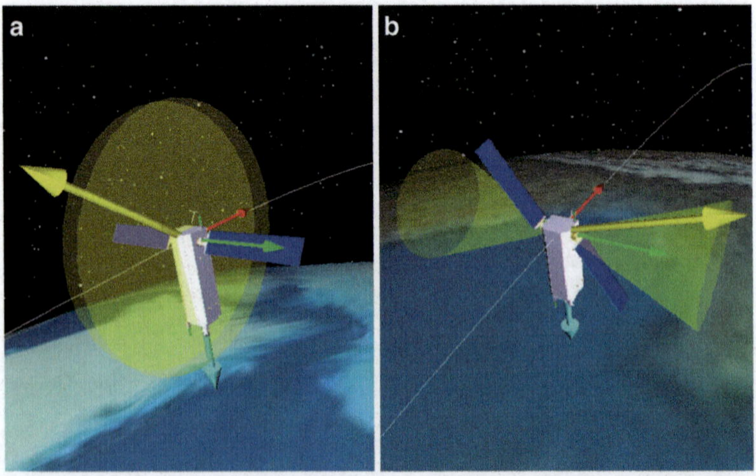

Figure 10.29. Iridium Sun and Earth Geometry Along the Seasons. (a) In a Season with the Sun Close to the Orbital Plane. (b) In a Season with the Sun Close to the Perpendicular to the Orbital Plane. In Each Season, the Solar Array Cant Angle Is Adjusted for Optimal Solar Array Illumination.

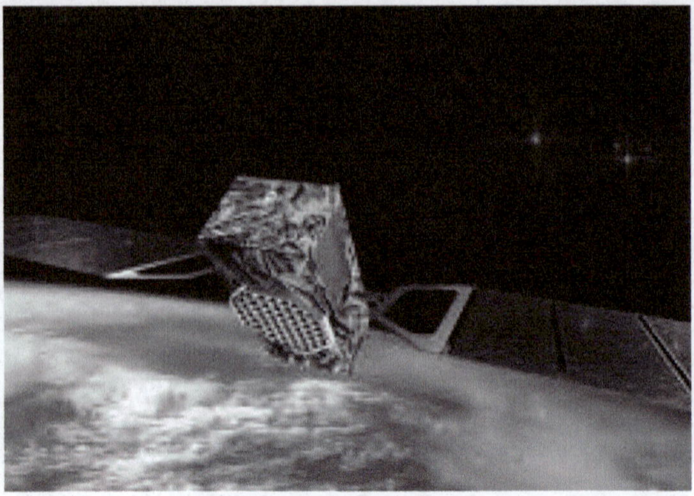

Figure 10.30. Galileo Configuration (ESA Achievements, BR-250, ESA 2005).

Cryosat measures ice thickness over the Artic and Antarctic and orbits at an inclination of 92°. At such an inclination, the satellite's orbital plane is almost fixed in inertial space, and the Sun orbits once per year around that plane. During some seasons the Sun will be perpendicular to the orbital plane and will illuminate the satellite laterally from a fairly constant position along the orbit. During the seasons when the Sun lies in the plane of the orbit, it will rise over the Earth's horizon and illuminate the satellite's trailing face; as the satellite proceeds along its orbit, the Sun will move toward the spacecraft's Zenith, and will finally set below the Earth's horizon while illuminating the satellite's leading edge side. To cope with this complex and varying geometry, Cryosat possesses fixed solar arrays located above the tent-like anti-Nadir face of the satellite. The radar altimeter and communications antennas occupy the Nadir-facing side. The angle and size of the two Zenith faces, where the solar arrays are installed, are optimized to provide similar solar energy production whether the Sun lies in the orbital plane or is perpendicular to it. The solar array's efficiency will never be very high but given the availability of space the energy produced is perfectly adequate to meet the satellite's needs. The NASA-DLR GRACE mission has the same orbit and satellite configuration.

The principal objective of MeghaTropiques is the study of the tropical water cycle. In order to maintain its orbit over the tropics, so maximizing its scientific mandate, it flies at an orbit inclination of 20°. The two wing solar array's rotation axis is perpendicular to the orbital plane and rotates uniformly at one turn per orbital revolution. The Sun revolves around the orbital plane, but being a low-inclination orbit the angle of the Sun with respect to the solar array will be as indicated in Figure 9.41 with the Sun angle always at a moderate angular distance of the orbital plane.

Iridium orbit inclination is very close to 90°, and the sunlight can reach the satellite from almost any direction. To permit communication between Earth and the satellite, Iridium must maintain a fixed geometry with respect to the ground; and it must maintain a constant geometry around Yaw in order that each satellite in the constellation can establish contact with the satellites flying around it. This means that it is necessary to provide a solar array with two rotation axes:

- One, which rotates from 0° to 90° to provide an optimal canting angle. This actuator operates in small increments to accurately track the seasonal variations of the Sun's position with respect to the orbital plane of the satellite.
- A second one that tracks the Sun by uniformly rotating 360º during each orbit.

Section 8.4.7 looked at missions in which the satellite can rotate freely with respect to the instruments' line of sight. Radar altimeter missions fall into this category. Galileo is a constellation of 27 European navigation satellites—plus three in-orbit spares – placed in three circular orbit planes at an altitude of 23,222 km and an inclination of 56°. Galileo emits navigation signals that allow receivers on the Earth to determine their geographical location by calculating the distance – determined from the delay time of the signal – from the ground receiver to the emitting satellites. The application allows rotation of the instrument around its line of sight that coincides with the Nadir direction. This results in a configuration with two solar arrays, each one with one rotation axis. The combination of this single rotation mechanism and the ability to orientate the satellite around the Nadir allows the solar array to be continuously oriented toward the Sun. Radiators are on the faces where the solar arrays are stowed and are always perpendicular to the Sun.

In non-Sun-synchronous orbits the Sun rotates seasonally around the satellite's orbital plane and successively illuminates each of the satellite's sides. To avoid sunlight reaching the radiators the satellite can perform seasonal yaw-flip maneuvers (Figure 9.40) around the Nadir direction when the Sun crosses the orbital plane. This prevents direct solar illumination of one face, and is a practical option for locating the radiators because the frequency of the flip is low and the resulting operational constraint is perfectly acceptable.

10.4.1.4. Low-inclination orbits

As depicted in Figure 9.41, in low-inclination orbits the angle between the ecliptic and the equatorial plane is moderate and solar arrays rotating on an axis perpendicular to the equatorial plane will be reasonably well placed with respect to the Sun along the entire orbit and throughout all the seasons. Given a cubic central body for the satellite, four of the six faces will be illuminated by the Sun (Figure 10.31) while two will remain in shadow. These two dark faces are the logical place for locating radiators. This is the case for all low-inclination orbits either LEO or GEO. Section 9.4.4 introduced geostationary orbits. The combination of these useful orbits with a Nadir-pointing attitude offers continuous coverage of extensive areas of the Earth. This combination is frequently employed for communication and Earth observation missions.

The instruments pointed to the Earth shall be designed to minimize the effect of the sunshine on them. This could be done by minimizing the scanning of the instruments outside the Earth's disk, by using an optical baffle to shadow the entrance of sunlight to the instrument or by providing shutters that will cut the light penetrating inside the instruments. When this shutter is activated the instrument cannot operate. This situation has been already depicted in Figure 9.43.

The configuration of the geostationary meteorological satellite GOESS (Figure 7.2) is a good example of this kind of satellite. GOESS's instruments are located on the Nadir face and are protected from sunlight by baffles. One of the two sides not illuminated by the Sun contains the radiators and the other carries a single-wing solar array. To compensate for the disparity produced by the pressure of sunlight on the solar array, the thermal radiator side features a deployable solar sail that is positioned by a boom far from the radiators, minimizing any thermal effects the solar sail might have on them.

Meteosat Third Generation (MTG) will be the future European geostationary meteorological satellite. The final design is not yet settled, but all configuration alternatives have elected a Nadir-pointing attitude. One design features a symmetric arrangement of solar wings (Figure 10.31). This is made possible by the use of active mechanical coolers that reduces the instrument detectors' working temperature without the need for a perfect view of cold deep space. The trade-off between complex mechanical coolers and complex configuration is treated with more detail in Section 10.5.4 (Figure 10.32). Most communication satellites that need their communication antennas to be aimed toward Earth share this Nadir-pointing GEO attitude. The most common configuration for communication satellites features two symmetric solar arrays. The Alphasat (Figure 10.33) provides a good illustration; it carries symmetric solar wings and the Nadir-oriented face is occupied by antennas which also occupy – when stowed – the two lateral faces that are also lit by the Sun. The two faces perpendicular to the solar arrays, which are never illuminated by the Sun, are used for the stowage of the solar array during launch and as radiators once in orbit. The bottom face of the satellite is used to attach it to the launcher.

10.4.2. Spinning satellites

The attitude control for spinning satellites was introduced in Section 9.3.3 and the resulting geometry around the satellite in Section 9.7.2. This section will examine the implications on satellite configuration of selecting a spinning satellite (Figure 10.34).

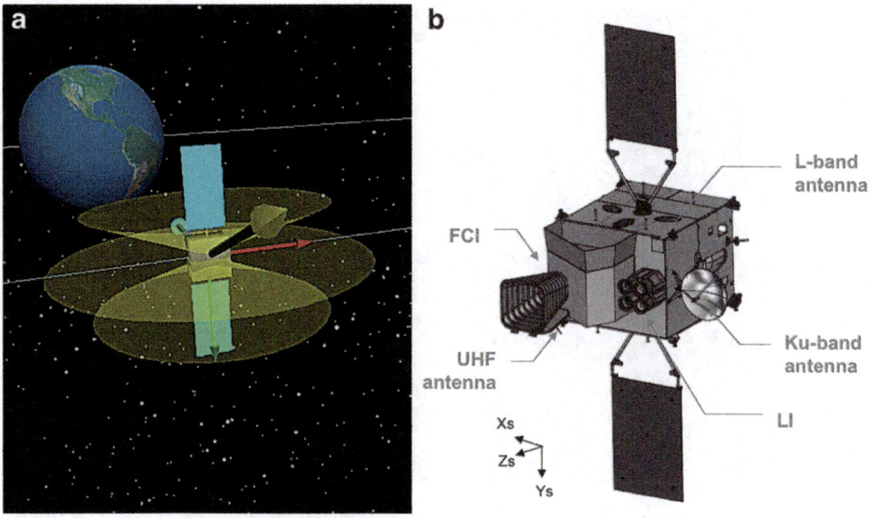

Figure 10.31. Solar Array-Sun Geometry (a) and Possible Configuration (b) for a Meteorological Geostationary Satellite (Image b) (Courtesy from Thales Alenia). This Configuration Has an Optimal Solar Array but the View of Cold Deep Space Is Hindered by the Solar Array.

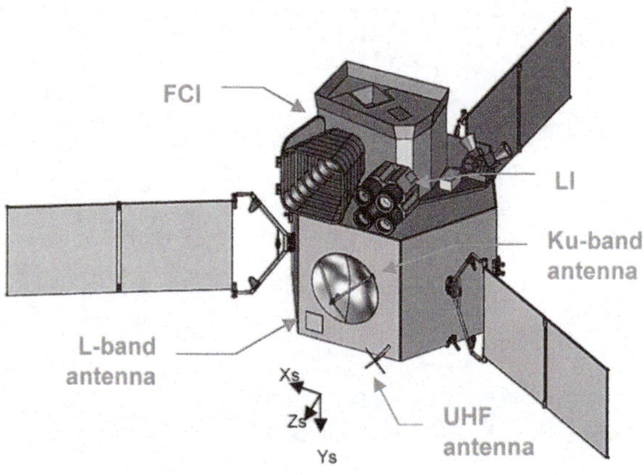

Figure 10.32. Possible Configuration for the Future European Meteorological Geostationary Operational Satellite MTG (Courtesy of Thales Alenia). This Configuration Has an Optimal View of the Deep Space for Cooling and Uses Three Fixed Solar Arrays.

Satellite faces perpendicular to the rotation axis will be fixed on an inertial reference frame, while faces parallel to the rotation axis will see the celestial sphere rotating rapidly around them. The most important element influencing the configuration of a spinning satellite is the centrifugal forces generated by the rotation. These forces will produce loads on any deployable structure required by the satellite. Deployable mechanisms are difficult to accommodate on spinning satellites unless they deploy

Figure 10.33. Communications Satellite Alphasat Configuration (Courtesy of EADS-Astrium).

Figure 10.34. Two Cluster Satellite Carried by the Fregat Upper Stage (The Cluster-II Mission, ESA Bulletin 102, Drawing Courtesy of NPO Lavotchkin, ESA 2000).

perpendicular to the rotation axis. The rotation axis also determines a cylindrical symmetry around itself. This cylindrical symmetry fits naturally into the cylindrical symmetry of the launcher fairing, leading to cylindrical satellites that will make efficient use of the volume provided by the launcher. The communication satellite Anik C (Figure 10.35); the meteorological satellite MSG (Figure 7.3); and the Earth magnetosphere mission Cluster (Section 11.4.1 and Figure 10.34) are examples using this configuration.

Spinning satellites are simple, stable, and as indicated above, make an efficient use of the volume provided by the launcher. Nonetheless, the cylindrical symmetry, the

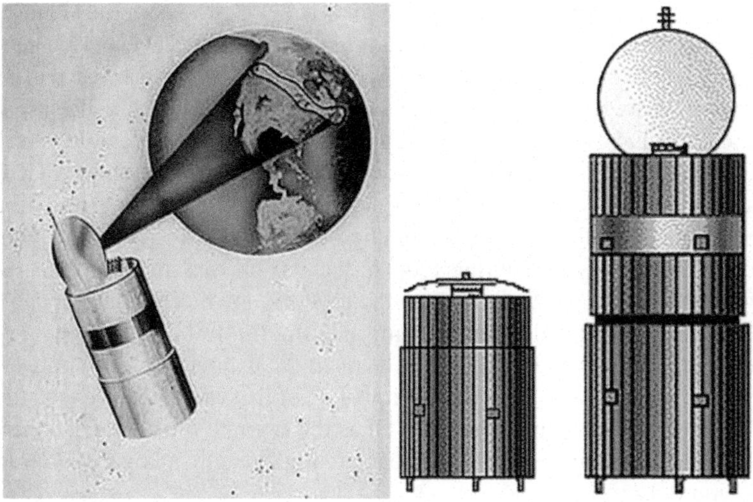

Figure 10.35. Anik C Configuration (http://www.boeing.com/defense-space) in Orbit, Stowed and Deployed.

lack of flexibility in pointing, and the high centrifugal forces allow designers little freedom. In addition, spinning satellites have difficulty overcoming limits on solar array size, scanning patterns, and thermal radiator geometry. But in cases where mission aims are compatible with these limitations, the configuration is very effective.

Satellites performing in situ observations are not constrained by the pointing direction of the instruments and a spinning configuration can be an attractive option. Ulysses (Section 10.3.2) is a good example of this kind of satellite. Ulysses has mechanisms deploying perpendicular to the rotation axis: the 70 m wire booms and the 5 m rigid hinged boom are deployed and stiffened by the centrifugal field. The 7 m axial boom is deployed along the rotational axis and any misalignment will be bent by the centrifugal forces, which increase as the deflection of the boom increases. This puts demanding requirements on the stiffness and alignments of this boom. In fact, during the mission's first days, the axial boom developed oscillations caused by centrifugal forces and the heat of the Sun. The oscillations produced dangerous nutation in the satellite's spin, but as soon as the Sun stopped shining on the axial boom the oscillations disappeared, and the boom has been behaving normally ever-since.

Cluster is another good example of a cylindrical spinning satellite performing in situ observations. It has two pairs of 100 m wire antennas and two 5 m long radial booms, all deployed perpendicular to the rotation axis. The configuration was designed to launch the four satellites of the Cluster constellation on Ariane-5's maiden voyage. Sadly, that flight failed and two Soyuz-Fregat rockets subsequently launched the constellation, each carrying a stack of two satellites. The satellites' ability to fit perfectly into the cylindrical launcher envelope is evident (Figure 10.34).

The Cluster constellation orbits Earth in an extremely elliptical high altitude orbit with an inclination of 90°. At this inclination the orbital plane is fixed in the inertial reference frame and the Sun rotates around the orbital plane once a year. The rotation axes of the satellites are perpendicular to the ecliptic plane, and the rotating cylindrical side of the satellite, which is covered by solar cells, is always perpendicular to the Sun direction. This is an optimal attitude for producing electrical energy. The top and bottom of each satellite never receive direct sunlight and the thermal radiators are located there where they can provide heat dissipation efficiently into deep space.

As mentioned above, the pointing of spinning satellites is inflexible, but there is one orbit where this is not a major problem: the geosynchronous orbit. A GEO spinning satellite, with the rotation axis perpendicular to the orbital plane and the instrument boresight perpendicular to the rotation axis, will scan along the Equatorial direction of the Earth's disk during 17.4° of the 360° of each rotation. This provides east–west scanning of the Earth without the need for any mechanism, though north–south scanning would still require a mechanism for full Earth coverage. MSG (Figure 7.3) is an example of this type. The main disadvantage of this arrangement is that observations can only be made during 17.4° of the satellite's 360° rotation; this radically reduces observations time as well as the strength of the signal collected by the instrument. To compensate for this reduction in strength, a larger aperture instrument is required.

From the point of view of satellite configurations and celestial sphere observing geometry, communication satellites are Earth-observing satellites. Their antennas must point toward Earth in order to receive and retransmit the corresponding data streams. Anik C is a good example of a communications satellite with a cylindrical configuration. Anik C was a Canadian communications satellite built by Hughes (now Boeing); three satellites were built and launched from 1983 to 1985. The satellite featured a cylindrical telescopic solar panel to increase the area available for the collection of solar power. It had a high-gain parabolic antenna on the topside that counter-rotated the satellite's spin and pointed continuously toward Earth. The configuration provided good pointing stability and efficient use of the volume offered by the launcher. The satellite's diameter was 2.16 m; its stowed height 2.82 m; and its deployed height 6.43. Its mass in the GEO position was 562 kg.

During the 1970s and 1980s this configuration was used extensively for GEO communications and meteorological satellites but today it is not common. Advances in attitude control and large deployable structures, coupled with the need for greater power, pointing flexibility, and payload performance have made this configuration nearly obsolete.

10.4.3. Inertially pointed satellites

Inertially pointed satellites are ideal for observing events located above the celestial sphere and in deep space; as such, inertially pointed satellites are generally astronomy satellites. All examples discussed in this section are astronomical missions. Angular accelerations in the deployed configuration are low and there is ample freedom for designs employing satellites with large, slender, deployable structures, if needed.

Inertially pointed satellites can point in any direction, and the positions of the Sun, the Moon, the Earth (for data downlink and commands uplink), and potential observation targets are highly variable. This makes the configuration of the satellite much more open than for Nadir-pointing or spin-stabilized satellites. In inertially pointed satellites, the configuration must be able to balance the need for accurate pointing; observation of the astronomical events being studied; adequate exposure to sunlight for electric power; exposure to deep space for heat dissipation; and Earth visibility for communications. The Sun's position must be considered not only for the location of the solar array and the thermal radiators but for the pointing of the instruments as well. Astronomy instruments are very sensitive and will not generally tolerate direct sunlight on them.

Section 7.2 provided an introduction to the interactions between the principal design drivers of space missions: observable, orbit, and attitude. Since all the missions examined in Section 7.2 are inertially stabilized, that chapter also provided an introduction to the configuration of inertially stabilized satellites. Specifically, Section 7.2.2 analyzed the fundamental design drivers for these kinds of satellites, and the premises of that analysis are applicable here. In those examples the configuration was governed by the telescope's need to withstand the launch load environment while providing a large telescope aperture and accurate pointing by the instrument in the satellite's deployed configuration.

10.4.4. Agile satellites

This is a special category of satellites whose outstanding characteristic is that they are able to rapidly maneuver into any direction required by the mission. All inertial satellites are agile in the sense that they can be pointed in any direction, but truly agile satellites are designed to allow swift attitude changes as required by rapidly changing targets. Agile satellites are especially appropriate for high-resolution observations of small areas of the Earth where events worthy of note are developing. Their agility allows them to cover numerous distinct areas of interest. This ability can also be employed to improve power production; for instance, by maneuvering from a Nadir-pointing attitude to an inertial Sun-pointing attitude so that the solar arrays are perpendicular to the Sun, thus increasing the amount of power generated by the array and to recharge the batteries in the process.

In agile missions the entire satellite points in the direction required by the observable and no dedicated instrument pointing mechanisms are needed, neither along- nor across-track. The satellite must be compact and rigid, minimizing inertia and avoiding dynamic couplings between the flexible deployables and the pointing of the instruments. The high-resolution Earth observation mission Pleiades (Section 9.6.4 and Figure 9.29) is a good example of this kind of mission and of the resulting configuration. The satellite configuration is compact, and the telescope is located in the center of the satellite's structure. The solar array is deployable but its panels are solidly attached to the satellite body and have supplementary stiffening stringers to maximize stiffness and minimize pointing perturbations. The compact configuration ensures lowest inertia and maximum maneuverability.

RapidEye[11] is a constellation of commercial medium–resolution, small satellites for Earth observation. The constellation is made up of five identical satellites flying in a Sun-synchronous orbit, at 620 km altitude and with a local time of 11:00 ± 1 h. It provides images in five spectral bands with a maximum resolution of 6.5 m. The spacecraft can maneuver across-track by ± 25° allowing a daily revisit over any part

[11]"The RapidEye" Spacecraft, Gary Buttner, IAA-B4-1105P.

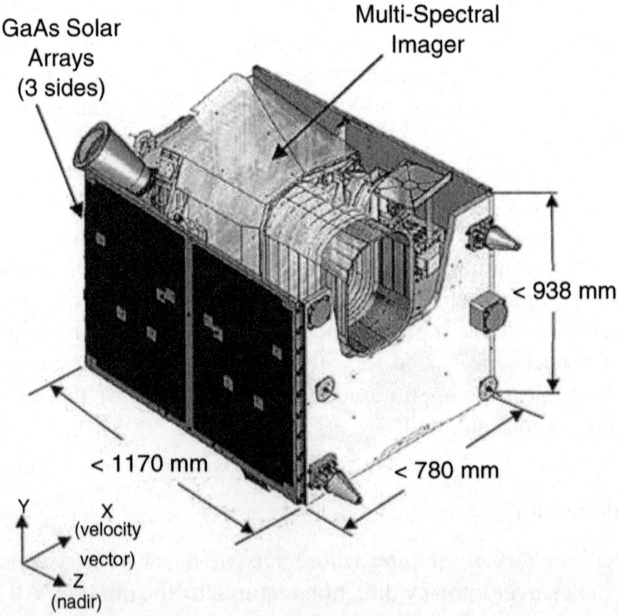

Figure 10.36. RapidEye: Example of Small Agile Satellite Configuration (http://directory.eoportal.org/presentations/6166/7466.html Courtsy SSTL).

of the Earth. The satellite's mass is 150 kg, with 35 kg allocated to the instruments. The instrument's aperture, is 150 mm. The satellite is a cuboid (Figure 10.36) without any deployable mechanisms. This minimizes inertia, maximizes agility, improves satellite reliability, and minimizes cost and risk. Being a Sun-synchronous satellite with almost 12:00 local time the Sun geometry is as in Figure 9.37a. The solar array is made up of three panels located in the velocity (+X), antivelocity (−X), and Zenith (−Z) sides of the satellite. The Sun, which due to its local time is always very close to the orbital plane, will illuminate sequentially the three panels along each orbit. The instruments and antennae are in the Nadir (+Z) plane pointing to the Earth. The arrangement does not provide an optimal use of the solar cells but is adequate for the mission and perfectly in line with a simple, robust satellite design approach.

10.5. Allocation of functions

This section analyzes the design decisions of this domain that imply to allocate functions to different mechanical elements and components. That means, who does what.

10.5.1. Primary structural shape

This section analyzes the different shapes that the primary structure of the satellite can take. The main design drivers for the primary structural shape are:

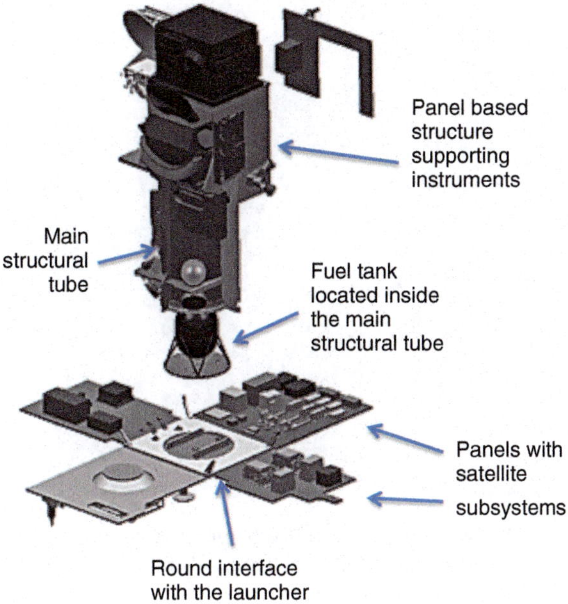

Figure 10.37. Exploded Views of Sentinel-3 Configuration with Lower Service Module (SVN) and Upper Instruments Module (PIM) (Sentinel-3 Definition Phase Final Report, Y. Baillion, Thales Alenia Space 2007).

- Launcher fairing volume limitations. Volume limitations will compel long satellites to adapt their shape to the – normally long – launcher fairing.
- Launcher payload mass limitation. Mass limitations will pressure short compact satellites to minimize load path and optimize use of structural mass.
- Minimizing axial center of gravity offset from the launcher's payload interface.

The following figure provides the internal structure of Sentinel-3. The satellite is divided into a lower part for the subsystems and the fuel tank and an upper part for the instruments. Launcher fairings are cylindrical-conical and longer than they are wide; to take advantage of the maximum volume permitted by the launcher, satellites must be elongated. Several examples of this have been discussed above: Figure 10.2 provided Sentinel-3's stowed configuration; and Figure 10.3 showed that of Snapdragon. In both cases the satellite is elongated and makes the best possible use of the volume allowed by the launcher. The configurations of GOCE in Figure 10.16 are also examples of a long satellite. Long satellites allow a natural split of the primary structure into two halves, upper and lower, each part with its own function: the upper half to carry the instruments and the lower half to carry the satellite subsystems. The lower part is usually called the "service module" and the upper part the "instruments module." This split design has a long history; TD-1, launched by ESA in 1972 already used it. Assembly can be done on both halves of the satellite in parallel and joined only when the majority of problems have been resolved in each separate half. This speeds up assembly. A long satellite divided into two parts is the most frequent configuration for LEO missions flying multi-instrument payloads. Figure 10.37 shows the internal structure of Sentinel-3. The lower service platform has a main structural

Figure 10.38. MARECS, Communication Satellite with "Short" Cubic Shape Optimize for Shared Launch to the Geostationary Orbit (ESA Achievements, BR-250, ESA 2005).

tube that is the primary load path to the launcher. The lateral panels accommodate the satellite subsystems and the large central tube provides rigidity, strength and space for the orbit control fuel tank. The upper-half primary structure of Sentinel-3 is based not on tubes but on panels. The panel-based primary structure allows for optimal accommodation of the large, roughly cubic instruments in independent compartments separated by load-carrying panels; this provides maximum use of the allowable volume.

Short satellites are of similar length and width. These satellites provide maximum internal volume for minimal cantilever distance. For the same structural mass, a short satellite will be stiffer and stronger than a long satellite of the same internal volume. This gives a clear structural advantage to short satellites. Short satellites, however, do not make efficient use of the launcher's internal elongated fairing volume. There are two situations in which this is not a shortcoming:

- When a single launcher is being used to launch more than one satellite
- For missions orbiting at or above GEO, where the launcher mass limitation would become more restrictive than the volume limitation

Many of the medium and large launchers optimized for GEO can launch more than one satellite at a time. In the case of GEO the launcher's elongated faring is split between two or three volumes, each for a different satellite (Figure 10.6). These volumes have similar longitudinal and transversal sizes and most medium-size satellites flying at GEO possess a cubic shape (Figure 10.38).

The high-energy levels needed to thrust a satellite far from Earth make mass limitation, not volume limitation, the principal design driver for the primary structure of deep space satellites. Deep space satellite Rosetta (Section 11.4.2), for example, is shaped like a cube; its central body measures 2.8 × 2.0 × 2 m and fits without any volume constraints into the faring provided by Ariane-5.

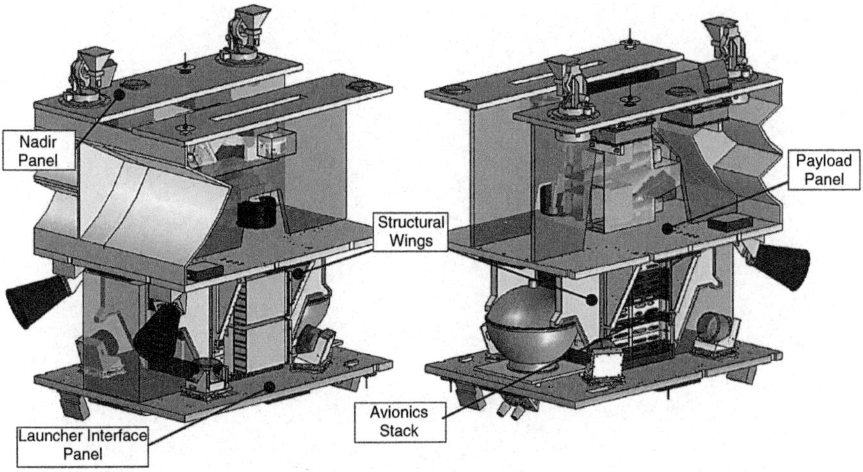

Figure 10.39. Front View (*right*) and Back View (*left*) of the Structural Arrangement for a SSTL Mini-Satellite (Courtesy of Surrey Satellite Technology Limited).

Structural requirements demand strength and stiffness. As satellite's shrink in size, their stiffness and strength is improved, permitting a reduction of structural material thickness. Furthermore, there are minimum practical thicknesses that cannot be further reduced. As a result of this small satellites fulfill quite easily structural requirements and it is feasible to use structures that do not look optimal but are perfectly able to fulfill the structural requirements while optimizing accommodation. The primary structure of the SSTL satellite provided in Figure 10.39 above is a good example for this type of structure.

10.5.2. Deployable structure and mechanisms: Fixed versus deployable

The primary structure defines the stowed configuration for launch and the deployable structure and mechanisms define the in-orbit deployed configuration. Large deployable structures and mechanisms are complex and expensive but they solve the contradiction between the need for a compact launch configuration and the eventual expanded in-orbit one. If an expanded in-orbit configuration is required, then a large deployable structure will be necessary. Most satellites have large deployable solar arrays for collecting the energy required for their operations. Deployable mechanisms are also crucial for providing the necessary unobstructed field of view needed by instruments or communication antennas. Chapter 8 made repeated reference to the need of sensitive instruments for large apertures to detect the frequently weak signals required for fulfilling mission aims. If the instrument aperture is larger than that allowed by the launcher, it will require the use of deployment mechanisms. The configurations studied in Section 10.3 encompass the compact GOCE satellite, with no mechanisms, to JWST that employs several large deployable structures and mechanisms, and each one crucial to the fulfillment of mission objectives. Section 8.4.1 alluded to the impact of observing frequencies on instruments and to how observations

Figure 10.40. Large 12 m Diameter Parabolic Antenna Deployment Test Under Its Gravity Compensation Ground Support Jig (European Large Deployable Antenna, F. Mini et al., Proceedings of EuCAP 2006, ESA SP-626). The antenna has 12 m of diameter.

at low frequencies demand larger instruments. Figure 8.12 compared the configurations of X- and P-band SAR satellites, and how the lower frequency P-band satellite required a very large antenna. Figures 10.12 and 10.40 offer also examples of large deployables, which are vital for fulfilling the demanding needs of P-band SAR observation. Ulysses (Figure 10.17), which observes electromagnetic waves at low frequency, also employs deployable mechanisms: the previously mentioned 7 m monopole axial antenna, a 75 m dipole radial antenna, and a 5.5 m radial boom.

To avoid the problems associated with complex deployment mechanisms, the use of larger launchers is advisable. Launchers like Ariane V, Delta-IV, or the Space Shuttle feature large volumes that will always make the task of accommodating large satellites easier; they are all, however, expensive. The choice between using large complex deployables or a large launcher is always an important trade-off in the design of the mission architecture. The Herschel satellite was presented in Figure 7.9; it has the largest rigid diameter structure able to fit inside an Ariane-5 rocket. Initial designs for the concept that eventually became Herschel – at that time called the Far Infrared Space Telescope – projected a deployable telescope with the primary mirror in segments, like the petals of a daisy, devised to fit inside the comparatively small volume provided by Ariane-4. The ultimate availability of the larger Ariane-5 allowed Herschel to fulfill its needs with a single element mirror. The Hubble Space Telescope's primary mirror size was reduced from an originally envisioned 3 m to a final 2.6 m in a trade-off occasioned by the need to fit it in one piece inside the Space Shuttle's cargo bay. JWST, on the other hand, includes complex deployment mechanisms needed to fulfill its very demanding mission requirements, which require an instrument with much larger diameter aperture than the Hubble's telescope, and the ability to operate in the infrared range without the life-span limitations associated with the use of liquid He for cooling (as it was the case with ISO (Section 7.2.1) whose life-span was limited to just 2 years). JWST's mechanisms include:

- Hinge mechanisms for the primary mirror to adapt its 6.5 m aperture to the 4.57 m diameter of Ariane 5.
- Mechanisms for situating the secondary mirror in its correct position with respect to the primary mirror.
- An 18 × 12 m deployable Sun-shield.
- Mechanisms for increasing the in-orbit separation of the instrument from the satellite platform. These mechanisms are necessary to minimize heat leakage from the satellite platform to the instrument.
- Deployable radiators, solar arrays and a pointable high-gain antenna, all located on the satellite's platform side.

These mechanisms can be seen in Figures 8.7, 10.20, and 10.21.

Large deployables also require complex testing procedures. In order to provide a credible estimate of the development costs of a large deployable, a detailed design must be provided not only of the mechanisms but also of test procedures and the required jigs needed for the testing. Figure 10.40 provides an indication of the complexity involved in the on-ground deployment of a large, storable parabolic antenna, which must simulate conditions in the weightless environment of space. This requires the use of special jigs to compensate, as far as is practical, for Earth's gravity. The design of these tests is complex, risky, and expensive. In most cases the solar array is by far the satellite's largest deployable. Most of the time the solar array also possesses tracking mechanisms to ensure that the largest possible area is exposed to the incoming solar radiation. Modern satellites need large solar arrays and in most cases that means they must be deployable. Fortunately, solar array deployment kinematics is far from complex, and designs can easily be adapted from one mission to the next, thus reducing the associated risks.

Recalling the missions described in Section 10.3: JWST and Iridium use deployable solar arrays; GOCE uses a fixed solar array; and Ulysses a radioisotope thermoelectric generator (RTG) in place of a solar array. GOCE cannot use deployable deployment mechanisms because their movements would introduce perturbations in the sensitive gravity-sensing accelerometers, though its fixed, rigid solar array is the largest of this type compatible with the launcher envelope. Ulysses orbits at a great distance from the Sun, so a radioisotope thermoelectric generator was selected to meet its electrical energy needs. Nevertheless, the use of solar arrays is also viable for missions flying far from the Sun. The Rosetta satellite – described in Section 11.4.2 – flies at 5 AU during the most important phases of its mission. At that distance the energy produced by its solar array is 25 times less than when flying near to Earth. To compensate, Rosetta uses a large, elongated solar array, which dominates the satellite configuration Europe lacks the US technological expertise in RTGs. Ulysses is a cooperative NASA-ESA mission and the delivery of the RTG by the USA was part of the mission's contractual conditions. Rosetta's configuration would have been more compact using a RTG, but its 32 m long array provides 400 W at 5 AU, sufficient to fulfill the mission's needs. The power delivered by this solar array compares satisfactorily with the 284 W provided by Ulysses's RTG.

10.5.3. Standard platform versus dedicated platform

It should be pointed out that a satellite configuration is not the same thing as a satellite platform. A satellite platform, either dedicated or standard, is the sum of a configuration – that is, a mechanical concept and an external shape – plus an electronics-avionics concept. It is perfectly possible to design satellites with totally different shapes but with the same avionics-electronics concept. The present chapter deals with satellite configuration, and will focus on standard platforms, meaning standard structural concepts and standard shapes.

Standard platforms with the ability to carry different instruments have been promoted as a way to reduce space mission costs for many decades; their actual performance, however, has been mixed. In some cases, they have indeed provided effective performance; while in others their performance has been disappointing. The reason is obvious: space missions are extremely diverse and the most difficult part of any mission is always the instrument configuration and performance. Standard platforms tend to reduce the cost of the simpler aspects of the mission, – the satellite structure and standard subsystems – while complicating the most expensive part of it – the instruments. Furthermore, the environment of space and the requirements of the mission are also exceedingly varied; so standard platforms almost always require minor or major modifications that counterbalance the advantages of the standardization. Section 4.11 reviewed the problems that NASA underwent in the 1980s during its drive to standardize planetary exploration.

Nevertheless, a standard platform provides a reference design for the satellite as to where the different instruments might be placed, and this can be helpful in many situations. Standard platforms are effective for satellite families flying very similar orbits and carrying instruments with similar demands, such as families of Earth observation satellites flying in LEO orbits. NASA[12] used a standard platform for the EOS (Earth-Observing System) Terra, Aqua, and Aura, and NOAA will use it for NPOESS, its future line of operational LEO meteorological satellites (Figure 10.41).

Both EOS and NPOESS are based on the T430 platform, a design by Northrop Grumman for missions orbiting Earth at a moderate altitude – around 800 km – and in Sun-synchronous orbits. The T430 platform has a composite lightweight structure that allows payload maximization and a titanium support surface for instrument accommodation flexibility. It provides modular interfaces and a standard data bus that facilitates easy accommodation and testing. The platform is compatible with orbiting at different local times with their different Sun and solar array geometries. This platform has already demonstrated its flexibility for carrying with different instrument combinations. In the case of Terra, the entire satellite weighs 5,200 kg; it carries six instruments with a total payload mass of 1,100 kg; its solar array provides 2.5 kW; and orbits at 705 km at a local time of 10:33. Terra uses TRDSS for data relay and its presence is quite conspicuous on its configuration. In the case of NPOESS[13] the satellite will operate in a Sun-synchronous orbit at 833 km of altitude at three different local times: 13:30, 17:30, and 21:30. Because they will carry different instrument

[12]The NPOESS Spacecraft and Payload Suite, H. Bloom, IEEE 2003.

[13]Payload Accommodations on Future Environmental Sensing Spacecraft – Lessons Learned from EOS and the Future with NPOESS, Leo Andreoli, SPIE 5659, Proceedings of SPIE Vol. 5659, SPIE 2004.

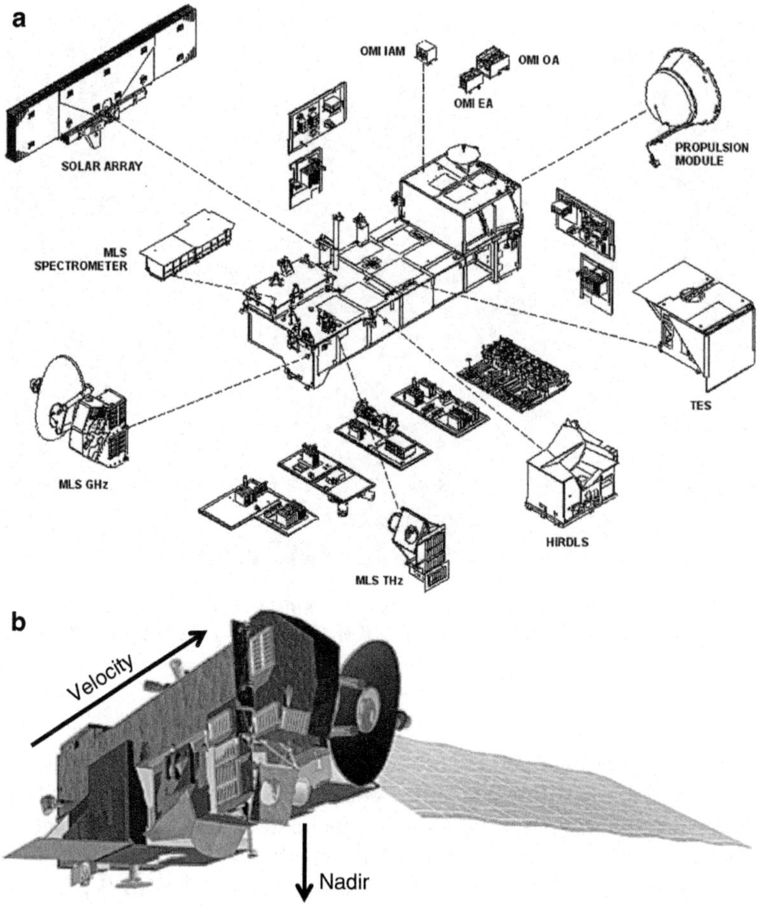

Figure 10.41. Aura on T 430 Standard Platform. (a) Details of the Payload and Satellite Subsystems Accommodation Inside and Outside the Standard Platform. (b) In Orbit Configuration of Aura (Configuration (a) NASA http://aura.gsfc.nasa.gov/spacecraft/index.html. Exploded view (b) NASA http://aura.gsfc.nasa.gov/spacecraft/images/equip_config.html).

sets, the mass of each of the three satellites will differ, though they will all fall into the 5,000 kg-plus category. The platform is designed to provide up to 7 kW of power. The different local times define different Sun and solar array geometries. The standard platform used by EOS and NPOESS has provided good service in space since 1999 and will continue to do so for another 10 or 20 years. This record is based on the strong similarity in the demands of the Earth-observing instruments carried by the standard platform implementations to date. Nevertheless, this standard platform has evolved in time: the payload space has increased from the 10 m^2 of EOS to the 14 m^2 of NPOESS; and its instruments data downlink approach has also changed. EOS satellites use the TDRSS data relay satellite while NPOESS will use a totally different downlink approach that will be discussed in Section 12.3.4. These are examples of the changes needed to accommodate the different needs of NPOESS and EOS over a theoretically "standard" platform.

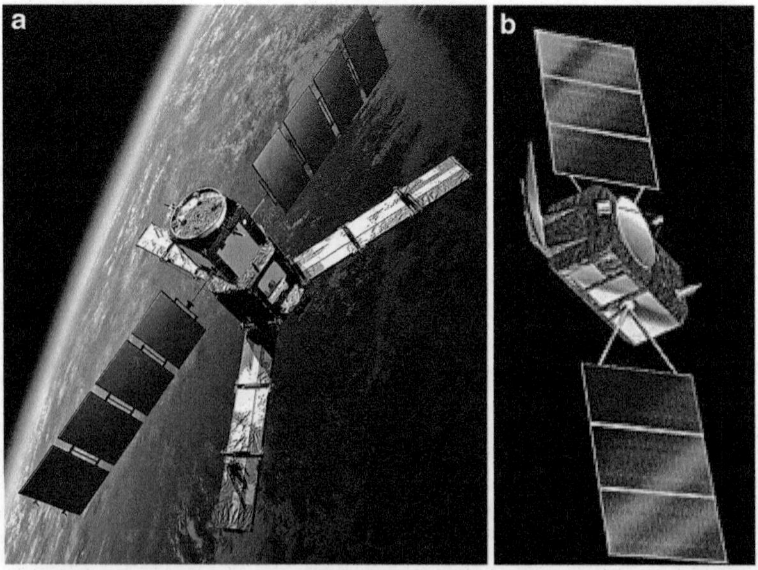

Figure 10.42. SMOS (a) and Jason (b) Satellites Using Proteus Platform (Image (a) ESA www.esa.int and image (b) CNES from http://132.149.11.177/html-images/HomeGB.html).

Another example of common platform is Proteus. This platform has been used in a variety of applications. Proteus, unlike our previous T430 example, was designed to operate at totally different orbit altitudes, inclinations, and Sun-Earth geometries. Proteus is small and can only accommodate a small number of instruments, in most cases just one. Proteus is an Alcatel-CNES concept designed for LEO payloads in the 300 kg/300 W class. Its name refers to the Proteus of Greek mythology, a monster able to change shape at will, and the Proteus platform has demonstrated the aptness of its name by operating in many different orbits and attitudes:

- In the case of Corot, the satellite is inertially stabilized with an orbit inclined 90° at 900 km of altitude.
- In the radar altimetry mission Jason it orbits at 1,336 km with 66° inclination.
- In the cases of SMOS and Calipso it operates in a Sun-synchronous orbit, the former at 755 km with a local time of 06:00 and the latter at 705 km and 13:30.

Proteus's capacity for flying in different orbits and attitudes has been demonstrated. Nevertheless, its efficacy with respect to the instruments is limited and in the three examples presented, an upper "instrument platform" was required to attach the instruments and which included data processing and data downlink services, vital to the instruments that Proteus' basic platform was unable to provide. The successful examples described here were made possible in exchange for one limitation each. T430 is able to provide extensive services to many instruments providing the orbit is LEO Sun-synchronous, and Proteus permits a variety of orbits providing the payload – preferably a single instrument – carries its own services (Figure 10.42).

10.5.4. Passive versus active thermal control

Satellites must dissipate the thermal energy they produce. The radiators provided for the purpose need clear exposure to cold deep space. Two conditions are possible:

- The need to cool a satellite's components to temperatures approximating normal temperature on Earth, around 290–300 K
- The need for cooling to very low temperatures, from 200 K down to near absolute zero, depending of the application

The preferred working temperature for most structures is ambient temperature, which minimizes thermoelastic deformation; ambient temperature also guarantees the smooth running of contact surfaces for mechanisms. Moreover, most surfaces not in continuous view of the Sun are capable of adequately dissipating the heat produced at ambient temperatures by the satellite's electronics boxes. Figures 10.23, 10.25, and 10.26 provided examples of varied radiator placement.

The situation changes completely when cooling to very low temperatures is called for. Very low temperatures are essential for reducing the noise on an instrument's detectors (Section 8.4.2). ISO (Figure 7.7) uses liquid Helium to cool the telescope's detectors to 2 K, while JWST (Figure 10.19) used a combination of passive radiators, Sun shields, and active coolers to reduce the temperature of the detectors to 7 K. JWST's complex thermal and mechanical approach facilitates a long mission life, which is a clear advantage for an operational astronomical observatory of these characteristics. The location of the radiators and the trade-off between using passive radiators or some kind of active cooler – mechanical or otherwise – is always an important decision that will have a major impact on the overall configuration of the satellite. Passive coolers increase reliability and mission life but complicate the satellite's configuration. Figures 10.31 and 10.32 compared two possible configurations for operational meteorological satellites, one with active and one with passive cooling. The choice between passive or active cooling has major implications, and the decision must be made after careful consideration of all the crucial issues and their impacts at all levels. This need for an end-to-end perspective plainly puts the decision into the hands of the mission architect.

10.5.5. Pointing by the instrument versus pointing by the satellite

To fulfill its mission the satellite's instruments must be pointed employing specified patterns. This function can be performed by the instrument or by the whole satellite. This issue was addressed on in Section 10.4.4 when discussing the topic of agile satellites and was also discussed in Section 8.6.1 and in Figures 8.20 and 8.3.

Allocating the scanning function to each individual instrument is the preferred solution for satellites carrying several instruments all with moderate apertures. In these cases, the pointing mirror or mirrors will be of moderate size, and the scanning function becomes the responsibility of each instrument, and thus no longer a major concern for the satellite's overall configuration. The pointing of each of the instruments will produce internal forces and torques that will alter the pointing of the other instruments, so it becomes necessary to specify the maximum level of perturbations that each instrument can introduce onto the satellite. The design of all mechanisms must take into account these maximum perturbation figures.

Large instrument apertures make pointing by movable mirrors or reflectors impracticable. It becomes necessary to point the whole instrument, and the most efficient way to do so is by pointing the entire satellite in the required direction. This demands major configuration adjustments. If pointing is the responsibility of the satellite, the satellite can only point in one direction at a time. This means that either the mission must employ just one instrument or that all the instruments can point in the same direction without detriment to the mission.

10.6. Allocation of performances

10.6.1. Mass budget

In the early stages of the design, it is necessary to establish an overall mass figure for the satellite and to provide preliminary allocations to the satellite's subsystems and instruments. In these stages, the mass of the satellite is not a fixed quantity but it should be considered in relation to the anticipated orbit and the launcher to be employed. As the design matures, the satellite mass and its orbit will converge into values that will become requirements for the space system. In fulfillment of its mission the satellite will require fuel for possible orbital changes. This will define two mass limitations: dry mass and wet. Wet mass is the total mass of the satellite that the launcher will put into orbit and dry mass is the total mass minus the mass of on-board fuel. The launcher will be selected to be compatible with the maximum "wet" mass of the satellite and the required amount of on-board fuel will determine the remaining "dry" mass. The resulting dry mass must be broken down to provide allocations to the instruments and subsystems on board the satellite. During the early stages of design the following measures could improve the mass margins:

1. Find a minimum mass configuration for the mission.
2. Analyze changes in the orbital path.
3. Consider a more powerful launcher allowing higher satellite mass.
4. Reduce the number and complexity of the instruments.

Measure 1 is the optimal one; coming up with a minimum satellite mass configuration does not mean reducing the mission's aims. If measure 1 is unfeasible, measure 2 could provide the solution; a judicious design of the orbit can still fulfill mission aims while allowing a heavier satellite. Measure 3 implies the use of a more expensive launcher and may not fall within the mission budget. Measure 4 will reduce mission output.

The weighing of options will lead to the identification of one or more architectures regarding orbit, payload, or launcher choice that could fulfill the mission's aims. These estimations can be used as reference points for the mission's preliminary design. In the early stages of design, the mass estimation will not be a bottom-up calculation but a top-down one, where the allowable total mass as determined by the launcher must be divided-up and allocated to the different satellite components. The adoption of measures 3 and 4 for reducing excessive mass in a mission's design phase – to either eliminate instrument capabilities or to choose a larger launcher – will have a major impact on the mission. These options will either reduce the scientific returns or increase costs. Nevertheless, they have the advantage of clarity. It is better to clearly state that the mission needs a more powerful launcher – and quote the associated supplementary cost – than to provide overly optimistic mass estimations that are likely to be proven impossible to meet during the industrial phase.

The relationship between dry and wet mass varies greatly depending of the mission. Appendix A of Wertz[14] provides the following relations (Table 10.2):

CNES,[15] on the other hand, suggests 6 % of propellant mass for Earth observation and 55 % for a communications satellite with a 15-year life expectancy. The difference arises from the different rules used by each source to define what is and is not included. Wertz often uses as examples satellites with dedicated upper stages that are not considered part of the satellite, which means that the propellant indicated is used only for orbit maintenance, while CNES bases its estimation on Ariane intermediate GTO's orbit (Section 9.4.4) and the 55 % of fuel figure includes the fuel needed to transfer the satellite from the highly elliptic GTO orbit to the GEO orbit plus the fuel needed or orbit maintenance. In the case of Earth observation, where the launcher puts the satellite in the final orbit, the estimations of Wertz and CNES are similar.

The next stage would be to provide estimations for the instruments and the various subsystems. The literature also offers estimations for preliminary mass allocation. Brown[16] and Wertz provide the estimations provided in Table 10.3. The figures provided by both authors are similar and they provide a good base for estimating the mass of the traditional subsystems, such as electrical power or structure. Instruments, however, are usually the most innovative part of any space mission and their mass cannot be allocated top-down, but must be estimated from the bottom-up, even in the earliest phases of the design. This should be possible because, as indicated in Chapters 7 and 8, even at the beginning of the design process there is enough of an instrument concept to allow a rough estimation of the instrument mass. On the other hand, as soon as the design begins to take shape, the top-down mass allocation must be contrasted with bottom-up estimations based on the increasingly more refined design. As the design matures and advances, the actual mass of a satellite always increases, and it grows more that any reasonable person would think it would.

Also according to the textbook of Brown, the mass growth of seven NASA missions from the authorization to proceed with a full-fledged project to launch was from 10 % to 36 % and the average of the seven was 27 %. Unavoidably, as the level of detail of the design increases the mass increases and the individual mass estimations should be balanced with maturity factors that take into account the uncertainties concerning the actual final mass of the satellite's instruments and subsystems. Maturity factors are values larger than 1 that multiply the best estimation of the mass for the various satellite components. As the design becomes better defined, the maturity factors will be reduced until they become 1 for a fully built flight unit. Different guidelines have been published by different sources on the figures to be used as maturity factors. For the early definition of totally new designs, the AIAA[17] uses the values provided in Table 10.4:

---

[14]Space Missions Analysis and Design, J. Wertz and W. Larson, Appendix A, in Page 894 of the Third Edition, Space Technology Library, Kluwer.

[15]Spacecraft Techniques and Technology Volume 1, Module 1–51, CNES, Cepadues Edition.

[16]Elements of Spacecraft Design, C. Brown, AIAA Education Series.

[17]Estimating and Budgeting Weight and Owner Contingencies for Spacecraft Design, ANSI/AIAA Publication G-020-1992.

Table 10.2. Mass of Propellent in Typical Satellites.

Category of satellites	Wet mass	Propellant mass	Propellant mass %
Communications	982 ± 550 kg	166 ± 180 kg	17 ± 7 %
Earth observation	1,475 ± 661 kg	63 ± 66 kg	4.3 ±2.4 %
Mini-satellites	140 ± 45 kg	70 ± 92 kg	50 ± 17 %

Table 10.3. Typical Satellites Mass Allocation per Subsystem.

Dry mass allocation in %	Communications		Meteorological		Planetary
	Brown	Wertz	Brown	Wertz	Brown
Structure	21	21	20	19	26
Thermal	4	3.6	3	2	3
Attitude control	7	7	9	4.5	9
Power + Harness	29	32	24	24	26
Propulsion	7	3.8	5	6.1	13
Telecom		4.8	4	3.5	6
Data handling	4		4		6
Payload	28	27	31	35	11

Table 10.4. Maturity Factors.

AIAA recommended minimum weight contingency %												
	End of phase 0			End of phase A			End of phase B			End of phase C		
Mass category	1	2	3	1	2	3	1	2	3	1	2	3
0–50 kg	50	30	4	35	25	3	25	20	2	15	12	1
50–500 kg	35	25	4	30	20	3	20	15	2	10	10	1
500–2,500 kg	30	20	2	25	15	1	20	10	0.8	10	5	0.5
Above 2,500 kg	28	18	1	22	12	0.8	15	10	0.6	10	5	0.5

In that table, class 1: is for a new spacecraft; class 2: for a next generation spacecraft based on a previously developed family, which expands in complexity or capabilities within an established design envelope; and class 3: for production-level development based on an existing design for which multiple units are panned and where a significant amount of standardization exists. The phases are as per Figure 1.6 and as per Section 6.3. The earliest mass estimations of some satellite components are consistently grossly underestimated: secondary structure and electrical harness are two examples of this. In addition to maturity factors, the satellite's design must include an overall reserve with respect to the maximum available mass injection capability of the selected launcher. The provision of comfortable maturity and safety margins is vital because mass estimations in the early design stages are always optimistic and because the real limiting factor for the satellite's overall mass (the capability of the selected launcher) cannot easily be changed. The need to change the initially selected launcher due to an increase in satellite mass would have a severe impact on the mission's overall budget and would jeopardize mission approval.

Advanced structural materials are lighter than conventional metallic ones. Carbon fiber, for example, is now a fully evolved technology and is no longer considered risky. The low thermal expansion coefficient of carbon fiber will also minimize thermoelastic deformations on satellites employing it, though it always requires metal inserts and special structural arrangements to avoid distortions and stresses when connected to secondary structures, instruments, or electronic boxes with different thermal expansion rates. This means that any reduction of mass in the primary structure will be partially offset by increases in masses in other parts of the satellite. The mass of connection elements such as nuts, brackets, inserts, and nodes is always larger than predicted and is very difficult to estimate in the early stages of the design, but the mass of all connecting elements is often comparable to the mass of the structural portion of the primary structure.

Some launchers require not only limiting the total mass of the satellite at launch, but also they specify a limited range of acceptable location for the center of masses of the satellite. The need to place the center of masses within this required range will also drive the configuration of the satellite. If the mass of the satellite is below the launcher acceptable limit, it will be possible to add balancing masses until the satellite center of masses is inside the specified range. If this is not the case the configuration of the satellite should be redesigned to make the concept compatible with both the total mass and the location of the center of masses.

10.6.2. Heat budget

In order for the satellite's components to function properly, the satellite must be kept within specified temperature limits. Satellites are divided into thermal enclosures and in each of these the temperature must be kept within the limits specified for that enclosure. To do so, the heat generated in each enclosure must be radiated from the components located within it into space at a temperature that ensures the maintenance of its specified temperature range. This balance between thermal production and thermal radiation must be guaranteed during all potential phases of the mission, including safe mode (Section 9.5). For each thermal enclosure this heat budget shall be balanced by providing a suitable radiative surface of adequate size with respect to the existing field of view toward space. The architect is responsible for designing the overall configuration. The thermal specialist will determine the size of the radiators, their thermo-optical properties and details concerning the internal thermal paths. The number of thermal enclosures depends on the configuration and on the need to provide clear interfaces between the different parts of the satellites. As a rule, each instrument is its own thermal enclosure and has its own radiator or radiators as well as its own heaters. In the case of a satellite divided into a service module and a payload module, each module is designed to be a separate thermal enclosure with a specified maximum heat transfer between them.

The working temperature for satellite subsystems is defined by guidelines for the preliminary design. General-purpose electronics units located inside the satellite are designed to operate within a 0–40 °C envelope. The batteries require narrower thermal margins requiring a 0–20 °C range. Components located on the outside of the satellite must be able to work within the wider temperature margins the exterior environment inevitably demands; solar arrays need to deploy in the −40 to +60 °C range and withstand −100 to +100 °C. These ranges, however, are "typical guidelines"; a mission's specific circumstances might require exceeding them; a satellite designed to rendezvous with Mercury or Uranus will need to operate in the especially hot or exceedingly cold thermal environments of these planets' neighborhoods.

Some instruments have much more demanding temperature control requirements. The ideal is for them to operate in orbit at, or near to, the temperature at which they were built and tested on-ground. These demanding requirements have implications not only inside each instrument but also outside them. To avoid thermoelastic distortions between different instruments pointing at the same observables, satellites carrying many instruments need to control the structure supporting them within a very narrow range around the standard ambient temperature of 20 °C, for example, 18 °C to 22 °C. The heat budget interacts with the power consumption budget because a radiator requires heating in cases when the visibility conditions change, such as during eclipses or in nonnominal attitude periods. Heat budgets require the use of adequate safety margins. The qualification testing of satellite components requires that they be able to operate in a temperature range including qualification margins to provide thermal safety margins and also to take into account thermal modeling uncertainty. As heat is produced by electrical power this margins will increase power production and the size of thermal heaters.

10.6.3. Power production budget

In traditional space engineering the power budget is normally a power consumption budget that allocates the power produced by the satellite to all the components that require it. Yet a power consumption budget of necessity implies the existence of a power production budget, upon which it depends. That is, a budget comprising the set of design decisions that ensures that the size, location, movement capabilities, and illumination conditions (including eclipse durations) of the solar array are enough to produce the required power and energy. All the main design decisions related to the fulfillment of this power production budget are satellite configuration decisions and will be treated here. The "conventional" power consumption budget will be studied in Section 11.6.1. The solar array is one of the most conspicuous components of the satellite structure. The solar array must be large enough to produce sufficient energy for the satellite to operate in all foreseeable conditions, and the spacecraft's configuration must be able to accommodate the required solar array size. The three crucial factors for balancing the power production budget are:

- That the solar array be positioned for maximal exposure to sunlight
- That the solar array is able to track the Sun's unavoidable changes of position relative to the satellite
- That the solar array possesses adequate area for energy production while illuminated, including the energy needed for battery charging throughout the mission's lifetime

The factors governing solar array size were discussed extensively in the previous section. Sun visibility and solar array geometry is thoroughly described Section 9.7.

The final size of the fully deployed solar array will depend on the satellite's stowed configuration, which in turn depends on the launcher size and the solar array's deployment concept. Solar arrays are generally standard components with clearly defined shapes and whose power delivery capabilities can be increased by providing additional panels within the same basic design while fulfilling the volume constraints presented by the launcher fairing. If both the design of the solar array and the launcher fairing volume permit, providing further power is as simple as adding one or several more panel to the same solar array design. Nevertheless, there are limits to the number of panels that a given design can carry, and to the number of panels and panel sizes that can be accommodated inside a given launcher fairing envelope. If these limits are reached, the addition of supplementary solar array area becomes very problematical because it will require either major changes to the design of the solar array, the satellite, or the use of a different launcher. This might be feasible depending on the specific situation, and the provision of energy would become a major design limitation. GOCE (Figure 10.14) is a good example of mission design limited by power availability. The orbit and configuration allows locating all the solar cells on a flat fixed surface that is roughly perpendicular to the direction of the Sun, thus optimizing electrical energy production. But GOCE's mission objectives rule out the use of mechanisms and its solar array shape and size are strongly limited by the space available within the selected launcher fairing. This limitation is absolute, and GOCE's orbit altitude, and hence its ultimate scientific performance, is a function of the launcher fairing size.

10.6.4. Alignment budgets

Alignment budgets are true complex budgets involving many aspects of the end-to-end space system – ground and space – and Chapter 9 has covered that area extensively. Section 9.8 introduced alignment, Section 9.10.6 analyzed pointing, and Section 9.10.8 analyzed co-registration. These budgets are related to the correct pointing of the instruments and of the attitude reference frames. However, absolute and relative pointing are also affected by configuration design decisions, which this section will examine. A list of all the error sources influencing alignment was given in Section 9.10.5. The configuration related error sources listed there include:

- Assembly and metrological test errors of structures and mechanisms
- Irreversible and permanent changes in alignment between the ground and the space environment
- Thermoelastic deformation produced by in-orbit thermal variations, outgassing or moisture desorption in composite structures, thermal gradients and by thermal expansion anisotropies

The use of highly accurate manufacturing and assembly procedures can reduce manufacturing and assembly errors but higher accuracy manufacturing is always associated with higher costs and a balance must be maintained between manufacturing accuracy and cost. Test equipment, such as metrology tools, will also produce its own errors and, as always, higher testing accuracy will incur greater costs. The realistic and accurate test of large deployable structures is especially difficult and expensive. The difficulty of designing proper test jigs for the functional testing of large deployable structures was touched on in Section 10.5.2. The test jigs depicted in Figure 10.40 is clearly complex and expensive. The jigs would become even more expensive if they were required to offer on-ground determination of the in-orbit shape of a large, complex deployable structure. In missions that feature large deployable structures, it is essential to provide an early detailed analysis of test methods that is detailed enough to provide credible estimations of testing costs and test performances including the expected final deployment accuracy. Strong structures with sufficient rigidity and strength safety margins will reduce or eliminate any irreversible changes in shape produced by the launch or deployment load environment. Metals do not change shape in a vacuum but organic materials, due to outgassing of water vapor, do.

Thermoelastic pointing errors can be minimized in three ways: by controlling the temperature of the satellite to minimize space and time variations; by using materials with low thermal expansion coefficients; and by avoiding mixing materials that have very different thermal expansion coefficients. Any alignment effect that can be properly analyzed can be eliminated as far as the quality of the analysis allows it; for example, the difference between in-orbit and on-ground satellite shape can be analyzed using structural analysis techniques and using the methods of comparisons between on-ground and in-orbit alignments, which were described in several sections of Chapter 9. The resulting results can be used to estimate – and compensate – the alignment errors induced by the lack of perfect knowledge of the deployed in-orbit satellite shape. Thermoelastic analysis can also be used to analyze the elastic variations in shape that the satellite will experience in orbit. The calculations can be used to compensate for these effects, so that the only remaining alignment knowledge error will be the uncertainty errors in the quality of the thermoelastic analysis. As in the case of permanent pointing errors, in-orbit validation of actual alignment can be used to verify and improve the quality of the thermoelastic analysis.

10.6.5. Volume budget

Mass budgets are discussed extensively in the literature of space system design; volume budgets are not. A mass budget document containing entries for the different masses is produced and updated throughout the process of design, manufacturing, assembly, and testing of any space mission; but no such document for volume budgets exist. Yet volume budgets are just as vital as mass budgets, and in the case of LEO missions are even more essential. Volume budgets are implicit in a satellite's blueprints as well as in its computer generated three-dimensional models. These models represent – and implicitly allocate – volumes and dimensions to the different components of the satellite. This volume budgeting becomes explicit when dimensions and envelopes are allocated to the different instruments and to any other components that require room on the satellite and that will compete for volume and visibility with all the other components around it.

The apportioning of this overall volume will be driven by the conflicting needs of the different instruments and subsystem components both inside and outside the satellite. All the section of this chapter discussed allocation of volumes and provided examples of satellite configurations that are adopted to optimize volume use. It is the responsibility of the mission architect to apportion the overall volume to the different components of the satellite.

Chapter 11

The Operational Data Flow Domain

Keywords Operational Data Flow, Satellite Ground Segment, Satellite Operations, Satellite Mission Operational Phases, Data handling, Satellite Autonomy, Systematic Operations, Interactive Operations, Power Budget, Link Budgets, Data Acquisition Delay

This domain was introduced in Section 7.1.4. This chapter starts with a brief introduction to the elements and components included in this domain that are related to the space "electrical space engineering" disciplines and with on-ground satellite operations. It continues with the analysis of mission phases and the different operational situations that they produce. The chapter continues with a section that studies this domain by analyzing the design decisions made for four space missions:

- The Earth magnetic field mission, Cluster; which features a constellation of four moderately complicated satellites with moderately complex operations and a low level of autonomy.
- The comet and asteroids rendezvous mission, Rosetta; which features a complex satellite with complex, autonomous operations.
- The global land and ocean operational mission, Sentinel-3; which features a complex satellite with systematic, simple, and autonomous operations.
- The Earth observation mission, SSTL Disaster Management Constellation (DMC); which features a constellation of six simple satellites, but with interactive operations.

These missions have been chosen to provide a rich selection of data handling scenarios: featuring both complicated and simple, and autonomous and nonautonomous operations. The last two sections analyze the allocation of functions and budgets included in this domain.

Figure 7.1 depicted the interrelations of this domain with the other domains:

- The on-board data handling is driven by the needs of controlling and monitoring the instruments and the satellite as a whole.
- The satellite orbit and the location and number of the ground stations will define: the ground to satellite contact patterns and the resulting opportunities for dialog between the satellite and their ground operators. This will drive the design of operational data flow.
- The nature of the events being observed – predictable versus unpredictable, fast versus slow—will define the way the observations should be organized and the resulting complexity of the instruments and satellite operations.

M.A. Aguirre, *Introduction to Space Systems: Design and Synthesis*, Space Technology Library 27, DOI 10.1007/978-1-4614-3758-1_11, © Springer Science+Business Media New York 2013

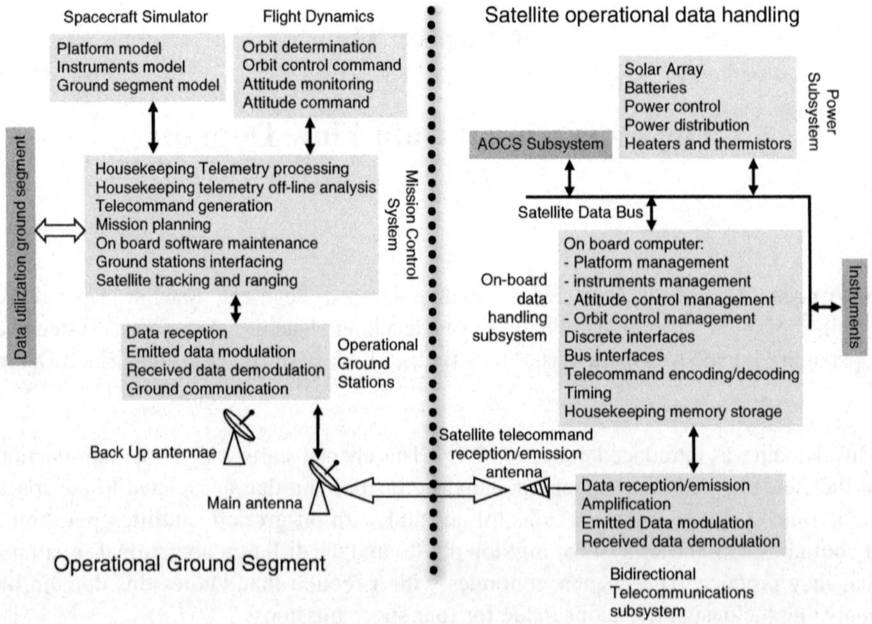

Figure 11.1. Operational Data Flow Design Domain Components and Connections.

Figure 7.4 described the end-to-end information flow of a space mission. That figure divided the flow into two domains: operational and instrument output domains; and each of these was divided into two parts: ground and space. Figure 11.1 is an expansion of the operational data flow aspects of Figure 7.4. Figure 11.1 also illustrates the connections to satellite and ground components that are discussed in other chapters: connection to the instruments, to the attitude and orbit control subsystem and to the instrument data utilization ground segment. Figure 5.8 of Section 5.6.1 divided mission effectiveness parameters in three classes related to: data quality, data quantity, and system responsiveness. The operational data flow deals mainly with system responsiveness.

The power subsystem will be discussed in Section 11.1.1, the on-board data handling will be discussed in Section 11.1.2, telecommunications in Section 11.1.3, the operational ground stations in Section 11.2.1, and the mission control in Section 11.2.2.

11.1. In-orbit components involved in the domain

Figure 11.1 provides the most important functions performed by each one of the in-orbit functional blocks of this domain. In orbit, this domain includes three classical subsystems:

- Power conditioning and distribution
- On-board data handling with the corresponding data buses and other data links necessary for the operation of the satellite
- The satellite's telecommand and telemetry communication subsystems

These three subsystems will be briefly described in the next headings.

11.1.1. Power

The power subsystem produces, regulates, stores, and distributes the electrical energy needed by the satellite. The power subsystem also directly controls power provision to heaters, pyro initiators, and other key elements that must be able to operate even when major components of the satellite are disabled.

A large majority of satellites produce electrical energy using the Sun. Solar cells of Si or GaAs mounted over solar panels are employed for this. The solar array frequently possesses a system for orientating it toward the Sun; this requires a dedicated mechanism. Some missions flying far from the Sun use the thermal energy provided by the decay of radioactive material; this is the case of Ulysses, depicted in Section 10.3.2. Short-lived probes or sub-satellites can use a battery as a single source of power.

Eclipses produce variations in solar illumination and in the corresponding production of power. The switching on or off of the satellite's various components produces varying power demands. To smooth out energy production and demand, the power subsystem stores and consumes electrical power from rechargeable batteries. There are several types of batteries available: NiCd, Hydrogen, and Lithium. Currently, Li batteries are the most widely used. Dedicated electronic units control the delivery of energy to the satellite, as well as the charging and discharging of the batteries. Payloads requiring high-pulsed power demand, such as a SAR, require particular care in the design on the power subsystem and harness, providing adequate capacitance to avoid sending current ripples to other electrical equipment.

The power subsystem also includes the harness needed to provide electrical power to all the components of the satellite.

The fundamental design requisite is the power demand profile, that is, which satellite components are operating and when. The power demand profile is matched by the power production profile, which depends on the solar array size and orientation, the frequency and the duration of eclipses, and the size of the batteries. Solar array size limitations have been discussed in Sections 10.1.4 and 10.6.3.

Power voltage and voltage ranges must also be selected. Normally, small- and medium-size satellites use 28 V, while large satellites use higher voltages. Higher voltages reduce losses in transmission but require the implementation of voltage reduction devices, DC-DC converters, for most electronic components.

As far as power distribution is concerned, a choice between regulated and unregulated buses must be made. Medium-size satellites normally work at 28 V if regulated, and in the range of 23–32 V if unregulated. Regulated buses simplify the design of satellite components and complicate the design of power subsystem. The situation is the reverse in unregulated buses. Earth-observing satellites normally use unregulated buses; science missions normally use regulated buses.

There are two main approaches for controlling solar arrays: peak power tracking and direct energy transfer. Peak power tracking gives a reduction in the size of the solar array. Direct energy transfer reduces the complexity of the power distribution electronics.

Solar array size depends on the solar illumination conditions, cell characteristics, mission life, and temperature; cooler batteries are more efficient.

Battery type, capability, and maximum allowable depth of discharge will be determined by the availability of qualified or about-to-be qualified equipment, the mission life and the corresponding number of charge/discharge cycles.

11.1.2. Satellite data handling

The heart of the satellite data handling subsystem is a general-purpose digital computer, which monitors and controls the satellite. In addition to the central computer, this subsystem may include other computers for performing specialized functions that include digital signal processors for the handling of the information produced by the instruments, a dedicated computer for controlling the attitude of the satellite, or a simpler, higher-reliability computer that can take control of the satellite in special emergency situations. To simplify interfaces, specialized computers can also be used for the monitoring and control of specific parts of the satellite. This is the case for multi-instrument missions that often have a specific computer to deal with the control and monitoring of the instruments. The control and monitoring of the satellite requires the storage of information; in the past, tape recorders were used, today solid-state digital memories provide this function.

To interchange information between the satellite's components, one or more data transfer buses and their corresponding bus controllers are necessary, depending on the requirements. As in the case of computers, general-purpose data buses like the 1553 or CAN are used. Special arrangements are provided to deal with functions beyond the capabilities of general-purpose buses, including the transfer of large amounts of data produced by some instruments, or to provide highly interactive real-time data transfer as needed by the satellite's attitude control. Discrete, point-to-point interfaces will also be necessary, for example, analog signals to command or to monitor electromechanical actuators. Some type of central clock will provide the mission's time reference.

The data handling subsystem also includes all the software required to perform all the necessary functions and to run in all the components mentioned above. The on-board software will include the software to perform the basic operations and the mission-specific software

Space missions can last a long time and the space environment is harsh, demanding high reliability for satellite hardware. For these reasons, demonstrating that an electronic digital component is adequate for this environment takes time, requiring a dedicated space qualification program, and therefore in-orbit computers are of older design and lower performance than their corresponding and more familiar ground-based equivalents.

The design decisions are driven by the complexity of the command and control functions that the satellite must perform, which are driven in turn by the number and operational profiles of the instruments and of the mission itself. The most important decision is the level of mission autonomy to be implemented as this will drive the number of operational data flow contacts between the ground and the satellite. In nominal conditions, the satellite must be able to operate without ground contacts for days or even weeks. At the same time, the satellite must be able to contact the ground as rapidly as possible in cases where an unexpected event has occurred.

As far as monitoring is concerned, it is also essential to determine how often the satellite's status and how many functional parameters are to be monitored and sent to the ground. This decision should be taken after considering in detail the failure modes of all the satellite's components and how they shall be monitored in order to differentiate between them. This information will be extremely valuable in case of a failure, as it will ease the designing of a recovery strategy for the satellite, allowing a successful continuation of the mission. This set of the satellite's functional parameters that define its status and that are sent to the ground is called "housekeeping data."

Modes have been already introduced in Section 9.5. It is recalled here that during its lifetime, the satellite will be confronted with widely different operational situations that require totally different approaches to how the satellite behaves. These are the satellite's operational modes. In the previous paragraph we identified two of these modes: the nominal one and an emergency mode when the "back-up" computer can take over. The satellite will also require a special mode for the period immediately after the launcher ejects it. In this mode the satellite will progressively wake up, maneuvre, and change configurations to reach its final nominal status. The execution of orbital correction maneuvers also requires a dedicated mode. The transition between modes can be performed either by ground command or automatically by the satellite.

The number and function of the computers on-board is another fundamental design decision. The satellite needs a central computer to control it in nominal operational circumstances, though supplementary specialized computers can also be used. A redundant computer, able to take over the main computer in case of failure, is also required. This computer must be simple and extremely reliable, and must be capable of controlling the limited number of functions absolutely necessary for the satellite's survival. Sometimes ground assembly and verification needs determine the number of on-board computers needed. A satellite that is often divided into two halves, one for the instruments and one for the satellite subsystems, would be easier to assemble and test if each half possesses its own intelligence. It must be decided what computer models and solid-state memory are going to be used. This selection is strongly influenced by the relative scarcity of adequate hardware. Due to the harsh radiation environment of space, achieving confidence in the adequacy of digital hardware for space use is not an easy task. In the USA, the power PC-based, radiation resistant ATIM RAD-6000[1] has been used on many missions; in Europe, the SPARC-based LEON[2] performs an equivalent role. To ensure added design robustness, the software in the redundant computer should be independent of the software in the main computer, avoiding the use of the same algorithms, and ideally supplied by an independent contractor. This is often difficult to realize in practice.

The number and selection of the data buses to be used is also crucial. In some cases a single bus, for example, 1553 or CAN bus, will be sufficient for all the satellite's needs. In other cases different busses will be needed for different purposes, for example, one bus for general control, one for high data throughput instrument output, data control, and another one for highly interactive attitude control hardware data interchange.

It must also be decided how to consider software: as either a separate subsystem under a central authority or divided into as many parts as the components that need software. A centralized approach will reduce the problems of interfacing between the different software modules but it will create interface problems between each software module and the hardware component it interacts with. This interface is especially complex because a component, which includes hardware and the software to operate it, needs to be approved as functional, and is paid for only after it has

[1] 32 Bit Radiation Hardened Computers for Space, J. Nedeau, IEEE Aerospace Conference 1998.

[2] Leon-1 Processor, Jiri Gailser, DASIA Conference, ESA SP-655, 2008.

demonstrated that it fulfills its requirements. This certification process and the allocation of responsibilities in the case of malfunction are especially complicated when different suppliers produce the hardware and the software.

The fundamental components of this subsystem (computers, clocks, memory, and buses) are selected from a normally small range of already-qualified or about-to-be qualified options that the specialist designer provides to the architect. Many factors influence the selection of the basic data handling architecture: from operational needs to verification needs; and from computer availability to the supplier's previous experience. Nevertheless, the number of basic possible architectures is limited, and there is a strong basic similarity between the data handling architecture for missions with widely different operational needs. The avionics architecture of the complex and expensive Sentinel-3 (Figure 11.9) and of the simple and cheaper SSTL-DMS (Figure 11.11) are strikingly similar.

11.1.3. Telemetry and telecommand data communications

The telemetry and telecommand data communications subsystem allows the interchange of information between the satellite and ground.

Antennas are necessary for the emission and reception of signals. The characteristics of these antennas will depend on their beam widths and frequencies. Antennas for telemetry and telecommand should be omnidirectional so they can provide satellite-to-ground contact when the satellite is in a nonnominal attitude in case of attitude failure, or if the satellite is agile during normal operations. Omnidirectional antennas are small, provide low gains, and are only adequate for receiving and transmitting at modest data rates. To cope with higher data rates, the satellite will also require antennas with larger dimensions and higher gains. Very high-gain antennas are appropriate for high data rate communication and for deep space missions. The signal shall be amplified and modulated at the frequency required. This will be performed by specialized components dedicated to these two functions. The characteristics of these units will also depend on the data rate and the frequency bands.

The uplink and downlink signals will be digitally processed: encoded or decoded and encrypted or decrypted. Dedicated digital processing modules perform these functions. With regard to the downlinking of information, these modules receive the digital information, produced by all of the satellite's components, merge them into a single data stream, and format them according to specific standards. As far as the uplink of commands is concerned, these modules receive the commands reaching the satellite and transform them into a list of commands that are validated as legitimate and then send them to the on-board data handling subsystem for execution. In every case the digital processing of the uplink and downlink data streams has three aims:

- To protect the data stream against possible errors in the transmission of information. This is the field of the error detection and correction codes.
- To allow the merging and separation of information by providing labels that indicate what information is related to what satellite function. This is the field of the packing telemetry codes.
- To encrypt the information so that only legitimate users can use it. This is the role of the encryption codes.

Each one of these functions is performed by its dedicated hardware.

The design decisions are driven by the necessary frequency of contacts between the ground and space, which is derived from the level of autonomy chosen by the mission.

The design decisions for this subsystem are strongly constrained by regulations. The International Telecommunication Union (ITU) regulates the frequency for all RF spectrum users. Therefore also the radio frequency uplink and downlink of information from ground to space will have to be implemented following the ITU Radio Regulations. These regulations allocate parts of the RF spectrum to satellite users. A subset of the space science bands is allocated to deep space science, defined as satellite operating at more than 2 million kilometers. To avoid interference, ITU also regulates the maximum power density allowed to arrive on Earth inside the allocated bands and, with much tighter requirements, outside them (unwanted emissions). The final result is that the range of possible design is limited by these regulations.

The power of the amplifiers needed for in-orbit data reception and emission will be directly derived from the selected communication speed, the chosen ground stations, and the chosen satellite antenna's gain. The power needed must be high enough to provide adequate margins to the communication link, but low enough to fulfill ITU regulations.

As an example the bands allocated with primary status to science and earth observation services are provided in the next table (Table 11.1).

There are also specific bands dedicated to meteorological or communications satellites.

Table 11.1. Frequency Bandwidth Allocation for Science and Earth Observation Satellite Communications (Radio Frequency and Modulation, ECSS-E-50-05, ESA 2003).

Primary allocated frequency bands			**(GHz)**
	Service	**Direction**	**Frequency**
S-band	All operations	Uplink	2.025–2.110
	Deep space operations	Uplink	2.110–1.120
	All operations	Downlink	2.20–2.29
	Deep space operations	Downlink	2.29–2.3
X-band	Deep space	Uplink	7.145–7.190
	Space science	Uplink	7.190–7.235
	Earth observation	Downlink	8.025–8.4
	Deep space	Downlink	8.4–8.45
	Space science	Downlink	8.45–8.5
Ka-band	Earth observation	Downlink	25.5–27
	Deep space	Downlink	31.8–32.3
	Deep space	Uplink	34.2–34.7

Coding schemes are also the subject of regulations by the Consultative Community in Space Data Systems (CCSDS).[3] Algorithms for the three types of encoding mentioned in the previous paragraphs must be selected: codes to improve reception, codes to interleave different types of data, and, if needed, codes to encrypt the information.

The most important design decisions are:

- What bands, within those permitted, will be used for uplink and downlink?
- Whether the information about satellite's status (housekeeping) is going to be mixed with the output of the payload, or if it will be downlinked using distinct frequencies.

The frequency band selection will be driven by the existence of satellite hardware and ground stations adequate for the different bands. Higher frequencies allow higher data rates but the availability of on-board and ground hardware is more limited, while higher frequencies suffer stronger attenuation by water vapor and rain. For the same gain, higher frequencies also reduce the size of the antennae. This can be an advantage for satellite accommodation.

The distance from the satellite to ground and the quantity of information produced by the instrument and monitoring of the satellite drive the decision to use one or two links. Earth observation satellites produce very high amounts of data and normally use two downlinks: one omnidirectional, moderate speed data link for satellite monitoring and command, and a second one, optimized for speed, for the instrument output. In the case of space science missions the ratio between the data produced by the satellite and that produced by the instrument is smaller; and in addition, contact with these satellites often takes place at great distances from the Earth. This promotes a single, very powerful data link that blends all the data produced by the satellite. Nevertheless, space science satellites are also moving toward separating these two functions, because of:

- The need for increasingly sophisticated instruments with expanding data production rates is pushing in this direction. One example will be JWST, as described in Section 7.2.1.
- Lower data throughput, higher pointing flexibility data links for critical events such as planet swing-bys, allows better command and monitoring. The multiple data links of Rosetta, described below in Section 11.4.2, are examples.

On the other hand this double approach requires the implementation in the satellite of two different communication channels. To avoid this it is also possible to implement one single frequency but working at two speeds: a lower speed for critical operations and a higher speed for normal operations, which includes also instrument data downlink.

The accommodation of the antenna is determined by its frequency and gain. Antennas for satellite operations should have a wide area of coverage; ideally there should be two antennas, each one with half-hemispherical coverage. Wide-field-of-view low-gain antennas are small and easy to accommodate. High-gain antennas, on the other hand, are large and might be dominant elements in the configuration of the satellite. The Rosetta configuration (Figure 11.6) and the Ulysses configuration (Figure 10.17) are good illustrations of this.

---

[3] http://public.ccsds.org/default.aspx

11.2. On-ground components involved in the domain

The aim of this book is to study the early design stages of space systems. To be able to provide a credible estimation of the overall mission cost, schedule, and risk, the satellite must be designed with a fairly high degree of detail. Nevertheless, satellite operations and ground segment physical facilities, like large ground antennas or mission control facilities, are almost never designed specifically for one particular mission. Ground control facilities are almost always already existing infrastructures requiring only adaptations for the particular needs of the mission under consideration. Moreover, these facilities are easier to implement than the space elements. Satellite operations are also a domain where standardized mission preparation and control methodologies are applicable. This would imply the use of multimission infrastructures, including common personnel, who can transfer know-how from one project to the next. For all these reasons, during the design phases prior to program approval, it is necessary to perform a thorough functional design of the ground segment, but it is not necessary to perform a physical design. Therefore, this section will be limited to a discussion of functional design decisions, while the physical implementation will remain uncovered. At any rate, the earlier the functional design of the ground segment is established, the better. A spacecraft designed without due regard to the ground elements is a deficient design. A correct early understanding of the end-to-end data flow and a correct apportioning of functions between space and ground is compulsory for a proper mission design.

11.2.1. Operational ground stations and data-relay satellites

These components relay the information to and from the satellite to the mission control centers. The data received from the satellite must be demodulated before being sent to the mission control center, and the commands received from the mission control center have to be modulated before sending them to the satellite. Considering the need for speed and for expert analysis of the information about the satellite's status, which is undertaken at the mission control center, the components under discussion act as fast conduits for the information, while experiencing only minimum delay or processing.

The fundamental design decision concerns the location and number of operational ground stations and the resulting time elapsed without contacts between ground and space. This decision will be driven by the operational approach and the responsiveness requirements regarding the mission's allowable reaction times.

Data-relay satellites allow much longer and more frequent contacts, almost continuous in cases where a sufficient number of data-relay satellites are available. This is a very powerful feature, which greatly enhances the mission's capabilities. On the other hand, a data-relay satellite or a constellation of data-relay satellites is an extremely expensive infrastructure. NASA is unique in possessing an operational data-relay satellite constellation. ESA operates just a single data-relay satellite: the geostationary Artemis.[4]

Another important design decision concerns the degree of centralization of the mission's operations function. Mission operations are highly centralized. The mission operational ground stations are capable of sending commands that could endanger the

[4] http://www.esa.int/artemislaunch/

satellite. This is a heavy responsibility, so there is almost always a single mission control center.

The availability of data-relay satellites is extremely limited. Their availability, or more likely, nonavailability, will be a system constraint imposed on the designer. The selection of the number and locations of the operational ground stations will also be very limited as a result of the high level of responsibility involved in commanding the satellite. These stations shall be selected from the small set of stations directly owned by the mission's operating entity. In practice this means that there are few design decisions to be made in this area, because most of the components to be used already exist, and their employment will be another design constraint.

11.2.2. Mission operations control centers

The functions performed by the mission operations and control center are many and complex. They can be divided into three main groups: (1) mission monitoring and control; (2) determination of the orbit and attitude of the satellite; and (3) maintenance of the know-how needed for the correct operation of the mission. Mission monitoring and control includes all the activities necessary for the operation of the mission:

- Reception, processing and analysis of the operational information sent by the satellite, allowing its monitoring.
- Generation of commands as required by the satellite, both in nominal and in non-nominal situations.
- Mission planning is the generation of all the commands that shall be sent to the satellite. This is a complex function that requires close coordination with the instrument's data processing center in order to acquire data as requested by the consumers.
- Maintenance and updating of the on-board software as required by the mission's evolution. All missions, even fully successful ones, require frequent software modifications to optimize the satellite's current behavior in orbit. This requires both the creation and uplinking of the software modifications, and continuous monitoring of the current status of on-board software.
- The monitoring and control of ground-based communication networks and of the operational ground stations themselves, including commands to the ground stations for receiving information from the satellite or for uplinking commands to it at the right moment and in the required direction.

Mission operations require the ability to reconstruct and predict the satellite's orbital position and attitude. Methods for orbit determination and control have been discussed in Section 9.1.2. Traditionally, the ground stations have performed "ranging" that measures the changes of the frequency signal received from the satellite. These changes are a function of the velocity of the satellite with respect to the ground station and to measure them allow deriving the velocity. To date, these functions have been performed on the ground, but are now being progressively located on-board. Satellites located in low Earth orbit can use GPS receivers to autonomously determine the orbit. Specialized ground-based flight dynamics calculation, however, will always be necessary for checking the autonomous on-board performance, and for determining and commanding exceptional, high-risk maneuvers, such as major orbital changes.

In order to carry out mission control functions, understand the satellite's behavior, and avoid the uplink of commands or software modifications that might be detrimental to the satellite, the mission control center must maintain databases containing crucial information related to the satellite:

- Satellite documentation and simulated computer models of the behavior of instruments and platforms. This will facilitate scrutiny of the satellite's health status and the simulation of any anomalous behavior.
- Information and computer models related to the ground segment itself.
- Records of all operational data flow between the satellite and ground segment, including the status of the on-board software and all its modifications.
- Records of the on-board monitored parameters to identify their status and determine any potential long-term evolution of the satellite's hardware.
- Access to satellite hardware experts in case of problems.

Proper processing of the data produced by the instruments will require knowledge of the values of numerous satellite parameters, such as the temperature of detectors or the attitude of the satellite. These parameters shall be provided to the instrument data processing ground segment to allow for the correct processing and calibration of the mission's data products. The mission control center will include components for performing this function.

The fundamental design decisions regarding the mission control center are directly related to the fundamental functional allocations of the overall operational data flow:

- The need for supplementary ground stations for special circumstances: for example, at the beginning of the mission, in case of failures or during especially risky operations.
- The most appropriate mode of data acquisition for the events under observation.
- Satellite autonomy versus ground intervention. This decision will allow the determination of the level of service to be provided by the mission control center: continuous; exclusively working hours; or only in case of unexpected events.
- Centralized versus decentralized. As explained above, mission control is almost always highly centralized and the use of existing facilities, adapted according to need, is almost always mandatory.

The mission control center requires a heavy investment that nearly always exists prior to the start of the mission design process, and the system designer must be able to count on the ground segment and operations specialists in order to understand the features of the existing mission control center. Mission control centers are more frequently acting as permanent multimission infrastructures that evolve and adjust themselves to new missions while maintaining a central core. The use of standard procedures, i.e., ESA SCOS-2000,[5] facilitates this progressive adaptation.

11.3. Mission phases

Satellite attitude and orbit control modes, as well as overall satellite modes, were presented in Section 9.5 as part of the discussion about the orbit and attitude design domains. The present chapter, which deals with the satellite's operations throughout

---

[5] Lessons Learned from the Integration of Hifly and SCOS-2000 in NASA GMSC, T. Beech et al., IEEE Aerospace Conference 2005.

the entire life of the mission, will discuss the related concept of mission phases. Once the satellite is launched it will pass through a number of varying situations; it must initialize itself; it must be characterized; it has to nominally acquire data and regularly perform specific maintenance functions; and finally, it needs to be disposed of. In each of these situations, the functions to be performed, and the requirements of the satellite and the ground system components controlling it will be different. Each one of these periods constitutes a distinct mission phase.

11.3.1. Launch and early operations

During the launch and early operations phase the satellite is ejected by the launcher, acquired by the ground segment, and reconfigured until it adopts first a safe configuration and finally its nominal in-orbit configuration. The launcher releases the satellite, which progressively switches itself on. It must reconfigure itself by acquiring the correct attitude, deploying its solar arrays and other deployables, and must contact Earth to confirm that it is ready and waiting. This is a critical period because the satellite release is always accompanied by some residual tumbling, and before producing electrical energy the tumbling must be stopped, the correct attitude acquired and the solar array deployed. All these functions have to be performed as quickly as possible using electrical energy from the satellite's batteries. During this early phase intensive interaction between the satellite and ground is necessary. This is why a large number of ground stations are rented for this brief critical period, which lasts several days. Some satellites perform the most critical actions – for instance, the deployment of the solar array – under ground control. In other cases the satellite performs the critical actions autonomously. Autonomous operation is risky but minimizes the time expended by the satellite in an unsafe configuration. In some cases there is a need for dedicated prelaunch activities of limited duration when there may be specific payload or platform-driven constraints, for example, transfer from external electrical power to satellite battery power, which will restrict the satellite battery life if the satellite is powered during launch.

11.3.2. Commissioning of the satellite

This commissioning phase begins when the satellite reaches its nominal initial orbit with its nominal configuration. During this phase each satellite component is studied and characterized to determine its actual performance. The various instruments are switched on and characterized. The on-board software is updated with parameters to make optimal use of the current performances of the satellite and its instruments. Optimal performance during this phase requires the involvement of a large number of experts on both the satellite's subsystems and its instruments in order to analyze their actual in-orbit performance. Depending on the satellite's complexity and its actual performance in orbit, this phase can last from a few weeks to a few months.

11.3.3. Nominal operation

The nominal operations phase is the main phase of the satellite's life. The satellite is fully functional and fulfilling its expected aims. The duration of this mission phase will be as long as possible, but at least as long as specified.

11.3.4. Safe mode and other dormant modes

The satellite needs a special attitude and satellite mode in case of a major malfunction. To cope with this, the satellite switches off the instruments and as many electrical components as possible. The satellite can also disconnect part of its intelligence. The satellite remains in a special easy-to-maintain attitude, which ensures a safe situation that permits any identified problem to be analyzed and solved without time constraints. A dormant mode can also be used to keep the satellite safe during the long travel periods needed for planetary exploration. Once the satellite's problem has been resolved, or once it reaches its planetary target, the satellite can be returned to its nominal mode. The safe mode requires minimum contact with the ground and can be of long duration, and theoretically of unlimited duration.

11.3.5. Nominal orbit correction maneuvers

This mode could potentially perform orbital changes. It might require the adoption of a special attitude and the switching off of the instruments. Once the orbital maneuver is performed, the satellite would return to the nominal mode. This mode should be as short as possible.

11.3.6. Decommissioning and disposal

At the end of the satellite's operational life, and in order to avoid adding to the continual increase of dangerous orbital debris, the satellite's orbital parameters could be changed to an orbit low enough that atmospheric drag will eliminate it within a reasonable period of time. Where this is not possible, i.e., in geostationary satellites, the satellite is moved to a "graveyard orbit" where it will not be a danger to other satellites. Any excess fuel is vented to avoid risk of subsequent explosion. Introducing provisos for allowing disposal is expensive because it demands the performance of a complex maneuver at the end of the satellite's life, when its reliability is at its lowest. Nevertheless, decommissioning is becoming compulsory in order to reduce the growing danger of in-orbit collision with orbital debris. This phase is brief but complicated, because the satellite could be partially nonfunctional at the end of its operational life.

11.4. Examples of data handling architectures

11.4.1. Cluster

11.4.1.1. Mission aim

Cluster[6] is a mission whose aim is to study small-scale structures in the magnetosphere of the Earth. To achieve this, Cluster is made up of four identical spacecrafts that fly in a tetrahedral configuration. This allows an analysis of the magnetosphere's small-scale structures because the simultaneous measurements provided by the four satellites permits a separation among the magnetosphere's variations in both space

---

[6] ESA Achievements, BR-250, ESA 2005.

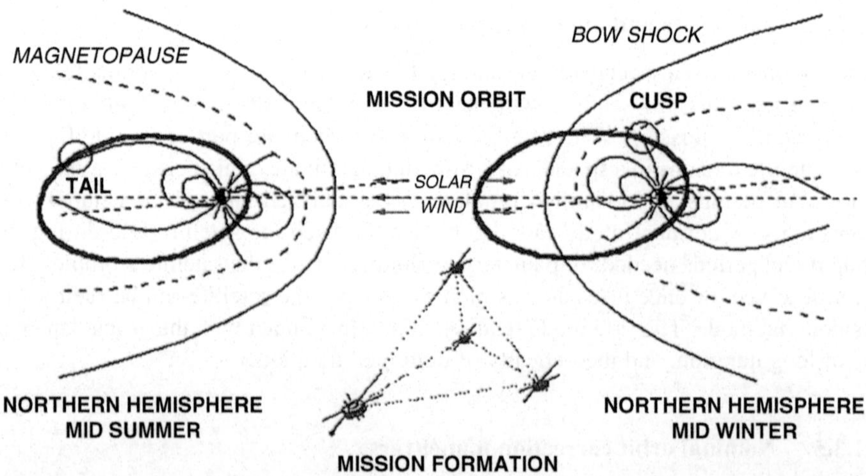

Figure 11.2. Cluster Mission Orbit (The Implementation of the Cluster II Constellation, J. Dow et al., Acta Astronautica 54 (2004) 657–669, Elsevier 2004). Allowing Testing of Different Parts of the Magnetosphere During the Different Seasons.

and time. The original mission life was 5 years. Cluster was launched in July 2000 and it remained in operation in 2009. The four satellites fly as a constellation along a highly elliptical orbit, with 19,000 km of perigee and 119,000 km of apogee; the orbital period is 57 h and the orbit inclination is 90°. The separation among the satellites varies between 40 km and 20,000 km, depending on orbital conditions. Cluster's orbit is essentially fixed in inertial space, so that in the course of a year the apogee moves from the tail region into the solar-wind region of the magnetosphere, thereby allowing all areas of the magnetosphere to be addressed (Figure 11.2).

The mission is now funded to continue operating until 2012.

11.4.1.2. Configuration

Each one of the four satellites is 1,200 kg, stabilized by spinning at 15 rpm and carry 12 "in situ" instruments for the detection of charged particles and magnetic and electrical fields. The total mass of each Cluster payload is 72 kg. The most sensitive instruments are located on long deployable booms to avoid electromagnetic noise produced by the satellite electronics.[7] In line with the spin-stabilized concept, the configuration is cylindrical, with the solar array located on the sides of the cylinder. The spacecraft's cylindrical design is shaped by the body-mounted solar array and also optimizes the fields of view available to the experiments, which are arranged around the rim of the main equipment platform on the upper side of the spacecraft. The height of the spacecraft's body has been minimized to make optimal use of the fairing volume offered by the launch vehicle (Figure 11.3).

11.4.1.3. Operational concept and mission phases

The satellite employs a single S-band set of components that is used for uplinking and downlinking and whose downlink includes both the satellite status "housekeeping"

[7] The Cluster Spacecraft, a Unique Production Line, Gustav Mecke, ESA Bulletin, ESA 1996.

Figure 11.3. Cluster Constellation Satellites Configuration (ESA Achievements, BR-250, ESA 2005).

and the instruments output data. Cluster's Mission Operations Centre is located in the European Space Operations Center (ESOC) in Darmstadt, Germany, with a single 15 m diameter S-band ground antenna station in Villafranca, near Madrid, Spain, and a Joint Science Operations Centre (JSOC) at RAL Didcot, UK, which establishes the observational strategy and provides it to the Mission Operations Centre in Darmstadt. Nevertheless, the 35 m diameter ground antenna located in New Norcia (Western Australia) is also used for critical passes.

The mission's initial phase, in which the four spacecraft are maneuvred to reach their final, highly inclined and highly elliptical orbits, lasted about 3 weeks. The final spacecraft configuration was then attained with the deployment of two rigid booms carrying the experiment sensors and four 50 m wire booms for electrical field measurements. The commissioning phase needed 10 weeks to properly characterize the satellites and their payload. The nominal science operations phase began at the end of the commissioning phase and has progressed until the present.

The observational strategy during the nominal phase was driven by the need for seasonal changes of observational strategy. The orbit is fixed in inertial space and the Earth's magnetosphere, which is determined by the Sun's orientation, which rotates once per year around the mission orbital plane. This implies that throughout the year, the orbit will move through different parts of the magnetosphere, hence requiring different observation strategies. This requires seasonal adjustments of the instruments' observing strategies and the volume of data production. As part of these seasonal adjustments of the observation strategy and in order to better study the magnetosphere, the relative distance between the four spacecrafts is varied every 6 months in a series of orbital corrections synchronized with normal orbit-maintenance maneuvres.

The single main ground station and the highly elliptical long-period orbit of 57 h combine to produce a ground control pattern in which a single long contact per day of around 28 h is shared between the constellation's four satellites. Cluster's satellites produce data at different rates, depending on their situation. When in contact with the ground station, the telemetry stream downlinks data in real time at 22 kbit/s, but also possesses a faster "burst mode" of 131 kbit/s for playback of previously recorded data. The Wide Band Data experiment, which operates during brief periods of each orbit, produces information in real time, and downlinks it at 220 kbit/s. On average, each satellite produces 0.8 Gbit per day.

The mission planning activities take place at different levels, each one having a different time horizon. The first level is long-term planning, which is based on a planning period of 6 months. The second level is the short-term operational planning, based on a weekly cycle. Both planning phases are created by interaction between the Joint Science Operations Centre at RAL and the mission operations center at ESOC. This interaction ensures that the science observation plan is compatible with the satellites' resources and positions.

It is worth mentioning that after 9 years of orbital life the spacecraft batteries are severely degraded, to the extent that during some eclipses some of the spacecraft are powered down and reactivated after the end of the eclipse. This is possible with Cluster since the spacecraft are spin-stabilized and their attitude is therefore inherently stable.

11.4.1.4. On-board subsystems

The power subsystem is made up of a fixed silicon-cell solar array covering the external cylindrical shape of the satellite, while five silver-cadmium batteries totaling 80 Ah provide power during eclipses, plus the necessary power control and distribution electronics.

The satellite on-board data handling OBDH subsystem is shown in Figure 11.4 and includes:

- A Central Data Management Unit (CDMU) with a central computer (CTU), a time reference Ultra Stable Oscillator (USO), and the command-decoding unit. The CDMU acts as the satellite's nerve center and directly controls the solid-state recorders (SSR), the data bus (OBDH bus), and the communications subsystem.
- A Remote Terminal Unit (RTU), which controls all experiments, the power subsystem, the thermal control and the satellite's attitude and orbit control subsystems.
- A solid-state recorder (SSR), that stores the data produced when the satellite is out of direct visibility with the single ground station. It is controlled by the CTU and it can store up to 5 Gbit of data.

The OBDH decodes and distributes commands, received by the telecommunication subsystem, which have been uplinked at a rate of 2 kbit/s, and acquires and encodes telemetry from payload and subsystem units. This telemetry is delivered either to the telecommunications subsystem for real-time transmission to the ground, or to the solid-state recorder (SSR) for later transmission. In order to reduce the duration of the downlink during the limited ground station visibility periods, dedicated high

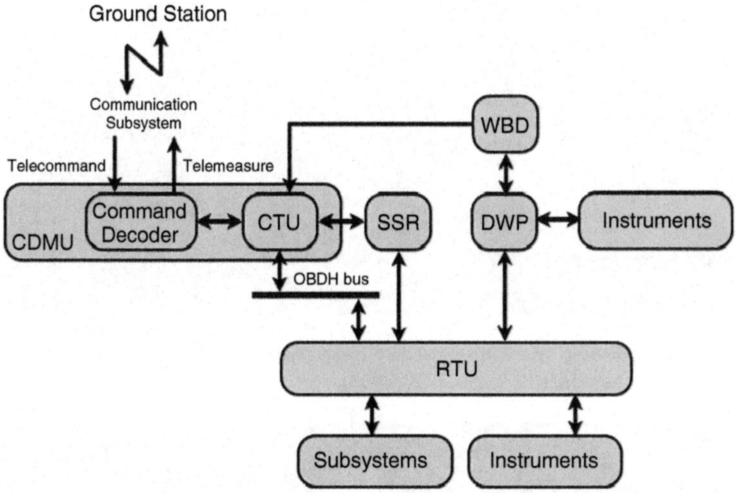

Figure 11.4. Cluster Avionics (The Cluster Spacecraft, a Unique Production Line, Gustav Mecke, ESA Bulletin, ESA 1996).

data rate interfaces are provided to the Wide Band Data (WBD) experiment. A dedicated digital wave processor (DWP) supports the real-time WBD experiment and other related instruments.

The S-band communications subsystem includes three low-gain antennas, a redundant set of transponders, a 10 W radio frequency amplifier, a distribution unit, and associated harnesses. Two hemispherical coverage, low-gain antennas are mounted on deployable booms attached to the upper and lower faces of the spacecraft. They ensure full coverage for uplinking and downlinking. A third antenna mounted on the lower side of the spacecraft is used until the in-orbit deployment of the lower antenna boom.

11.4.2. Rosetta

11.4.2.1. Mission aim

Rosetta[8] is a deep space science mission whose aim is to study the origin of comets, the relationships between comets and interstellar material and their implications with regard to the origin of the solar system. Rosetta consists of an orbiter and a small sub-satellite lander called Philae. If all goes as planned, Rosetta will rendezvous[9] with comet 67P Churyumov Gerasimenko in 2014. Rosetta will orbit and observe the comet, while the Philae lander is released from Rosetta to land on the comet's surface. In order to rendezvous with the targeted comet, Rosetta must fly a complex orbit, including one Mars and three Earth swing-by maneuvres to increase the satellite's speed. On its way to the target comet, Rosetta will also visit the asteroids Lutetia and Steins. The mission will last 10 years.

---

[8] Description of Roseta on ESA Bulletin and Brochures.

[9] Mission Operations for the new Rosetta, Paolo Ferri, International Astronautical Congress, IAC-03-Q.5.01, Bremen 2003.

Figure 11.5. Rosetta and the Sub-satellite Philae Approaching the Targeted Comet (ESA Achievements, BR 250, ESA 2005).

Figure 11.6. Rosetta Configuration Is Dominated by Its Long Solar Arrays and Large Parabolic Antenna. (ESA Achievements, BR 250, ESA 2005).

11.4.2.2. Configuration

The satellite is an inertially stabilized cubic box with the Philae lander attached to one of its sides. The satellite configuration is dominated by the large high-gain parabolic communication antenna for contacting Earth, and by the very large solar arrays needed to receive energy from the distant Sun. The satellite's mass is 3,000 kg, including 1,600 kg of fuel. Philae weighs 100 kg. There are pointing mechanisms for steering the solar array toward the Sun and the parabolic antenna toward Earth (Figure 11.5).

The payload of Rosetta includes four remote sensing instruments: one providing panoramic and detailed color images, and three spectrometers. There are also three instruments for in situ chemical composition analysis and four others for studying the dust and electromagnetic environment around the satellite. The Lander carries ten experiments, which will provide images and in situ data on the comet (Figure 11.6).

11.4.2.3. Operational concept and mission phases

Rosetta's mission operations center is in Darmstadt, Germany, with a primary ground station located in New Norcia, Australia, which features a large 35 m diameter antenna. For emergencies, the mission can make use of the American Deep Space Network's antennas, up to 70 m in diameter. Perth provides up to 12 h of coverage per day, though only around 9 h will normally be used. New Norcia is one of the three

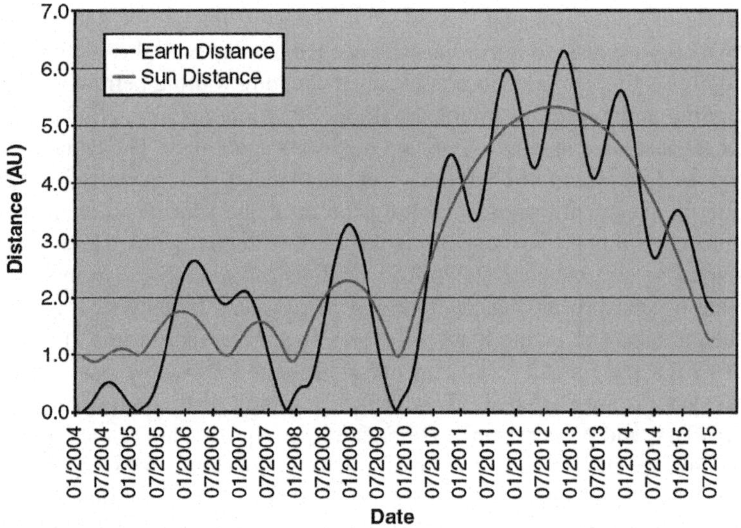

Figure 11.7. Varying Rosetta Satellites Distances to the Sun and to the Earth (Mission Operations for the New Rosetta, Paolo Ferri, International Astronautical Congress, IAC-03-Q.5.01, Bremen 2003).

European Deep Space Network stations. The second one is in Cebreros Spain and the third one – under construction in 2009 – is in Malargue, Argentina. There is also a Lander Control Centre and a Science Operations Centre for use by investigators. Given Rosetta's complex orbit, its operations are also complex:

- In order to reach its intermediate and final targets, the satellite's projected life is very long.
- The long distance to the final target obliges long periods of satellite hibernation in dormant mode.
- Hibernation will be interrupted by special high-activity events including orbital maneuvres, planetary swing-bys, and encounters with the two asteroids to be approached and studied during the satellite's journey.
- The mission's latter stages, including the release of the Philae lander and the orbiting of Rosetta around the comet requires high levels of activity.

The widely varying distances from the satellite to the Sun and the Earth further complicates the mission:

- The varying distance to the Sun changes the satellite's thermal and power environment. This compels changes in energy management strategies to cope with the varying electrical energy production and heat dissipation profiles.
- The varying distance from the satellite to the Earth requires the use of different types of communication links employing low, medium, and high-gain antennas, each of them specialized for specific distances and purposes.
- The long distance between the Earth and the satellite – up to 6 AU – entails long waiting times between the satellite monitoring, satellite commanding, and commands execution instants. Data needs up to 50 min to travel from the satellite to ground (Figure 11.7).

Rosetta's operations manage to be simultaneously interactive and autonomous. Interactive because ground intervention is needed to cope with the complex scenario of the satellite's life, and autonomous because the long waiting period between command sending and execution, coupled with the long hibernation periods, means that the satellite must have its own significant decision capabilities. The launch in March, 2004, and the first period of operations went as planned. The spacecraft, its payload, and the ground segment were successfully commissioned during an intense commissioning phase, split into two parts during the flight's first 6 months. The routine cruise phase started in mid-October, 2004. The satellite had already performed Mars and Earth swing-by maneuvers and observations. In between, the satellite was in hibernation. When in hibernation, the spacecraft spins once per minute while facing the Sun, so that its solar panels can receive as much sunlight as possible. Almost all of the electrical systems are switched off, with the exception of the radio receivers, command decoders, and power supply. Hibernation is required due to the large distance from the Sun reached by the spacecraft in its heliocentric orbit (up to 5.3 AU). Although the spacecraft has large solar panels (64 m^2 – 32 m tip to tip), the generated power at those distances does not allow to keep the spacecraft active. So it will be spun up in early 2011, and then most systems deactivated. A timer will switch them on back in January 2014, when the distance to the Sun will be below 4.6 AU. During the rest of the cruise the spacecraft is not hibernated, but kept in a low-activity mode and contacted once per week

When Rosetta reaches its final target, it will carefully orbit the weak-gravity comet at a distance of a few kilometers, keeping its instruments pointed toward it, for a period of 22 months. The 100 kg Philae lander will be ejected from Rosetta and navigate to a soft landing on the comet's surface, and will relay to Earth the information it gathers for as long as it manages to remain operative. Philae will be released from a distance of 1 km from its target and the descent will take 30 min. Philae will be totally autonomous, and plans are for the lander to remain alive on the comet for at least one week.

11.4.2.4. On board subsystems

Rosetta[10] possesses a power subsystem, a telecommunication subsystem, the attitude and orbit control subsystem, and the data management subsystem. A harness subsystem connects the individual units to the power and data handling units. The solar array area is 70 m^2 and measures 34 m tip to tip. Special low intensity, low temperature solar cells supplemented by Li-Ion batteries provide the spacecraft with electrical power.

The power is conditioned and distributed by the power subsystem via a regulated 28 V voltage bus. Any two of the four MA31750 processors are able to act as the satellite's brain, providing for all its functions, including attitude control, avionics control, and satellite and instruments data management. Data is stored in a 25 Gbit solid-state mass memory. A navigation camera provides optical target detection and can be used in a special mode to control one axis of the spacecraft during the fly-by of an asteroid. In proximity of a small body, like the asteroid Steins in 2008, images

[10] The Rosetta Misssion, D. Kolbe, R. Best, Acta Astronautica, Vol 41 Nos 4-10, Elsevier 1997.

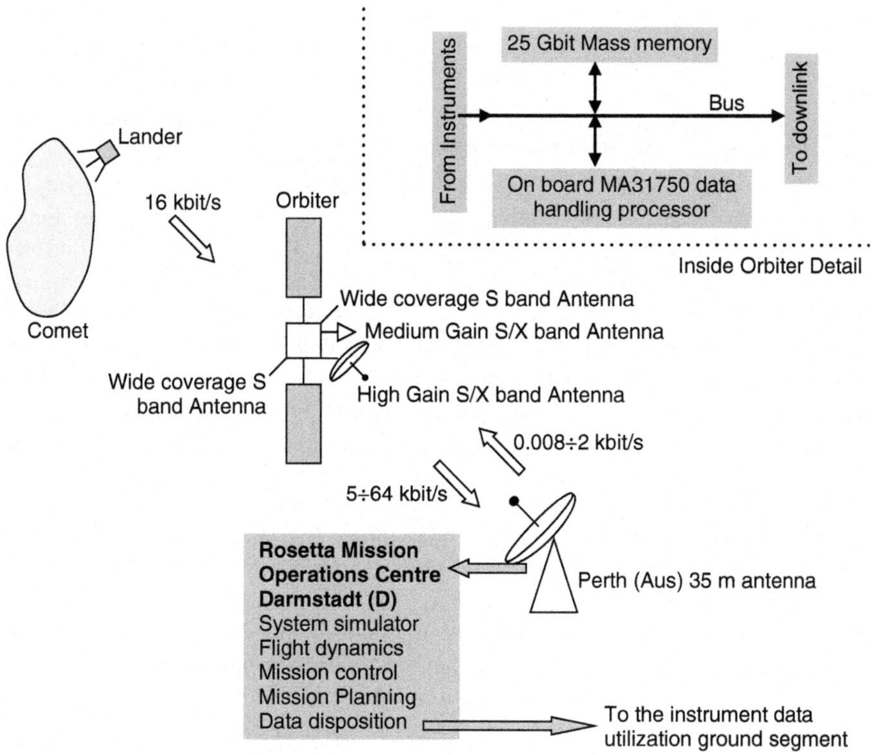

Figure 11.8. Rosetta Data Flow.

taken by the navigation camera were processed on ground to refine orbit determination. The communications components[11] are also complex. The distance from Earth to the satellite can reach up to 930 million km. To cope with this scenario, there is a high-gain X-band link with a large 2.2 m diameter parabolic antenna with 28 W of transmission power. This link is able to provide from 64 kbit/s to 5 kbit/s depending on the distance. During the critical rendezvous with the comet the distance will be about 5 AU (Figure 11.8) and the data will be 5 kbit/s. Uplink telecommand rates vary from 0.008 to 2 kbit/s, also depending on the varying distance. The Philae lander has a 700 Wh nonrechargeable battery that can provide up to 60 h of operation. It has also a 68 Wh rechargeable battery fed by Si solar cells providing a maximum power of 10 W. Philae sends information to Rosetta at 16 kbit/s. In addition of the parabolic high-gain antenna there are two other sets of antennae:

- Two medium-gain antennas; 9° beam-width in X-band and 30° beam-width in S-band, for operations at intermediate distances
- One pair of omnidirectional S-band antennas for operations near Earth

[11] Rosetta Ground Segment and Mission Operations, M. Warhaut, P. Ferri, and E. Montagnon, Paper Presented to SpaceOps 2000.

11.4.3. Sentinel-3

Sentinel-3's mission aim and configuration have been already introduced in Section 8.4.3.

11.4.3.1. Operational concept and mission phases

Sentinel-3's operations are simple. Once the satellite is installed in its final orbit, followed by separation, reconfiguration, and attitude acquisition, the commissioning phase of the mission will start. When the satellite is fully characterized and in-orbit validated, the routine operational phase of Sentinel-3 will begin. As mentioned above, this routine phase will last at least 7 years, extendible to 12. After 12 years, to prevent it becoming dangerous orbital debris, the satellite will be deorbited. During nominal operations all the instruments are in continuous operation, but there is no acquisition by the visible light channels at night, and radar altimeter modes differ when observing different surfaces such as open ocean, coastal zones, sea ice, or inland waters. The satellite, which performs its own orbit determination and carries a digital model of the Earth, autonomously changes the observation mode of the instruments as required by the surface under observation. The satellite also autonomously prepares itself for downlinking to the instrument ground station or for bidirectional communications with the operational ground station located in Kiruna, (Sweden). During nominal operations, the satellite can remain without contact with the operational ground station for up to 2 weeks, but the "housekeeping" satellite information is downlinked to instrument ground stations during every orbit and is immediately forwarded to the Mission Control Centre located in Darmstadt for analysis.

11.4.3.2. On-board subsystems

The power subsystem include a deployable GaAs solar array of 10 m^2 area, orientated by a solar array drive mechanism (SADM), a Li battery, and a power control and distribution unit, which controls the solar array, the battery's charge and distribution of power to the satellite platform and instruments. The satellite's power consumption is 1,100 W (Figure 11.9).

Sentinel-3 possesses a satellite management unit (SMU), which includes a central computer that acts as the satellite's nerve center, controlling all the components needed for the handling of commands, including the decoding and validation of uplinked commands and the coding of "housekeeping" satellite data for downlink. The SMU also includes a data storage memory. The SMU controls the satellite through two mil-1553 buses. One supplementary "Spacewire" high data throughput bus connects the SMU to the payload data handling unit (PDHU). This bus is used for providing the necessary data interchange needed for the monitoring and control of the instruments. The SMU also includes point-to-point discrete connections to the attitude and orbit control subsystem (AOCS) components.

Sentinel 3 telecommunication components include a fully redundant, low-speed S-band up- and downlink communication subsystem and a high-speed, 450 Mb/s X-band subsystem for downlinking of the satellite's instrument output, which is part of the payload data flow. The S-band subsystem has two hemispherical low-gain antennas, which provide communication with the ground, whatever the satellite's attitude.

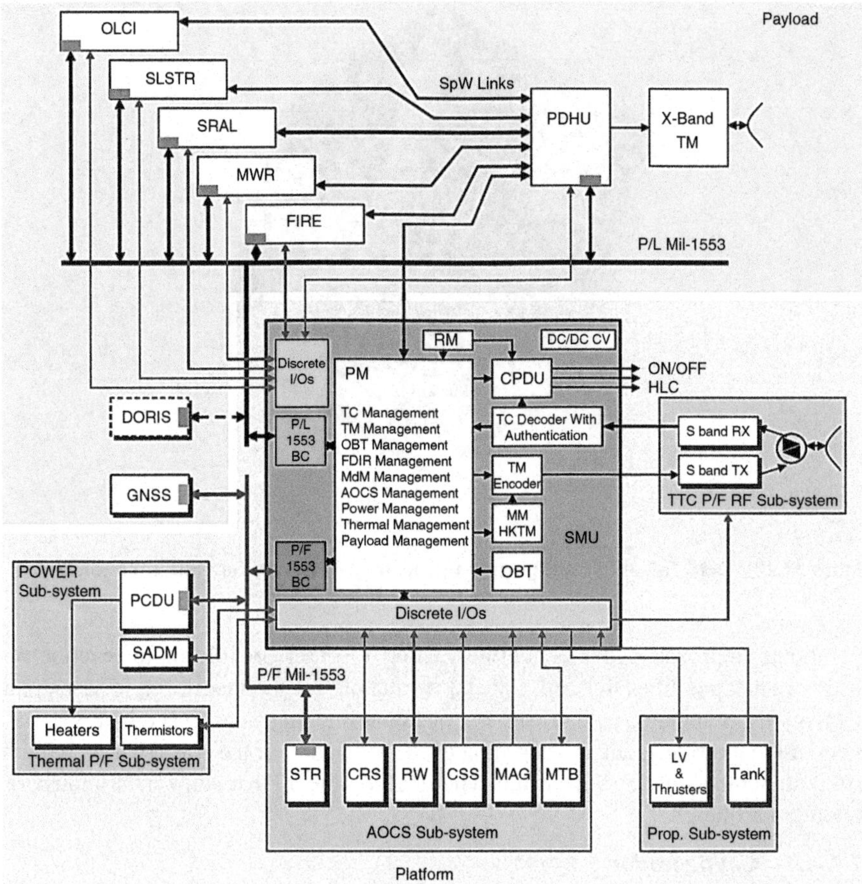

Figure 11.9. Avionics of Sentinel-3 (Sentinal-3 Definition Phase Final Report, Y. Baillion, Thales Alenia Space 2007).

11.4.4. SSTL-DMC

11.4.4.1. Mission aim

The Surrey Satellite[12] (SSTL) disaster management constellation (DMC) of multipurpose Earth observation satellites consisted at the beginning of 2010 of six satellites built by SSTL (Section 7.4). Each member of the consortium has its own independent satellite imaging capability, which under the consortium agreement is made available to support common objectives during limited periods of time. All members signed a nonlegally binding agreement to make available up to 5 % of satellite imaging capacity for the other satellite owners. Such an arrangement allows a cooperative

[12] Overview of SSTL Small Platforms for ESA Surrey Satellite Technology, GG-009296-1.01, SSTL 2007.

Figure 11.10. DMC Satellite Configuration (Courtesy of Surrey Satellite Technology).

operational approach. All the satellites acting together provide rapid reaction and observation capabilities that are especially valuable for the monitoring disasters and other swiftly changing events. DMC International Imaging Ltd., owned by SSTL, acts as central marketing agent and coordinates the utilization of the constellation with the different members of the consortium. The satellites are all in a Sun-synchronous orbit at 640 km altitude.

11.4.4.2. Configuration

Each SSTL-DMC satellites is a three-axes-stabilized cube measuring approximately 0.7 m side. They weigh ~90 kg and their average power consumption is 50 W. All the satellites in the constellation provide Earth images in three bands with a spatial resolution of 32 m, though several of them carry a supplementary panchromatic camera with a resolution of up to 4 m. The swath of the 32 m resolution channel is 640 km (Figure 11.10).

11.4.4.3. Operational concept and mission phases

Each satellite can be operated either individually or as part of the overall constellation. Individually, each owner has its own ground station and mission control center to interact with its own satellite. The ground stations are S-band with an antenna 3.7 m in diameter for telecommand and telemetry, as well as instrument downlinking. Possessing identical communication interfaces, any ground station in the system is capable of addressing any of the satellites, but in practice this is done only in emergencies or for experimental purposes. Nonetheless, all the satellites can act together by interchanging their tasking plans to allow the central DMC center to request images taken by one satellite as required by the owner of another. Finally, each individual owner will generate and uplink the tasking plans for its own satellite. That is, the different satellites collaborate but the uplink of commands is handled exclusively by each owner.

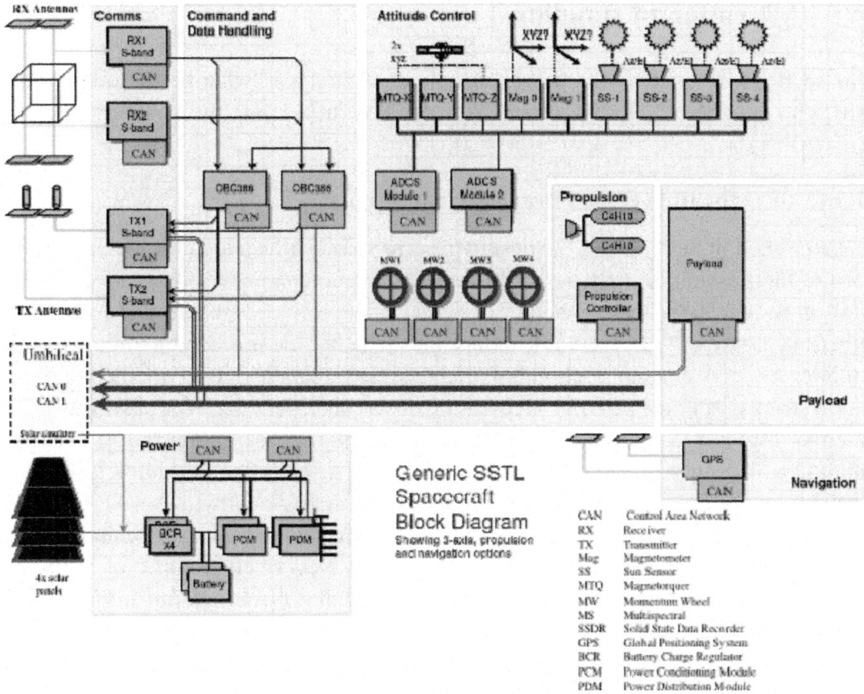

Figure 11.11. Avionics of a DMC Satellite from SSTL (Utilization of DMC Experience and the Potential Usage of DMC Services to Provide Additional Support to an European Global Monitoring System, M. Cutter et al., Small Satellites Systems and Services Symposium 2008, ESA SP-660, ESA 2008).

Acting together, these satellites are able to observe any part of the Earth with a mean delay of 1 day. They can be rescheduled within 24 h (resources available), and imaging to downlink times range from 10 min to 24 h depending on the location of the target and the ground station. The time from data reception to product release varies from 4 h (not orthorectified) to 48 h (orthorectified using ground control points to produce subpixel accuracy). All ground stations are connected via the Internet at 1 Mbit/s.

11.4.4.4. On-board subsystems

The power subsystem includes 4 GaAS body-mounted solar panels feeding power to an unregulated bus (Figure 11.11).

Besides the difference in mass and cost, the electronics architecture of DMC and Sentinel-3 are almost identical. In the case of DMC, two Intel 80386-based OBC386 computers manage the satellite; the main one and a second one providing redundancy in case the main computer fails. The computers control the satellite through main and redundant CAN buses, which provide up to 388 kbit/s and point-to-point connections when higher speeds are needed. A GPS receiver provides autonomous navigation and a time reference. There are two redundant communication links: one low data-rate S-band for up and down satellite command and control, and a high data-rate S-band, 8 Mbit/s, for the instrument output. The uplink is done at 9.6 kbit/s.

11.5. Allocation of functions

This section will analyze the design decisions of the type "what needs to be done?" and "who does what?" in the operational data flow of the satellite.

11.5.1. Systematic versus interactive operations

As mentioned in Section 7.1, the design process starts with an idea for the operational need of the mission directly derived from the phenomena under observation. Hence, there is no decision about whether the mission is to be systematic or interactive; rather the nature of the observable determines whether the mission operation shall be interactive or systematic. Section 8.4.5 analyzed systematic versus interactive observation strictly from the point of view of the observable. The need for observation can be either highly predictable, which favors a systematic observation mode, or highly unpredictable, requiring an interactive mode. In addition to the mission's fundamental aim of acquiring information, other activities must be performed as well, including the satellite's launch, in-orbit configuration, commissioning, and decommissioning. In the past all of these activities are interactive. Nevertheless some of them, for example, solar array deployment and acquisition of the correct satellite attitude could be handled autonomously by the satellite. Satellites must also perform regular activities such as calibration or orbital corrections that can interrupt observations. If these operations are frequent, they will push the mission design toward autonomous and systematic execution of these activities; if they are irregular and depend on unpredictable factors, they will require interactive operation. We will now analyze in depth how the systematic versus interactive alternatives function in the four missions outlined above.

Sentinel-3 will provide regular and consistent information about Earth's entire surface, including the open ocean, coastal zones, sea and land ice, inland waters, and dry land. Each surface requires distinct operational modes. Over the ocean the radar altimeter will operate with a lower resolution, while over coastal areas and inland water, it will operate in a higher resolution, along-track "synthetic aperture" mode. As far as the optical instruments of Sentinel-3 are concerned, during the night, or when the Sun is low over the horizon of observation, images that require substantial illumination by the Sun will be curtailed, though thermal infrared images, which do not depend on solar illumination, will continue to be gathered. The changing of the seasons, determined by the varying geometry of the Sun with respect to the Earth's surface, and the presence and absence of ice, must also be taken into consideration when determining instrument modes. While it is evident that the issue of changing instrumental modes is complex, they can all be programmed as functions of the orbital position, the time of the year and the Earth's geography; that is, the changing from one operating mode to another depends on the data that can be programmed into the satellite's intelligence; hence, modes can be made autonomous and systematic.

Cluster systematically analyzes regions of the magnetosphere: the bow shock, the magnetopause, and the polar cusp, according to the yearly cycle of relative positions among the satellite constellation's orbits, the Sun, and the Earth, as illustrated in Figure 11.2 and its accompanying text. Each magnetosphere region possesses unique characteristics, which require distinct modes from the observing instruments.

Nevertheless, the boundaries between regions is less defined and fixed than the boundaries encountered by Sentinel-3. The Sun-magnetosphere interaction, not being fully predictable, will entail adjustments of the operational profile to account for changes in the exact position of the different boundaries. The distances maintained among the four Cluster satellites must occasionally be changed to enhance analysis of the fine structure of the phenomena under observation. To cope with special events, for example, sudden changes in solar activity, sudden modifications in the overall pattern of operations must be made, such as those undertaken during a strong solar storm that took place in November 2000.[13] Another significant factor is the adjustments that need to be made as a result of the knowledge gained over the magnetosphere as the mission proceeds, modifications of the operational plan aimed at refining the analysis of the phenomena being investigated. That is, as in the case of Sentinel-3, Cluster's observing behavior depends on its orbital position and the changing seasons; but unlike Sentinel-3, this overall pattern requires regular corrections. This entails a higher level of human interaction and ground segment modifications to the regular overall planning. This explains the double loop of mission planning: the weekly and 6-monthly cycles described in Section 11.4.1.

SSTL-DMC, as its name implies, is a constellation designed to deal with disasters, which are localized and unpredictable. This requires frequent changes in the satellites' image acquisition plans to accommodate the changing requests of the consumers. It is always possible, however, to define "background" acquisition plans where images over large areas are gathered in a systematic way. This systematic mode will be adopted when there are no urgent or specific requests. This defines DMC as an interactive mission with some elements of a systematic mission. In general, each ground segment creates the acquisition plan for its own satellite. In situations where a rapid response is crucial, an image acquisition request can be sent to the central SSTL control center, which receives, consolidates, and forwards the request to the first satellite in the constellation to reach the area of observation. Requests for acquisition of images also incorporate different levels of priority, as some acquisitions will be more urgent than others. These factors, coupled with the project's large number of ground stations (one per satellite), results in very fast reaction times. Even so, final authorization for the uplinking of commands will always rest with the owner of each individual satellite, who can refuse to cooperate at any time, since the cooperation agreements are not legally binding. DMC operations are strongly interactive, and dominated by the need to create and update changing operational plans. All these activities are performed on the ground. This high level of on-ground activity permits the designing of a relatively simple satellite with basic computing capacities and minimal on-board autonomy. This is an example of a complex, interactive, and non-autonomous environment.

Rosetta's orbital track is calculated with a high level of accuracy, and the exact moment of its rendezvous with the comet, as well as the planets and asteroids it will encounter along its path are all known beforehand. Nevertheless, Rosetta must be responsive to what it encounters when it makes its different rendezvous. Rosetta faces a constantly changing environment and must adopt its behavior to what it encounters

[13] Cluster 1 Year in Space, C. Escoubet, ESA Bulletin, August 2001.

on its journey. Furthermore, the prolonged travel time needed for information to reach Earth from the satellite requires the satellite to be able to make decisions on its own. This is an example of an environment that is complex, interactive, and autonomous all at the same time.

11.5.2. Autonomy versus ground intervention

The previous section discussed systematic versus interactive operations. It was explained that there is no trade-off between systematic and interactive operations: the nature of the events under observation decides beforehand what option should be chosen. In this section we will discuss where the "intelligence" dealing with the predictable or unpredictable operational changes should be located: to what extent it should be on-board an autonomous satellite, or on the ground, requiring ground-based decision making and frequent ground to space contact. The designer's autonomy versus ground intervention strategy will be influenced by two conflicting factors:

- The need to increase mission robustness by allowing ground-based, human intervention to correct anomalies and to cope with unexpected events
- The desire to reduce ground-based operation costs and to foster mission flexibility by increasing satellite autonomy

Independent of the care taken during testing, all missions must confront numerous on-board problems. These problems are normally identified early in the mission and are solved by the use of on-board software updates and extensions. These software problems are generally associated with data handling or attitude control. To cope with these issues the commissioning phase of any space mission includes a high level of ground intervention. Anyway, already commissioned space missions could continue experiencing nonnominal situations that would entail human intervention.

The most perilous situation is a loss of attitude with respect to the Sun, with the resulting power run-down and subsequent lack of contact with the ground. Satellites are very resilient, and ground intervention is capable of recovering a theoretically lifeless satellite, as occurred in 1991[14] when Olympus remained in a nonoperational state for 77 days, including 3 weeks of total silence. The satellite, including its especially sensitive fuel and batteries, withstood temperatures well outside its design specifications. Nevertheless, the satellite was recovered and put again in operation. This is a clear demonstration that no matter how autonomous a satellite is, it should permit maximum ground intervention during all phases of the mission, in order to facilitate proper handling of unexpected events. The increasing pressure to reduce ground operations costs, combined with the growing sophistication of computers, encourage the provision of autonomy to satellites. This trend is supported by the possibility of increasing mission performance and flexibility by providing functions that allow autonomous and intelligent satellite decisions.

This need for autonomy becomes obligatory for satellites operating in a changing environment far from Earth, as we saw with Rosetta. These two conflicting but legitimate

[14] ESA Achievements, BR 250, ESA 2005.

influences push in different directions. Satellite operation specialists will advocate a low level of satellite autonomy, and will point to numerous examples in which prompt human intervention made the difference between the success or failure of a mission. On the other hand, the increasing cost of ground operations and the recent sophistication of on-board data handling equipment encourage increasing autonomy. Many sources praise the virtues of autonomous satellite control as a way to cut costs, and argue that the largest enemy of operational autonomy is tradition. They are right, but traditionalists also have a valid point; a single mission saved by human intervention compensates numerous cost reductions by the application of autonomy. SOHO was nearly lost for good because they tried to reduce ground support with a software patch. Nevertheless, there is a clear tendency toward higher autonomy underscored by the need to reduce costs and the increasing performance of on-board data systems. In-orbit autonomous orbit generation, image acquisition, orbit corrections, or the execution of programmed calibration are all increasing. Special missions aimed at fostering confidence in on-board autonomy have been undertaken: in Europe with PROBA satellites (proof of on-board autonomy); and in the USA with several "New Millennium" projects including Deep Space One[15] and the Autonomous Science Experiment on board Earth observation satellite EO-1.[16] EO-1 can autonomously detect and respond to special events such as volcanic activity, flooding, and water freezing or thawing.

In the case of Sentinel-3 all the functionalities required for nominal instruments operations are located on the satellite. To make this possible, the satellite must carry hardware and software that permit it to independently determine its position over the Earth as well as the position of the Sun. Sentinel-3's specifications require it to operate autonomously for at least 14 days. This level of autonomy allows the use of a single ground station and infrequent operational contacts with the satellite. Nevertheless, housekeeping telemetry, monitoring the satellite's status, is included as part of the instrument data stream, which is downlinked on every orbit. This permits fast reaction times by mission operations in cases where an anomaly forces contact at the first possible opportunity. This approach allows low operating costs, especially important in a mission like that of Sentinel-3, which specifies an unusually long operational life of 15 years. The accumulated cost reductions associated with this type of simplified operations are significant.

Cluster's orbit allows long periods of continuous contact with the Earth; nevertheless, these periods of contact with the single ground station must be shared by the four satellites making-up the Cluster system. In addition, at certain times of the day, poor alignment between the satellite and ground station prohibits proper contact. These two factors combined entail a moderate level of satellite autonomy, allowing it to operate during intervals without ground contact. Originally, Cluster was to have been served by two ground stations, but the mission was eventually altered to operate with just one.[17] Two ground stations would have permitted almost continuous monitoring of the satellites, but the single ground station meant a significant reduction in the cost

[15] Mission Operations with an Autonomous Agent, B. Pell et al., IEEE Space Conference 1998.

[16] ST6 the Autonomous Science Experiment, R. Sherwood et al., IEEE Space Conference 2005.

[17] Cluster II. Evolution of the Operations Concept, M. Warhaut, ESA Bulleting, August 2001.

of operations. The increase of on-board data processing and storage capabilities, combined with the pressure for low-cost operations, prompted this change.

DMC operations are strongly interactive, and dominated by the need to create and update changing operational plans. These are all activities performed on the ground. As mentioned above, this high level of on-ground activity enables a simple satellite with a simple computer and without too much on-board intelligence to be used. All complex functions are located on the ground. Nevertheless, there is not a heavy up-front investment, because each new component of the satellite constellation comes with its own ground segment, which is included in the mission's overall budget. So the ground segment expands naturally in step with the space segment.

Rosetta must be highly interactive and at the same time highly autonomous. It requires a high level of intelligence both on the ground and in-orbit, and the design reflects that. One of Rosetta's cameras forms part of the attitude and orbit control subsystem, allowing the satellite to identify the surrounding environment so the satellite can make autonomous decisions regarding the satellite's attitude and orbit. These interactive and complex operations are compatible with long periods of little activity during dormant phases. The satellite performs the necessary housekeeping activities during these periods independently.

It is easy to conceive of increasing levels of autonomy for in-orbit operations. One possibility is a meteorological observation system able to alter its behavior according to how the weather evolves; for example, to autonomously sample in more detail a region with a developing storm.[18] The system could also react to requests for acquisition patterns Also, in an emergency situation, in situ staff could make direct requests for the satellite to acquire and download images directly to them, thus bypassing the mission control center. This is not only a sharing of information but also a sharing of command authority, which would necessitate the establishment of priority criteria for the satellite to choose autonomously between conflicting requests. Such a system would need to develop not only new components with enhanced capabilities but would also entail overcoming "political" barriers, as it would disperse control over a very expensive asset. Sharing authority over a satellite is as much a political issue as a technological one, and some consideration should be given to development of collaborative environments that ease the difficulties of communicating across organizations. Indeed, a system such as this presents numerous challenges regarding the evolution of organizational systems, and it involves technological advances as well. But there is no doubt that fostering high levels of operational autonomy and the sharing of operational responsibilities will strongly improve the performance of space systems.

11.5.3. Fast versus slow commanding

Figure 8.17 depicted three time-related figures of merit that the data produced by a space mission can have: responsiveness, revisit, and data latency. The same figure also defined acquisition delay as responsiveness minus latency. Revisit or temporal

[18] Architecture Vision and Technologies for Post-NPOESS, Michael Kalb, SPIE Vol. 5659, SPIE 2005.

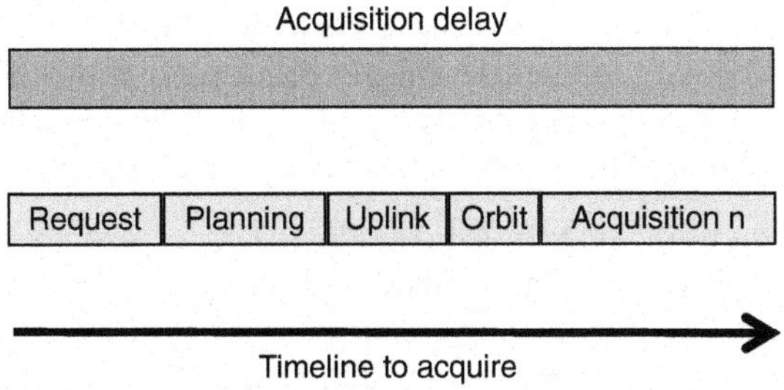

Figure 11.12. Data Acquisition Timeline.

resolution was introduced in Section 8.4.2 and it is related to the satellite orbit selection that was analyzed in Section 9.9.3, Data latency is related to instrument data processing and will be discussed in Section 12.4.2 of the next chapter, and acquisition delay is related to the operational ground segment and will be discussed here.

In some missions, where the delay in the delivery of the information provided by the satellite is important, it is necessary to minimize the interval between the consumer's request regarding a specific item of information such as an image, and the moment he actually receives the item. This interval can be divided into two parts: the time needed for the item's acquisition, and the time needed for delivery of the acquired item to the consumer. The former will be discussed here, while the latter, as part of the instrument data processing domain will be studied in detail in Section 12.4.3 of the next chapter.

The need for rapid response to commands, depending on the urgency of the information, is maximized in missions studying dynamic events involving military and civilian security, dynamic astronomical events, or in manned missions requiring complex, interactive, and real-time operations. The process to start the acquisition of commanded observations requires the steps indicated in Figure 11.12.

The acquisition time forms part of the satellite operations cycle studied in this chapter. The acquisition of an image involves the following steps:

1. The consumer needs time to contact the mission and make a request. The time needed for this is a function of the tools provided by the mission operator for facilitating requests: i.e., 24-h availability versus normal office hours, the use of web-based request forms, etc.
2. The operator needs time to plot and program the set of commands needed by the satellite. This will depend on the number and nature of received requests, their priority status, and the availability of tools for optimizing the acquisition plan and minimize waiting times, and for providing flexibility in the accommodation of high-priority last-minute requests. The outcome of this process is a list of commands to be uplinked to the satellite(s), and which optimizes the operations of the satellite and the instruments.
3. The satellite receiving the uplinked list of orders must be visible to the relevant ground stations. The time needed is a function of the orbital trajectory of the satellite and the number and location of the commanding ground stations. As the number of satellites, their orbits and the number of ground stations and their locations are established, they directly determine delay times.

4. Once the command has reached the satellite there will likely be an interval before the satellite, following its predetermined orbital path, achieves visibility of the requested region or item of observation and acquires the requested image. This is also a direct function of the orbits and number of satellites.

As far as the design of the operational data flow is concerned, to reduce acquisition times would entail an increase in the number of operational ground stations, or the use of data-relay satellites for relaying orders to the relevant satellite or satellites. The geosynchronous mission IUE was described in Section 7.2.1; the orbit and the number of ground stations were chosen to allow continuous contact between the satellite and one of the two ground stations. For low Earth orbit satellites, the most common case, the number and location of ground stations needed for continuous contact between satellite and ground is impractical. This is why data-relay networks are sometimes used. One data-relay satellite, located in the geostationary position, is able to provide permanent links to any satellite located in low Earth orbit for about 1/3 of the orbital time. The European Artemis[19] mission is an example of this. One constellation of three or more well-distributed geostationary data-relay satellites, as with the American TRDSS,[20] permits almost permanent links from any satellite in low Earth orbit to anywhere on Earth. Due to their equatorial geostationary position, and the their resulting geometry, even a full constellation of data-relay satellites loses contact for brief periods with satellites flying at very low altitude over the poles. Geostationary constellations for data relay are expensive to design and maintain. As they represent such a heavy investment, they would likely be one choice among other, possibly more practical infrastructure options. Hence, the possible employment of data-relay satellites as part of the design process will be among the system constrains provided by the mission sponsor. The actual use of data-relay satellites is in fact rare, and none of the four missions discussed in this chapter uses data-relay satellites for commanding.

Data-relay satellites allow almost real-time command and data delivery but the only existing operational data-relay satellite constellation, TDRSS, was designed with manned flights in mind, and all the other missions have lower priority. This has produced a rather uneven scenario for the use of data-relay satellites for commanding (Figure 11.13):

- The Hubble mission (Section 7.2.1) uses TDRSS for information downlink but not for satellite commanding.
- Landsat 7 mission operations control center, located at NASA Goddard near Baltimore, USA, can command the satellite using the TDRSS network but downlinking is done directly to Landsat-7's own ground stations.
- Landsat 5, on the other hand, downlinked data through the TDRSS. The TDRSS network is also used as instrument downlink for Terra, Aqua, and Aura, but instrument data downlinking will be discussed in the following chapter.

It will always be possible to reduce the response time by providing a large number of well-placed ground stations. This will be studied in the next section.

---

[19] ESA Achievements, BR-250, ESA 2005.

[20] www.spacecomm.nasa.gov/spacecomm/programs/tdrss/default.cfm

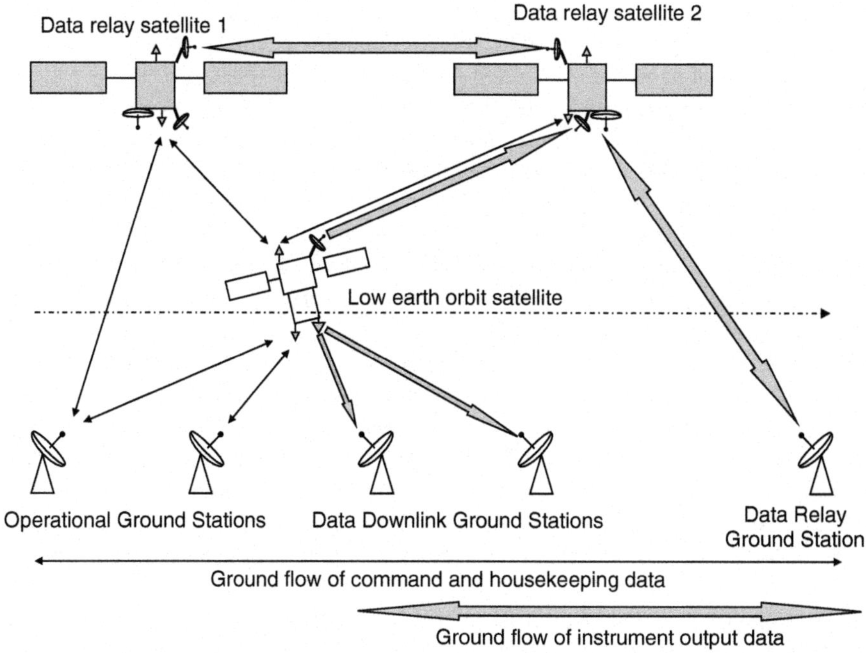

Figure 11.13. LEO Data Flow Using Data-Relay Satellites. Thin Lines Indicate Operational Data Flow and Thick Lines much Higher Throughput Instrument Data Flow.

11.5.4. Number and location of operational ground station

With the exception of the critical and brief early stages of a mission, when a large number of operational ground stations are used, the number of operational ground stations is normally quite small, and in a majority of cases only one station is used. This accords with the desire for highly centralized satellite operations discussed at the end of Section 11.5.2, and with the need to reduce operational costs discussed in Sections 11.5.1 and 11.5.2. This tendency toward a small network of operational ground stations has been steadily reinforced by technical advancements, permitting a higher degree of on-board autonomy.

Cluster performs all its communications, operational and instrument data downlinking, using a single station in Villafranca, near Madrid, Spain, which operates in the S-band for communications, in the 2,025–2,111 MHz band for uplinking and in the 2,200–2,290 MHz band for downlinking.

Sentinel-3 uses different stations for operational and instrument data. The uplinking and downlinking of operational data is done using a single S-band ground station in Kiruna, which will be used rarely due to the mission's high level of autonomy.

Rosetta uses also a single ground station, located in Perth (Australia) as its main operational center, but can resort to NASA's deep space network in emergencies. Two bands are used: the deep space S-band for the uplink and the deep space X-band for the downlink.

DMC employs the same approach. Each satellite's owner has their own ground station, and apart from the sharing of resources, the use of a centralized facility to coordinate acquisitions, satellite owners have exclusive command of their own satellites through their own ground station, though the technical capacities for cross-commanding is at hand.

Recent missions employing a single operational ground station are in the majority. In the past, before reliable on-board autonomy was common, a larger number of stations were the norm, but the trend toward single stations is clear. The older ISO and IUE had stations in both Europe and America; Cluster was originally designed for two operational ground stations but was finally implemented with just one. NASA's deep space network of three ground stations (in California, Spain, and Australia) offers continuous contact but, as explained in Section 7.2.1, JWST was designed to operate with a single 4-h downlink contact per day.

The adequacy of operational ground stations will be proportional to the number and duration of the opportunities of communications they provide. The number and duration of these opportunities will be a function of the satellite orbit ground track and the location of the ground station. For example, low Earth-orbiting satellites with very high orbit inclination will fly over the poles twice per orbit and ground stations locate at the poles will provide opportunities for contact with every orbit. The ground track of the orbit will define what are the best geographical locations for the ground stations of each mission and Section 12.4.5 will study the best location in a systematic fashion. Nevertheless, the location of an operational ground station is not just an issue of geographical optimization. Operational ground stations send commands to the satellite that ensure its survival or that could produce its destruction. Therefore, most operational ground stations are located in places under close supervision by the owner of the satellite. Most of the time, the locations of the operational ground stations are a constraint provided by the mission sponsor.

11.5.5. Orbit determination and control functions allocation

Orbital determination and control were previously examined in Section 9.1.2 as part of the chapter dealing with Orbit and Attitude Domain. That section introduced the various kinds of orbital maneuvers and different factors, such as injection errors, necessary orbital changes, or unwanted orbit perturbations that make them necessary. Also described were the tools necessary for determining the orbit, and it was noted that, traditionally, orbits were determined via range-ranging in the ground-to-satellite communications link. This allowed the ground control to determine the satellite's velocity with respect to the satellite-to-ground station line of sight. Nowadays, for satellites in low Earth orbit, it is also possible to determine orbit by including a GPS or Galileo navigation receiver onboard the satellite. Nevertheless, the standard practice, even today, is for Mission Control to:

- Calculate the satellite orbit using ranging data and any other available information.
- Compare the actual orbit with the nominal one.
- Calculate and command the necessary orbit control maneuvers, large or small.
- Verify the outcome of the maneuvers once executed.

As indicated in Section 9.1.2, for low Earth orbits it is perfectly possible for the satellite to independently determine its orbit using navigation satellites such as GPS or Galileo. Once the orbit has been determined it is also perfectly feasible to implement on board algorithms allowing the satellite to autonomously calculate, program, execute, and check the necessary orbit correction maneuvers. The advantages and shortcomings of autonomous operations were discussed in Section 11.5.2. Clearly, autonomous operations reduce costs, but they also increase risks.

Autonomous orbit determination and control is especially advisable in cases requiring frequent but minor orbit corrections, such as constellations that require flying in either tight or loose formation, or for satellites demanding particularly accurate control of their orbital paths. Opting for the autonomous execution of many small maneuvers will reduce operations costs, while any individual minor autonomous orbit correction is unlikely to represent a major risk for the satellite. Major, sporadic corrections, however, are better performed under close ground control and rare orbital corrections are better performed under close ground control.

During the early stages of mission definition it is crucial to establish a basic orbit determination and control philosophy and to adjudicate this function to either the satellite, to the mission control, or to both.

11.6. Allocation of performances

11.6.1. Power budget

Each of the electrical components on board a satellite consumes power. Moreover, each of them will have a power consumption profile; some, like the central computer, will operate continuously, while others will feature different modes requiring different power levels. Electronic units might have zero power consumption when off, low power consumption during "stand-by" mode prepared for swift reaction, and one or more operational modes with different power needs when performing different functions. All these factors must be taken into account when determining the overall power profile for the satellite as a whole. This variable power demand must be provided by the satellite's primary power source, that is, the solar array, and by the secondary power source, the batteries. Ultimately, two budgets will be needed:

- An instantaneous power budget to ensure that both sources of electrical power (the solar array and battery) are always able to deliver enough electrical power to fulfill the needs of all the units operating at each moment
- An energy budget to ensure that the primary source of power, the solar array, is able to produce enough energy during each orbit to provide for all the satellite's energy needs while engaged in fully recharging the battery during the illuminated portion of the orbit

It will be necessary to calculate as many different power and energy budgets as required by the varying power demands and power production situations of the satellite during all the different phases of its operational life.

- It must be verified that the satellite can assume the right attitude with respect to the Sun, and can deploy and operate its solar array, before the battery drains below its permissible minimum charge level.

- It must be verified that the satellite will have enough power and energy during special situations such as major orbital changes that could require the solar arrays to point away from the Sun.
- It must be verified that the satellite has enough power and energy during the special geometry and power demands of the safe mode.

As in the case of masses (Table 10.4 of Section 10.6.1), power demand estimations must include factors that take into account our ignorance regarding the actual final demands of the electronic devises that will eventually be manufactured and put into orbit. Early predictions of the power demands of electrical systems are highly unreliable, much more so than predictions about mass, and the use of very large margins is crucial.

Different guidelines have been published by different sources outlining figures to be used as maturity factors for deriving reliable power budgets for the different stages of the project. For early definition, the AIAA[21] applies the following guidelines (Table 11.2).

As already explained in Section 10.6.1, class 1: is for a new spacecraft; class 2: for a next generation spacecraft based on a previously developed family, which expands in complexity or capabilities within an established design envelope; and class 3: for production-level development, based on an existing design for which multiple units are planned and where a significant amount of standardization exists. The phases are as per Figure 1.5 and as per Section 6.3.

As far as energy production is concerned, its evolution over time will be a function of the varying orbital geometry, which can be determined to a high degree of accuracy, and of the aging of the electrical subsystem's components, which is covered by the application of known mission-life degradation factors. Solar arrays degrade over time, and designs for the end-of-life phases of lengthy missions require a corresponding increase in power budget.

Various engineering specialists will contribute to power production forecasts, which, once the different safety and maturity factors have been calculated, will determine the solar array's size. The accommodation of the required solar array is a satellite configuration issue and it has been discussed in Section 10.6.3.

Normally, large power consumers cannot reduce their power needs; active instruments like SARs need power. There is no easy way to reduce power demands. In the

Table 11.2. AIAA Recommended Minimum Power Contingency %.

	End of phase 0			End of phase A			End of phase B			End of phase C		
Class category	1	2	3	1	2	3	1	2	3	1	2	3
0–500 w	90	40	13	75	25	12	45	20	9	20	15	7
500–1,500 w	80	35	13	65	22	12	40	15	9	15	10	7
1,500–5,000 w	70	30	13	60	20	12	30	15	9	15	10	7
Above 5,000 w	40	25	13	35	20	11	20	15	9	10	7	7

[21] Estimating and Budgeting Weight and Power Contingencies for Spacecraft Systems ANSI/AIAA G-020-1992.

early stages of a project it is crucial to always design a power subsystem able to provide the satellite's power needs with ample margins. In balancing this budget we have the following degrees of freedom by order of preference:

- Increase the solar array size to provide an energy budget within the desired margins. This is by far the best way to balance the power budget. This is only possible when the satellite configuration, solar array qualification limits, attitude control limits, and launcher capacity permit the size increase.
- Increase solar array size by extending it beyond its qualification envelope. This introduces some risk and will entail supplementary qualification testing, but the original design might allow the increase.
- Improve the technology by employing more efficient solar cells or batteries, fully regulated buses, or a more efficient control of the solar array.
- Try to run the solar array colder, which increases efficiency. This requires a more sophisticated thermal design around the solar array.
- Identify subsystems that require too much power. Two possible sources are the thermal subsystem and the harness. In large satellites, increasing bus voltage will reduce mass as well as losses in the harness.
- Reduce instrument consumption. This will likely involve degradation of mission performance and should be avoided.
- Employ a larger launcher to accommodate a larger solar array. Serious cost impacts make this a last resort.

Designing a solar array that meets mission requirements usually presents no difficulties. If problems occur, however, they are among the hardest to solve because the maximum size of the fixed solar array is determined by the chosen launcher capacity, and the power demands by the required satellite performances and both are very difficult to change.

11.6.2. Communications link budgets

Link budgets are calculations that verify that the power of the signal arriving to the receiving antenna is strong enough for the information encoded in the signal to be recovered.

The following factors will enhance the link budget:

- The power of the signal emitted by the satellite or the ground station. In the case of the satellite this factor is limited by the availability of technology for power amplification, by the capability of the satellite power subsystem to satisfy the emitter demands, and by the International Telecommunication Union (Section 11.1.3) regulations on the maximum energy density arriving on Earth from satellites. In the case of the ground stations the power will be limited by the availability of ground stations with powerful transmitters.
- The gain of the satellite antenna. High-gain antennas concentrate the signal into a narrow beam, which improves the strength of the emitted signal on-ground. The larger the gain the larger the antenna and the gain will be limited by the resulting accommodation constraints of a large antenna on the satellite. The accommodation of Rosetta's antenna, depicted in Figure 11.6, is a good example of this. Moving to higher frequencies permits smaller antennas, thus reducing accommodation

problems; nevertheless, higher frequencies require complex advanced hardware that could call for dedicated development.

- The gain of the ground station antenna. The larger the antenna the higher the gain. The large ground stations used for Rosetta's communications, described in Section 11.4.2, are prime examples. NASA DSN stations have a diameter of 70 m. The limiting factor here would be their availability and cost, since ground stations with large antennas are rare, and expensive to build, maintain, or rent.
- The ground station receiving quality expressed by its system gain to noise temperature ratio G/T noise figure. This quality factor behaves in a way similar to antenna gain; better antennas improve the situation but are more expensive to operate and are less common.
- The use of enhanced coding schemes to protect against bit error. This allows the correct recovery of transmitted data with low receiving powers. The use of Reed Solomon or convolutional codes is now widespread. CCSDS regulations recommend specific codes to be used.

The following factors degrade the link budgets:

- The speed of data being recovered. The larger the amount of information, in bit/s, being transmitted, the larger the amount of power needed to recover it with a low bit error rate. Operational data flow are normally modest. The problems associated with the transmission of very high data rates, which is a problem specific to the downlink of instrument data, will be covered in Chapter 12.
- Losses in the satellite emitting subsystems. Choosing the best quality hardware can reduce this loss; however, it is generally small.
- Losses due to mispointing of the satellite antenna. This factor is important where high-gain antennas are employed. To reduce this loss it is necessary to provide a high-accuracy antenna pointing mechanism and a correspondingly accurate attitude pointing system for the satellite. High-accuracy pointing mechanisms for high-gain, large-sized antennas are complex and expensive, as are high-accuracy attitude pointing systems.
- Free space progressive attenuation, that is, the unavoidable dispersion of the signal beam as it travels along its path from the satellite to the ground station, is by far the dominant loss in the link budget. It depends on the distance between the satellite and the ground station and is thus directly determined by the orbit selection, which is directly driven by the needs of the consumers. The orbit chosen and the resulting satellite-to-ground distance will govern the link budget.
- Attenuations due to the atmosphere, including attenuations due to water vapor and rain. These are unavoidable factors that the design must incorporate. They are greater at higher frequencies and when the satellite is low over the horizon. These attenuations limit the satellite's minimum elevation above the horizon needed for successful data interchange.
- Losses due to mispointing of the receiving antenna. The pointing accuracy of any ground station is always commensurate with its dimension; the larger the antenna diameter, the larger the antenna gain, and so the highest pointing accuracy. Cost is the main limiting factor, as large diameter, high-accuracy ground antennas are rare and expensive to operate.

The distance between the satellite and Earth varies widely according to the nature of the mission and the type of orbit used. This will be discussed further below.

11.6.2.1. Flying near earth

An important design consideration in operational communications is that the link shall be possible regardless of the satellite's attitude, that is, even with the satellite out of nominal orientation or while tumbling. This implies that the on-board operational communication antenna must be omnidirectional. Two hemispherical coverage antennas normally provide this. Sentinel-3, SSTL-DMC, Cluster, and Rosetta all employ a pair of hemispherical antennas providing contact with the satellite regardless of its orientation. The data stream speed necessary to take control of a satellite is modest; that means, the low antenna gain associated to the hemispherical antennas is not a problem.

Housekeeping data streams are normally small. Nevertheless, if housekeeping and science downlinks are mixed, as they are in space science, the data streams will be larger. Near to Earth, very high data streams are compatible with moderate gain, moderate-size antennas. Sentinel-3 downlinks information at 450 Mbit/s but uses wide-field-of-view antennas. Cluster's antennas, as described in Section 11.4.1, also provide hemispherical coverage and are fully capable of downlinking all the information, including both housekeeping and instrumental, produced by the mission. In missions flying near to Earth the data speed is compatible with wide-field-of-view antennas.

11.6.2.2. Deep space missions

In this type of mission the large losses associated with the long distances between the satellite and Earth presents an entirely different scenario. The power arriving at the receiving antenna decreases with the square of the distance, and this degrades the link budget. To compensate for this, it is possible to increase the antenna emitting power, though any improvement will be limited by the resulting power demands of the satellite. In deep space missions, it is essential to use the largest ground antennas, with the highest gain, pointing accuracy, and quality. There are few stations suitable for this purpose, and building new ones is a major and costly endeavor beyond the means of single projects. This is a general infrastructure investment for use by many projects. Mission designers are forced to choose from existing facilities, which means that selecting the ground stations for a deep space mission is not a design decision but rather a mission constraint. The only really free design parameter is the satellite antenna's gain. Large, high-gain communication antennas dominate the configurations of deep space missions like Rosetta, in Figure 11.6, or Ulysses, in Figure 10.17.

Satellites at great distances from Earth will use with advantage long contact times with ground stations because the amount of data produced by a mission will be higher in cases where longer downlink times are permitted. A limiting factor is that it is impractical to fully book a ground station for a single satellite. The few ground stations capable of receiving data from remote satellites must share their time among all the deep space probes needing their services. Hence, the time available for downlinking is not a technical decision, but rather a programmatic constraint resulting from negotiations between the various space missions simultaneously requiring the ground station's services.

Whatever the tools used to improve the link of a deep space mission, the amount of data transmitted will be small in comparison with the consumer's wishes. The scientific community would appreciate submeter resolution images of the large areas of Mars, and cameras exist with that capability, but there is no way to transmit the hundreds of Mbit/s[22] of resulting data throughput to Earth. This figure contrasts with the few kbit/s that is currently feasible. Inevitably then, the science to be performed will be limited by data downlink's unavoidable limitations. To alleviate this problem, efforts at all levels are needed, most fundamentally antennas with improved gain both in orbit and on the ground. On the ground it is possible to use arrays with a large number of medium-sized antennas.[23] Nevertheless, the heavy penalty in mass and volume inherent to deep space missions, and the heavy investment needed for large arrays of ground antennas, makes the solution to this problem especially difficult.

In the previous discussion about near-Earth satellites, reference was made to the problems associated with out-of-control satellites or those in any other emergency situation. This also applies to deep space satellites. To cope with these problems deep space missions also include wide-field-of-view links. As shown in Figure 11.8, Rosetta includes a pair of full coverage S-band antennas and a medium-gain S-/X-band antenna. NASA's mission Galileo[24] was a complete success despite the failure of its high-gain antennas. All the data produced by the mission were slowly down-linked to Earth through a wide-field-of-view communications link not originally designed for this purpose. This is a good example of the capacity of a mission to provide service beyond its specified range, i.e., a robust system design.

11.6.3. Computer load budget

The computing power of processors on board the satellite must be compatible with the processing load, which is a function of the level of autonomy, and of the complexity of the satellite. The factors driving the number and functions of the on-board computers were discussed in Section 11.1.2. Given the large degree of uncertainty in the early stages of a project, very large safety margins for computer processing power, software size, satellite monitoring, and control memory are crucial. During the early design phases, safety factors of at least two are perfectly reasonable. This budget normally does not present complicated interfaces and can be left safely in the hands of the specialist.

11.6.4. Operational on-board storage

Operational mass memory for housekeeping data needs is normally moderate. The memory required for on-board storage of the data produced by the instrument is much larger and will be discussed in the next chapter. The operational memory must store operational on-board software and operational procedures for long periods of time as

[22] Evolution of the Communications System and Technology for Mars Exploration, Ramon de Paula et al., Acta Astronautica Vol 51, Pergamon Elsevier 2002.

[23] Low Cost Architectures for Large Arrays of Small Antennas Dayton Jones, IEEE AC paper #1199, IEEE 2005.

[24] http://solarsystem.nasa.gov/missions/profile.cfm?MCode=Galileo

required to perform the satellite's functions. The operational memory also stores the commands sent to the satellite by the mission control center. The storage capacity shall be determined by the frequency of updating of commands plus a safety margin in case of failure in the uplink. The mass memory must also store all the data related to monitoring the satellite's status and the instrument and housekeeping data. This data will be downlinked at each opportunity to the operational ground station, though it could also be downlinked together with the data produced by the instruments during the normally more frequent instrument data output downlinks. Given the high degree of uncertainty in the early determination of the sizing parameters, on-board storage is calculated with safety margins of at least two, as described in the previous section. In some cases it is desired that all housekeeping data be downlinked.

11.6.5. Data acquisition delay budget

The total time delay between image acquisition request and image delivery was discussed in Section 8.4.6. The system acquisition delay part that is relative to the operational data flow has been discussed in Section 11.5.3. The mission specification will apportion the overall delay between all their contributors. This apportioning will establish a requirement for the response time necessary for the end-to-end system to start the acquisition of the data requested. The specified delay time shall be divided and apportioned to the factors contributing to it as indicated in Figure 11.12:

1. The time allocated for reception of the consumer's request, which can be reduced by computerized fast handling of the requests
2. The time allocated for the planning of the mission that can be improved by enhanced computerized handling supplemented with human supervision
3. The time needed for the ground station to uplink the information to the satellite, which is fully determined by the orbit, the number and location of the ground stations and the possible employment of data-relay satellites for commanding
4. The delay in acquiring the information, which depends on the first opportunity for observing the relevant location offered by the orbital path plus the time necessary for the satellite to finish acquiring the data requested

Steps 3 and 4 are fully determined by the orbital geometry and the definition of the area to be acquired and do not require safety margins. The overall delay of steps 1 and 2 will depend on the total time allocated for all the steps minus the time needed to perform 3 and 4. The time remaining for the execution of step 1 plus 2 will be used for the specification of the handling of the consumer's request and for the mission operations including the planning of the observations. As mentioned above in Section 11.2, the early designs for a space system do not include a detailed physical design of the ground segment, but instead focuses on the functional design. Ground processing experts will be able to advise the architect if the values allocated provide comfortable margins and are compatible with the existing and likely evolution of the ground segment capabilities. These delay values will also be crucial to the estimation of operational costs, needed to establish a credible cost estimation for the end-to-end mission because fast and permanent handling of consumer requests is expensive to implement. The provision of cost estimations will also be in the hands of the ground segment specialist.

11.6.6. Level of service and availability budget

The objective of satellite missions is to acquire the information requested by the consumers. The organization of satellite operations, including interfaces with the consumer, must maximize the time dedicated to this fundamental task. This factor will determine the level of service offered by the mission, and has two aspects:

- What percentage of time can consumers request specific observations?
- What percentage of the total time are the instruments actually acquiring data?

Offering a high degree of availability to consumers entails the creation and maintenance of structures that permit them to request information at any moment, without limitations. The satellite requires time to perform procedures such as orbit correction, maintenance, or monitoring that may interrupt data acquisition. Some missions require a high level of availability, while others possess varying degrees of tolerance to interruptions in data gathering. Generally, interactive missions require fast reaction times as well as high levels of availability, while systematic missions studying non-dynamic events are more flexible regarding temporary data interruptions. However, systematic missions studying dynamic events, such as meteorological satellite missions, also demand the highest level of availability. Availability requirements higher than 95% for meteorological mission are perfectly reasonable.

The requirements on system availability and level of service must be apportioned between all the factors contributing to interruptions of the services. The apportioned values will determine the design of numerous elements of the mission, including orbit control tools, orbit control frequency, the need for special calibrations forcing the interruption of the mission, or the possible need for a permanent mission control team available 24 h per day and 7 days per week.

It is evident that design decisions necessary for providing high availability levels will have a major impact on operation costs and must be taken into account if a credible cost estimation for the mission as a whole is to be formulated.

Chapter 12

The Instrument Output Data Flow Domain

Keywords Instrument Data Flow, Satellite Ground Segment, Satellite Data Processing, Satellite Data Distribution, Satellite Data Storage, Satellite Users Interfaces, Data Volume Budget, Users Data Request, Users Data Delivery, Data Latency

This domain was introduced in Section 7.1.5. This chapter starts with a brief introduction to the elements and components – ground and space – included in this domain. As it is strongly related to the previous chapter, the present chapter will use as examples three space missions already discussed in Chapter 11: Rosetta, Cluster, and Sentinel-3, as well as a new one, NOAA POESS. NOAA POESS produces very large amounts of data that need to be processed at high speed, and uses four different paths for data downlink and processing. This makes it an interesting mission to analyze.

In Chapter 11 the key point of discussion was how autonomous the operations of the mission should be. In this chapter and recalling the three categories of mission effectiveness of Figure 5.8 – data quality, data quantity, and system responsiveness – the main points of discussions will be the data quantity and the latency aspect of system responsiveness.

The topics covered in this chapter are those indicated on the right-hand side of Figure 7.4 expanded here in Figure 12.1.

The instruments data handling will be discussed in Section 12.1.1, the data downlink in Section 12.1.2, the payload ground stations in Section 12.2.1, and the payload processing and users services in Section 12.2.2. The next section will analyze the four above-mentioned mission examples. As Chapter 11, this chapter will finish with two sections analyzing the allocation of functions and allocation of budgets included in this domain.

12.1. In-orbit components involved in the domain

12.1.1. Instrument output data handling

These components capture, digitize, store, and process the output of the instruments. The instrument data handling must also process supplementary data, for example,

M.A. Aguirre, *Introduction to Space Systems: Design and Synthesis*, Space Technology Library 27,
DOI 10.1007/978-1-4614-3758-1_12,

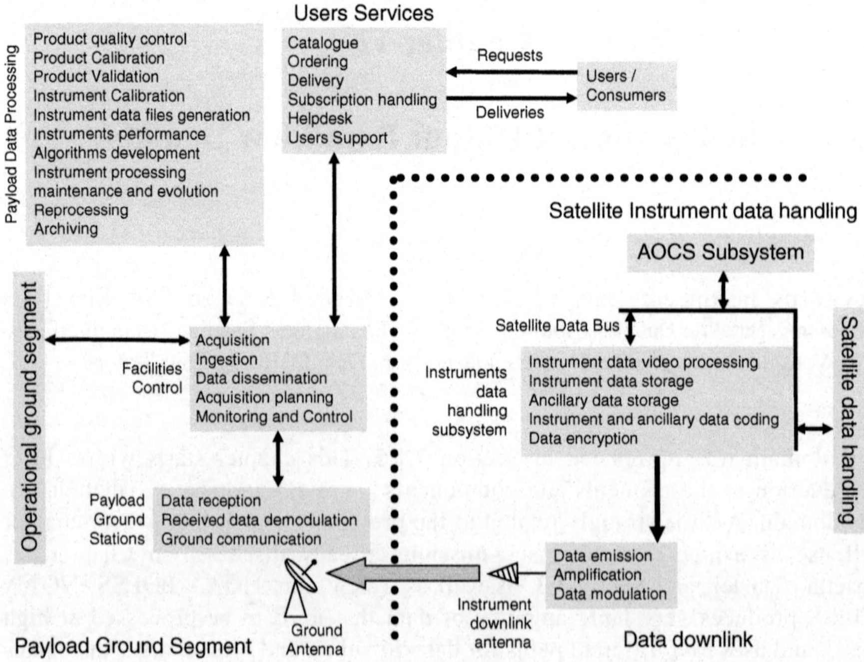

Figure 12.1. Instrument Data Flow Design Domain Elements and Interactions.

satellite attitude, position, or time of acquisition, indispensable for the correct interpretation of the acquired data.

The necessary components are the same as those described in Section 11.1.2: general-purpose and dedicated processors, storage memories, and data buses. The difference is that modern instruments can produce huge amount of data; one modern SAR instrument can produce several hundred Mbits per second of operation. This high data rate requires specific solutions for data processing distribution and storage. Application-specific integrated circuits (ASIC's) and fast analog-to-digital converters are developed for the capture and digitization of the information flow produced by the instruments. High-speed data buses like Spacewire,[1] based in IEEE 1355, can provide data transfer speeds as high as 400 Mbit/s. Digital signal processors or dedicated general-purpose computers can also be used for processing the instruments' output. The on-board storage of large amounts of data requires the use of large solid-state mass memories. It is essential that these storage devices be capable of working in the high radiation space environment with high reliability. This requires the development of radiation-hardened modules, which include error detection, correction components and redundancy.

The fundamental design decisions are all connected with the amount of data produced by the instruments, which is driven by the requirements of the observables. These

[1] SpaceWire Nodes Links and Networks, ECSS-E-50-12, ESA 2003.

design decisions include: the speed of the data handling and processing hardware, the capabilities of the data bus, and the storage capability of the on-board memory.

As in Section 11.1.2, the main problem is to achieve the right balance between low-risk conservative design and high performance. These decisions are especially important for the design of the mass memory, which can be a heavy item requiring large amounts of mass and power. Newer memories always offer larger data storage capacities using smaller and lighter components, but they cannot be flown without the necessary evaluation and development to verify their adequacy for working in space.

The implementation of on-board instrument data processing is also an important decision of the highest level of significance. On-board processing, including data compression, can reduce and simplify the data storage and transmission problem, but not all applications can tolerate data compression. Irreversible data compression algorithms should be avoided for scientific and advanced applications. All new space development results in the discovery of unexpected applications, so access to the uncompressed raw data is essential. The use of interferometric SAR images for deriving the displacement of the Earth's crust before and after earthquakes is a good example of an important application that was not considered when the satellite was being designed.

Irreversible on-board data processing could destroy information that eventually might become crucial. On the other hand, irreversible data compression can be tolerated for well-characterized applications, where the amount of information lost during compression is known. This is often the case in operational missions.

The hardware shall be selected from a small set of available "space compatible" components either already ready or in the course of development, while design decisions will normally be in the hands of the specialist engineers. However, the decision on the overall architecture and the sharing of functions between the different components of the architecture is in the hands of the architect.

12.1.2. Instrument data output downlink

The data output downlink components code, encrypt, modulate, amplify, and transmit to Earth the information produced by the instrument and satellite and are required for the correct processing of the instrument data stream.

The components used for the instruments downlink are the same as those described in Section 11.1.3: encoders, amplifiers, modulators, and antennae. The difference is that the very high data rate calls for the use of specific design solutions and further limits the availability of hardware adequate to the relevant functions. In line with the frequency limitations explained in Table 11.1, the bands allowed for instrument data downlink as well as the allowable bandwidth and the maximum power flux density arriving to the Earth surface are specified by ITU regulations and must be respected. This will introduce further design limitations. High data throughput downlinks must use efficient modulation schemes capable of carrying as much information per Hz of used bandwidth as possible. In cases where the information produced by the instruments is of a sensitive nature, for example, high-resolution images with intelligence potential, the inclusion of special components for the encryption of information becomes essential.

One way to increase the data throughput is to increase the frequency of the carrier of the data stream. Earth observation satellites use the X-band but could use the Ka-band in cases where higher speed was needed, because there is a higher bandwidth available in that range. It is also possible to use optical data links. Optical data links could be used to beam information to ground stations or for sending it to data relay satellites in the geostationary position. The Earth observation mission Spot 5 uses an optical link to the geostationary communications satellite Artemis. The problem with optical links to ground is the high atmospheric attenuation and that clouds or haze will interrupt the communication links. Inter-satellite links, on the contrary, are free of this problem and greatly benefit from optical links

It is also common with this type of application to use different downlink chains for different types of consumers. NOAA POESS currently uses four different data downlinks to send data from the satellite to ground. Each one features its own dedicated encoding, modulation, and amplifications, as well as antenna. This is described in Section 12.3.4.

The first and most important design decision is how many and how fast the data downlinks shall be, and what the resulting amplifier power and antenna gain of each one of the downlink paths should be. The amplifier power will be limited by the availability of equipment. The availability of amplifiers and the distance from satellite to ground station will also determine whether a high-gain antenna is needed or not. If pointable high-gain antennae are used the pointing concept must be selected: mechanical or electronic. This will drive the mechanical accommodation design solution for the antennae.

Another important decision is whether or not to use data encryption. This is a programmatic constraint that is provided to the designer as a compulsory guideline. The domain specialist selects the coding and encryption algorithms. The use or not of encryption will also impact the transmission power required.

There is not a great variety of hardware capable of fulfilling these demanding functions. Furthermore, many design features in this area are strongly regulated: the frequency to be used for the downlink, density of power in the downlink, bandwidth allocated, algorithms for coding, and the standards for the packing of the data streams. All these factors combined limit the freedom of the designer. Nevertheless, these limits are counterbalanced by the freedom to select the overall data downlink architecture and allow quite different sets of solutions for very similar missions. This is a design area where the architect has to be proactive. The specialists will always provide the best design solutions compatible with the chosen architecture, but it is the duty of the architect to provide the architecture with the number and speed of the different data paths.

12.2. On-ground components involved in the domain

As in the case of operational data flow, the instrument on-ground components are normally not designed for the particular mission being designed, but are selected within an already existing infrastructure that has been implemented for use by several space missions. This is made possible by the use of widely accepted standards for the linking of information from space to ground. A supplementary factor encouraging the use of existing elements is the existence of many ground stations that can provide services to the satellite. The reception and delivery of payload data is much less critical than the commanding or monitoring of the satellite, that is why it is considered perfectly acceptable to use ground stations and data networks not under the direct control of the mission owner. One interesting exception to the use of existing infrastructure is the SafetyNet™[2] concept of NOAA POESS that was created for the special needs of this operational meteorological project. This system is discussed in Section 12.3.4.

As detailed in the previous chapter, during the preliminary design phase prior to program approval, it is vital to perform a functional design of the ground segment, but unnecessary to perform a physical design; therefore, only functional design will be covered in this subsection.

12.2.1. Instrument downlink ground stations and data relay satellites

Ground stations and data relay satellites transfer the large amount of information produced by the instruments from the satellite to the processing centers. In the case of centralized instrument data processing, the level of processing performed in the stations is usually small: the data are demodulated, decoded, and separated into data streams and each stream is sent to its specific processing center. It is perfectly possible to perform these functions with the data encrypted until it arrives at the final processing center. This would be advisable in cases where the owner of the space system does not control the ground stations under discussion.

In some cases the instrument data processing is decentralized and each ground station will possess its own data processing and mission products derivation components. The NOAA POESS Direct Readout Mission is a good example of this approach, and will be described in Section 12.3.4.

The fundamental design decision on this area encompasses how many ground stations and how many classes of ground stations are required, and if data relay satellites will be used. The design will be driven by the amount of data produced and the resulting data throughput requirements, in addition to the acceptable data latency. It is necessary to recall that maximum data downlink speeds will be finally limited by the frequency allocated by the ITU regulations as provided in Table 11.1.

Another important decision is what tools will be used to transfer the data from the ground station to the processing centers. Ground stations are often in difficult-to-reach places like Antarctica or Svalbard in the Arctic, which are very well situated for downlinking data from high-inclination Earth observation satellites. Svalbard has optical data links connecting it to mainland Europe, but the only way to relay the data from Antarctica is via communication satellites.

Determining the downlink data flow and the number of ground stations is another design decision where the architect has to be proactive. Downlink speeds are limited by regulations, the use of a large number of ground stations is expensive, data relay

---

[2] Latency Features of SafetyNet™ Ground Systems Architecture NPOESS, James Duda et al., Proceedings of SPIE Vol. 5659, 2005.

satellites are rare, and consumers may require downlinking of very large amounts of data. This is the typical situation of conflicting needs that must be balanced by the architect after taking into account all the conflicting factors.

The number of ground stations and the use or not of data relay satellites is also strongly influenced by nontechnical program constraints. Ground stations and data relay satellites are expensive infrastructures built to last for many years and to service many satellites. It is inevitable that the final design decision shall be made based strongly on the use of existing infrastructures. Cost considerations become a major factor in defining the final architecture.

12.2.2. Payload data segment

The payload data segment processes the data produced by the satellite and distributes it to the mission's data consumers. The essential functions include:

- Processing of the information, to derive the required mission data products
- Control and management of facilities related to the instrument's data flow
- Long-term storage of instrument output and all other information relevant to the correct interpretation of this output
- Interfacing with consumers, receiving requests and delivering mission products

Processing the data produced by the satellite requires the calibration and validation of the output of the instrument. The Committee in Earth Observation Satellite (CEOS) defines calibration[3] as "The process of quantitatively defining the system responses to known, controlled signal inputs" and validation as "The process of assessing, by independent means, the quality of the data products derived from the system outputs." To calibrate means testing and adjusting the accuracy of the measuring instruments. To ensure their accuracy the instruments carry devices that monitor their environment, and can correct the sensitivity of the instrument by checking the temperature of its detecting elements, or by identifying deviations in the instrument's behavior by comparing the instrument's output to well-characterized inputs. This has been already discussed in Chapter 8 when analyzing instruments. Further qualitative improvements can be obtained by comparing the instrument's results with results provided by similar instruments in other satellites observing the same events. In some cases it is possible to perform "in situ" determination of the parameters under observation that can be compared with the read-outs provided by the instruments. System validation will involve activities that include the use of references external to the space system itself. Validation shall be used to provide absolute references of quality. Each instrument's measurement must also be geo-referenced to a specific location (Section 9.9.3) on Earth or on the celestial sphere. To perform this function the payload data segment needs data on the satellite's position and attitude at specific times used by the system's space and ground segments. After calibration and referencing, the data will be processed according to special algorithms to produce information

[3] http://wgcv.ceos.org/wgcv/wgcv.htm#defs

with an increasing level of relevance to the analysis of the physical reality of the events under observation.

The data processing will normally have two levels:

- A first set of basic data processing that produces calibrated and geo-referenced (Section 9.10.4) "basic" data. This processing will be common and necessary to all applications that will use the data produced by the mission.
- A second set of data processing that will be different and optimized for each possible use of the mission data. This second level of processing would normally require the use of large amounts of data external to the system that will increase the value of the resulting products.

Data processing sometimes requires the acquisition of external information. The payload data segment should be able to acquire any needed external information. All data provided to consumers must be quality controlled. The calibrated and located data, once controlled for quality, must be provided with the speed and in the mode required by the consumer.

The payload data segment requires the handling and management of large amounts of data from many sources, some of them external to the space system. This requires the establishment and maintenance of adequate interfaces. Of special relevance is the payload data center interface with mission operations. Mission operations will provide information on the status of the satellite relevant to improving mission products quality to the payload data center. In addition, the payload data center, by controlling the quality of mission products, could provide information on possible malfunctions to the mission operations center. The payload data center is also normally the place where instrument use and image acquisition plans are established; nevertheless, this function should be shared with the mission control center, which commands a better view of the overall status and capabilities of the satellite. The data quality control function may sometimes be allocated to a specific facility, for example, where there is the specialist expertise available to monitor and check for consistency, errors, trends, and quality.

The payload data segment shall also provide long-term storage of the data produced and allow for the proper archiving of the documentation associated with the processing of data. The algorithm and methods used for calibration are developed in conjunction with the space mission design; nevertheless, they can be improved after launch, as the behavior of the satellite in orbit becomes better understood. Hence, facilities for the reprocessing of old data with the updated algorithms must also be provided.

The payload data segment's last group of functions involves interfacing with consumers: the delivery of mission products and reception of requests. This interfacing may take place in two rather different environments:

- In payload data segments for advancement of science missions (Section 4.5.3) that entail frequent interacting with a select set of researchers usually called "principal investigators." Cluster and Rosetta are examples of this.
- In payload data segments for operational missions, which require the delivery of data to a large and varied community. Sentinel-3 and NOAA-POESS are examples of this.

The "astronomical observatory" missions described in Section 7.2 share both environments: they normally feature a restricted core of principal investigators with privileged access to the mission, and a larger set of general consumers that must follow more complex procedures to obtain data. The type and form of delivery of mission products affects the interface with the consumers. Missions that routinely deliver data of a single or few types will require simple interfaces with the consumers. On the other hand, missions that have a strong service request character, where different users will have very variable needs requiring the delivery of different products and where different products can have different time delivery priorities and different quantities or throughput, will require the design and implementation of complex interfaces with the consumers. Fulfilling consumers' multiple needs requires the implementation of dedicated organizational structures for both the reception of their requests and delivery of the pertinent mission products. In addition, the mission may require different levels of service quality. One extreme will be operational services needing to ensure high levels of availability, which will require permanent services and guaranteed delivery of data under any circumstances. This requires not only adequate interfaces with the consumers receiving these permanent services, but also the use of communication links able to provide absolute reliability as regards availability. Operational meteorology or defense-related ventures are examples of missions requiring this guarantied availability of services.

As a rule the principal investigators have a high level of knowledge about the mission because they would have been deeply involved in its implementation; they are project insiders. As regards general consumers, for either scientific or operational undertakings the establishment of networks capable of receiving requests and delivering products is crucial. The implementation of these functions entails a heavy investment. The Space Telescope Science Institute of Maryland, described in Section 7.2.1, provides a clear illustration of the extent of these necessary investments.

It is clear that the definition of the payload data processing is important and that adequate effort must be dedicated to an early definition of its functions and architecture. The following decisions must be made:

- Identification of the fundamental building blocks for the payload data center, and determining if the existing infrastructure will be adequate for providing the required services or if new or additional infrastructures are needed.
- Where to locate the instrument processing: on the ground or in orbit. This will be discussed in Section 12.4.4.
- Centralized versus decentralized data processing. This will be discussed in Section 12.4.6.
- The identification of any special long-term calibration, external data validation, or reprocessing needs.

The role of the architect in the definition of the payload data segment is strongly dependent on the mission constraints. For advancement of science missions handled by authoritative and sophisticated "principal investigators," the latter will assume the leading role, with the architect supporting them in the definition of the payload data segment.

In operational systems, responsibility for the definition of the architecture payload data segment might be in the hands of different organizations. On both sides of the

Atlantic there are space research and operational agencies: NASA and NOAA in the USA and ESA and EUMETSAT in Europe. In cases of collaboration between a research and an operational agency, the share of responsibility might have been decided before the design process even started. Nevertheless, the architect's view will always be valuable because it provides an end-to-end understanding of the mission. This perspective is essential, since it will ensure a properly designed space mission, not just a properly designed satellite. On the other hand it is the specialist who will always advise the architect on what architecture is possible, as well as on instrument data processing costs. The cost aspect is especially important given that the cost estimation of the ground data processing shall be based on a functional design, without the greater certainty offered by a physical design.

12.3. Examples of architectures

12.3.1. Cluster

Cluster is a mission led by principal investigators (PI) that have been involved in the mission from its origin and who are familiar with the satellite and the instruments. As with many space science missions, the downlink and processing of both operational and instrumental data is accomplished in the same data stream to the same ground station, and all the data – instrument data and satellite status housekeeping data – is processed by the mission operations center. The specialized function of the mission operations center for the processing of mission products is known as the Central Data Disposition System (CDDS) and is located at the Mission Operations Centre in ESOC, Germany. There is also a Joint Science Operations Centre, which helps the mission operations concentrate on instrument mission planning and authorizes non-PI researchers to submit requests for obtaining data to the CDDS. All housekeeping telemetry and other mission data will be made accessible, in addition to the relevant science telemetry and timing information. There are two paths for accessing the processed data:

- Quasi-real-time access via electronic network
- Off-line access via data media (CD-ROMs)

Quasi-real-time access will be available only to the principal investigators and to the Joint Science Operations Centre.

The CD-ROM's contain the same data accessible electronically through the CDDS, but it is produced and shipped by mail. Scientists interested in doing research with Cluster but who are not PIs receive a copy of the CD-ROMs mentioned above. An average of two CD-ROMs per day are needed to store all the data produced.[4] The volumes of data for distribution – around 0.8 Gbit/day – and they way they are generated, stored, and downlinked, have been already discussed in Section 11.4.1 (Figure 12.2).

[4] From P. Ferri CNES 195 Workshop on Mission Operations.

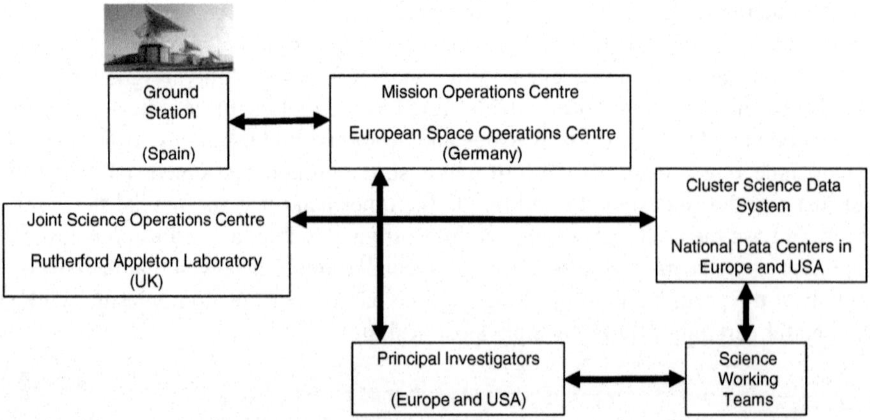

Figure 12.2. Cluster Scientific Data Distribution (The Cluster Mission Operations Concept, P. Ferri et al., Paper to CNES Mission Operations Workshop 1995).

12.3.2. Rosetta

As in the case of Cluster there is a single data stream that includes instruments and operational data, which is downlinked to the ground station and processed together at the same mission operations center.

Rosetta mission operations are divided in two parts:

- Rosetta mission operations center (RMOC) for satellite operations
- Rosetta science operations center (RSOC) for science operations

The RSOC works in close cooperation with the RMOC and are both located in Darmstadt (Germany).

The principal investigators will support RSOC during the critical periods when the spacecraft reaches the comet, and RSOC will be the focal point for all inputs generated by the principal investigators. It is the responsibility of RSOC to coordinate all inputs and to propose a strategy for overall payload operations near the comet. With the help of the PIs, the RSOC and the RMOC will establish the operational planning and the "near-real-time" science analysis to cope with the complex and interactive operations required near the comet's nucleus. The RSOC will also perform the scientific data processing distribution and archiving of the data produced by the mission. The scientific operations will be performed during the available 9–12 h of daily contact between the single ground station in Perth and the satellite. A separate ground segment will be established for the operations of the Lander Philae. The Lander PI's will also generate their operations requests that will be sent to an independent Lander Science Center and from there to a Lander Control Center. The consolidated operational plan for the Lander will finally be directed to the Rosetta mission operations center for uplink to the Rosetta satellite, and from there to the Lander Philae when needed.

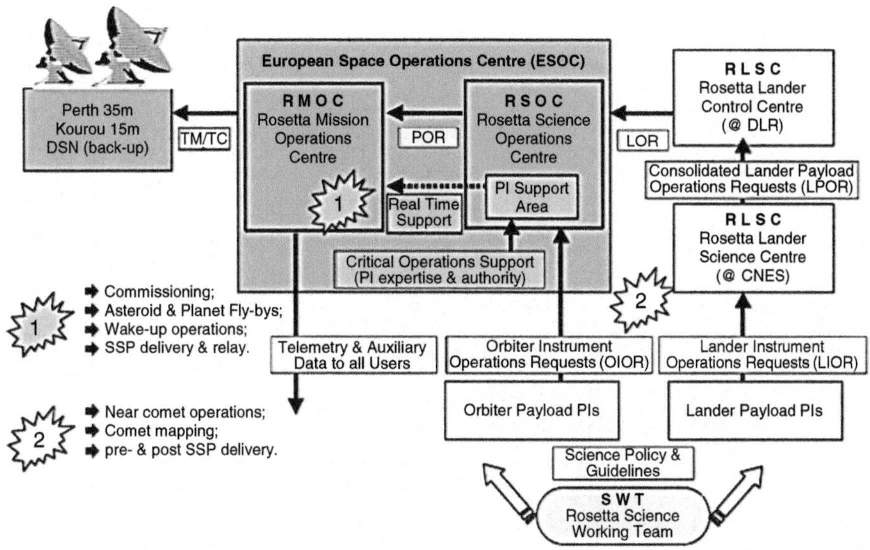

Figure 12.3. Rosetta Instrument Data Flow and Science Data Distribution (Science Operations Implementation for Rosetta P. Ferri, Paper to SpaceOps Workshop 2000).

The non-PI researchers are federated in Science Working Teams and will produce requests for observations that will be judged by the PIs using previously established guidelines (Figure 12.3).

Rosetta has a storage memory of 23 Gbits that is used as buffer to even out the continuous data acquisition with the discontinuous periods available for data downlink. The buffer mass memory also allows for the production of data at fast speeds, while downlinking it slowly to Earth. After proper processing, the data generated by the instrument will be delivered to the PI and to the other scientist through the data disposition system (DDS), which also forms part of the RMOC. The DDS will support direct delivery of data through the network and physical delivery of data through physical media (CDs or DVDs).

12.3.3. Sentinel-3

The relative simplicity of the data distribution approach of the missions described in the two previous sections, whose products are only of interest to a small group of scientists contrast to the complexity of Sentinel-3, which addresses itself to a large community of users interested in global ocean and land information. Figure 12.4 provides the overall data flow of Sentinel-3. Figure 11.9, in the previous chapter provides details of the satellite's electronics and also the instruments data flow. The data produced by the instruments are transmitted by high-speed Spacewire links and stored in the large on-board mass memory located in the Payload Data Handling Unit (PDHU). The data is downlinked by X-band communications, which uses a low gain, wide field-of-view antenna, which does not need to be pointed to provide adequate link strength. Sentinel-3 produces a large amount of information that has to be

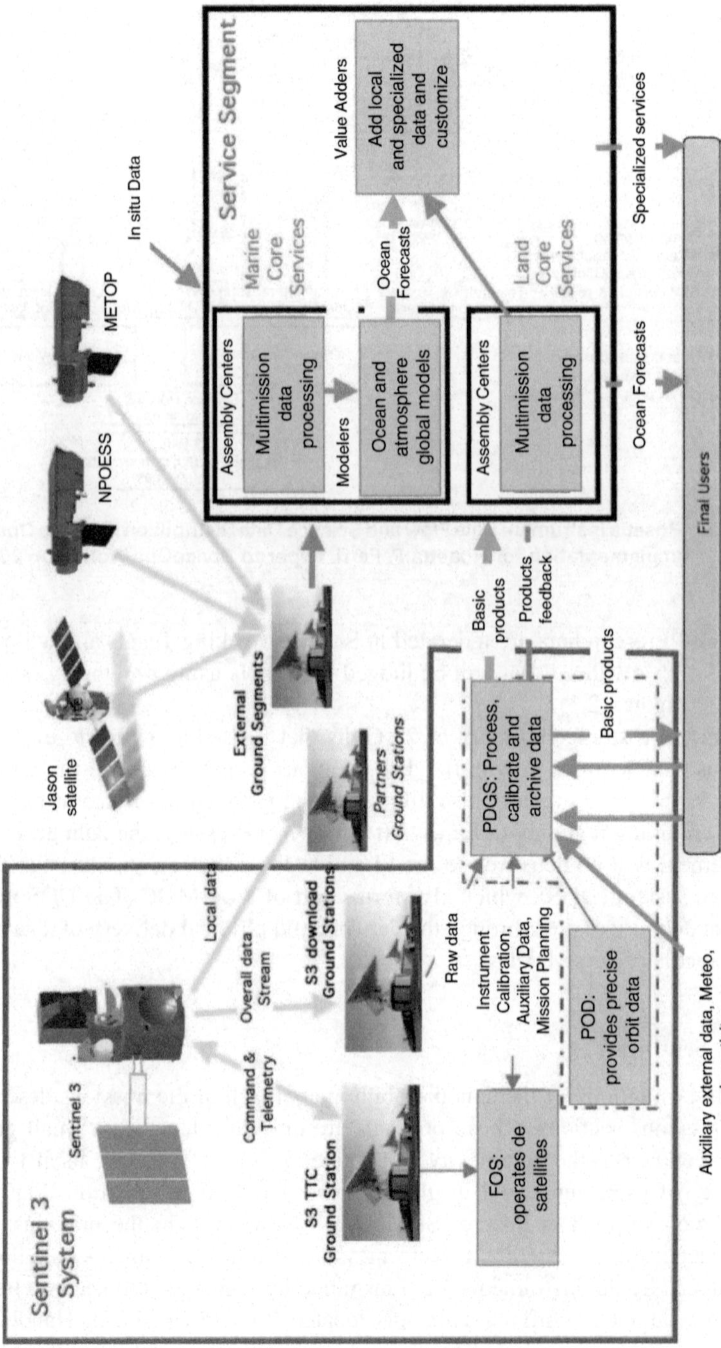

Figure 12.4. Sentinel-3 Overall Data Flow (GMES Sentinel-3 A Long-Term Monitoring of Ocean and Land to Support Sustainable Development, Y. Baillion, M. Aguirre, IAC 2007).

routinely delivered in a highly reliable fashion. Sentinel-3's three instruments continuously observe the Earth, producing 200 Gbits of information per orbit. This large amount of information can be downlinked at 450 Mbit/s; 8 min of contact is sufficient for sending the data generated in one orbit to Earth.

Sentinel-3's main ground station is located in Kiruna, Sweden, but uses Svalbard for contact during the two or three contiguous orbits when there is no contact with Kiruna. The data received is sent from the ground station to the Payload Data Ground Segment (PDGS). The processing of Sentinel-3 is complex and requires the use of external information. See level anomalies are used to derive ocean currents, and are observed by Sentinel-3's radar altimeter, but this only provides the distance from the satellite to the surface of the ocean; to be able to provide sea surface altitude, the data needs to be corrected by determining with, with high accuracy, the orbit of the satellite, and this requires precise orbital data from GPS satellites. This activity is called "precise orbit determination" (POD of Figure 12.4). Sentinel-3's ground segment also includes a Flight Operation System (FOS) to handle the operations of the Sentinel-3 satellites. All the calibrated and corrected data produced by the PDGS are eventually delivered to some final users that will prefer to use them directly, and to the "Service Segment" that is external to the Sentinel-3 system. This Service Segment includes:

- Ocean and Land Specialized Assembly Centers that will analyze and compare information provided not only by Sentinel-3, but also by other satellites delivering similar information. This will permit the derivation of multi-instrument, multimission integrated products, and services. This multimission information will further improve the quality of the final data and of the derived predictions.
- Modelers, who will use Sentinel-3 and other data to create ocean, land, and atmospheric models for the elaboration of predictions about the evolution of the atmosphere, oceans, and vegetation.
- The data produced by the Ocean and Land Specialized Centers and by the modelers will be provided to "value adders" that will further process the data including local information – like population density or economic value at risk – and customize the results according to the needs of the different specialized users.
- These "value added" products will be finally delivered to end-users that will use them as help in decision making.

All the mission products shall be distributed in near-real-time in less than 180 min from acquisition. Sentinel-3's orbit has a period of ~90 min and uses a single X-band downlink ground contact per orbit; either with Kiruna or with Svalbard ground stations. This single contact allows one downlink to ground per orbit. This approach allows a maximum of 90 min from acquisition to downlink and another 90 min for on-ground data processing and distribution. Actually Sentinel-3's Payload Ground Segment does not provide a single data stream but three (Figure 12.4):

- The above-mentioned Near-Real-Time (NRT) data stream, which has to be delivered with a data latency shorter than 180 min.
- Slow Time Critical (STC) data that has to be provided in less than 24 h and that is used mainly for oceanographic operational monitoring. The supplementary time

permits improved corrections to be made – mainly improved precision orbit determination – and better products accuracy.
- Non Time Critical (NTC) data that includes the most accurate correction methods and that is used for long-term monitoring of the sea level and for global change application.

Sentinel-3 is also capable of providing real-time data to its X-band Partner's Ground Stations. The downlink data rate is the same as that used for the main stations Kiruna or Svalbard, but this service can only be provided a few times per orbit.

12.3.4. NOAA – POESS

12.3.4.1. Mission aim

The National Polar-orbiting Operational Environmental Satellite System (NPOESS) is the backbone for the operational provision of meteorological and environmental data.[5] This program is a fusion of the civilian POES program of NOAA and of the military DMS program of the American Department of Defense (DoD). It was established in 1994 as a common endeavor of NOAA, NASA, and the DoD. NPOESS will be a series of satellites and the system will be implemented progressively. The satellites will carry multiple instruments for the determination of atmospheric and surface data relevant for operational meteorology and environment monitoring. This is a very large endeavor and its instrument compliment has changed, with some instruments being disembarked or substituted. In its final shape NPOESS will fly as a constellation of three satellites with instruments providing horizon-to-horizon coverage. Each satellite is able to provide whole Earth coverage in 12 h and the three together are able to provide full Earth coverage every four hours. To do so, the local times of the ascending nodes of the three satellites are regularly distributed at 13:30, 17:30, and 21:30 local time. Figure 9.15 provided the orbital geometry of these three satellites. This will allow a frequent updating of the parameters under observation. All satellites will fly in a Sun-synchronous orbit at 833 km of altitude and with an inclination of 98.7°. The mass of each satellite is variable because they do not carry identical instrument sets but they will be all in the 5,000 kg-plus category. The first satellite will be launched in 2013. This constellation will be reinforced by the European MetOp satellite that will fly in a complementary orbital plane with a LTDN of 09:30 (Figure 12.5).

The design life for the platform is 7 years with 0.9 spacecraft reliability. It can provide up to 7 kW of power and it has a very high stability optical bench to ensure optical co-alignment between all the instruments.

12.3.4.2. Instruments output data flow

The atmosphere changes rapidly and users of meteorological information vary enormously. To fulfill the need for swift delivery of the different communities of consumers, the architecture of the NOAA-POESS data distribution is necessarily complex. It possesses four parallel data downlink systems (see Figure 12.6).

[5] NPOESS A Program Overview and Status Update, F. G Hinnant, 0-7803-7031-7/01, IEEE 2001.

Figure 12.5. NOAA POESS (http://www.st.northropgrumman.com/media/presskits/media-Gallery/npoess/photos/media1_4_16358_16359.html#).

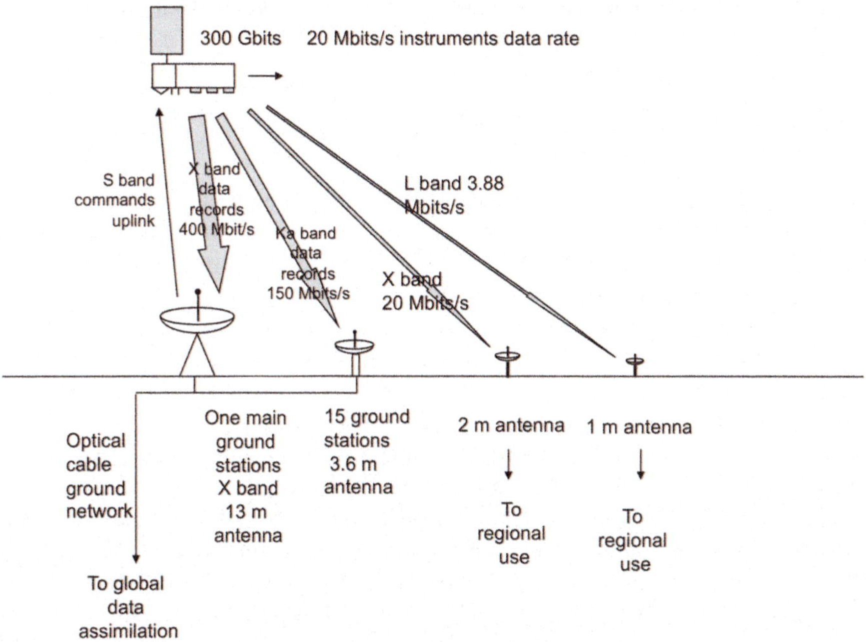

Figure 12.6. NPOESS Instruments Data Flow. Different Users Will Be Served by Different Downlink Ground Stations with Different Antennae Size.

To satisfy the most important consumers' needs for fast delivery of global atmospheric information, NOAA-POESS has implemented the SafetyNet™[6] data routing concept, which is able to provide 95 % of the data produced by the instruments in less

[6] Latency Features of SafetyNet™ Ground Systems Architecture for NPOESS, James L. Duda et al., Enabling Sensor and Platform Technologies for Spaceborne Remote Sensing, SPIE Vol. 5659, WA, 2005.

than 28 min after acquisition. The concept is based on the implementation of 15 Ka-band unmanned ground stations more or less uniformly distributed around the globe and providing an average of five contacts per orbit between ground station and satellite. When a satellite enters a SafetyNet station visibility circle, all stored data it has recorded is burst to the ground at a speed of 150 Mbit/s. Once all stored data is dumped, the satellite downlinks real-time data until it leaves the station visibility circle. For use in the gaps between station circles, NPOESS has a 300 GBit on-board solid-state recorder. When NPOESS comes within the field-of-view for the next SafetyNet™ station, the process is repeated. These stations are linked by a fiber-optic system to the central data processing centers where the modeling of global weather patterns is undertaken with the aim of providing whole Earth meteorological predictions.

The second system is an X-band downlink at 400 Mbit/s, downlinking once per orbit to a large, 13 m diameter antenna in Svalbard, in the Arctic. This download architecture is similar to Sentinel-3's: one contact per orbit that is downlinked at a speed rapid enough to empty the stored data produced during the last orbit.

To cope with the needs of consumers interested in regional meteorological predictions, the Direct Readout Mission[7] (DRM) broadcasts, that is, continuously downlinks data as the instruments produce them. DRM will service users in the field providing regional environmental coverage with two direct broadcast links:

- High rate data (HRD) in X-band at 20 Mbit/s compatible with a 2 m Ø antenna
- Low rate data (LRD) in L-band at 3.88 Mbit/s with a 1 m Ø antenna

The signal sent by the satellites to the field terminal is amplified, down converted, demodulated, decoded, and decrypted. Later it is processed and the environmental data records are generated. Ancillary data needed for the production of the data records, such as satellite position or attitude, is included in the same downlink. HRD and LRD ground stations are provided with their own data processing algorithms able to produce a large number of data products. The resulting data records can be produced at high speed, in 2–15 min, by computers of moderate performance. The small size of the requisite antennae, the availability of data processing software, and the modest processing power required guarantees that consumers interested in local meteorological information can acquire HRD or LRD components easily. Consumers with more demanding requirements will need to employ HRD, while consumers with less demanding requirements can acquire the much simpler and cheaper LRD. These two systems provide a fully integrated and easily acquired solution. This ensures maximum operational use of NOAA-POESS, optimizing the social benefits produced by the mission. This approach continues the legacy of previous NOAA operational meteorology missions, which also provided direct broadcast of mission products by the HRPT and LRPT subsystems. In HRD, 50 mission products (environmental data records) can be downlinked, while in LRD the limit is 29 with data compression.

[7] NPOESS Direct Readout Mission, John A. van de Wouwa et al., Atmospheric and Environmental Remote Sensing Data Processing and Utilization, SPIE Vol. 5890, 2005.

12.4. Allocation of functions

12.4.1. Large versus small amount of data

Instruments data output quantity has been already discussed in Section 8.4.4, which analyzed the factors driving the data rate. Depending on the nature of the observables, the data rates could vary widely: Cluster obtains data in situ, over a single point, and its data rate is just of 0.8 Gbits per day; Sentinel-3 produces 3,000 Gbits per day. High spatial resolution, large-swath instruments can produce several hundreds of Mbit/s. Spectrometers with very fine spectral resolution also have very high data throughputs. Synthetic Aperture Radars are notorious for the amount of data they produce. The transmission to Earth of such a quantity of data is difficult and the downlink capacity of the satellite can turn into a bottleneck for the whole mission. Providing high data flow requires:

- High-power data emitters in the satellite
- High-gain antennae in the satellite
- Higher downlink frequencies able to code more information on the available bandwidth
- High-power and high-gain antennae on the ground
- The use of data compression techniques
- The provision of long contact times by providing data relay satellites or using higher orbits
- The provision of long contact times by provision of a large number of ground stations
- The provision of long contact times by a judiciously locating ground stations in places with optimal satellite visibility

All the aforementioned tools have limitations: high-gain antennae are large in size; data compression techniques might be unacceptable for some applications; data relay satellites may be unavailable and are expensive, as is the use of numerous ground stations. The downlink speed is limited; once the downlink frequency is chosen, the maximum speed of the downlink is rigidly determined by the existing allowable bandwidth standards (introduced in Table 11.1 of the previous chapter). This implies that there are often absolute limitations in the data rate, forcing a reduction of consumer expectations. During Rosetta's rendezvous with the comet, scientist's employing the satellite would have welcomed larger data rates than the meager 5 kbit/s mentioned in Section 11.4.2; nevertheless, the large distance between Earth and the comet at the time of the observation imposed the limitation. Also due to satellite data throughput limitations, high-resolution Earth observation payloads can often only acquire images during a brief period of the satellite's complete orbital time. The rest of this chapter will focus on the advantages and disadvantages of increasing data throughputs of missions by various means.

Sentinel-3 provides regular and consistent information over the whole of the Earth. That means the instruments are continuously acquiring information. The only exception to this constant 100% duty cycle is the obvious one of visual imagers not being able to acquire information at night. Being global, the optical instruments have a high swath width and provide information in many bands. The radar altimeter also possesses a high data rate synthetic aperture mode that produces large amounts of data, as well as a conventional low data throughput mode. The tools used to moderate the data rate are the instrument's resolution and the percentage of time that the synthetic aperture mode of the radar altimeter is allowed to be operational. The instrument's resolution is limited to 250 m, which fulfills the expressed needs of the consumers; nevertheless, if higher spatial resolutions had been possible, the consumers would have chosen them. In the case of the radar altimeter modes, a somewhat artificial half-and-half division between the high and low data throughput modes has been adopted. This allows for observations in the better resolution synthetic aperture mode during half of the time. This is sufficient for characterizing all the coastal zones, ice-covered areas, and inland waters. The chosen solution provides a data flow architecture compatible with proven techniques for downlinking and with just one station contact per orbit. Here we see a clear example of the interests of the consumer being modulated by the need for an affordable system.

Cluster studies the magnetosphere by measuring the electric and magnetic fields of the area it is flying through, by means of in situ instruments. At each instant in time the area of observation is a single point. This brings the data throughput to 20–130 kbit/s. The daily average is just 0.8 Gbit. Due to the orbital geometry, the average contact times with the ground are lengthy, and they must be shared between the four satellites of the constellation. The 5 Gbit on-board memory acts as a buffer for the data produced by the instruments, and can store up to six orbits of data. The modest data rates ensure that the mission can be built with comfortable margins, using low risk technology, and requiring just one ground station to serve the mission.

The long distance between Earth and the comet at the crucial scientific periods is the dominant constraint on Rosetta's instrument data flow. This distance oscillates between 4 and 6 AU (Figure 11.7). Rosetta carries a set of varied instruments, including in situ and imaging instruments. The ground station for data delivery is the 35 m diameter ESA ground station in Perth, Australia. The orbital geometry provides a single contact per day with a contact length of about 12 h. Rosetta carries a large high-gain antenna that dominates the configuration of the satellite. The data throughput is determined by the 12-h daily contact and the 5 kbit/s of data rate. The mass memory can store 25 Gbits of data and allows the observation of complex events that can be stored and downlinked slowly to Earth. Both the emitting and ground-receiving antennae are the largest and most powerful currently compatible with affordable technology. The NASA DSN network has 70 m diameter antennae, while those of the ESA equivalent stations are just 35 m in diameter.

NOAA-POESS produces 20 Mbit/s of data, the equivalent of 110 Gbits per orbit. As with Sentinel-3, the operational meteorological data that this mission produces is acquired continuously, except that visual optical data is not gathered at night. The instrument complement of NOAA-POESS is quite varied, and gathers highly accurate data over many bands as well as having wide swaths. Nevertheless, the spatial resolution of the instruments is moderate and the resulting data rates, while large, are compatible with existing technology. The use of an average of five contacts per orbit, provided by the 15 SafetyNet ground stations (which allow a high speed data downlink of 150 Mbit/s) provides a scenario in which the handling of the instrumental data quantity is not critical. NOAA-POESS's criticality is within the limits of data delivery speed that will be discussed later.

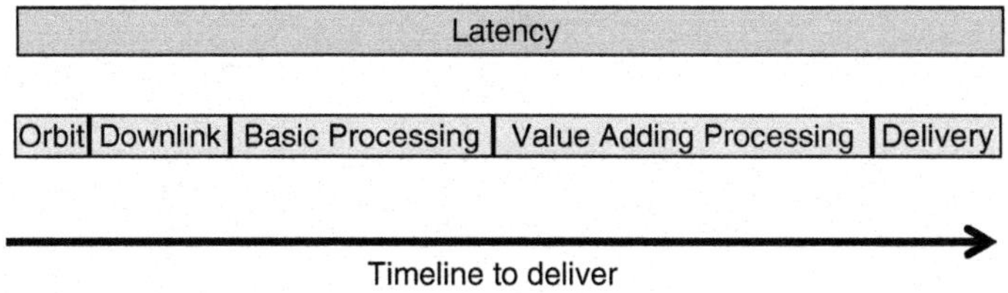

Figure 12.7. Data Delivery Timeline.

12.4.2. Short versus long data latency

This section continues the discussion of fast-versus-slow data request and delivery started in Section 8.4.6 and continued in Section 11.5.3 of the previous chapter. The discussion in Section 11.5.3 focused on the "time to acquire" part of the end-to-end timeline. The discussion here will focus on the "time to deliver" and on the resulting "data latency" that was defined in Section 8.4.6 as the time elapsed between the acquisition of data by the instruments and the delivery to the final user of the products based on these data. Data latency includes the time necessary for the satellite to be able to downlink information plus the time necessary to process and to deliver the data. Figures 12.7 and 12.8 provide consumer-centered perspective of the same end-to-end timeline already depicted in Figure 8.18.

The specified latency will be more demanding for missions that study fast-changing phenomena, and less demanding for those studying more slowly changing ones. The delivery of mission products with data that are as recent as possible is of extreme importance for operational missions because the information is used to make decisions that could be detrimental if the data used to derive the decisions were too old. The latency period will be especially critical for missions that require fast decision making, as in cases of military and civilian security. Science missions also prefer fast delivery of data, as it allows a far more interactive study of the phenomena under observation; this is especially so for analyzing the observations of transient events such as frequently occurs in X-ray astronomy. In the other hand, missions that are pushing the state of the knowledge will deal with production of data that requires deliberate processing and a long time for correct understanding. This type of missions will not put a heavy emphasis in short latencies. Cluster provides reasonably fast delivery of data to the principal investigator and lower delivery to the secondary consumer. This is an example of a science mission with rather relaxed overall data latency requirements.

Rosetta would have benefited from fast timelines in order to allow for the adaptation of observational strategies to the phenomena being detected by the satellite. However, Rosetta is limited by the long delays required for the signal to travel to and from the Earth, and by the slow data uplink and downlink speeds. This forces long waiting times for the delivery to ground of the acquired data, which are stored in the satellite's 25 Gbit solid-state mass memory until they can be sent to Earth.

Sentinel-3 and NOAA-POESS have similar architectures and requirements because both produce global data to be assimilated into large computer models, but

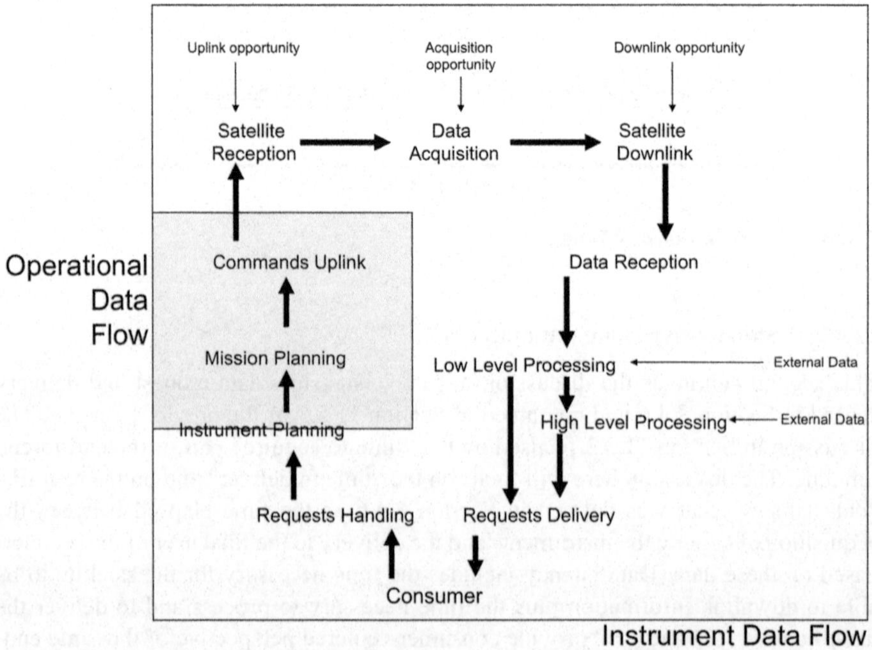

Figure 12.8. Step-by-Step Image Acquisition and Delivery Process. It Includes the Steps Related to the Uplink/Operational Timeline and to the Downlink/Instrument Timeline.

Sentinel-3 provides data for ocean and vegetation modeling, while NOAA-POESS data is used for modeling the much faster changing atmosphere. The need to monitor phenomena such as hurricanes, tornadoes, or Mediterranean autumn flash floods requires the rapid and increasingly accurate knowledge of the current weather and the reduction of data latency. The fast-changing nature of meteorology explains the sophistication – and associated cost – of the NOAA-POESS data downlink architecture that allows the reduction of data latency from 3 h to 30 min.

SSTL-DMC, discussed in Section 11.4.4 above, possessed an instrument data processing system capable of providing the mission products in a period between 4 h to 2 days, depending of the level of processing and quality of the data products.

The steps necessary to deliver the acquired data are the following:

1. The satellite needs time to reach a point in the orbit which offers a data downlink opportunity either through contact with a ground station or, in the rare cases where they are available, with a data relay satellite. This period is dependent on the orbit, the number of downlink ground stations, and the use or not of data relay satellites.
2. Once the data has arrived on Earth it must undergo a first processing step that normally involves the delivery of calibrated and well-referenced data, with respect to the Earth or sky; for example, if the instrument is a radiometer, the processed data will be calibrated radiometric fluxes entering the instrument arriving from clearly

identified locations. This data will still be more related to the physics of the instrument than to the physics of the phenomena under observation. Sophisticated users familiar with the instrument will know how to take full advantage of this data.
3. High-level processing will convert the data to physical parameters directly related to the phenomena under observation, for example, the radiometric fluxes could be transformed into temperatures of the terrestrial surfaces under observation. All potential consumers, including those not familiar with the instruments, will be able to use this data. Its reception requires further processing and supplementary delay.
4. Finally, high- and low-level processed data can be delivered to the consumers. This can be done through the physical delivery (i.e., by post) of media such as CDs or DVDs, or via the Internet by ftp or http. For users requiring systematic and large datasets, dissemination through communication satellites using small antennae, 1–2 m in diameter, is another, and affordable, possibility.

The delay for steps 2 and 3 can be reduced by the use of sophisticated algorithms and powerful computers, step 4 can easily use the possibilities provided by the Internet and by the newer broadband communications systems, but step 1 can only be reduced by providing a large number of ground stations, as done by NOAA-POESS or by providing the use of data relay satellites, as done by ESA with Envisat and Artemis. Both alternatives are complex and expensive.

The use of data relay satellites located in the geostationary position increases the data throughput because it amplifies the allowable contact time from the 10 min per contact per ground station to 30-40 min per orbit and per data relay satellite. Data relay satellites provide clear advantages for heavy data producers. The NASA NOAA Landsat has used and continues using the operational TDRSS constellation of data relay satellites. The constellation of NASA Earth science satellites (Terra, Aqua, and Aura), also use TDRSS. ESA and CNES have also used the demonstration data relay satellite Artemis: CNES to downlink Spot V data through an optical data relay link and ESA to downlink Envisat (Figure 12.9) data through a Ka-band microwave link. The use of Artemis has allowed more extensive employment of the instruments by Envisat and Spot. The distance from low Earth orbit to the geostationary position is large, and a data relay link will require pointable high-gain antennae in both the low orbit and geostationary satellites. Data relay satellites are an infrastructure that service many users, hence will inevitably be shared with other satellites.

12.4.3. Existing, to-be-acquired and subscribed products

Four types of consumer request situations are possible:

1. The request is for a not-yet-performed observation that shall be acquired, returned to ground, processed, and delivered.
2. The request is a subscription for regular and systematic delivery of a certain type of product, for example, all the observations over a specific area of the Earth or sky.
3. Requests are unnecessary. The data processed by the instruments are broadcast and can be received by anyone after some type of registering process and with the use of adequate equipment.
4. The request is for already acquired observations stored in the mission's archive.

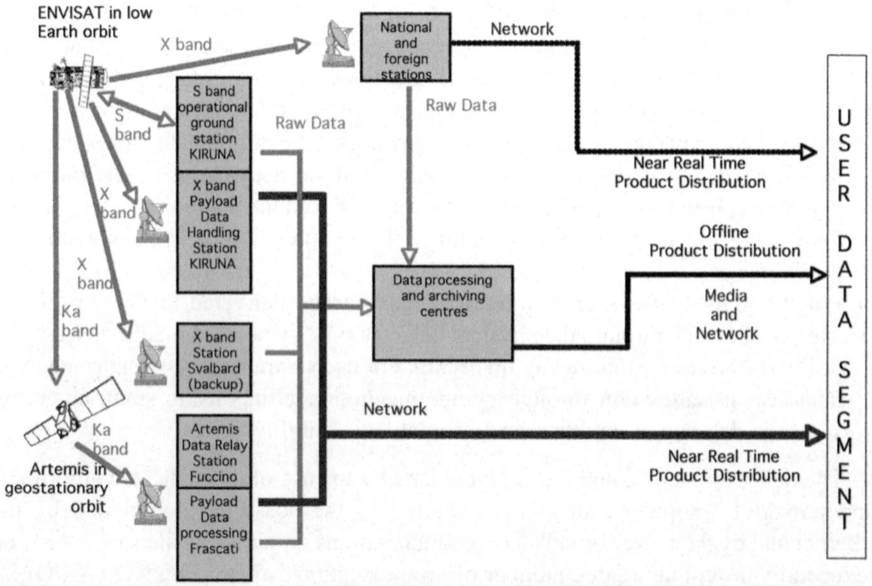

Figure 12.9. Use of Artemis Data Relay Satellite in Envisat Data Downlink (Image Courtesy ESA).

Each of these delivery modes has implications for different aspects of the end-to-end space system.

Type 1 is the broadest, it has implications affecting every element of the space system, and was analyzed in Sections 8.4.6 and 11.5.3. Figure 12.8 depicts the process from the point of view of the consumer and Figure 8.17 from the point of view of the satellite. This is the commonest type of situation in advancement of science missions. The astronomical observatory missions described in Section 7.2 are all of this type. High-resolution Earth observation missions, like SSTL-DMC, described in Section 11.4.4, and Pleiades described in Section 9.6.4 are also of this type. These missions study a small area within a large area in great detail. This is the reason it is necessary to command the satellite to point in a special direction. In these missions instrument operations are interactive (Section 8.4.5).

Type 2 is simpler; it requires no continuously changing mission planning, but impacts all the other parts of the end-to-end system, from in-orbit instrument data production to data distribution to consumers. This situation occurs in Earth or sky survey missions such as that of Sentinel-3, discussed in Sections 11.4.3 and 12.3.3. Operational Meteorological missions like NOAA POESS, described in Section 12.3.4, are also of this type. An extreme case is provided by global meteorological assimilation centers that provide whole Earth meteorological predictions. These centers require the regular and timely delivery of all the data produced by one or several space missions. The architecture capable of providing this type of service does not require a high degree of flexibility but must be optimal to ensure timely and reliable delivery of data to the assimilation centers producing the global meteorological predictions. These missions provide synoptic regular views of large areas. In these missions, instrument operations are systematic (Section 8.4.5).

Type 3 situations are best suited to ensuring the widest use of the data produced by the satellite. The two NOAA-POESS DRM systems, described in Figure 12.6 of Section 12.3.4, which broadcasts information to regional users, are good examples. In order to provide higher-level, superior quality mission products, this kind of data distribution also requires the provision of algorithms to the consumers, permitting them additional ground processing of the broadcast data. NOAA-POESS, for instance, provides this type of algorithm delivery and maintenance service.

Type 4 requests are easy to satisfy because the delivery to consumers of already existing data only affects the payload data segment. This delivery mode is common in studies of slow changing physical phenomena, like climate change. Providing this type of services entails the long-term maintenance of consistent quality data and the documentation relative to the algorithms used for data calibration throughout the duration of the mission. It should be noted that this long-term maintenance of consistent data can be expensive.

12.4.4. In-orbit versus on-ground processing

It is clear from the previous sections that the amount of data to be downlinked by the satellite can pose problems and that when in-orbit processing is used to reduce it, the mission's design will be simplified. Synthetic aperture radars and interferometers produce enormous amounts of data that can be reduced by on-board processing. The amount of data produced by high-resolution high-swath imaging instruments can also be reduced by data compression. There are four possible scenarios:

- Complex data processing that can significantly reduce data throughput. This is the case for on-board SAR or on-board Fourier transform spectroscopic processing.
- Complex data processing that destroys data recognized on-board as irrelevant. One example is the nondownlink of images covered by clouds.
- Data compression that preserves information and only moderately reduces the data throughput. This is the case of lossless compression of images.
- Data compression that destroys information and that significantly reduces data throughput. This is the case of lossy compression of images.

On-board computers are progressively more powerful and synthetic aperture and spectroscopic processing is progressively becoming within the capabilities of existing systems and, the near future will see increased possibilities for this type of on-board processing.

Also are rare space missions that allow satellites to autonomously decide whether the data produced by the instruments is irrelevant and could be deleted. To unequivocally decide that clouds cover part of an image is more difficult than it looks. Despite a great deal of research, as yet there are no algorithms capable of fulfilling this function with no mistakes being made, but as is the case of interferometric and spectroscopic on-board processing, this is another area where future improvements will allow for this type of data elimination. A second problem associated to this approach is that the downlink of variable amounts of information will require variable data rates and that the downlink system shall be anyway dimensioned for maximum data throughput. That means, this approach rather complicate than simplify the data downlink.

The use in satellites of data compression, both lossy and lossless, is extensive and simple to implement. It is used in all missions that produce large data throughput. In the case of lossless compression, the reduction of data throughput is modest: a guaranteed minimum compression factor of 1.8. This minimum value applies to data containing a lot of phase information, for example, that produced by spectroscopes or synthetic aperture radars. Algorithms able to significantly reduce data throughput with only moderate reduction of image quality are used for high-resolution optical information. The Pleiades lossy data compression algorithm allows data to be reduced by an average minimum factor of four, and QuickBird by a minimum factor of 5.[8] Despite the advantages that extensive use of on-board data compression provides for mission performance, it must be used with extreme caution. In general, advancement of science missions opt to avoid data compression, while operational missions tolerate it far better. The deletion of information, even of apparently small amounts of it, is dangerous in science missions where the actual value of the information being destroyed is unknown. On the other hand, operational missions benefit far more from aggressive data compression approaches. Consumers have the final say on whether to allow on board data processing, with the risk that valuable information might be destroyed, or whether they prefer to reduce the mission's output in order to accommodate a lower data throughput, for example, by allowing coarser spatial or spectral resolution, or a smaller swath.

12.4.5. Number and location of ground stations

The ground stations considered in this section do not send commands to the satellite; they only receive data from the satellite. Therefore, there is a much higher level of freedom in the selection of their location. It is very common to subcontract data reception to ground stations that are not under the direct control of the authority that owns the in-orbit satellite. This gives the architect freedom to choose the location of the stations that are the best placed to provide a service to the satellite.

The number and location of ground stations must be optimized to allow downlinking of all the data produced in the satellite at the speed allowed by the chosen design. The number and use of ground stations will depend on the frequency and time periods that ground stations are within range of the satellite and can establish a link with it.

There are three possible scenarios:

- The satellite is not orbiting Earth. The contact pattern is governed by the Earth's rotation.
- The satellite is orbiting Earth but its period is long, comparable with Earth's rotation. The orbital period and Earth rotation combined govern the contact pattern.
- The satellite is orbiting the Earth at low altitude and with a short orbital period. The contact pattern is governed by the orbital period and inclination.

We will now analyze these three scenarios in depth.

[8]Observation of the Earth and Its Environment, H.Kramer, Springer 2002.

12.4.5.1. Satellites not orbiting the Earth

Satellites located far from Earth, such as Rosetta, appear as fixed over the celestial sphere and the visibility pattern is governed by the Earth's rotation. Earth's rotation allows ground stations to scan the heavens; the satellite becomes visible when located above the ground station's celestial horizon. One ground station provides visibility during a single period of approximately 10 h per day, while three uniformly distributed ground stations can provide permanent contact. That is why NASA's Deep Space Network has stations in California, Australia, and Spain. ESA is implementing a similar network of stations located in the same countries. The stations located in Spain and Australia are now operational and the third one is under construction.

The use of these systems is expensive and the few stations able to provide communications to far-away planets must be shared by all the existing space missions requiring them. In order to reduce costs, satellites are designed to minimize the need for ground contact. A single ground station (in this case located in Australia) serves Rosetta. As described in Section 7.2, the three ground stations of NASA's DSN would be capable of continuously serving JWST; but JWST's data flow architecture was designed to require just a single contact of 4 h per day. This is a further demonstration of the progressive tendency to satellite autonomy, permitted by the improved capabilities of computers and the need to reduce costs.

12.4.5.2. Long period satellite orbiting the Earth

Satellites with periods similar or equal to Earth's rotation will have long periods of contact followed by periods of noncontact (longer or shorter depending on the orbit) with their ground station. An extreme example is a geostationary satellite with a period of 24 h, which allows permanent contact between the satellite and a well-located ground station. IUE was launched in 1978, and as discussed in Section 7.2 had an orbit of 32,050 km of perigee and 52,254 km of apogee, with an orbital period of 24 h. It used two ground stations, one near Washington DC and the other near Madrid. The plane and eccentricity of the satellite's orbit were such that the European station had visibility for 8 h a day and the American one 16 h a day.

Cluster (Section 11.4.1) has an orbital period of 57 h, equivalent to 2 days and 9 h; its orbital inclination is 90°, the apogee altitude is 119,000 km and the perigee altitude is 19,000 km. The pattern of one Earth orbit and the visibility from the Madrid ground station can be seen in Figure 12.10. The orbital period is long with respect to the 24-h rotation of the Earth, and when the satellites are near the apogee they are nearly fixed in the sky and Earth's rotation dominates visibility. This provides in every orbit a first long contact period of 11:23 h and a second one of 10:25 h. All together, Cluster satellites are visible approximately 38% of the total orbital time.

As we have seen with the two examples provided, satellites orbiting the Earth with long orbital periods permit long periods of contact with a single, wisely chosen ground station. The percentage of time with full ground contact will oscillate from 30 % to 100 % in the case of geostationary orbits.

This visibility pattern allows the design of a communications architecture in which one ground station allows a high percentage of contact time as well as the downlinking of large amounts of data at moderate speeds. This geometry can also produce long

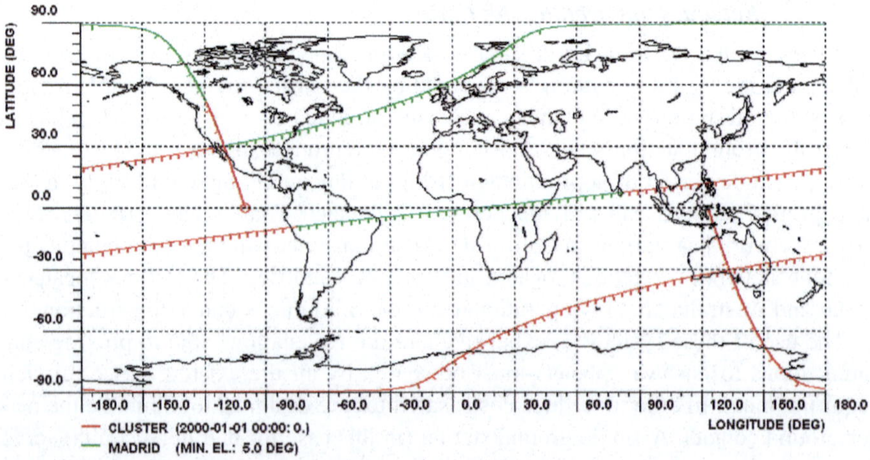

Figure 12.10. One Cluster Orbit Visibility from Madrid (*green*).

periods without ground contact. The length of this blind time can be reduced by judicious choosing of the orbit inclination, the eccentricity, and the location of the ground station. The satellite will require on-board memory for the storage of data acquired by the instruments while the satellite is out of range with its ground station.

12.4.5.3. Short period satellites orbiting the Earth with high inclination

This is the traditional orbit for Earth observation and delivers observation coverage of the entire Earth from pole to pole. Satellites in this orbit pass over both poles on each orbit, and ground stations near the poles ensure contact on every orbit. Figure 12.11 provides the visibility of a typical Earth observation satellite contacting the Svalbard ground station on each orbit. This ground station provides contact lengths varying from 6 to 12 min. The station allows downlinking of data 2:38 h per day, equivalent to 11 % of the total time.

This alternative of a single ground station and one contact per orbit provides a reasonable data flow architecture implementation that is satisfactory for many types of missions. Considering that the typical orbital period for a low altitude orbit Earth-observing satellite is about 90 min, if one downlink per orbit is foreseen, then at the moment of the downlink the oldest data transmitted will be one and a half hours old. With another 90 min allowed for data processing, calibration and quality checking, the information delivered to the customer will be less than 3 h old. The consequences of missing a contact with such a ground station, for example, due to a maintenance issue on-ground can result in permanently lost data

However, there are many situations where this 3 h delay will be insufficient. For situations requiring faster data processing it is necessary to provide a more efficient and complex data flow architecture. One possibility is to provide stations not only near the North Pole but near the South Pole as well. Due to the harsh climate, the use of stations and their associated infrastructure at the poles requires heavy investment.

This visibility pattern allows the design of a communications architecture in which one ground station allows a high percentage of contact time as well as the downlinking of large amounts of data.

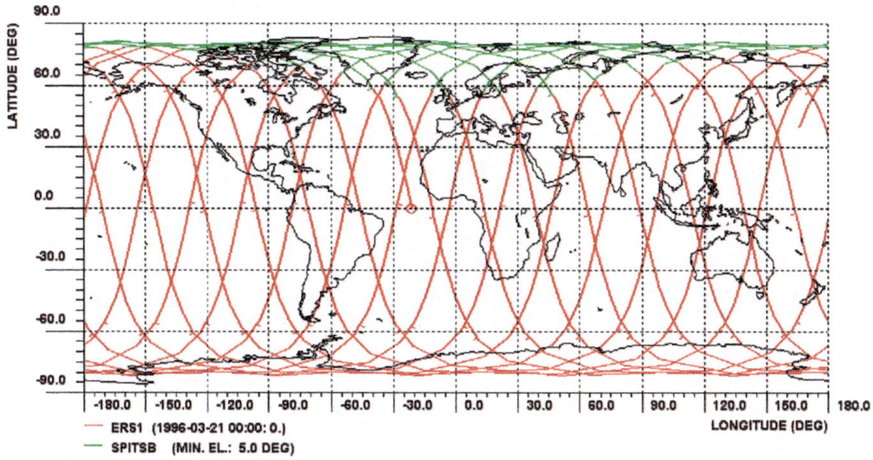

Figure 12.11. ERS 1 Day Orbit and Visibility from Svalbard (*green*).

12.4.5.4. Short period satellites orbiting the Earth with low inclination

The most efficient way to put satellites into orbit is to place them in low-inclination low-altitude orbits around the Earth, and most satellites not requiring a specific orbit end up in low inclination orbits around the Earth. The most efficient way to communicate with a satellite in a low Earth orbit is to locate a ground station with latitude similar to the inclination of the satellite's orbit. Figure 12.12 provides the contact pattern for an orbit with an inclination of 40°, using a ground station near Madrid.

In the most likely case that this type of orbit was chosen for satellites with less demanding requirements, one single ground station is normally able to provide adequate ground contract. Madrid can provide 1:17 h of contact per day distributed in six or seven contacts of around 11 min each. In cases that it will be necessary to provide longer contact time, more stations – always located within the area overflown by the satellite – will be necessary.

12.4.6. Centralized versus decentralized processing

Processing of the data produced by the instrument is crucial. This processing requires the use not only of the instrument output but also of supplementary data provided by the satellite, and frequently of external data used as a reference. Processing will produce data of increasing levels of sophistication and value. This progressive processing can be performed at a single site used for distributing mission products to the consumers, or alternatively, different consumers can independently perform their own processing. NOAA POESS provides examples for both cases.

SafetyNet and the centralized X-band downlink provide data to a select group of data assimilation modelers that will use them to produce global operational meteorological predictions. This centralized approach works well because centralized global meteorological predictions can use the most sophisticated data and the most powerful computers while providing information valid for the whole Earth. On the other hand,

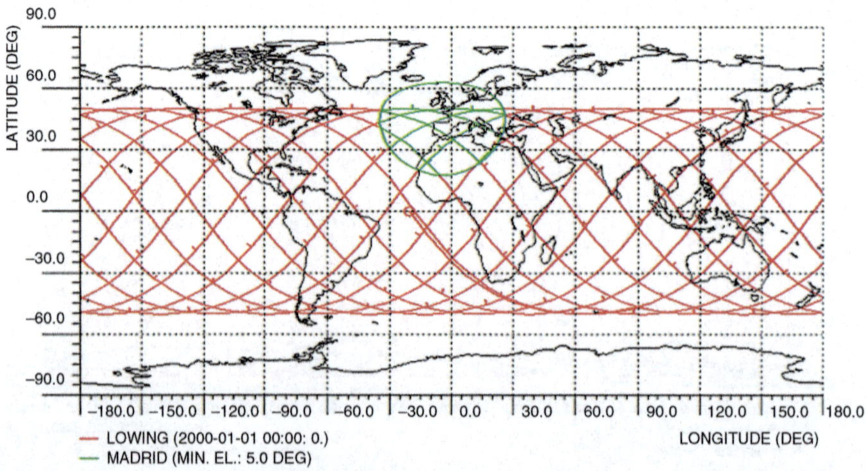

Figure 12.12. One Day 40° Inclination Earth Orbit and Visibility from Madrid (*green*).

NOAA-POESS DRS system produces data broadcast directly from the satellite to numerous regional users. These regional users will perform their own processing. This ensures the widest and fastest use of the information produced by the satellite. The existence of these two totally independent systems for data downlinking and processing is made necessary by the different needs of different consumers. Some will prefer the quality and global scope of centralized data processing; others will prefer the speed and independence of the decentralized data processing. Sentinel-3 has a more centralized data downlink and processing architecture. The central PDGS processes the data and provides it to the modelers and assembly centers that also use data produced by other satellites and by in situ instruments. All the data assimilated together provides predictions on ocean status that are delivered to the users.

Cluster and Rosetta provide mixed scenarios. The data processing is performed by a selected group of principal investigators, but the data can be distributed to external researchers who prefer to do their own processing.

Centralized data processing centers allow a tight control of the production and dissemination of the information produced by the satellites. An additional benefit of centralized data processing is that it allows the long-term storage of well-documented data. This rigorous data archiving is crucial to many scientific applications.

12.4.7. Science operations separated or as part of overall operations

The satellite's operations have as an important subset the operations of the instruments. The mission control center, which prepares and uplinks the commands to the satellite, must take into account the consumers' data acquisition requests. These requests are filtered and processed by the science operation center, which considers them according to their scientific merit, scientific mission adequacy, and mutual compatibility. The resulting requests must then be filtered from the perspective of view of their technical adequacy: to what extent are the request compatible with the availability of satellite resources, and to what extent are they compatible with the satellite's operations, (for example, the need for an orbit correction maneuvre). Hence, the establishment of the observation plan has dual components: technical and scientific.

In the case of the Hubble Space Telescope (Section 7.2) we discussed the functions performed by Goddard , which performs the technical role, and the Space Telescope Science Institute, which performs the scientific role. The description of Rosetta's scientific operations in Section 12.3.2 also offers a good example of the collaboration of satellite (RMOC) and instrument (RSOC) specialists together establishing observation plans.

12.5. Allocation of performances

12.5.1. On-board storage memory budget

The instruments produce data at varying speeds depending on their operating modes. The instrument downlinks the data produced when visibility with the ground stations permits, or when permitted by the link with the data relay satellite. An on-board mass memory is essential for providing a buffer storage between data production and data downlink. All the missions studied in this chapter employ mass storages of differing sizes: Cluster uses 5 Gbit; Rosetta 25 Gbit; Sentinel-3 uses 200 Gbit; and NOAA-POESS 300 Gbit. The amount of memory must be high enough to allow the buffer storage of all the data produced by the instrument with the largest data production, combined realistically with the amount produced during the longest period without downlinking opportunities. The time allowable for actual data downlinking is generally small with respect to the overall mission time; consequently, the downlink speed must be higher than the data production speed. In the case of NOAA-POESS, data is produced at 20 Mbit/s and the maximum downlink speed is 400 Mbit/s. The on-board accommodation of the required mass memory size is normally not a problem. The longest period without ground contact is the crucial determining factor and is a function of the orbit and the downlink geometry, but the data production rates of the instruments cannot be predicted with a high degree of accuracy during the earliest stages of design. In line with the general tendency to use high safety margins in the early dimensioning of digital hardware, safety factors of 1.5 to 2 are used in the early estimations of mass memory size. Is it worth stating that early in the design stage an average system data circulation analysis is produced, but this early analysis can be inaccurate, and pending the determination of the platform and payload data production rates, orbit, ground stations number, and locations it is all that can be done. However, a more accurate dynamic data circulation analysis must soon follow as data became more defined. It is necessary to avoid situations where the mass memory or the downlink rates could exceed the originally allocated values. This will require the use of safety margins.

12.5.2. Data downlink budget

The link budgets and its associated problems were discussed at length in Section 11.6.2 of the previous chapter, that section studied with detail the problems associated to the downlink of information from satellites located far away from the Earth. The problems depicted there must also be considered here, but this section will provide supplementary material applicable to cases featuring large volumes of instrument data, an issue particularly associated with the downlinking of instrumental output from near-Earth orbits, and not properly covered in Section 11.6.2.

In near-Earth orbits, the distance from satellite to ground station is small, and theoretically, it should be possible, using reasonable communication components, to downlink at enormous data rates. Nonetheless, to avoid interference between distinct users, and as explained in Section 11.1.3, the international Telecommunication Union (ITU) regulates the communications environment. These regulations introduce two supplementary constraints on designers:

- The signal's maximum allowable power flux density is limited. Mission designers cannot increase the signal power arriving at the ground station beyond the regulated limit. Because it is a power flux density limitation, it restricts antenna gain as well as the total power radiated by the satellite.
- The spectrum bandwidth itself is limited, as it must be shared between all possible users. This is a severe limitation for Earth observation satellites that would benefit from large data throughputs. Therefore, the diversification of the ground station location and the increase of the receiving antenna directivity are preferred methods to the frequency diversification.

In near-Earth orbit missions, the power reaching the ground antenna is usually enough for transmitting the necessary data throughput without the use of large power amplifiers or high-gain antennae. Almost all Earth-orbiting space missions, as well as those downlinking the maximum allowable data permitted by the bandwidth, use low-gain antennae. Cluster uses omnidirectional antennae, while Sentinel-3 uses a cardioid antenna that moderately increases the gain when the satellite is low over the ground station's horizon, and reduces the gain when the satellite is flying directly over the ground station and the distance between them is minimal.

Cluster is a good example of a relatively simple data downlink system. Cluster only carries in situ instruments that produce moderate data rates, and its orbit allows long contact periods. This makes fulfilling the link budget easy, and leaves ample margin for the allocation of requirements and the design of all components.

NOAA-POESS DMS system, described in Section 12.3.4, provides mission products to a large number of small antennae, as little as 1 m diameter in the case of the L-band component. The L- and X-band DMS downlink broadcast continuously information using low-gain antennae, and the system was designed to simplify the ground station hardware, and make the system accessible to as many distributed users as possible. This creates a situation in which the apportioning of requirements reduces the difficulties of the ground segment, making it easy to design, and minimal for the space system, forcing it to use as much emitting power as allowed by ITU regulations for the corresponding bands.

Sentinel-3 does not continuously broadcast information. It downloads and empties its entire memory when flying over its main download ground stations: Kiruna in Sweden and Svalbard near the North Pole. The principal limiting factor in the design of Sentinel-3 was not the availability of hardware, but the stringent regulations. As Section 12.4.5 made clear, Earth observation missions benefit from stations near the poles. Many Earth observation missions would prefer to downlink information to

Svalbard, near the North Pole; they would also prefer to use maximum data throughputs and to downlink at every visibility opportunity. It is no accident that both Sentinel-3 and NOAA POESS have an X-band 400 Mbit/s downlink system and that both use Svalbard; this is indeed a very attractive solution; the problem is that it could generate interference between the satellites.

Data relay satellites provide one possible solution to the problem of limitations on bandwidth and power density. Data relay satellites allow the transmission of similar amounts of data during longer times, reducing the need for high speed and high bandwidth use, and permitting downlinks via geostationary satellites with good visibility from Earth.

The SafetyNet system of NOAA-POESS is another method for radically solving the regulatory problems indicated above. The use of numerous ground stations increases the allowable contact time and reduces the required data downlink speed. The ground stations are uniformly distributed around the world, reducing the problems caused by congestion over the Poles. A supplementary tool for reducing regulatory problem is that SafetyNet works at Ka-band. ITU regulations allow 1.5 GHz in Ka-band with respect to the 0.3 GHz that they allow in X-band. That means Ka-band provides a much larger data throughput capability.

12.5.3. Data latency budget

This section is a continuation of Section 12.4.2 on short versus long data latency. We have already extensively analyzed the major factors affecting the timeline for data delivery. The four periods discussed were:

1. Downlink time, which is fully determined by the orbit, by the location and number of ground stations and by the use or not of data relay satellites for downlinking.
2. Low-level processing time. This can be optimized through the use of fast processing algorithms and computers.
3. High-level processing time. This can also be optimized through the use of fast processing algorithms and computers. However, in the frequent instances that processing requires the use of external information, for example, near real-time, high-accuracy orbital data from GPS satellites, punctual delivery methods for the processing of this data must be determined.
4. Distribution time. As in 2 and 3 above, this can be reduced by the use of computerized handling, but might also require the use of permanent or semi-permanent human intervention.

As explained in Section 11.6.5, period 1 is deterministic, and can forego the use of safety margins, but periods 2, 3, and 4 are functions of the processing algorithms, computing power, ground communications and infrastructure, the amount of human support needed, and the delay in the delivery of the external data. This requires the use of comfortable safety margins. Also as explained in Section 11.6.5, the early design of ground processing and distribution will focus on functional design and cost estimation, but will not include a detailed physical design. This is an additional reason for including comfortable safety margins in the planning of time allocated for steps 2, 3, and 4. Ground processing specialists must determine if the resulting time allocated for 2, 3, and 4 is realistic and what costs would be associated with implementing a system capable of providing the requested performance. Special care should be given to the identification of external data and what kind of arrangements are necessary for ensuring the availability of this external data within the time delays allocated to it. Very short data latency is of extreme importance for applications such as security, defense, or meteorology. To reduce latency it is necessary to reduce the times required by steps 1–4 of the list above. This will require fast processing and data distribution networks but it will also require a large number of ground stations and/or the use of data relay satellites. Section 12.3.4 introduced the SafetyNet™ set of 15 ground stations for reducing the data latency of the data used for operational meteorology. Such a complex structure was implemented in order to provide the quickest delivery of meteorological satellites and the swiftest delivery of weather forecasts. This is especially useful when dealing with fast and extreme weather conditions such as hurricanes or floods. The following figures provide the latency figures for different ground station networks for a single satellite in Sun-synchronous orbit at an altitude of 800 km, the most frequently used orbit for operational meteorology missions. The following figures depicts progressively more complicated architectures that will reduce the data latency

A single ground station in Svalbard provides a downlink time that fluctuates between a few minutes – for observations acquired just before downlink – to 70 min for observations acquired just after losing contact with Svalbard (Figure 12.13) and that are delayed until renewed contact is made nearly one orbital period later. Three supplementary ground stations uniformly distributed along the Equator only moderately improve these latency figures (Figure 12.14). A dramatic improvement would be provided by the use of data relay satellites (Figure 12.15). One single data relay satellite would reduce the latency over roughly one-third of the Earth, while three would afford rapid delivery above the entire planet. The infrastructure investment for a comprehensive data relay satellite network is costly. The SafetyNetTM network described in Section 12.3.4 also dramatically reduces latency times (Figure 12.16) over the whole globe but implementation of the necessary 15 ground stations also requires heavy investment.

Satellites often produce large amounts of data, and sending data to Earth is just the first step; it must also be delivered to processing centers. Overall latency calculations must also include transfer speed of the data from the receiving ground station to the processing and distribution centers. Transfer speeds to processing and distribution centers will vary according to the geographical location of the different ground stations. Figure 12.17 provides the estimated transfer speeds from different ground station locations. The figures vary considerably depending on the nature and availability of communications infrastructures. The network evaluated here belongs to a possible next generation European operational meteorological satellite system.

Svalbard and other European stations can use special high-speed networks able to provide up to 10–20 Gbps. Equatorial stations, however, have much slower rates. A data relay satellite can deliver directly to the processing center. McMurdo ground station in Antarctica, has, like Svalbard, very good downlink contact patterns with polar satellites. Nevertheless, delivery of the data from the station to the processing center is slow. It is evident that the space mission appraised in the figure must share the communication links with other users, which means the data throughput available for the mission is only a fraction of the total. A thorough end-to-end design perspective should take into account all these “nonspace” factors.

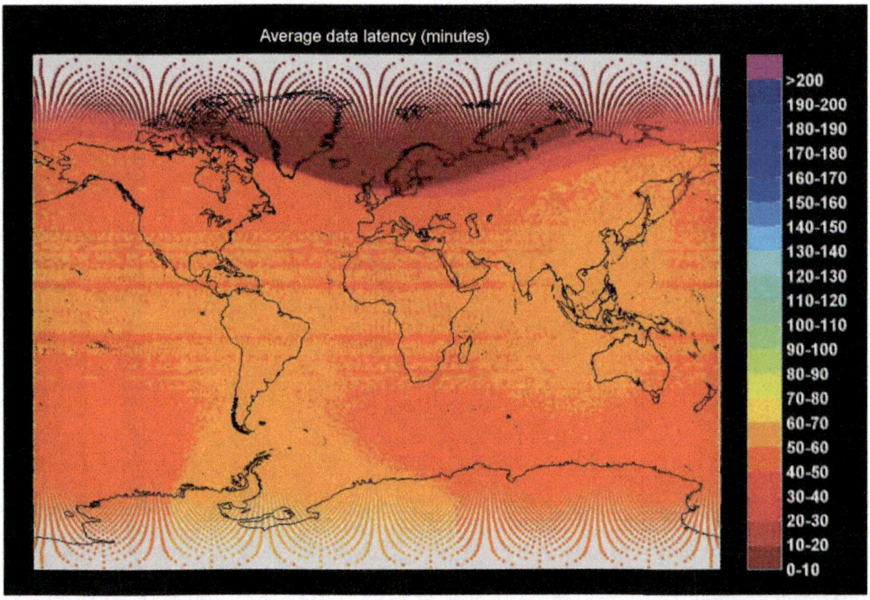

Figure 12.13. Latency in Minutes for a Single Station in Svalbard (Image Courtesy Astrium and GMV).

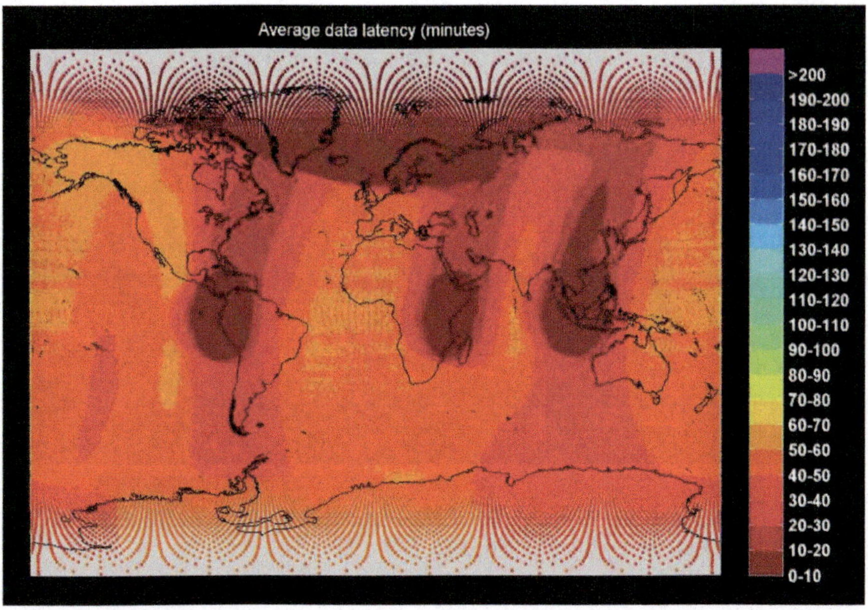

Figure 12.14. Latency in Minutes for a Single Station in Svalbardand with Three Stations Guayaquil, Malindi, Singapore (Image Courtesy Astrium and GMV).

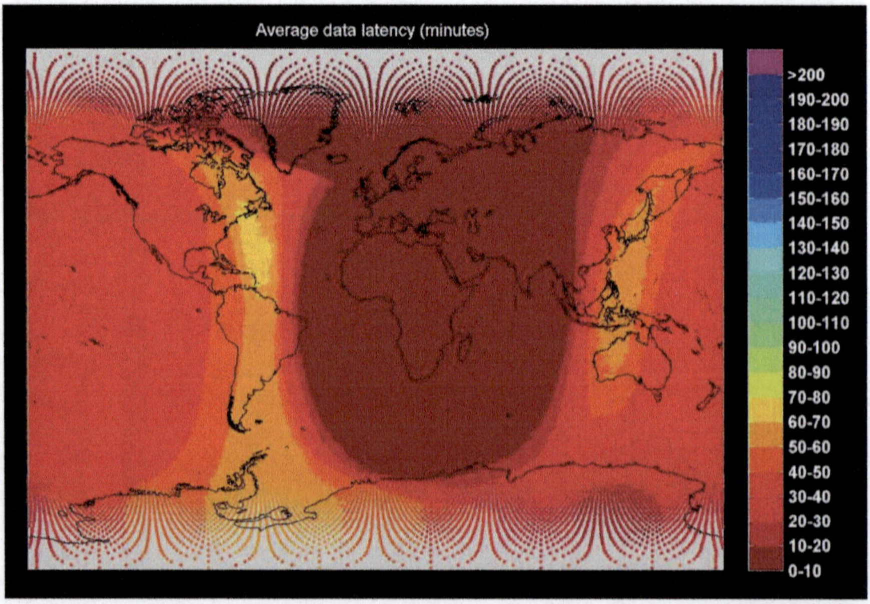

Figure 12.15. Data Latency in Minutes Using a Single GEO Data Relay Satellite (Image Courtesy Astrium and GMV). The Use of One Data Relay Satellite Allow Very Brief Latencies for the Part of the Earth Covered by the Data Relay Satellite.

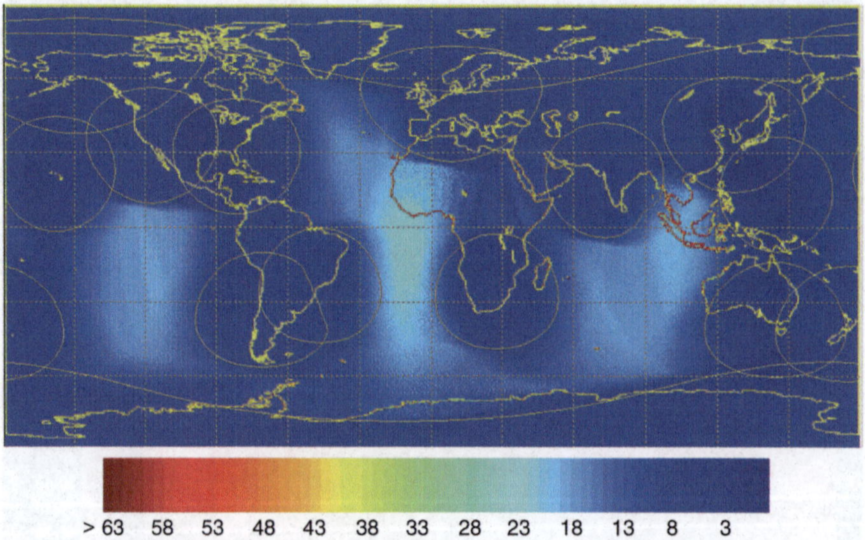

Figure 12.16. Data Latency in Minutes for NOAA's SafetyNet Network of 15 Distributed Automated Ground Stations (Image Courtesy Astrium and GMV). Due to the Large Number of Ground Stations the Waiting Time Needed to Download the Data Is Shorter than 20 min over the Whole Earth.

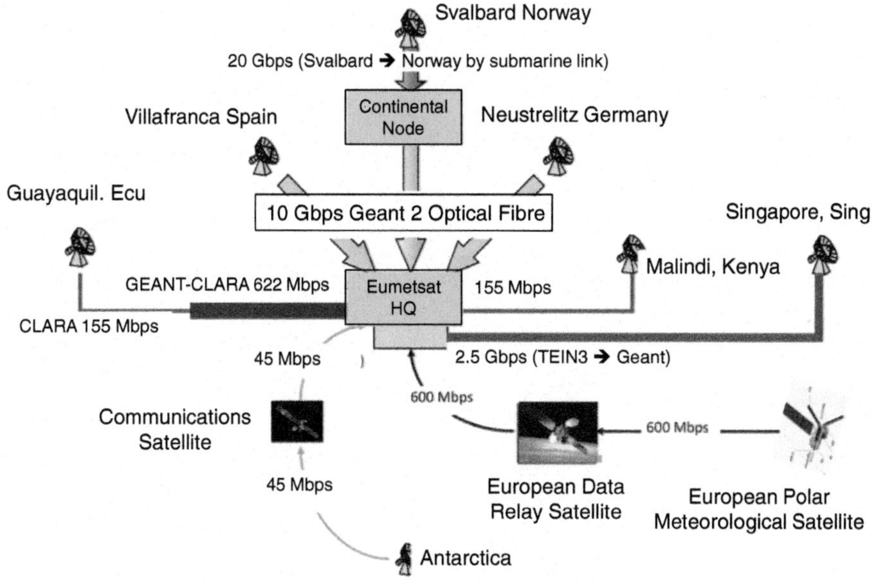

Figure 12.17. Data Transfer Speeds from Possible Ground Stations to a Central Processing Centre for a European Polar Low-Earth-Orbit Meteorological Mission (Image Courtesy of Astrium).

Chapter 13

Space Missions Cost and Alternative Design Approaches

Keywords Small Satellite Missions, Space Missions, Cost Reduction

Mission costs have been discussed several times in previous chapters. Sections 4.5 and 4.6 described the different kinds of space projects; Section 4.8 analyzed costs and how to evaluate them; Section 4.9.3 examined the development approach and model philosophy; Section 5.3.1 looked at low cost designs. The following chapter will again consider the topic, providing a synthesized view, and examining cost reduction approaches proposed in the recent and not so recent past.

We will begin by reviewing which technical and nontechnical factors contribute most to the high cost of space missions. This will be followed by a section examining methods of cost reduction and evaluating both their benefits and limitations. Three previously studied methods will be examined further, and the potential of small satellites as cost savers will be discussed at length. The next section will analyze possible cost reductions in situations when the sponsor and the consumer are the same, while the last section will look at missions with a low novelty level that can be implemented without a customer.

13.1. The space mission and cost

To begin with, space missions are expensive because they are highly complex and difficult to implement. There are obvious physical reasons that make a satellite an expensive piece of hardware: space is an unfriendly environment: a vacuum is incompatible with most common lubricants, and makes convective heat transfer impossible, while space radiation degrades electronics and other satellite components. The weightlessness of space, on the other hand, favors large, slender structures that are impossible in the Earth's atmosphere, though for the same reason testing them on the ground is complex and expensive. Launchers, used to put satellites in orbit provide a brief but demanding ride that requires satellite structures to be strong, rigid, and compact. Launchers are expensive and satellites must be light to fly on them. The intrinsic physical difficulties provided by launch conditions and the space environment are compounded by the fact that satellites carry advanced scientific instruments likely to be on the leading edge of technical developments. Another particular feature of space missions is the need for the engineering to be reliable, as fixing a failed spacecraft is

M.A. Aguirre, *Introduction to Space Systems: Design and Synthesis*, Space Technology Library 27, DOI 10.1007/978-1-4614-3758-1_13, not easy and usually not possible. Hence, the technology and components used must be of high quality and provide the necessary redundancy in its architectures and detailed design. Further, the provision of the design and associated verification of the design requires the use of highly trained personnel and specialized facilities, both of which are costly to provide and use.

There are two supplementary nontechnical factors that make satellites expensive:

- Many – indeed most – satellites are unique.
- Space missions are usually paid for by someone other than those who most directly benefit from it.

The satellite's uniqueness suggests that its realization is the result of original, technically difficult design processes that can rarely be copied from well-known precedents. Furthermore, being unique, failure of the mission means consumers completely lose the opportunities offered by the mission in the first place. A scientist or operational user – for instance an operational meteorologist – would garner little satisfaction from knowing that the mission was insured if it should fail; he wants his data, not his money back.

Satellites must operate correctly, with an extremely high level of reliability, the very first time it reaches its intended environment. This means its design must incorporate ample safety margins, conservative testing and thorough analysis, combined with a project philosophy of "failure is unacceptable." All these factors result in high costs.

The fact that space missions must answer to two masters, the sponsor and the consumer, is another factor affecting costs. The consumer's priority is to achieve a mission that fulfills all his aims with an unprecedented level of quality. The sponsor possesses the final authority to commit the funds required by the mission, though he will not be directly involved in the day-to-day aspects of mission definition, and his only real choice is to either cancel the mission or demand drastic reductions in performance criteria to satisfy the available funds. Both these alternatives seem extreme, but real-life experience shows that most satellites are overspecified, overdesigned, and overcost. This situation repeats itself in nearly all missions attempting to advance our knowledge of the physical world. In these kinds of missions, problems such as this are nearly impossible to solve, but they can be alleviated by prompt and energetic debate about mission requirements relaxations between the customer, supplier, and the consumer.

Most discussions about cost reduction focuses on reducing the expense of the satellite platform, but a healthy approach to the issue of space mission costs should center on a balanced reduction of all the relevant aspects: ground segment, launcher, instruments, and satellite platform. The standardization of satellite platforms offers a good opportunity for reducing expenditures. Nevertheless, the issue of reducing the cost of instruments is a complex one, because instruments differ so greatly according to their applications. Each new mission requires new instruments, and new as well as ambitious ones require ambitious new instruments. There are few reliable launchers and these are usually expensive. Efforts to design an affordable launch method, capable of putting a satellite into orbit have been constant for more than 40 years; the space shuttle, introduced in the 1970s, was supposed to reduce the cost of putting payloads into orbit by a factor of 10. In the 1990s, private initiatives tried to further reduce launch costs, but failed. The lack of small, reliable, and inexpensive launchers is hindering the development of satellites with these same characteristics.

Ground-related mission costs: satellite operations and mission data processing and distribution constitute a sizable portion of the total mission cost. Furthermore, as these activities are ground-based as well as strongly based on information technologies, we could have assumed that costs related to this area would be decreasing dramatically, in parallel with the development of ever more powerful and inexpensive information technology. This is not the case, however. This can be explained as a result of the enormous amounts of data that space missions are now providing, and the increasingly large consumer communities that make use of this data. This requires massive storage and processing capabilities, flexible interfacing with consumers, and flexible satellite operations to adequately respond to the different needs of each member of the consumer community. While it is true that advances in information technology are transforming the way space missions are implemented, the changes tend toward better performance levels at moderately lower cost, than in the direction of constant performance levels at dramatically lower cost.

Another way to reduce costs is to seek funding from a third party. Although financing the mission's instruments might ensure real savings for the mission's principal backer, overall costs are not actually reduced, they are just transferred. Taking advantage of free launch opportunities or being launched piggyback beside a primary mission that is paying for the launch can provide real cost reductions, but this approach is only feasible for a small percentage of all missions; obviously, the entire space effort cannot fly piggyback.

Given the points examined above, it is important to realize that not all space missions are equal and that cost-saving measures might be advisable for some missions but not for others. This is often forgotten, and cost reduction strategies are frequently presented by their defenders as the whole solution instead of what they actually are; part of the solution.

13.2. Methods of cost reduction

This section is divided into subsections examining different cost reduction methods. The first three methods were discussed at length in previous chapters; the following provides a final synthesis. These are:

- Cost reduction by proper architectural definition
- Cost reduction by hardware solutions
- Cost reduction by programmatic solutions

The final subsection is devoted to small satellites and introduces several new topics.

13.2.1. Proper architectural definition

As indicated in the previous section, the architect must answer to two clients: the sponsor and the consumer, whose interests, far from similar, are contradictory. The consumer's priority is ensuring superior mission performances, while the sponsor's is for containing costs. The architectural definition phases must reconcile this contradiction to deliver value to the consumer while containing costs in line with the sponsor's desires. In order to reconcile the two conflicting aims it is crucial to view the design process not as the fulfillment of the requirements but as evolving together with the requirements until they reach a mutual balance. Achieving this balance necessarily reduces costs because it results in the least expensive mission still capable of meeting the fundamental needs of the consumer.

Figure 1.1 shows that missions with a well-funded architectural definition phases have small cost overruns, thus reducing overall costs. Section 5.3.1 examined the apparent contradiction that increasing costs at the beginning of a mission reduces costs at the end. This is not surprising. A deliberate and well-funded architectural definition stage prior to project approval permits:

- Negotiating a reduction of the consumer's requirements in as optimal a manner as possible
- Agreement on an optimal design that fulfills the consumer's final set of requirements while minimizing the sponsor's costs and both of their risks
- The production of a comprehensive and detailed system specification that is completely free of open or undefined points
- Providing maturity to key technologies required by the mission

Chapter 3 analyzed the establishment of requirements and the creation of specifications, and made clear that establishing requirements was an interactive process, involving constant feedback between the consumer's needs and the hard technical realities those needs entail and which the supplier brings to the attention of the architectural definition team. This process (see Section 3.2) produces a chain of specifications with increasing levels of concreteness. Simultaneously, the mission's final design is being realized through a process of production, analysis, and selection of alternatives (see Chapter 5). The mission's architectural definition process advances step by step through a series of milestones (see Chapter 6) until both the specification and the mission design are finally defined together.

In order to achieve an ideal trade-off between requirements and the design responding to them, several factors are essential:

- A mathematical model able to provide end-to-end mission effectiveness for the alternative concepts under consideration. The distinct concepts might be fulfilling different sets of requirements for different levels of difficulty. Mission effectiveness models were discussed in Sections 3.4, 5.6, and 8.5.
- A flexible cost estimation tool able to provide realistic cost forecasts for each alternative. This was discussed in Section 4.8.
- Common sense, in order to clearly understand that the reality behind the final performances and costs is far more complicated and subtle than the figures provided by the cost and performance models. This was discussed in Section 5.6.2.

Section 5.4 ended with a reference to the heuristic that states: "**The last 10 % of the performance sought generates 1/3 of the costs and 2/3 of the problems**." According to that heuristic, obtaining a reduction of this last 10 % of performance from the consumer before receiving a fully funded project approval will mean a substantial decrease in overall costs and schedule. An additional tool for convincing consumers to accept a nominal relaxation of requirements is the fact that – due to the existence of explicit as well as hidden safety margins – real space missions that do not fail catastrophically will usually perform better than specified. This was discussed in Section 5.6.3.

Finally, it should be remembered that the mission's existence is due to the consumer, and he shall be the ultimate authority on the degree of requirements relaxation; after all, the consumer has the right to mandate an unaffordable mission!

These methods for reducing costs are independent of all others: they are equally applicable to missions with high or low risk acceptance, small or large satellites, and expensive or less expensive missions. The only case to which they are not applicable are those where there is an extremely low level of novelty; if there is an existing satellite that adequately fulfills the consumer's needs within the cost constraints defined by the sponsor, then there is no need for an architectural definition phase. Then it is simply a question of purchasing the satellite, as is often the case for standard communication satellites.

13.2.2. Hardware optimization

The history of space travel has shown that it is possible to reduce the costs of missions by designing standardized or reusable satellite hardware. Section 4.11 discussed several trends that have been influential in the past; of these, the following were hardware related:

- In the 1970s, use of reusable launchers, namely, the space shuttle
- In the 1980s, using replaceable orbital units allowing in-orbit servicing by astronauts
- In the 1980s and 1990s, using standard platforms;
- Standard plug-and-play electronic subsystems

The biggest push has been toward standardization, including entire standard satellite platforms. This was discussed above in Section 10.5.3, where it was shown that due to the diverse payloads carried by satellites and the varied environments in which they fly, successful standardization is not easily accomplished. Moreover, standard satellite platforms force payloads to adapt themselves to the previously designed platform. This inevitably complicates the most difficult and expensive part of the mission – the payload or instruments – to gain a cost reduction in the cheapest –the satellite platform.

On the other hand, standard platforms have shown success in missions featuring satellites flying in similar environments and with similar instruments, such as those making meteorological or environmental observations from low Earth orbit or from the geostationary position, as well as for communication satellites. Most communication satellites fly in the same orbit, and are functionally quite similar, acting as pipelines providing communications bandwidth, and their payloads are mostly similar from one mission to the next. This success of standardization for communications missions is demonstrated by the fact that almost all communication satellites are built using standard platforms that can accommodate different types of payloads, depending on the application.

Standardization is also sometimes advantageous for science satellites. Herschel and Planck are two ESA science satellites flying in the same orbit but with totally different payloads. However, both payloads have similar basic shapes and their requirements are not dramatically different. They both use a common satellite platform, thus reducing overall costs. The savings is also shared by satellite operations because both communications and commanding can be similarly handled.

In the late 1980s ESA tried to reduce the cost of its Soho and Cluster satellites by using common components for both of them, but analysis demonstrated that imposing common elements actually increased the cost of the satellites, because the shape, instruments, requirements, and orbits of both missions were completely different. The only thing they had in common was that they were to start at the same time and both were going to study the Sun and Sun-Earth interactions.

The conclusion to be drawn is that cost reduction through hardware standardization or commonality is mainly effective when missions are similar. In most cases cost reductions associated with these methods will not be dramatic, because the hardware must still be able to cope with the demanding environment of space, and the instruments will be new because they must be able to provide the new data and mission products demanded by the consumer. Nevertheless it is possible to reduce costs moderately by standardization of instrument, components, interfaces, or software.

13.2.3. Organization optimization

The organization of space projects is a complex undertaking. Traditional projects involve an industrial consortium of suppliers with a prime contractor responsible for the satellite, and with various levels of specialized subcontractors for the different subsystem assemblies and parts inside the satellite. The industrial consortium is directed by a permanent project team at the space agency that acts as the customer. This permanent team is supported in turn by technical experts that offer advice on specific issues. The industrial contractors provide the assemblies under their responsibility, supported by data packages substantiating their adequacy and allowing integration into larger assemblies on up to the satellite level. Both the satellite and instruments have to be properly documented for the purpose of satellite operations. The establishment of the infrastructure and training needed for satellite operations are normally projects carried out independently from the satellite itself. Product assurance consists of an independent set of experts that make sure that the progressively larger assemblies have been produced and tested according to the agreed-on procedures. Last but not least, there is a chain of review milestones where experts external to the project certify the adequacy – or lack thereof – of the work done up to the milestone. All this is very complicated, generates a large amount of documentation, and involves large numbers of people checking each other and using complex procedures and communication channels. This cumbersome approach was developed early in the history of space endeavors because space exploration was rightly judged as an especially difficult undertaking. Indeed, today the most likely source of mission malfunction is launcher failure, and most successfully launched satellites have demonstrated a high level of reliability, thus demonstrating the effectiveness of this cumbersome methodology.

Claims that this complex scheme can be greatly simplified for all types of projects with a low impact on the actual quality of the final product have long been heard. So too have claims that different projects should be handled with different levels of thoroughness, and that the most exhaustive project organization should be limited to high-priority projects with an extreme level of complexity, those providing unique opportunities, requiring extended mission life, and that push the frontiers of the state of the art. Section 4.5 presented various missions according to their aims and how they had different levels of tolerance to risk. Section 4.6 discussed projects according to their criticality and how they can be classified using objective criteria along a continuous scale of risk tolerance, and how projects of lower levels of criticality could reduce costs by relaxing quality assurance standards. The following organizational changes can be used to reduce project costs:

- Simplification of the model philosophy. This was discussed in Section 4.9.3.
- Reduction of the formal quality of components.
- Reduction of the number of formal project reviews.
- Reduction of the formality and deliberateness of formal project reviews.
- General simplification of the communication channels and reduction of the amount of documentation to be produced.

This approach, which advocates a progressive but cautious reduction in project complexity, is now firmly established. Nevertheless, during the 1990s, a more revolutionary than evolutionary approach to project simplification was proposed. This radical approach was associated with management trends of the period such as "faster, cheaper, better" or reengineering, which recommended a fundamental review of the procedures used to implement and operate complex systems. This trend, examined in Section 4.11, did sometimes work, but failed as well, mainly when applied to complex, ambitious missions (Figure 13.1).

We can conclude that a simplification of project organization reduces costs but unavoidably increases risks, which is certainly not advisable for difficult missions requiring the implementation of new techniques that advance the state of the art. However, projects managed by a lean organization have been able to deliver moderately complex satellites with tighter costs and schedules than would have been possible using conventional approaches; ESA's Giove-A is a good example of this.

13.2.4. Organization and hardware centered: small simple satellites within a lean project organization

The previous section concluded that a lean project organization increases risk. Surrey Satellite Technology pioneered an approach that used a lean project organization for satellites that were designed to be tolerant to the resulting risk increase. This required that the mission's aims were modest and that the satellite design was robust. Robust design was defined in Section 5.3.1 as simple, employing well-proven and fit-for-purpose technologies. The modest mission aims ensured that the instruments required moderate levels of novelty or risk. The DMC series of satellites (Section 11.4.4) is the best example of this approach, which almost always uses small satellites. The use of small satellites has been promoted as a way to deliver large, even dramatic cost

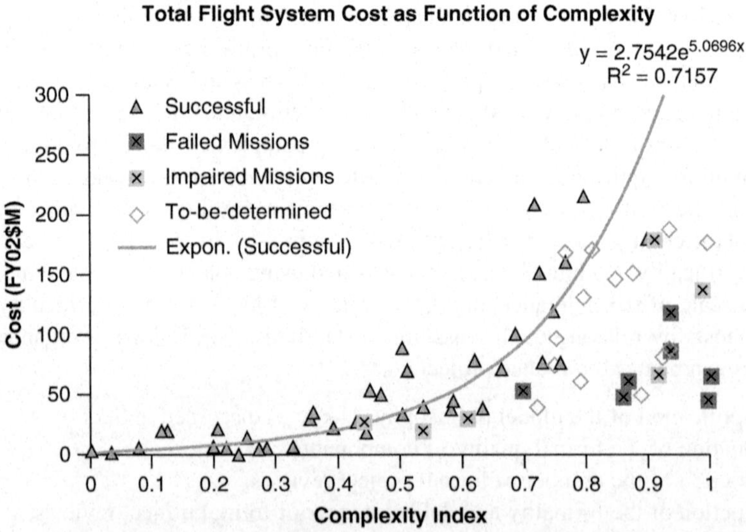

Figure 13.1. Faster Cheaper Better Mission Track as a Function of its Complexity as Expressed by a "Complexity Index" Depending of Mission Requirements (When Is a Satellite Mission Too Fast and Too Cheap? David Bearden, The Aerospace Corporation 2001 MAPLD International Conference).

reductions. Small satellites provide real potential for large cost reduction. Of course, small satellites are nothing new. The first American satellite, Explorer-1, launched in 1958, weighed just 14 kg. So small satellites are not a "new paradigm"; but recent missions featuring small satellites have been able to deliver reasonable performance at very low prices. There are a few main reasons that small satellites cost less:

- Small satellites can be built by small teams, and small teams have simpler communication channels. This allows a major reduction in the tasks associated with the creation, maintenance, and distribution of documentation.
- In small teams every member knows the problems experienced by other members. This allows fast and easy solutions to interface problems where engineering area A could be helped by actions in engineering area B.
- Small satellites often, though not always, have modest aims. Modest aims entail less demanding requirements, simpler instruments, lower risk, and simpler design.
- Small satellites need smaller facilities with lower overhead costs.
- The loss of a small satellite would not be a total catastrophe because a new one can be built at a reasonable price. This means that each satellite can be built with lean organizational approaches and lower levels of product assurance and reliability.
- Missions employing small satellites often fly several in a constellation. The failure of a single satellite will not finish the mission, but only reduce the amount of data produced. This permits lower levels of product assurance, redundancy, and numerical reliability.

This means that small satellites are naturally adapted to the use of a lean project organization. There is a balance between all the mission's conditions: hardware, organization, and aims that favors reduced costs.

Missions employing small satellites are often technology demonstration missions, but there are no intrinsic reasons preventing small satellites from addressing new scientific uses or fulfilling relevant operational functions. This is especially true for missions using a constellation of small satellites. Constellations were discussed at length in Sections 7.3.1 and 9.10. It is evident that for some applications many small, simple satellites can provide performances comparable with one large complicated satellite. Furthermore, the use of synthetic aperture techniques could solve the limitations that physical laws impose on the resolution of instruments with small real apertures. Synthetic aperture techniques, however, have limitations and Section 8.6.3 discussed the strengths and weaknesses of synthetic aperture techniques. Optical aperture synthesis imposes extremely demanding parameters in the relative position between small satellites, which must fly in tight formation. Optical aperture synthesis via multiple satellites is likely to increase not reduce costs. Again, it is necessary to contrast management approaches with the laws of physics to ensure that the proposed solution will actually work. It can be concluded that a constellation of small satellites can perform some, but not all, of the functions performed by large complicated satellites.

Hence, while it is true that small satellites are economical, it is untrue that by simply making a satellite smaller, while maintaining all its requirements will necessarily make the satellite cheaper. It has been said: "**The cost of any system correlates well with mass, but to decrease the mass of a given system increases costs**." [1]

It is obvious that to fulfill a given set of requirements while being severely mass constrained is more difficult – and more expensive – than fulfilling the same set of requirements without being mass constrained. Not being mass constrained also permits the use of standard, nonoptimal, off-the-shelf components that are not only cheaper to acquire but also allow a reduction in testing due to the good characterization and wide safety margins they provide. A robust small satellite is indeed quite heavy when compared with alternative mass-optimized designs, performing the same functions. In extreme cases, the use of nanosatellites and nanotechnology has been proposed as a promising and dramatic way to reduce the design costs of space missions. This has created a push for miniaturization employing the innovations of information technology. Nevertheless, the translation of concepts between engineering domains can be misleading. As G. Helmeier, director of Darpa, once said: "If the automotive industry had progressed during the last two decades at the same rate as microprocessors have, Rolls Royces would today cost just $3.00 and cars would measure just a quarter of an inch a side." Helmeier's statement is an interesting metaphor, though as a literal statement it is inaccurate because a family car needs to be large enough to fit a family, of humans, not amoebas. It is clear that the laws of physics do not scale well, and shrinking a satellite has intrinsic limits that will not disappear.

[1] Augustine's Laws, Norman R. Augustine, AIAA 1983.

Besides the above-mentioned limitations regarding aperture limits, reducing a satellite's size produces changes in the relative complexity of the satellite's different subsystems. When a satellite is scaled down, its mass decreases at a greater rate than the strength and stiffness of its structural elements. This seems to imply that small satellites will possess strong, light structures. Nevertheless, extremely thin structures tend to suffer malfunctions such as buckling that thicker structures avoid, and extremely thin structures are incompatible with human handling or conventional manufacturing methods. The use of nanotechnology can indeed make feasible some missions, such as interplanetary probes, where mass and volume have a strong premium, but it is unlikely to provide any far-reaching revolution to the space mission field. To continue growing, the space community needs to reduce costs and this can be achieved by designing satellites to be as light and small as possible, but not 1 g more.

Two supplementary problems associated with small and nanosatellites that limit their actual usefulness are those related to launching and disposal. The widespread use of very small satellites would require practical ways to place them in orbit and once their mission has been completed, practical ways of removing them from those orbits where they would become dangerous flying debris.

Most current small launchers have injection capabilities of around 1,000 kg to low Earth orbit. It is extremely unlikely that this situation will change in the foreseeable future, because the development of reliable new launchers, small or large, is a long, complicated, and risky process. So there are no extra-small launchers optimized for extra-small satellites, and it is not likely that this situation will change anytime soon. Nonetheless, there are two practical ways to put small satellites into orbit:

- By piggyback on a larger satellite that employs – and pays for – most launcher costs
- Multiple small satellites that share a launcher dedicated to all of them

While both these possibilities are viable and practical, neither is fully satisfactory.

In cases where a small satellite flies piggyback on a larger one, the latter will dictate the injection orbit and the former will end up in an orbit that is unlikely to be an optimal one for its mission. This limits the usefulness of the piggyback approach to technology demonstration missions whose only aim is to demonstrate the feasibility of the small satellite or a specific technology, and does not require a specific orbit; once feasibility has been proven, each new "real" application would call for a particular orbit. In some cases the large satellite's orbit might coincide with one that is suitable for the small satellite, but this means designing the smaller satellite's mission around a lucky accident.

The launching of a constellation of mini-satellites by a small launcher would appear to be a sensible technical solution for fulfilling the requirements of a demanding mission. They would be launched into an orbit suitable to all of them, and would use the small launcher's capacity efficiently. However, packing a large number of mini-satellites inside a launcher fairing (see Figures 10.4 and 10.5) is no simple matter. For very small satellites, the use of some kind of central structure where they can be attached is required (see Figure 10.4), as well as a special mechanism for releasing the satellites one by one in a controlled fashion. All of these elements require mass and space under the launcher fairing. Hence, it is not in fact possible to launch ten 100 kg satellites in a launcher designed to put 1,000 kg in orbit, and even accommodating four or five 100 kg satellites is no easy task. The reality, then, diminishes the apparent attractiveness of small satellites. The problem only increases when going from small to mini-satellites; the number of 1 kg satellites that can be practically stowed inside a 1,000 kg launcher fairing is very unlikely to be above 100. Furthermore, designing and manufacturing a structure capable of holding all of them during launch and releasing into orbit would be neither cheap nor easy.

The problem of satellite disposal is especially critical for very small satellites:

- Any practical use of very small or nanosatellites will entail the use of a large number of them. Once their mission was complete, they would become potentially dangerous space debris.
- Small satellites are unlikely to be able to carry on-board fuel or other components for changing their orbit altitude in order to remove themselves from the region where they pose a danger. The problem would be compounded because the deorbiting hardware would have to be extremely reliable to ensure that most, ideally all, the satellites shall indeed be removed.

The problem of small satellite disposal remains unsolved and it will be difficult to find a comprehensive solution, thus calling into question any proposal for the implementation of space missions employing a very large number of mini or nanosatellites.

Small satellites with modest aims and managed by lean teams have demonstrated their ability to provide exceedingly good value for the money invested. They perform well in technology demonstrations, and can provide adequate services to consumers with moderate aims. Small satellite constellations can also provide alternative solutions for some, though by no means all missions. Hence, very small or nanosatellites might enable a number of special missions, but they will not be used extensively until the difficult problems associated with them can be solved, even though they provide dramatic cost savings.

13.3. Projects without the duality sponsor/consumer

This book is based on the four principal players involved in the process of undertaking a new space mission. These four players, the consumer, the sponsor, the customer, and the supplier (Figure 1.2) are implicit throughout the entire book. This basic framework underlies all classical space missions with large-scale aims, handled by a space agency and financed with public funds. Not all space missions use this organization, however. For instance, the dichotomy between sponsor and consumer is absent in privately funded space missions. As a principal objective for owners of a privately funded satellite is to generate profits, they will avoid expensive features not likely to produce an adequate return on their investment. This will automatically eliminate the supplementary costs associated with the customer's wish for ambitious mission objectives when there are limited funds available. Privately funded missions typically allow for a faster and less-convoluted establishment of mission requirements and the subsequent design process.

Moreover, as private owners can insure their mission, they can reduce the level of mission reliability, with the knowledge that they will be fully compensated in case of failure. This means that issues associated with space mission uniqueness are reduced. It is the responsibility of the insurer to determine the level of premiums according to the mission's product assurance and reliability. Missions lacking the sponsor-consumer duality permit cost reduction methods not easily applied to those in which the duality exits.

13.4. Projects with a very low level of novelty, projects without customer

The subject of this book is proper mission architectural definition. It is appropriate then that this last section will serve to point out that not all projects need such a process. As we have seen, the customer's task was to interpret the wishes of the consumers, while taking into account the constraints and guidelines provided by the sponsor, as well as analyzing the supplier's design. This scheme assumes that the customer needs something new that requires interpretation. However, consumer needs that can be fulfilled by an already existing, reliable satellite that fits the budget determined by the sponsor do not require any architectural definition. The resulting mission does not need a customer. The supplier will simply provide the existing hardware able to fulfill the specified needs, for which he will be paid.

In cases where the sponsor and the consumer is the same person, the situation is even simpler. Imagine a buyer that wants to acquire an off-the-shelf system from a supplier at a firm, fixed priced. This can occur for situations requiring fairly standard communication satellites or to fulfill Earth-observing needs that can be met by already existing instruments and satellites. It is obvious that an existing space system with a proven track record of reliable in-orbit performance can be delivered at greatly reduced cost. In cases such as this, all activities related to design can be eliminated, and those associated with ensuring product quality, including testing, can be minimized. It is also sometimes possible to pay for the satellite only after it has been put into orbit and demonstrated that is functioning properly.

13.5. Cost engineering as art and science

It shall always be recalled that cost estimation is as much an art and not a science, and its output is always subject to interpretation. A "cost" is in fact most likely to be the basis of negotiations between the supplier and customer. The assessment of a "cost" is always subject to the commercial pressures of the specific circumstances that apply at a particular time; it will be affected by the commercial entities involved, and must be treated with caution. Further, the cost estimates are produced by "company confidential" procedures that are anything but universal, and may be the subject of quite different assumptions and assumed boundary conditions or constraints by the supplier organizations bidding for a program. Hence, comparisons require yet further judgments being applied before an assigned cost is approved. Therefore, when costs are calculated there is much scope for error being made and commercial risk for all concerned. The space agencies have the benefit of collating program costs for a multitude of programs over many years from many suppliers and these can form the basis of making generic cost estimation tools. However, this process is still fraught with potential sources of error, which must be taken into account, such as ensuring that the evolution in technologies, system performances, and even fundamental changes in architectures are not grouped inappropriately. In effect, a program's cost is part of a legally binding commercial agreement between the parties involved, and as such represents a judgment on the part of the individuals concerned, with the benefit of past experience playing a major role.

Index

M.A. Aguirre, *Introduction to Space Systems: Design and Synthesis*, Space Technology Library 27, DOI 10.1007/978-1-4614-3758-1, © Springer Science+Business Media New York 2013

E

J

K

L

M

Q

R

U

V

W

X

Y

Z

Printed by Printforce, the Netherlands